1992-93

Merry Christmas Dad and Todi
    I hope you enjoy This
insight into my life in Antarctica...
It was one of The most significant
experiences I've had to date. There
is no place on This planet like
it.
        Enjoy! Much Love,
                Kim

P.S. This book focuses on
The imediate area around
McMurdo Base... called Ross Island...
It includes photos of many
places I flew to in helicopters as
well as the very aircrafts.

Pictures really don't do justice
to The awe I felt in
these various areas.

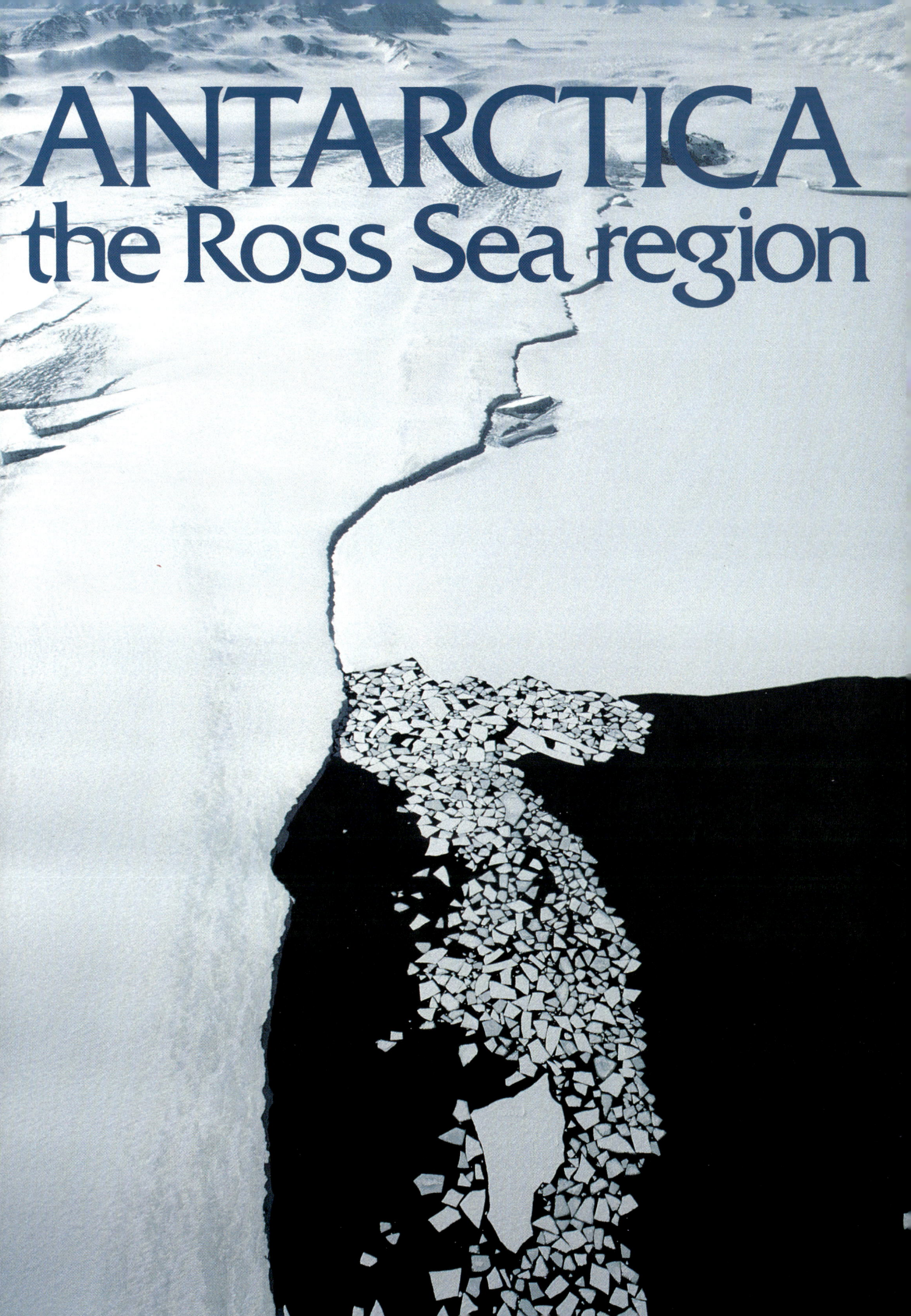

ANTARCTICA
the Ross Sea region

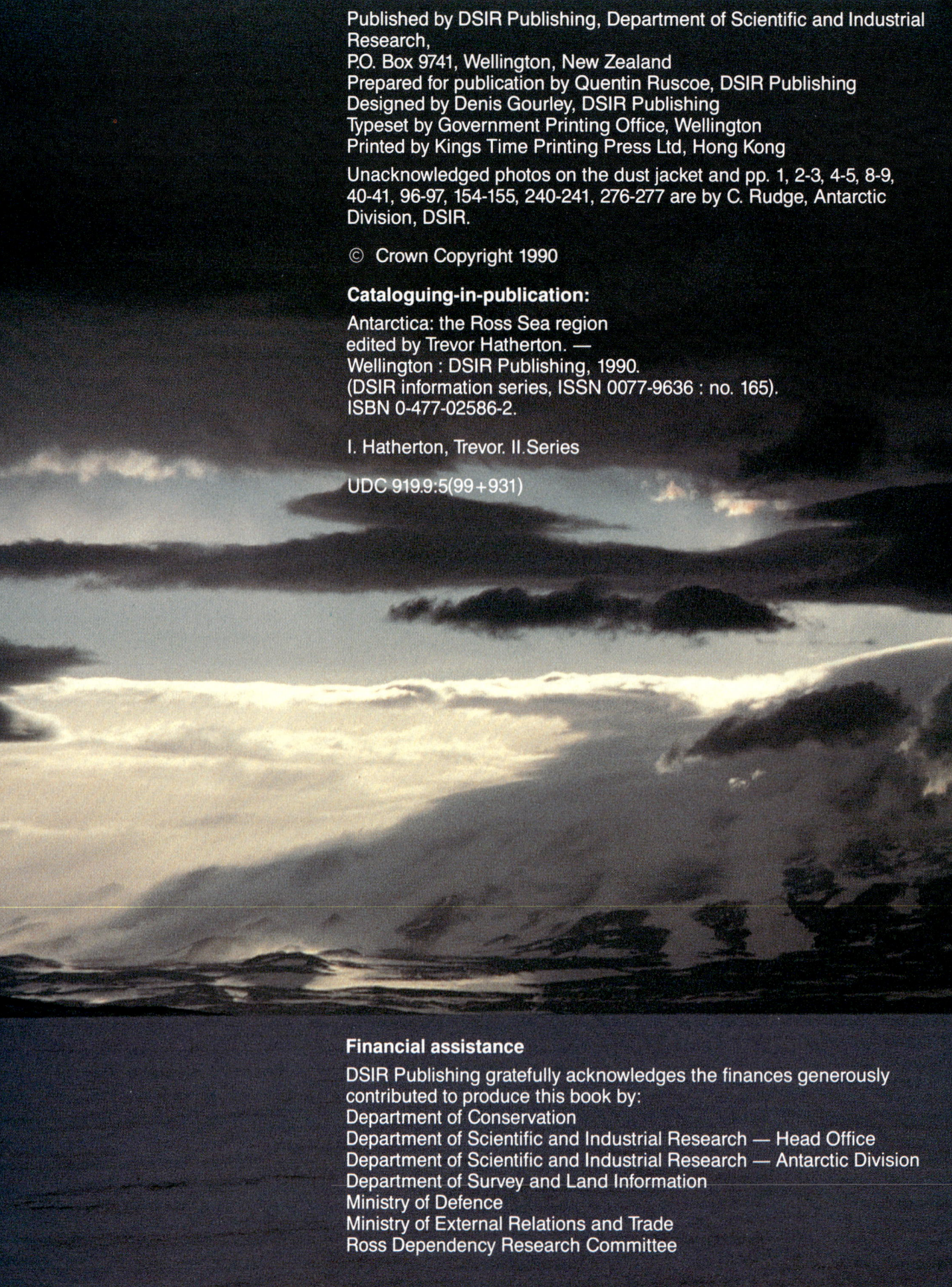

Published by DSIR Publishing, Department of Scientific and Industrial Research,
P.O. Box 9741, Wellington, New Zealand
Prepared for publication by Quentin Ruscoe, DSIR Publishing
Designed by Denis Gourley, DSIR Publishing
Typeset by Government Printing Office, Wellington
Printed by Kings Time Printing Press Ltd, Hong Kong

Unacknowledged photos on the dust jacket and pp. 1, 2-3, 4-5, 8-9, 40-41, 96-97, 154-155, 240-241, 276-277 are by C. Rudge, Antarctic Division, DSIR.

Cataloguing-in-publication:

Antarctica: the Ross Sea region
edited by Trevor Hatherton. —
Wellington : DSIR Publishing, 1990.
(DSIR information series, ISSN 0077-9636 : no. 165).
ISBN 0-477-02586-2.

I. Hatherton, Trevor. II.Series

UDC 919.9:5(99+931)

Financial assistance

DSIR Publishing gratefully acknowledges the finances generously contributed to produce this book by:
Department of Conservation
Department of Scientific and Industrial Research — Head Office
Department of Scientific and Industrial Research — Antarctic Division
Department of Survey and Land Information
Ministry of Defence
Ministry of External Relations and Trade
Ross Dependency Research Committee

ANTARCTICA
the Ross Sea region

edited by Trevor Hatherton

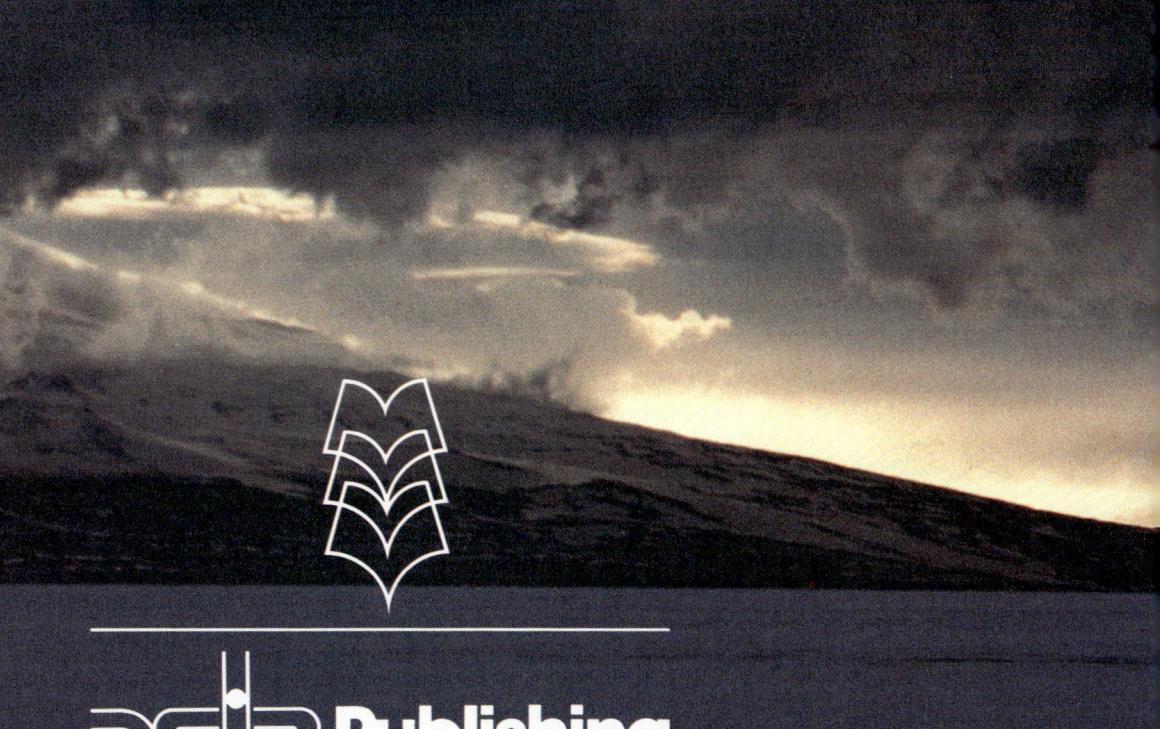

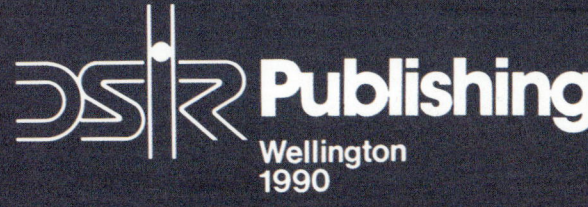

DSIR Publishing
Wellington
1990

Contents

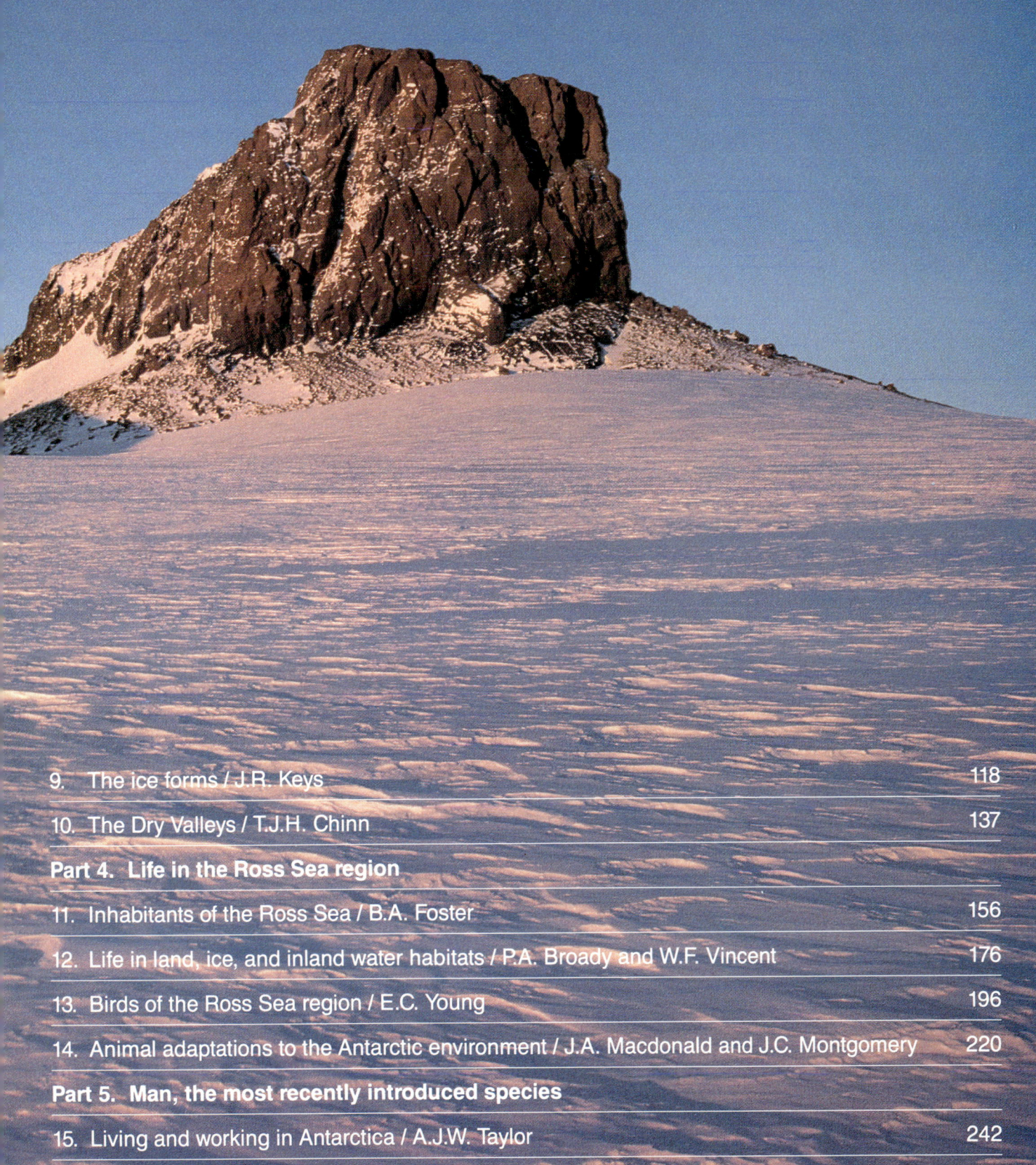

Dedication

This book is dedicated to the late
Sir Holmes Miller, Knight Bachelor;
Officer of the Most Excellent Order of
the British Empire; on whom was
conferred The Polar Medal;
B.A.; D.Sc. (Hon. Causa); Fellow of the
New Zealand Institute of Surveyors;
Honorary Member of the Royal
Institution of Chartered Surveyors;

by profession—surveyor,
by inclination—explorer,
by conviction—conservationist,
by nature—gentleman,

for 30 years, doyen of New Zealand
Antarctic activities; a member of the Ross
Dependency Research Committee since
its inception in 1958, and its Chairman
from 1971 to his resignation in 1983.

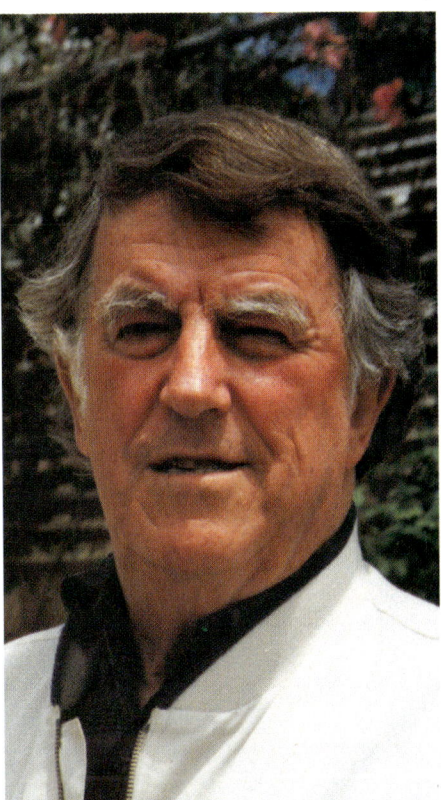

The history of human association with Antarctica can be divided into four periods. The first began with the conception of a cold south polar region following the discovery, attributed to Aristotle, that the earth was spherical. But no attempt was made to exploit the new knowledge in the two millenia which followed this, the most important of all geographical discoveries.

The "Great Age of Discovery" initiated the second phase. In the sixteenth century the famous sea captains, Magellan and Drake, gave the world its foretaste of Antarctic conditions during their circumnavigations; and, incidentally, they became the first to exploit Antarctic wildlife when they killed thousands of penguins to replenish their food supply. They were followed by many other mariners, searching for a fabled "Terra Australis". Eventually, James Cook, between 1772 and 1774, demonstrated that if a continent did indeed exist it must lie in a region of perpetual snow and ice.

The third period, which opened up the continent for exploration, resulted from the voyage of James Clark Ross. In early January 1841, he penetrated for the first time the pack ice which had protected the continent from human access for the previous three centuries. Moreover, the large marine embayment which he discovered beyond the pack ice proved to be the gateway for the attempts on the South Pole by three men whose names are now legend.

In the fourth period exploration, greatly facilitated by modern transport technology, is now complete and has given way to science. The importance of Antarctica as a major laboratory for investigating and monitoring changes in the earth's environment is now widely accepted. This book tells us what we know about the region which Ross on the sea and Amundsen, Scott, and Shackleton on land, opened up—a region in which the intellectual adventure of science is still accompanied by the physical adventure of working in a landscape which is dramatic, harsh, and often dangerous.

Sir Edmund Hillary

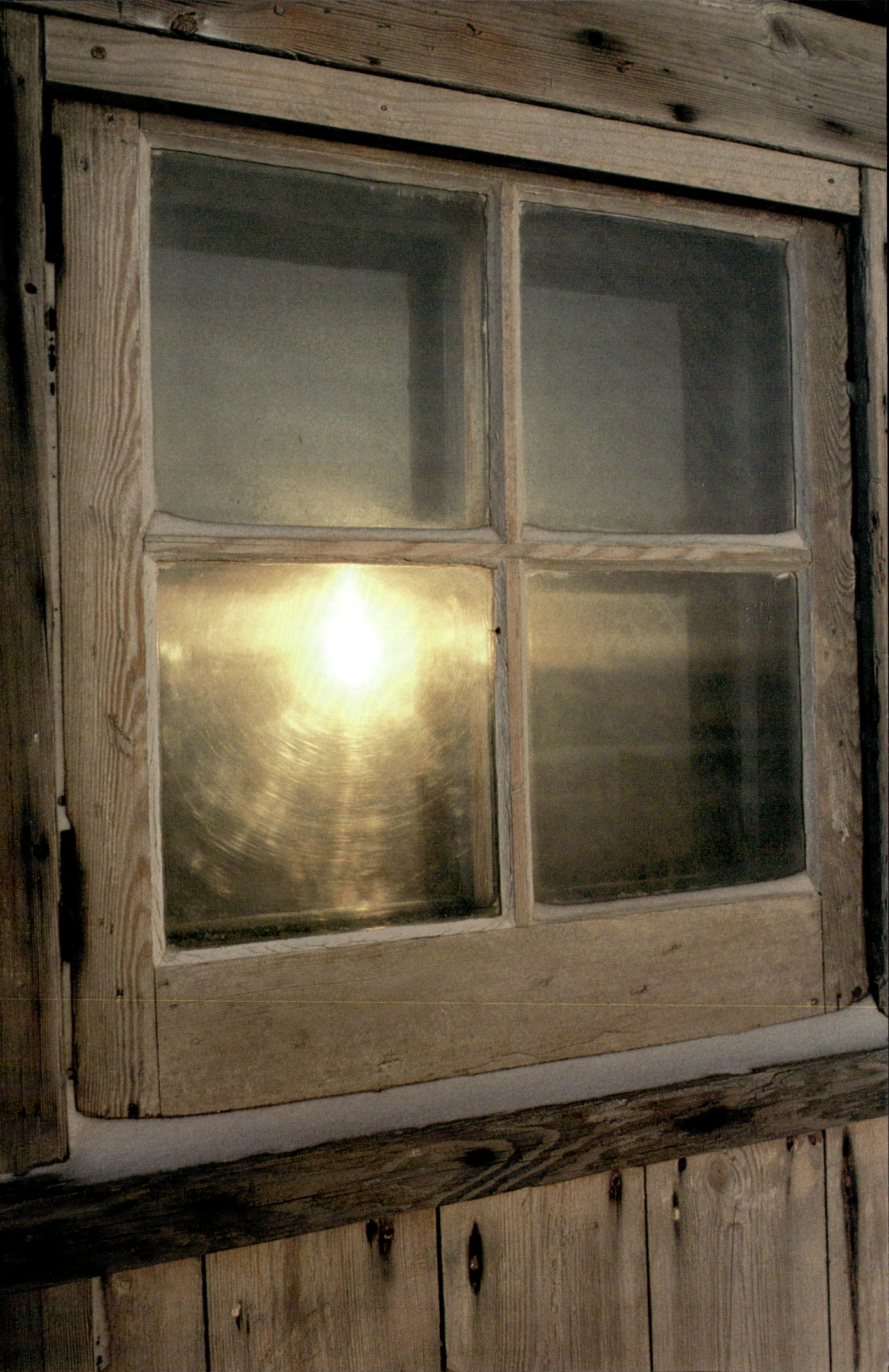

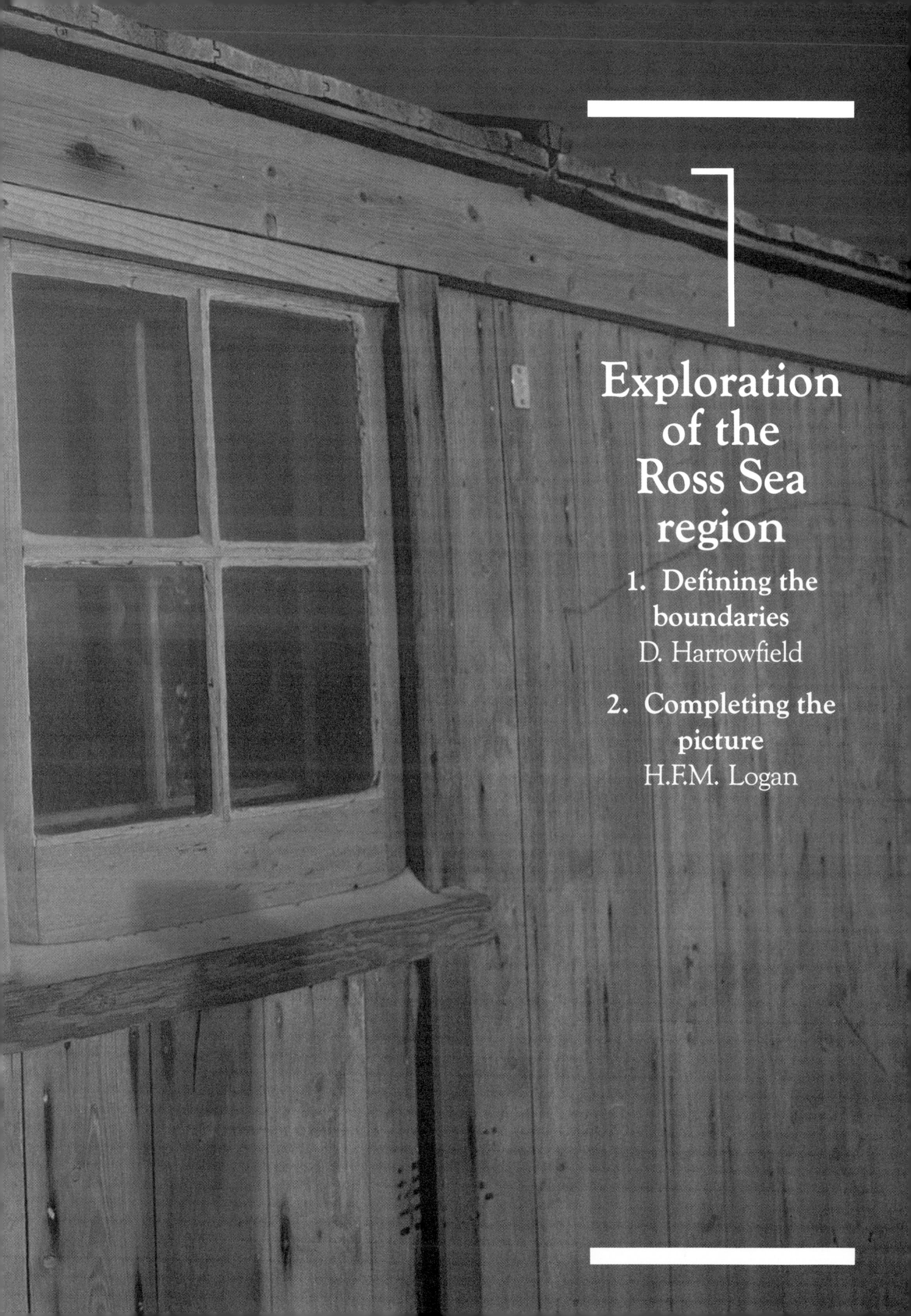

Exploration of the Ross Sea region

1. Defining the boundaries
   D. Harrowfield

2. Completing the picture
   H.F.M. Logan

1. Defining the boundaries

The existence of a southern polar region was predicted by the early Greeks, for Aristotle is credited with discovering the spherical shape of the earth in the fourth century B.C. The name given to the region was "Antarktos"—opposite the Bear, the northern constellation which contains the pole star Polaris. One hundred years later Eratosthenes calculated the earth's diameter quite accurately and thus speculation about the nature of "Antarktos", and the possibility of reaching it, naturally followed (Figure 1.1).

Motivation and essential technology are the mainsprings of exploration. For nearly 2 millenia after the Greeks, mankind had neither the motivation nor the technology to explore any southern land. In the later Middle Ages the westward transmission of Greek and Arabic knowledge, the strengthening and centralisation of power in the states of the Iberian Peninsula, the location of these states on the shores of the Atlantic Ocean and developing shipbuilding, sailing, and navigation techniques all led to the "Great Age of Discovery". There arose from the many consequent voyages the notion of a great southern continent—"terra australis nondum cognita". But over 2 centuries were to pass before two great navigators were to limit the boundaries of this unknown and unobserved continent.

The first was James Cook who, with his ships *Resolution* and *Adventure*, in what has been described as "the greatest voyage ever made", proved that if land did exist then it lay in latitudes higher than 60°S along which he made his circumnavigation in 1772–75. Forty-five years later the Russian Thaddeus von Bellinghausen carefully filled in the gaps left by his great predecessor. Both explorers considered that to penetrate further south would be both difficult and dangerous.

Exploration of Antarctic waters did not cease however, for the many reports of whales and of fur seal colonies soon led to an influx of small commercial expeditions, notably those of Bransfield, Palmer, Weddell, and Balleny. The first sightings of the land were made in 1820, but it was not until 1839 that the first landing was made south of the Antarctic Circle (66°30'S). In that year John Balleny, a captain with the British sealing firm of Enderby Brothers (in command of the vessels *Eliza Scott* and *Sabrina*), discovered the islands which now bear his name, and Captain Freeman of the cutter *Sabrina* landed on a spit at the north-west corner of Borradaile Island.

The route opens

At that time one focus of scientific research was the study of the earth's magnetic field. The North Magnetic Pole had been located and reached in 1831 by a British naval officer, James Clark Ross (Plate 1.1). Three, separate national expeditions were dispatched within 2 years with the objective, partially or wholly in mind, of reaching the South Magnetic Pole which the great mathematician and physicist Gauss had calculated to be 25° south of Tasmania. They were a French expedition (1837–40) led by Dumont d'Urville, who gave his wife's name to Adélie Land; a United States expedition of six ships (1838–42) commanded by Charles Wilkes, whose orders, hardly scientific, began "The Congress of the United States having in view the importance of our commerce embarked in the whale-fisheries and other adventures of the great Southern Ocean . . . "; and a British expedition.

The British Antarctic Expedition of 1839–43 arose from a meeting of the British Association for the Advancement of Science in 1838 which adopted the resolution; "That this Association views it as highly important that the deficiency, yet existing of our knowledge of terrestrial magnetism in the southern hemisphere should be supplied by observations of the magnetic directions and intensity, especially in the

Figure 1.1. *The earth as drawn to the ideas of Macrobius (about 410 A.D.) following Cicero (first century B.C.). The known world in the Northern Hemisphere, centred on Jerusalem, is balanced by a large southern continent separated by two hot areas (perusta) and an equatorial ocean.*

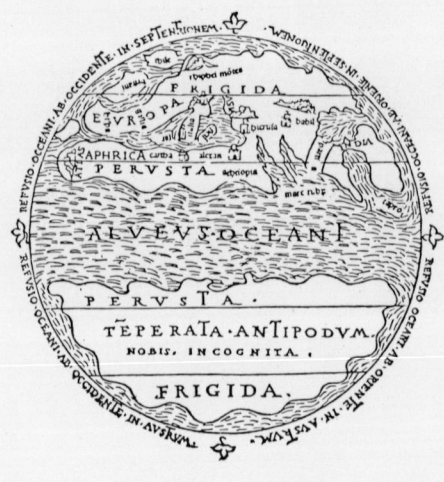

Plate 1.1. *Sir James Clark Ross in Captain's undress uniform of 1833–56. Photo: National Maritime Museum, London, from a portrait by H. W. Pickersgill.*

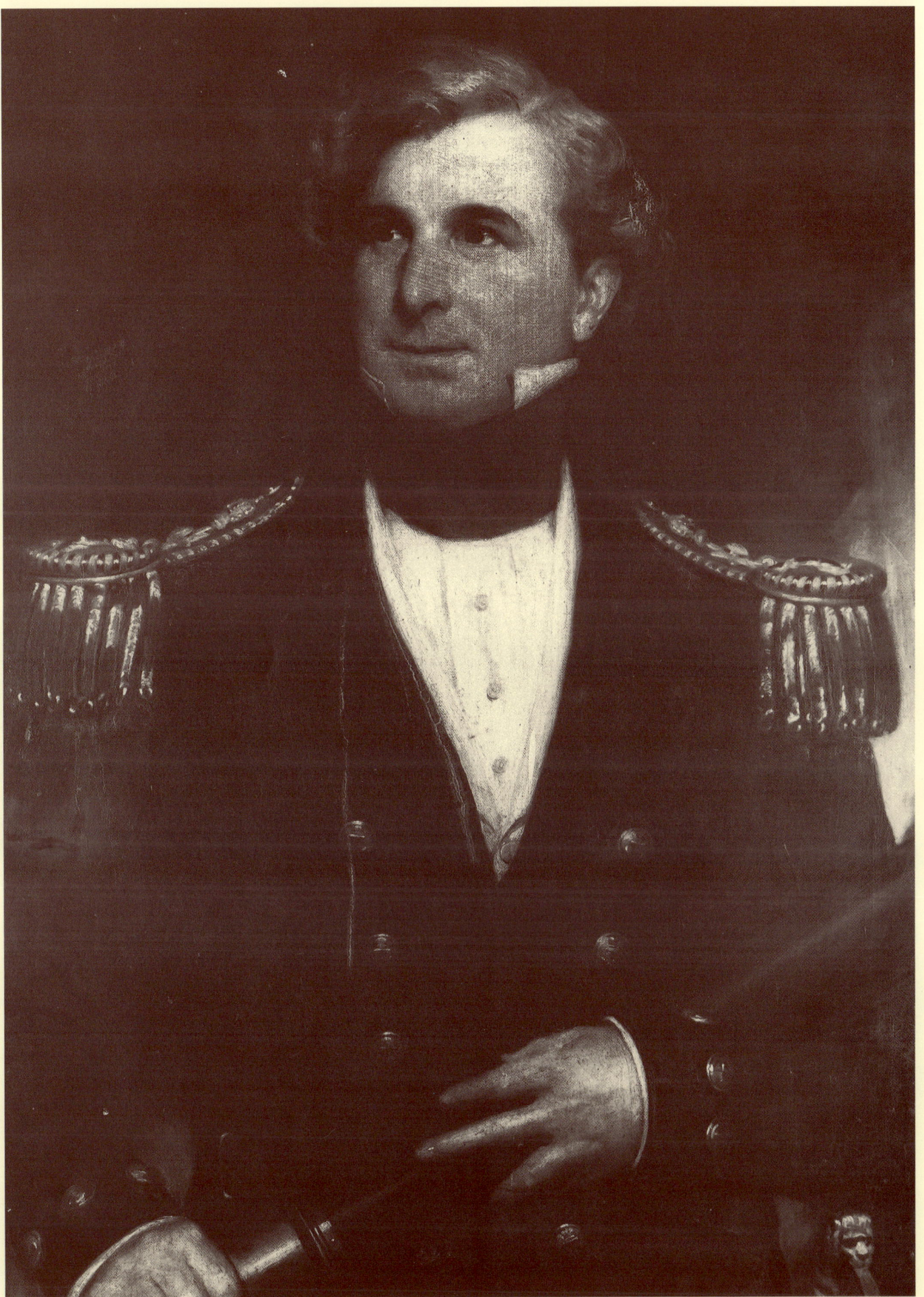

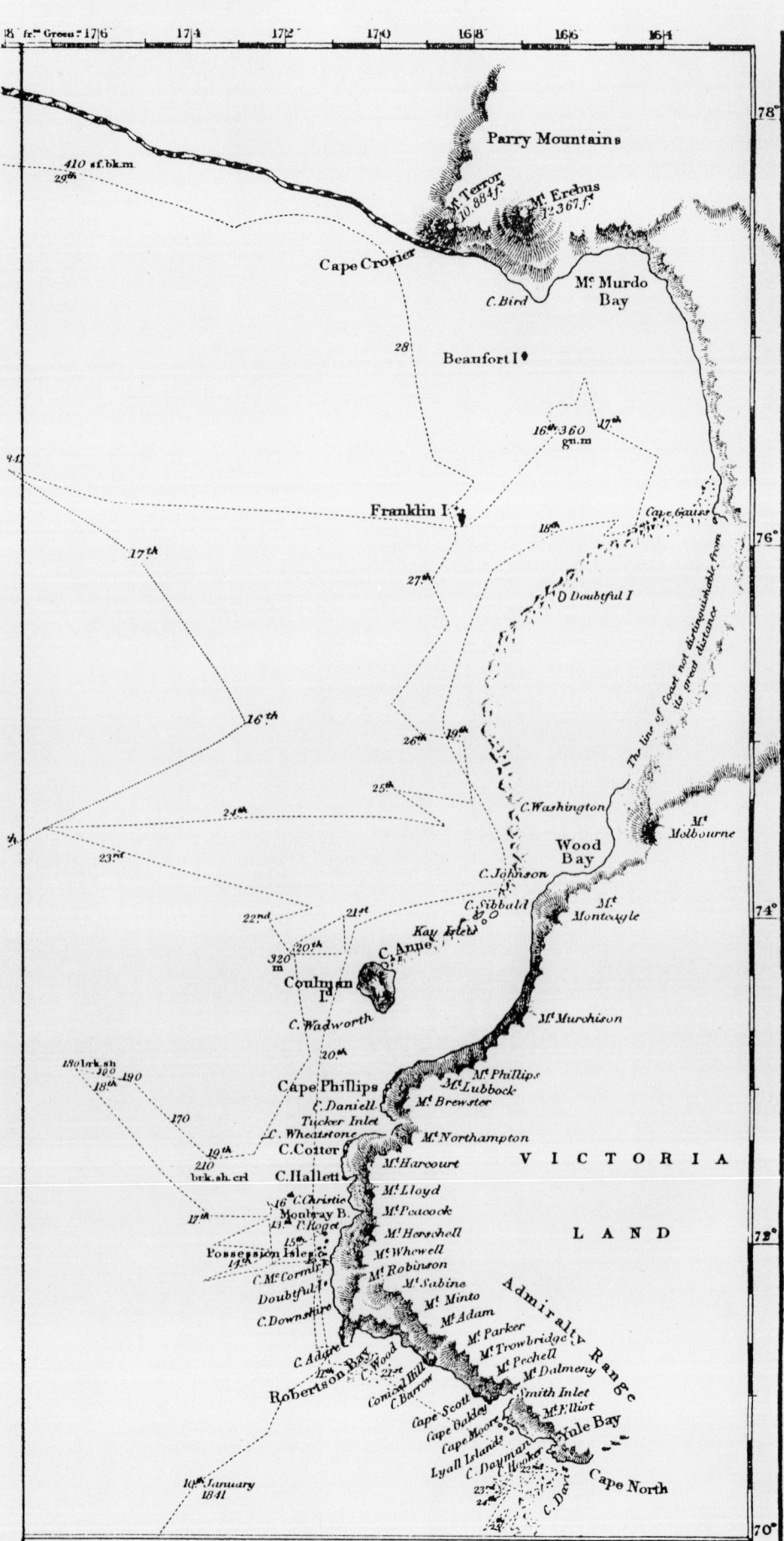

Figure 1.2. *Part of Ross's chart of the western Ross Sea, published by the Hydrographic Office of the Admiralty in 1846.*

high latitudes between the meridians of New Holland and Cape Horn; and they desire strongly to recommend to Her Majesty's Government the appointment of a naval expedition expressly directed to that object".

The British Government accepted this recommendation and appointed James Clark Ross to the expedition's command on 8 April 1839. No such appointment could have been better justified. Between 1818 and 1836 Ross had spent 8 winters and 15 summers in the Arctic, earning an impressive reputation as explorer, seaman, and navigator. In addition, he was an acknowledged authority on terrestrial magnetism and a capable naturalist.

The British expedition was the last of the three and Ross, benefitting from the experience of d'Urville and Wilkes, determined to make his high latitudes further to the east than theirs. On 5 January 1841, after cruising along the edge of the pack ice for several days, Ross decided to head into it. He was soon to find the first "Ross's seal". On 9 January, "We had a most cheering and extensive view; not a particle of ice could be seen in any direction from the masthead". The icy defences to the south had been breached. On 11 January the officer of the watch reported land ahead which "rose in lofty peaks, entirely covered with perpetual snow" and Cape Adare, Cape Downshire, and the Admiralty Range became the first names (with the exception of the previously discovered Balleny Islands) in the Ross Sea region. Next day a party landed on an island which, because of the ceremony of taking possession of the newly discovered lands in the name of "our most Gracious Sovereign, Queen Victoria", was named Possession Island. The officers celebrated with "a glass of excellent sherry"; the men had rum. Surgeon McCormick shot a skua and left his name to ornithological history.

Ross, with his two ships, continued his triumphant way southwards along the coast of Victoria Land discovering, plotting, and naming the features of the continent and the offshore islands—Coulman, Franklin,

Plate 1.2. *The first buildings on the Antarctic Continent were Borchgrevink's huts at Cape Adare, constructed in February 1899. Photo: Alexander Turnbull Library, Wellington.*

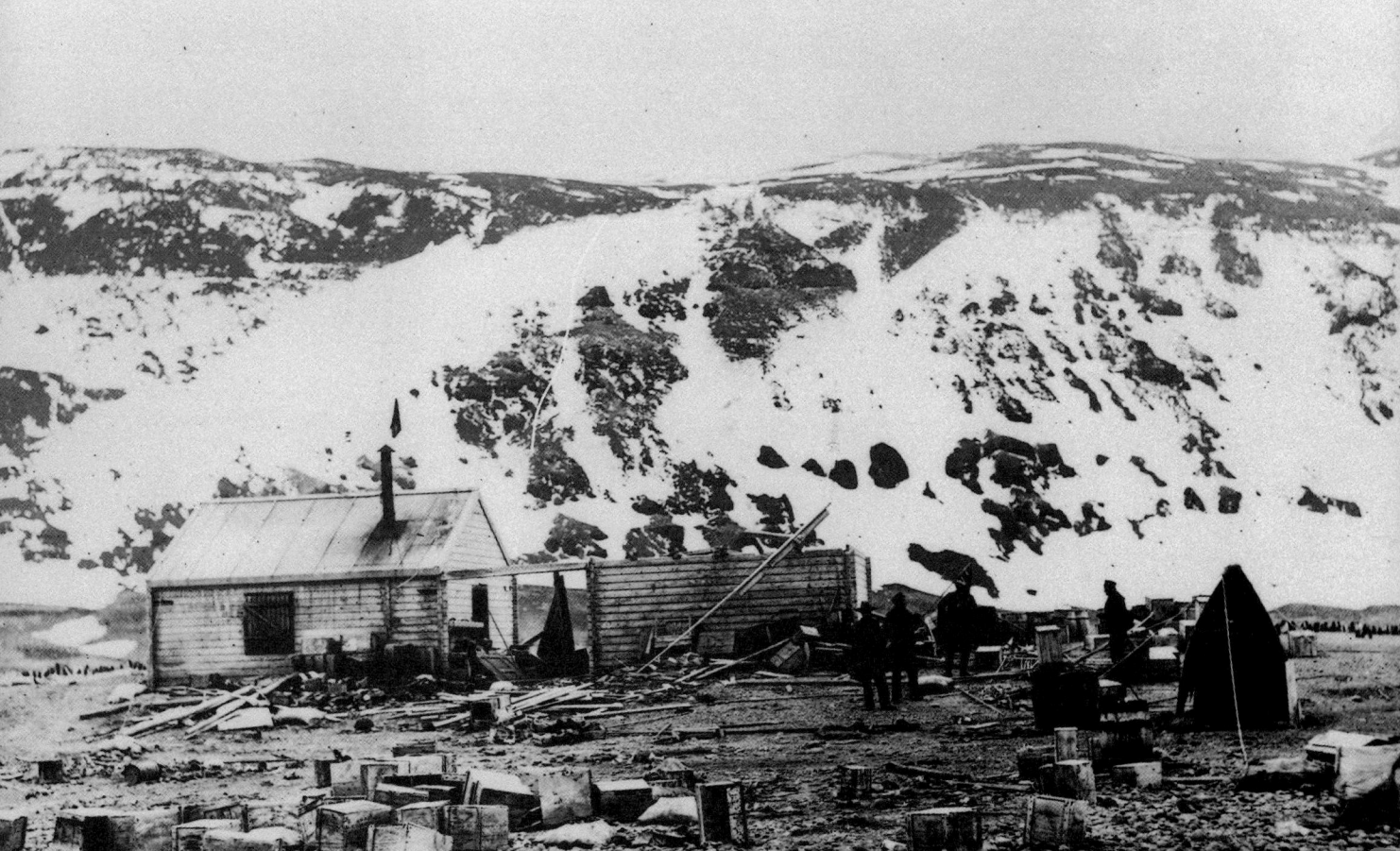

and Beaufort (Figure 1.2). After observing an active Mount Erebus with
wonder and estimating its height with astonishing accuracy, he sailed
along the length of the "Great Ice Barrier" before returning to Cape
Adare and continuing his survey westwards as far as Yule Bay and Cape
North. His greatest disappointment was not "to have found a place of
security for the ships in which to pass the winter; we might then with
ease have travelled the short intervening distance (to the South Magnetic
Pole) and reached the summit of those noble mountains. Mount Erebus
might also have been ascended in the spring . . . ". Thus, he was
thwarted of his long cherished ambition "of being permitted to plant the
flag of my country on both the magnetic poles of our globe".

Although the objective of the voyage, the South Magnetic Pole,
remained unvisited the discoveries of Victoria Land, the Ross Sea, Ross
Island, and the Ross Ice Shelf not only set the seal on Ross's
distinguished career but also opened the pathway along which Scott,
Shackleton, and Amundsen would attempt to reach the Pole 60 and
more years later. During that long hiatus British attention was directed
back to the Arctic in numerous searches for the lost Franklin expedition
and then to the exploration of Africa; colonialism was endemic among
European powers; and the United States suffered the trauma of its Civil
War and subsequently concentrated on its internal exploration and
development.

The first foothold

Ross, like Cook, had made sightings of whales and this stimulated
an Australian of Norwegian birth, H. J. Bull, to organise a whaling
expedition to the Ross Sea. Bull obtained a steam whaler, the *Antarctic*,
and after it was refitted in Melbourne he sailed south in 1895. During
sealing operations about Macquarie and Campbell Islands the propeller
was damaged and the ship turned north to New Zealand. Here repairs
were made and four new hands were taken on at Stewart Island as the

Plate 1.3. Discovery in Winter Quarters
Bay, 1902. Observation Hill is in the
background. McMurdo Station now occupies
the area between Discovery and
Observation Hill. Photo: Alexander
Turnbull Library, Wellington.

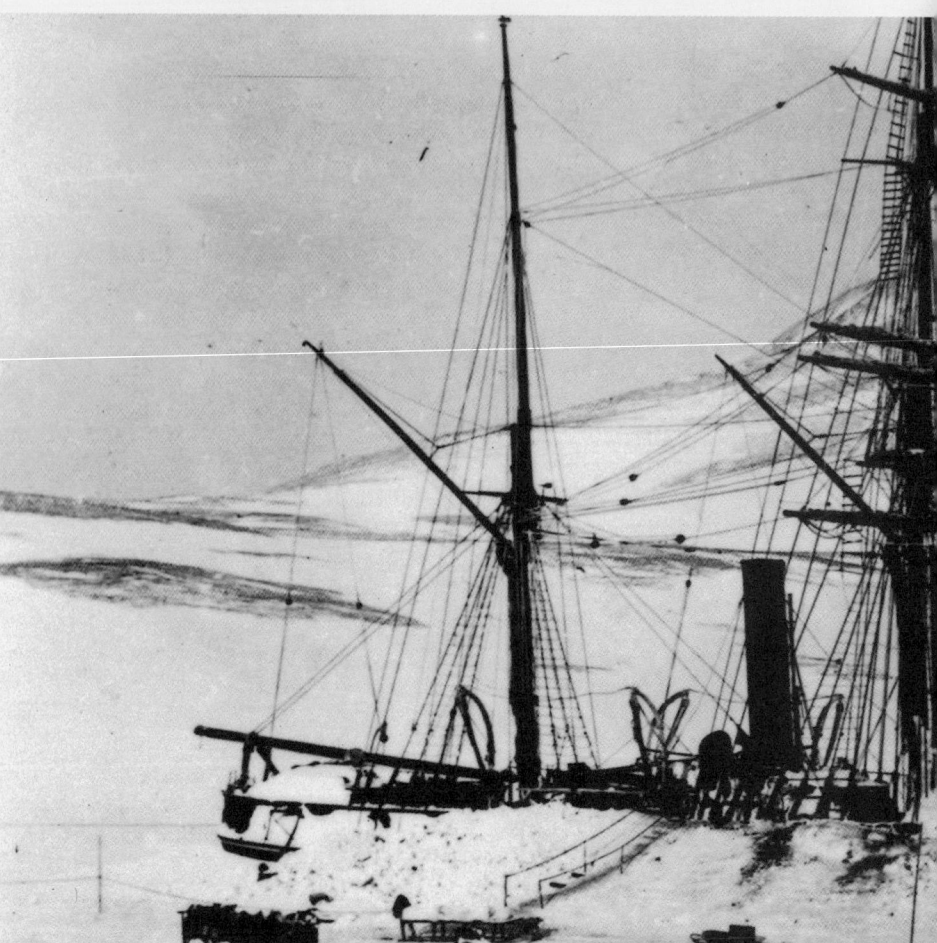

Antarctic headed south. One of them was 17-year-old Alexander von Tunzelman, who subsequently maintained that he was the first person to step ashore on the Antarctic mainland, at Cape Adare when he leapt out of the boat to hold the bow steady. The crew of the *Antarctic* had included another Norwegian–Australian, Carsten Borchgrevink. Borchgrevink (Plate 1.6), in spite of having a blunt manner, was dynamic and although failing to organise an expedition to exploit guano deposits on Possession Island managed to obtain the support of a wealthy British publisher, Sir George Newnes, for a more ambitious project.

With £40 000 now at his disposal he purchased a former whaling vessel the *Pollux* and renamed it the *Southern Cross*. A small but competent group of scientists was enlisted and 90 Siberian dogs, the first to be used for Antarctic work, were obtained. However, although the expedition received many good wishes it was not officially recognised.

In February 1899, two 4.5 × 4.5 metre wooden huts were erected on a large cusp-shaped foreland at Cape Adare to form Camp Ridley, named after Borchgrevink's mother's maiden name (Plate 1.2). The huts, the first to be erected on the continent, were lined with seal skins and canvas and the walls were insulated with papier mâché. Ten men made up the shore party. During winter a series of short sledging trips were made over the sea ice to the head of Robertson Bay, and in October the zoologist Nicolai Hansen died and was buried on top of Cape Adare. Late in January 1900 the *Southern Cross* returned and the expedition proceeded south into the Ross Sea. A party landed on the Ross Ice Shelf and made a short sledging trip to latitude 78°51'S, thus attaining the farthest south man had been until then. The expedition, however, had not been a very happy one and only a little scientific work was accomplished though this included magnetic observations, a detailed meteorological record, and new zoological discoveries. Also, the physicist Louis Bernacchi suggested Ridley Beach was either the result of glaciation or was merely a raised beach—these were the first theories of

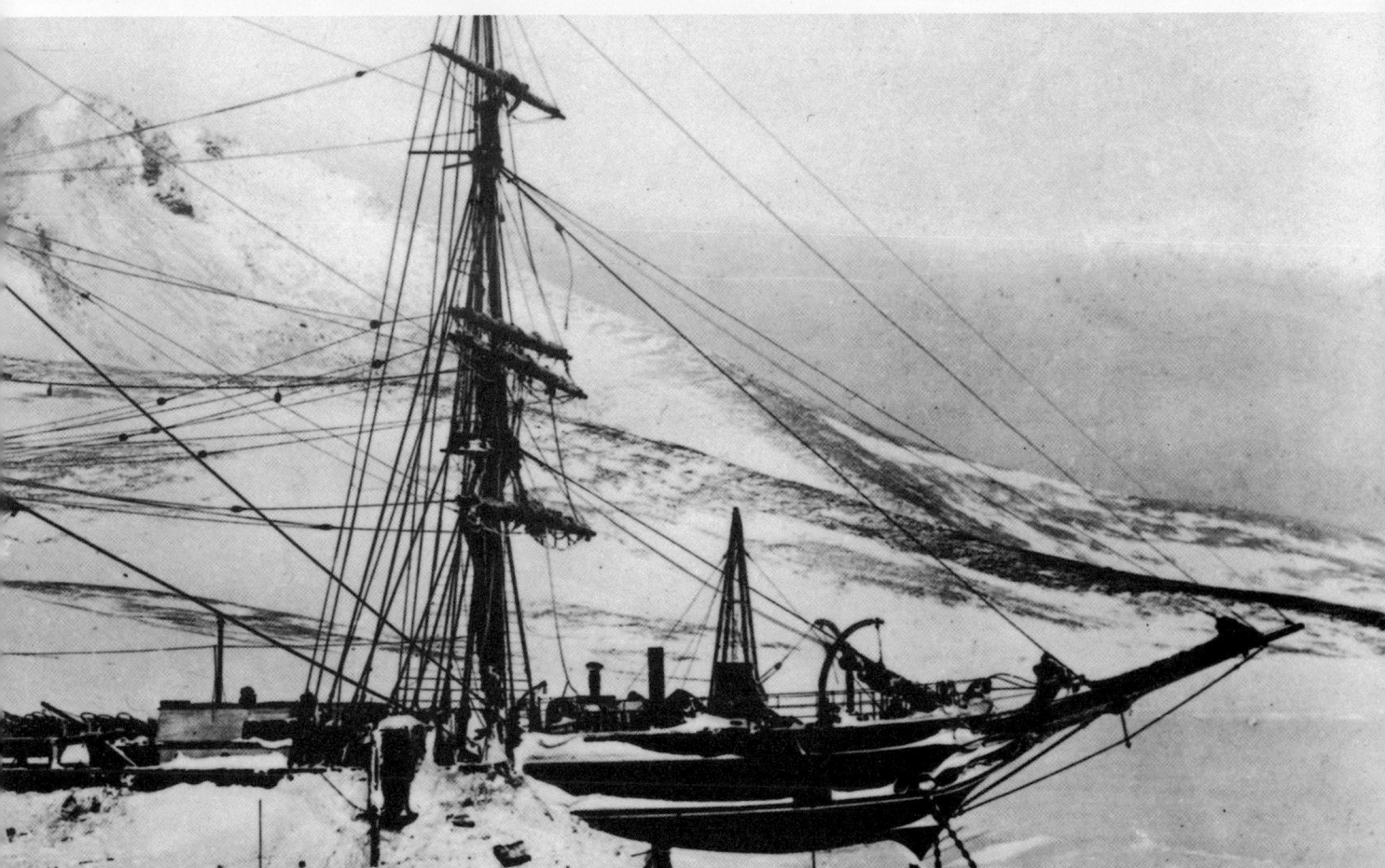

Antarctic coastal geomorphology.

Scott's first expedition

By now a more extensive British national expedition was being organised. With support from the Royal Geographical Society and the backing of its President Sir Clements Markham, a new barque-rigged, steam-powered vessel the *Discovery* was built at a cost of over £50 000. Robert Falcon Scott (Plate 1.6) was appointed Commander. There was also an able scientific staff of which 3, including Bernacchi, had polar experience and a crew of 17 (mostly from the Royal Navy). An amazing variety and quantity of supplies were taken, including 23 dogs, an 11 × 11 metre hut from Australia, a balloon, and a windmill for power generation.

On 21 December 1901 the expedition left its final port of call in New Zealand and after breaking through pack ice landed on the Ross Ice Shelf. Here, using the balloon, Scott made the first aerial ascent in Antarctica and took photographs from a height of 250 metres. As the expedition moved southwards into McMurdo Sound the unfolding landscape created much interest. On 8 February 1902 Bernacchi recorded "Mts Erebus and Terror are undoubtedly on an island, which is a most surprising and totally unexpected discovery. The cause of the Great Ice Barrier is more than ever difficult of solution." The *Discovery* was now allowed to become frozen in at Winter Quarters Bay (Plate 1.3) and a hut was erected on a small promontory from what then became, appropriately, Hut Point Peninsula. The hut was used for various purposes including storage, dramatic productions, and scientific observations but it was never used much for accommodation until subsequent expeditions.

Following the first winter, during which the windmill was demolished, sledging began in earnest (Plate 1.4). In November 1902 Scott, Wilson, and Shackleton set out across the Ross Ice Shelf with 18

Plate 1.4. *Although Scott used dogs and ponies, he is best known for his manhauling journeys. Here we see: (a) the first sledging party of the Discovery expedition with Shackleton in the lead and Wilson and Ferrar in the rear; (b) a more professional approach as a party sets out to the Pole from 80°S Depot; and (c) the sheer hardship of manhauling on deep snow. Photos: Alexander Turnbull Library, Wellington.*

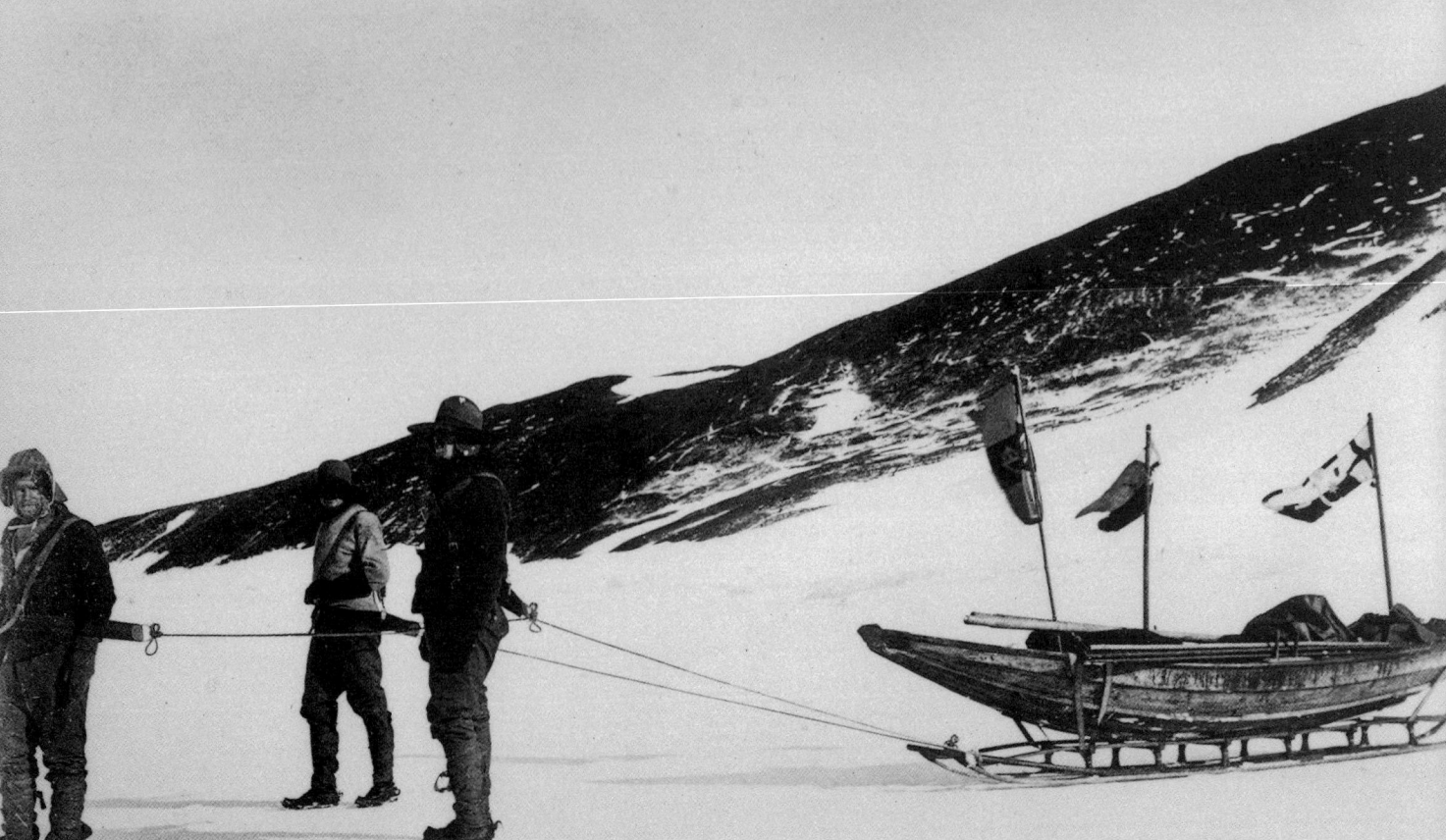

dogs, which soon began to fail. Fifty-nine days later, after reaching latitude 82°17′S, having travelled over 600 kilometres from their point of departure, and with Shackleton suffering from scurvy, they had to turn back. Meanwhile, a party led by Armitage had crossed the Sound and sledged up the Blue and Ferrar Glaciers to become the first on the Polar Plateau, reaching an elevation of 2800 metres. On their return Armitage's party discovered the ice-free area now known as the Taylor Valley.

In January 1903 the relief ship *Morning* arrived under the command of Captain Colbeck. Departing with Shackleton and several other expedition members it left the *Discovery* and the remainder of the party to winter-over for a second year. With the arrival of spring, sledge parties again went out. One which included Scott, Skelton, and Ferrar, following Armitage's route of the previous year up the Ferrar Glacier, travelled 480 kilometres westwards from the ship. In January 1904 the *Morning* and a second relief ship the *Terra Nova*, under Captain Mackay, arrived. The *Discovery* was freed using explosives and the three ships departed for New Zealand.

Apart from the tragic loss of Able Seaman George Vince in March 1902 the expedition had been remarkably successful and set a precedent for the combination of exploration and science which was to be followed by subsequent British parties. New geographical discoveries included the Blue, Ferrar, Koettlitz, and Taylor Glaciers and the Polar Plateau; the Ross Ice Shelf was confirmed as a floating ice mass and the many geological finds included carbonaceous material and plant fossils. An extensive meteorological record and much geomagnetic data were also obtained.

Shackleton returns

Shackleton (Plate 1.6), disappointed at his breakdown in health, was determined to go south again. By 1906 he was enlisting support for an

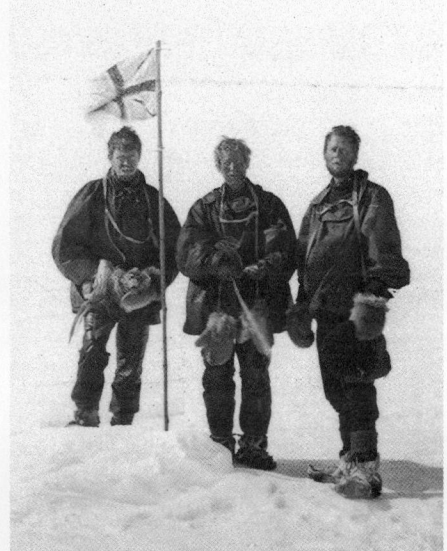

Plate 1.5. *Ross's goal realised, as Mackay, Edgeworth David, and Mawson reach the South Magnetic Pole at 3.30 p.m. on 16 January 1909. Photo: Alexander Turnbull Library, Wellington.*

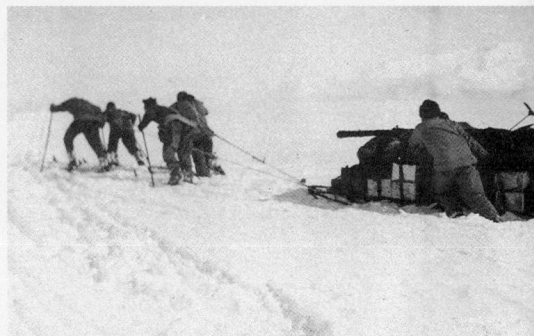

(c)

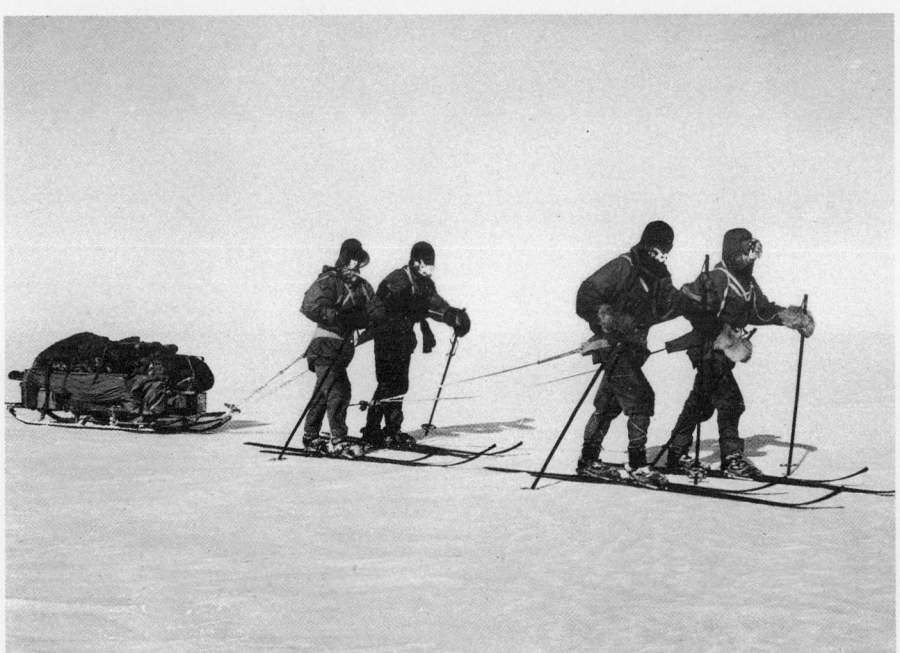

(b)

expedition thats main purpose was to make an attempt on the South
Pole, and he managed to raise £44 000 largely from private sources. A 40-
year-old vessel the sealer *Nimrod* was obtained for £5000 and converted
to a barquentine. The 15 personnel selected for the shore party included
2 from the *Discovery* Expedition (Joyce and Wild) and "equipment"
included a 12–15 h.p. motor car made by the Wolseley Company (see
Plate 1.9), 15 Manchurian ponies, and 9 dogs which were descended
from 75 left by Borchgrevink on Stewart Island, New Zealand. A 10 ×
5.5 metre prefabricated hut insulated with granulated cork was to be
their home, but very little furniture was taken as the men were expected
to make most of their own.

On New Year's Day 1908 the *Nimrod*, with only 1.8 metres of
freeboard, left Lyttelton and was towed for 2400 kilometres to the pack
ice by the *Koonya*. It was a rough trip. The Second Engineer, J. G.
Rutherford, noted "The 9th (of January) was the worst day I have ever
experienced, terrific squalls and mountainous seas sweeping everything
before them. The *Nimrod* was continually swept with solid water from
stem to stern . . . " Six days later the pack ice was sighted and the
Koonya returned to New Zealand.

Shackleton's intention had been to establish his base on the eastern
side of the Ross Ice Shelf but this was not possible. The *Nimrod* then
steamed into McMurdo Sound and a sledging party was sent to Hut
Point where the *Discovery* Expedition hut was found to be practically
free of snow. Eventually, a good site for their own hut was found at
Cape Royds on 3 February.

Living within sight of Mount Erebus soon presented the challenge
first noted by Ross. Professor Edgeworth David, one of the two
geologists with the expedition, wrote "For us, living under its shadow,
the longing to climb it and penetrate the mysteries beyond the veil soon
become irresistibly strong". The first ascent of the mountain was
achieved by a party of six between 5 and 10 March (see p. 79). During

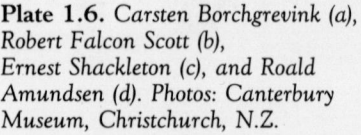

Plate 1.6. *Carsten Borchgrevink (a),
Robert Falcon Scott (b),
Ernest Shackleton (c), and Roald
Amundsen (d). Photos: Canterbury
Museum, Christchurch, N.Z.*

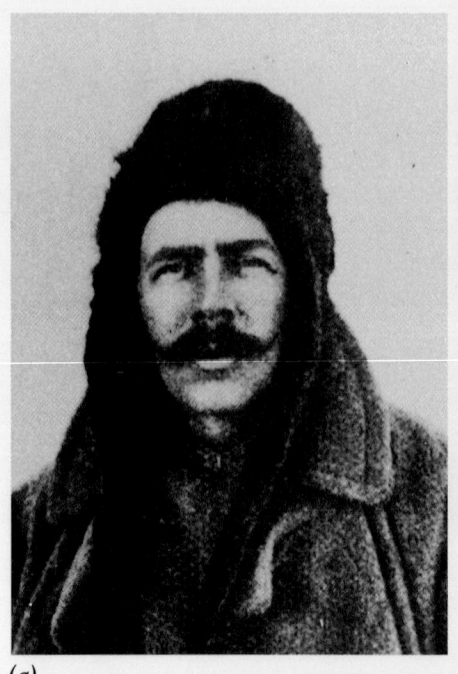

(a)

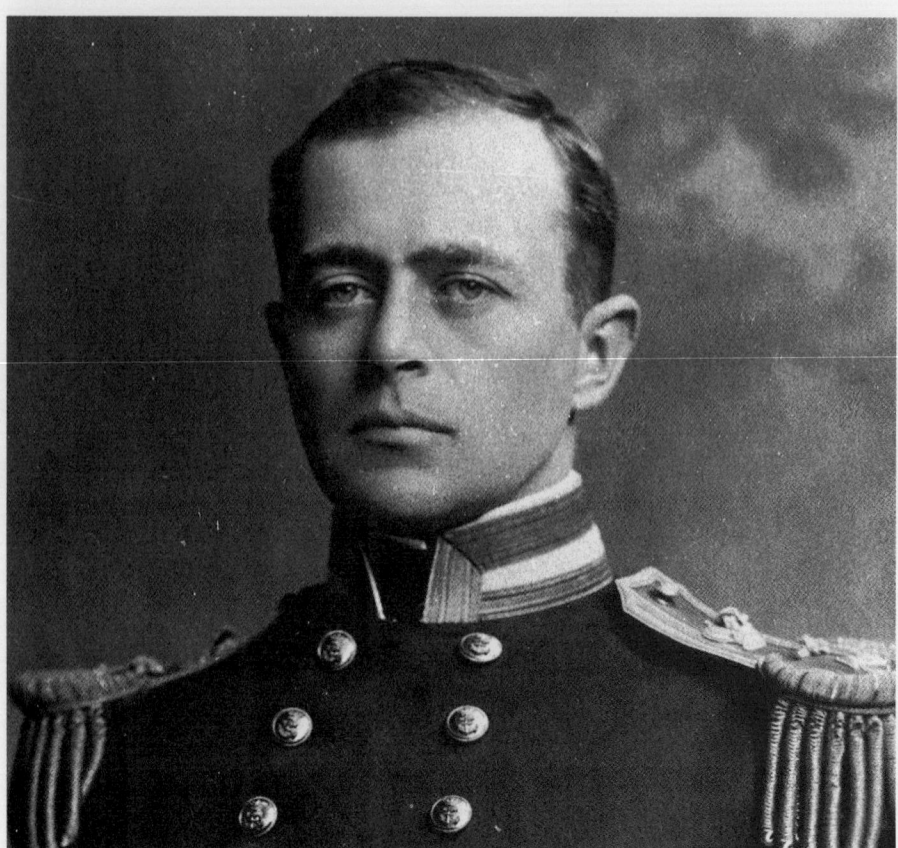

(b)

the winter, in the light of acetylene lamps, the first book to be printed and bound in Antarctica was produced. Titled "Aurora Australis", only about 90 copies of the 120-page edition were printed and bound with wood from packing cases. After August a series of short sledging trips took place and in October Shackleton, Adams, Marshall, and Wild set out with four ponies towards the South Pole. The journey across the Ross Ice Shelf, up the Beardmore Glacier, and across the Polar Plateau was not without incident. Crevasses were a major problem and resulted in the loss of a pony and almost the entire party. The remaining animals were shot and cached for food. The members of the party trudged on but were soon suffering from a shortage of food and the difficult decision had to be made to turn back when they were only 148 kilometres from the Pole. They had now been marching for 72 days. The return trip was "touch and go" but on 1 March they were safely aboard the *Nimrod*.

While the southern party was away, a second group, comprising geologists T. W. Edgeworth David and Douglas Mawson, and Forbes Mackay the physician, had on 16 January 1909 achieved Ross's original ambition and reached the South Magnetic Pole (Plate 1.5). The 2000 kilometre return journey of 122 days took them north from Cape Royds to the Drygalski Ice Tongue and up the Larsen Glacier. From here they continued manhauling to the north-west across the Polar Plateau until the South Magnetic Pole was reached. Although at the time rather overshadowed by Shackleton's southern venture, the journey of David, Mawson, and Mackay remains one of the great sledging trips in the Ross Sea region.

Scott's last expedition

Shackleton's efforts had barely ended when Scott announced plans for his second expedition, the prime objective again being to conquer the South Pole for the British Empire. However, the expedition had other goals which included an exploration of King Edward VII Land and an

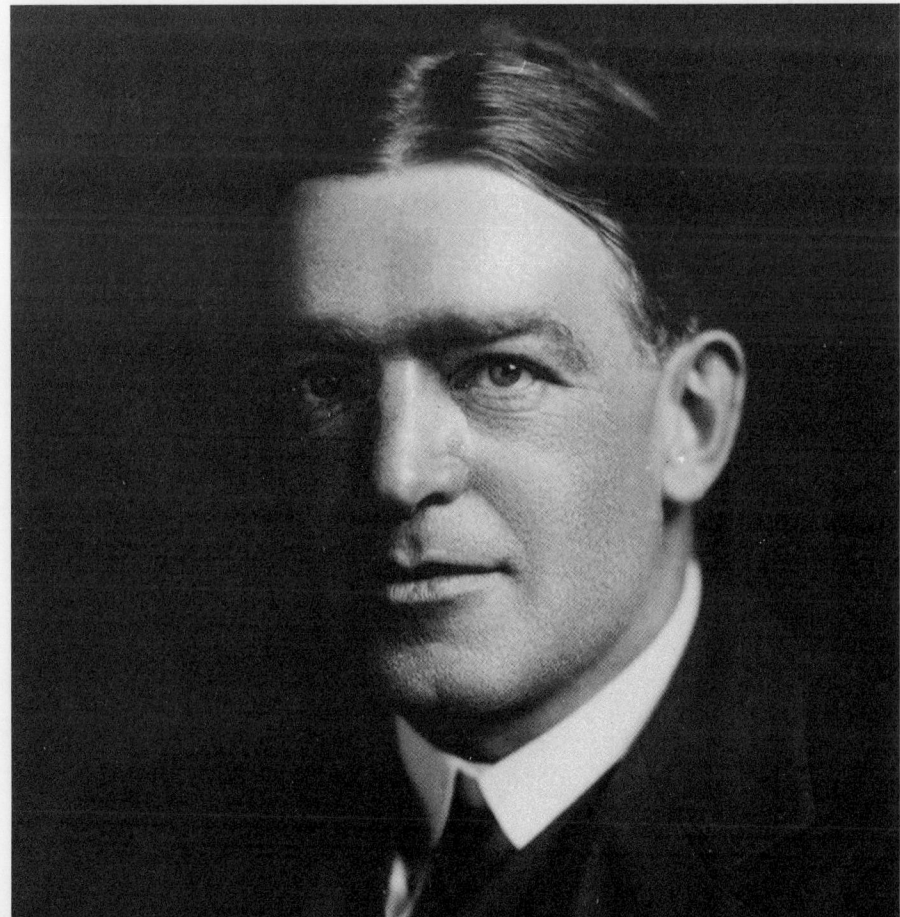

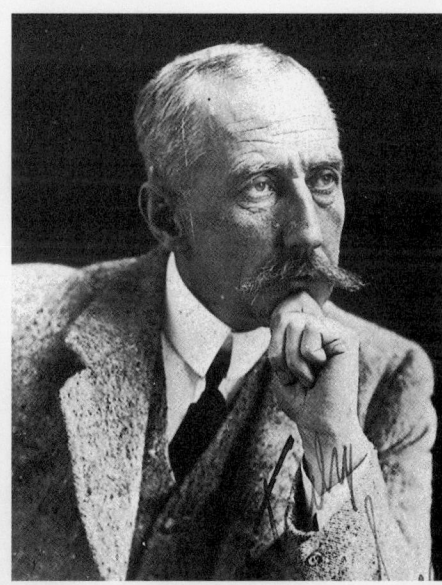

(d)

(c)

extensive scientific programme. But there were other rivals: Japan, Germany, and Norway had also expressed an interest in the Antarctic.

Scott's expedition was to cost an estimated £40 000 and the former Discovery Expedition's relief-ship *Terra Nova* was acquired for £12 500. Problems were encountered in raising finance and New Zealand contributed 600 tonnes of coal, 180 sheep, and 1585 kilograms of butter. The *Terra Nova* departed from Cardiff on 15 June 1910. On board were three, 14 h.p four-cylinder crawler tractors (see Plate 1.9), the first such machines to be used in Antarctica, a 15.5 × 7.7 metre prefabricated hut, and a vast quantity of stores which included 1 tonne of oatmeal, 17 tonnes of self-raising flour, 6 tonnes of dog biscuits, and 1 tonne of pemmican (40% beef and 60% fat). To assist with sledging operations, 19 Siberian ponies and several dogs were procured by Cecil Meares and shipped from Vladivostok to Lyttleton. Twenty-six men, many with Antarctic experience, would make up the shore party.

When Scott reached Melbourne, on his way to New Zealand, he learned that there was another expedition under way. Amundsen (Plate 1.6) having been beaten by Peary to the North Pole, and now equipped with Nansen's famous vessel *Fram*, had decided to make an attempt on the South Pole. On 26 November 1910 the *Terra Nova* departed from Lyttleton. A brief stop was made at Port Chalmers and the expedition was then under way for the Antarctic.

The voyage south was far from comfortable and after 20 days, when they were held up in the pack ice, Ross Island was reached and a base established at Cape Evans only 22 kilometres north from where the *Discovery* was icebound during 1902–03. By late January 1911, sledging parties had begun to lay depots for the following season's operations and the first western geological party, led by Griffith Taylor, had completed a successful spell of field work. Meanwhile, Campbell, unable to establish a base in King Edward VII Land, had found Amundsen at the Bay of Whales (Plate 1.7) and subsequently erected a hut for the northern party

Plate 1.7. *Amundsen's base, Framheim. Photo: Alexander Turnbull Library, Wellington.*

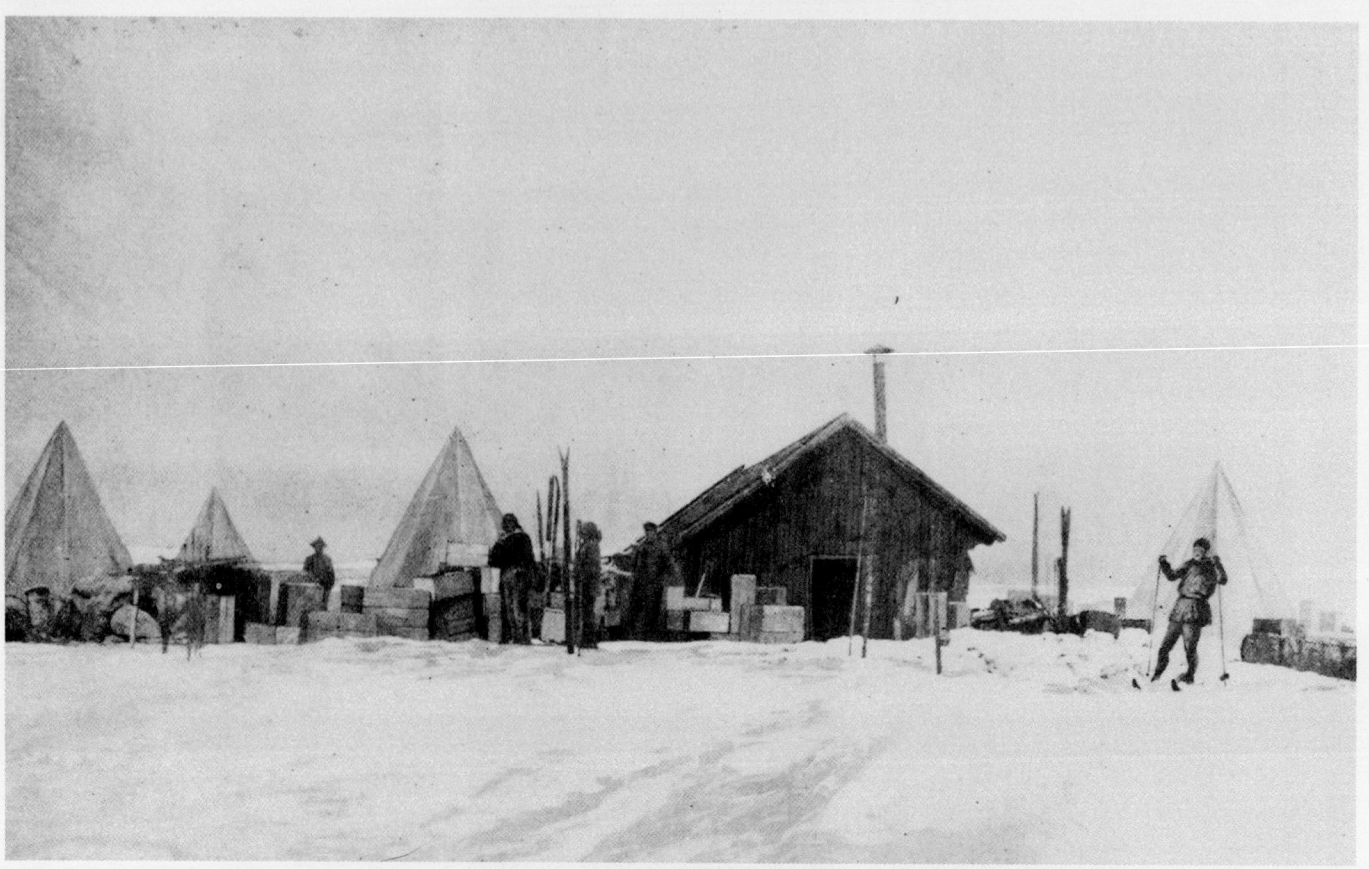

at Cape Adare (where Borchgrevink had wintered 12 years previously).

About this time a Japanese expedition led by Lieutenant Choku Shirase with the ship *Kainan Maru* also visited the Bay of Whales. Apart from a short sledging trip on the Ross Ice Shelf little was achieved on this, Japan's pioneering Antarctic expedition.

At Cape Evans the shore party settled down for the winter and a comprehensive scientific programme was begun. In mid-winter Wilson, Bowers, and Cherry-Garrard made an epic journey to Cape Crozier to observe the breeding of the emperor penguin. Pulling 270 kilograms each, on some days only 3 kilometres were achieved during 8 hours of sledging, and a temperature of –60°C was recorded. By comparison, life at Cape Evans was very comfortable (Plate 1.8).

On 3 November 1911 the main southern journey began when a party of 16 with the remaining 2 motor tractors (Plate 1.9), 10 ponies, and 2 dog teams departed from Cape Evans. Amundsen had started 2 weeks earlier. Soon the tractors were abandoned, the ponies were later shot, and the dog teams turned back from 83°S. On 18 January 1912, Scott, Wilson, Bowers, Oates, and Petty Officer Evans reached the South Pole to find Amundsen with his dog teams had arrived one month before. The party then began the desperate dash for home, only to perish on the way.

While the southern party was away, a second geological party was making exciting new discoveries in Victoria Land. In February the last polar supporting-party returned and the remaining 13 men prepared for a second winter at Cape Evans. Increasing concern was felt for the northern party which the *Terra Nova* had been unable to pick up, and as winter drew near Campbell and his men were forced to excavate a cave in ice on Inexpressible Island. This was to be their home for 6 months and in the spring they were able to sledge 400 kilometres down the coast. They picked up food depots on the way until Hut Point was reached, where a message informed them of the death of Scott and his party (Plate 1.10). In November the second ascent of Mount Erebus was achieved and on the 12th a search party, using Indian mules taken south the previous summer, located the tent containing the bodies of Scott, Wilson, and Bowers. With the return of the *Terra Nova* on 18 January 1913 the expedition ended.

The South Pole reached

Meanwhile, what of Amundsen? The *Fram* had reached the Ross Ice Shelf at the Bay of Whales on 11 January 1911. Depots were laid as far as 82°S and such was Amundsen's mastery of polar travel that on one journey the teams raced back to Framheim from the 80°S depot at a speed of almost 80 kilometres a day.

The polar journey began on 19 October and for Amundsen too the first part of the assault lay across the Ross Ice Shelf. Amundsen was accompanied by Hanssen, Wisting, Hassel, and Bjaaland. Fifty-two dogs pulled four sledges and from the base to 85°S the party travelled at speeds of up to 9 kilometres an hour. As they reached this latitude they, like Shackleton, found mountainous country barring the way. But, sometimes using as many as 20 dogs to a sledge, the party surmounted the steep gradient with its nightmare of crevasses.

Passing Shackleton's farthest point south on 9 December they reached the Pole 5 days later and, to ensure that they had reached its exact location, remained there until the 17th. They erected a tent and raised the Norwegian flag which was to be found one month later by Scott. On 25 January 1912, only 39 days after leaving the Pole and 99 days after their journey began, Amundsen and his party sledged triumphantly into Framheim.

While they were away a party of three under Prestrud had sledged east to explore King Edward VII Land. In the field for 64 days, the men were the first to set foot on the eastern land bordering the Ross Sea.

The final Antarctic ambition

There remained one great journey to be accomplished—a crossing of the Antarctic Continent, and in 1914 Shackleton announced plans for his second expedition. Two parties would be involved. A depot-laying party using the ship *Aurora* would proceed to McMurdo Sound from where food and fuel dumps would be laid across the Ross Ice Shelf to the bottom of the Beardmore Glacier. On the other side of the continent, under Shackleton's command and using the ship *Endurance*, the main crossing party with dogs and motorised sledges would set out. Both parties were to meet with disaster. In January 1915 the *Endurance* became trapped in the pack ice of the Weddell Sea and, after drifting during the winter, sank on 21 November. The entire ship's company camped on ice floes until it was possible to launch the ship's boats and reach Elephant Island. From here Shackleton, after an open boat voyage of 1300 kilometres, reached South Georgia and eventually returned to rescue the rest of the party. The Ross Sea party that had become established at McMurdo Sound also experienced problems. In May 1915, after the preliminary autumn-depot laying, the *Aurora* which had been frozen in at Cape Evans was taken out to sea in a blizzard and drifted in the pack ice for 9 months before reaching New Zealand. A. L. A. Mackintosh, the captain of the ship and leader of the Ross Sea party, and nine others were marooned on Ross Island. Few stores of food, fuel, or clothing had been landed safely ashore. By drawing on Scott's leftover supplies, and by manufacturing their clothing, the party was able to subsist and during the following spring and summer laid the planned final depots across the Ross Ice Shelf. On the return journey, however, the Reverend A. P. Spencer-Smith died and Mackintosh and Hayward,

Plate 1.8. *Scott's Cape Evans hut and Mount Erebus on the first warm, sunny spring day (−26°C), 1911. Photo: Alexander Turnbull Library, Wellington.*

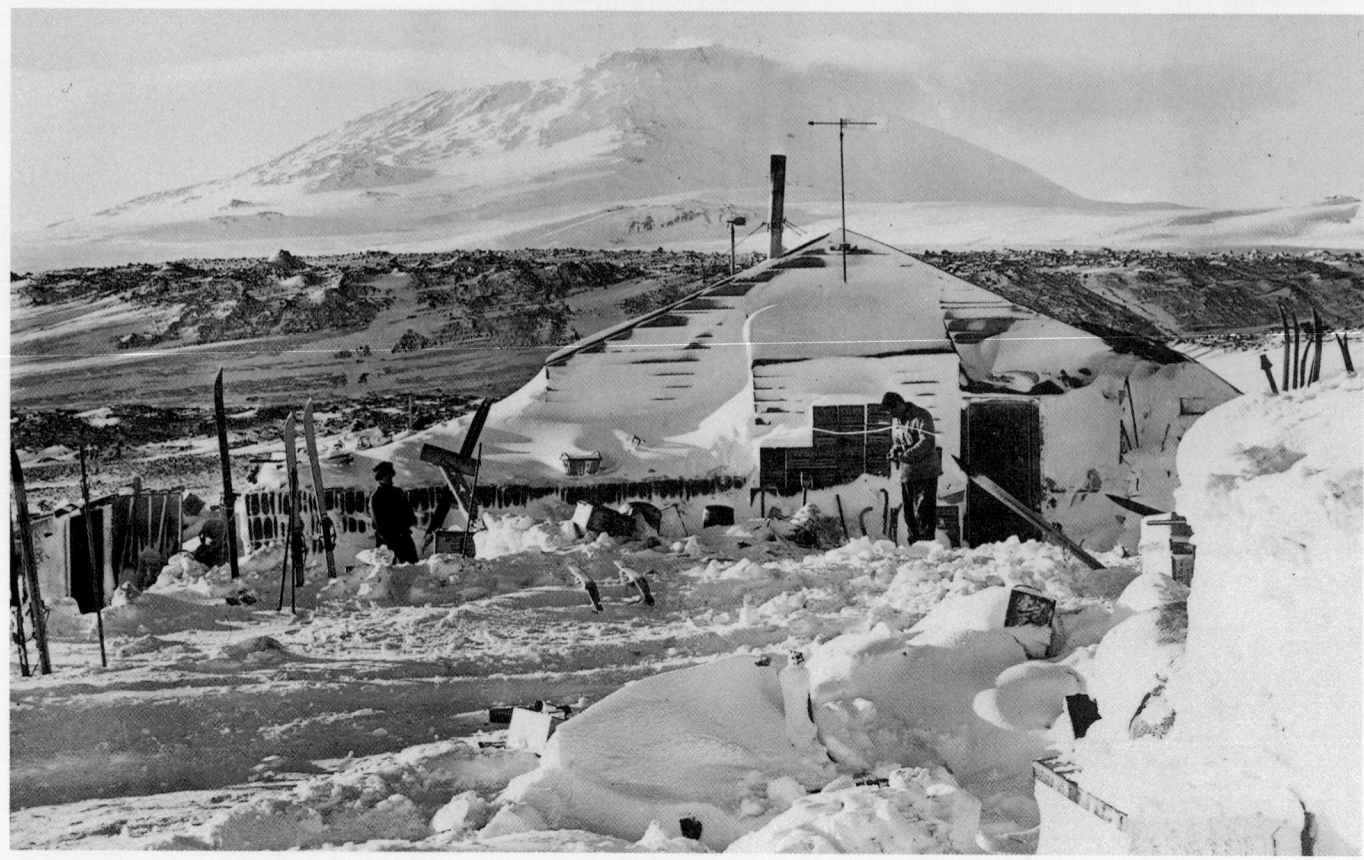

(c)

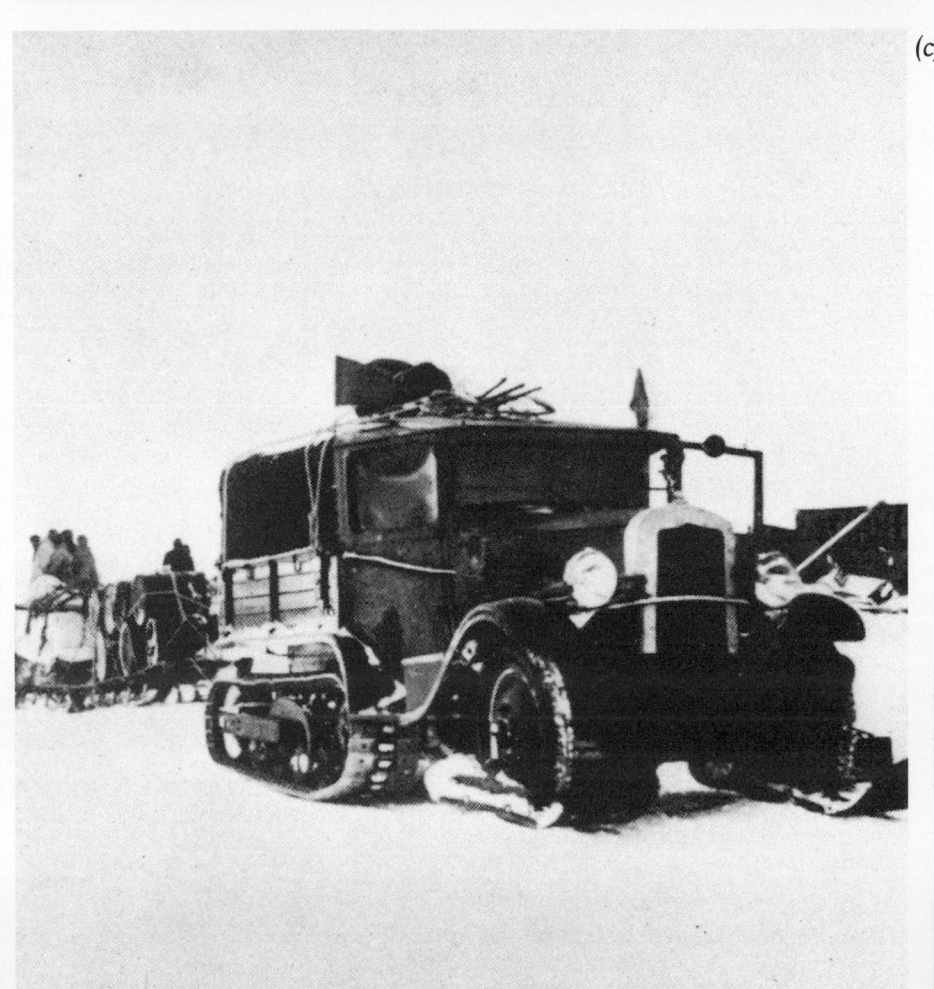

Plate 1.9. *First experiments with mechanical transport: (a) Shackleton's Wolseley car (1908); (b) Scott's crawler tractor (1911); (c) Byrd's Citroen vehicles with rear track-drive and skis under the front wheels (1934). Although Byrd's vehicles travelled several hundred kilometres, it was not until after World War II that mechanical transport was reliable enough for travel over snow. Photos: Alexander Turnbull Library, Wellington.*

(b)

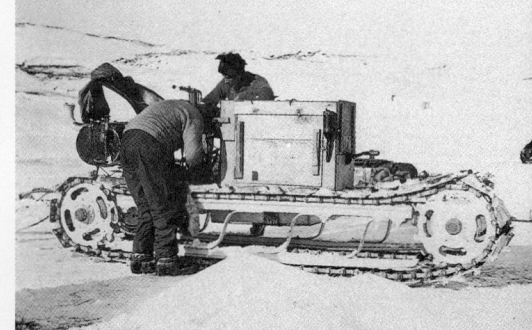

(a)

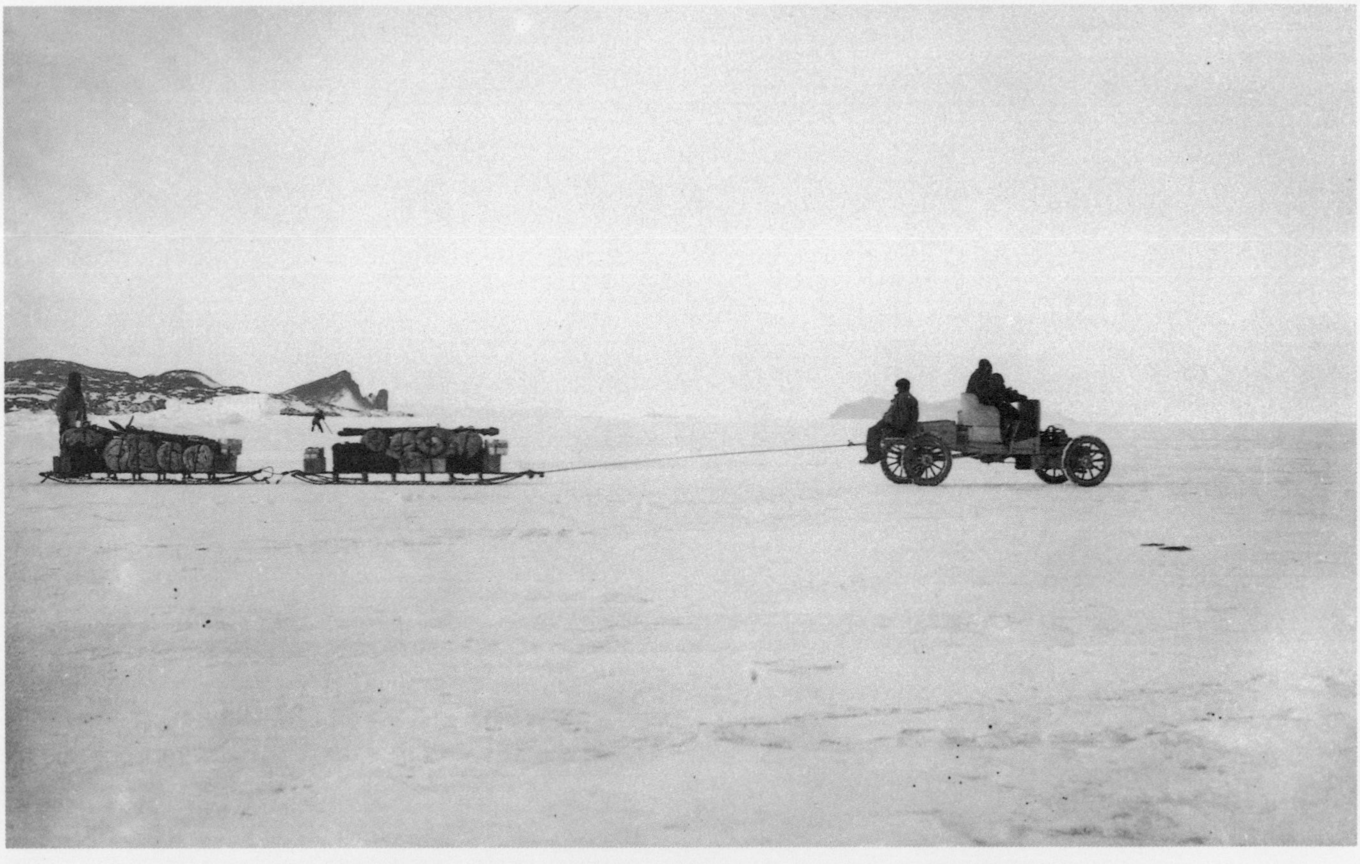

who were also recuperating from scurvy, were lost when they tried to cross the sea ice from Hut Point to Cape Evans. Eventually, the *Aurora*, after repairs in Port Chalmers, and with Shackleton as a passenger, was able to rescue the seven survivors in the summer of 1916–17.

"There be of them, that have left a name behind them" (Ecclesiasticus, xliv, 8)

Ross, out of the desolate Antarctic prospect "wrested an open sea, a vast mountain range, a smoking volcano and a hundred problems of great interest to the geographer". Scott "had a tale to tell of the hardihood, endurance and courage of my companions" which stirred the hearts of every Englishman and many besides. Shackleton had breached the Transantarctic Mountains via the mighty Beardmore Glacier, his men had ascended Mount Erebus and also reached the South Magnetic Pole, he had achieved one of the great small boat voyages in history and his utter loyalty to his men was repaid by the agonising efforts of the *Aurora* party. Amundsen, the *professional* polar traveller reached the South Pole, in what has been called "a model of technical performance", only 13 years after Borchgrevink established human habitation on the continent.

To these leaders and their men, living, working and travelling under conditions unimaginable today, we owe the framework on which we base our knowledge of the Ross Sea region of Antarctica. Everywhere we travel we are among features that they named. The earliest topographical and geological maps were produced by them, as were the first climate and geomagnetic compilations, and the first glaciological theories. Although whaling ships were at work in the 1920s in the Ross Sea, more than a decade was to pass after the *Aurora* party's departure before men were to live there again. With them came superior technology—aeroplanes, radiocommunications, and better diets.

Plate 1.10. *Scott's Northern Party on the return from their enforced wintering in the ice cave on Inexpressible Island. From left to right: Seaman Dickason, Petty Officer Abbott, Petty Officer Browning, Lieutenant Commander Campbell (leader), Raymond Priestley, and Surgeon Levick. Photo: Canterbury Museum, Christchurch, N.Z.*

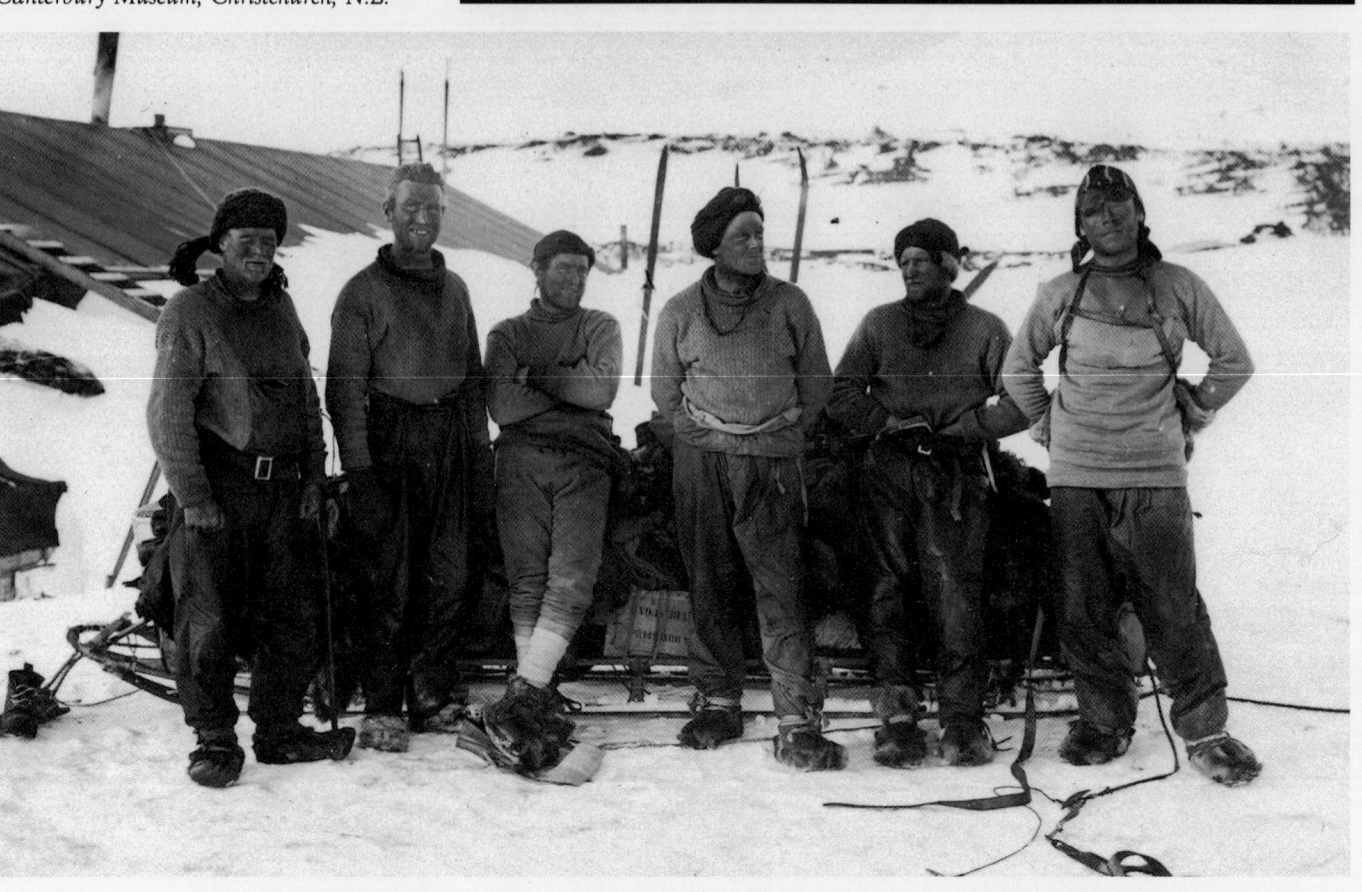

When the *Aurora* sailed out of McMurdo Sound in 1917, taking the survivors of Shackleton's Ross Sea party north to New Zealand, only the debris of the last 16 years of expeditions remained behind. The huts slowly filled with snow. The crosses on Observation Hill, at Hut Point, and at Cape Evans were left unvisited. An era of intensive land-based exploration in the western Ross Sea had ended. It was to be 30 years before anyone would again set foot on Ross Island. Instead, man looked to the eastern side of the Ross Sea, and to the resources of the Sea itself.

Whales and politics

Antarctic waters have been known for some time to be a haven for the large blue and humpbacked whales. Climate and distance had protected them until the early twentieth century when harpoon guns, steam-powered whale catchers, and factory ships ended that happy state. The first onslaught on these mammals occurred around the Antarctic Peninsula, where several ice-free anchorages for mooring factory ships were available.

The Ross Sea was another matter. There were few suitable anchorages and the nearest ports were in New Zealand, over 3000 kilometres away. In 1923, however, Norwegian whaling pioneer C. A. Larsen took a factory ship (appropriately named the *Sir James Clark Ross* (Plate 2.1)) and five whale catchers into the Ross Sea. Larsen anchored his ship against the shelf ice in Kainan Bay on the eastern Ross Ice Shelf and during the summer 221 whales were killed and processed, proving that it was possible to return a profitable catch from Ross Sea whaling. Other companies followed his lead and the increasing whaling interest in the Ross Sea acted as a spur to political interests. In 1908 Britain had claimed the Antarctic Peninsula as British territory. In the early 1920s British authorities hoped to extend this claim to all of Antarctica, and consequently approached British Dominions such as New Zealand and

Plate 2.1. *The factory ship* Sir James Clark Ross, *the first of two ships of this name, which exploited the whales of the Ross Sea. Photo: Alexander Turnbull Library, Wellington.*

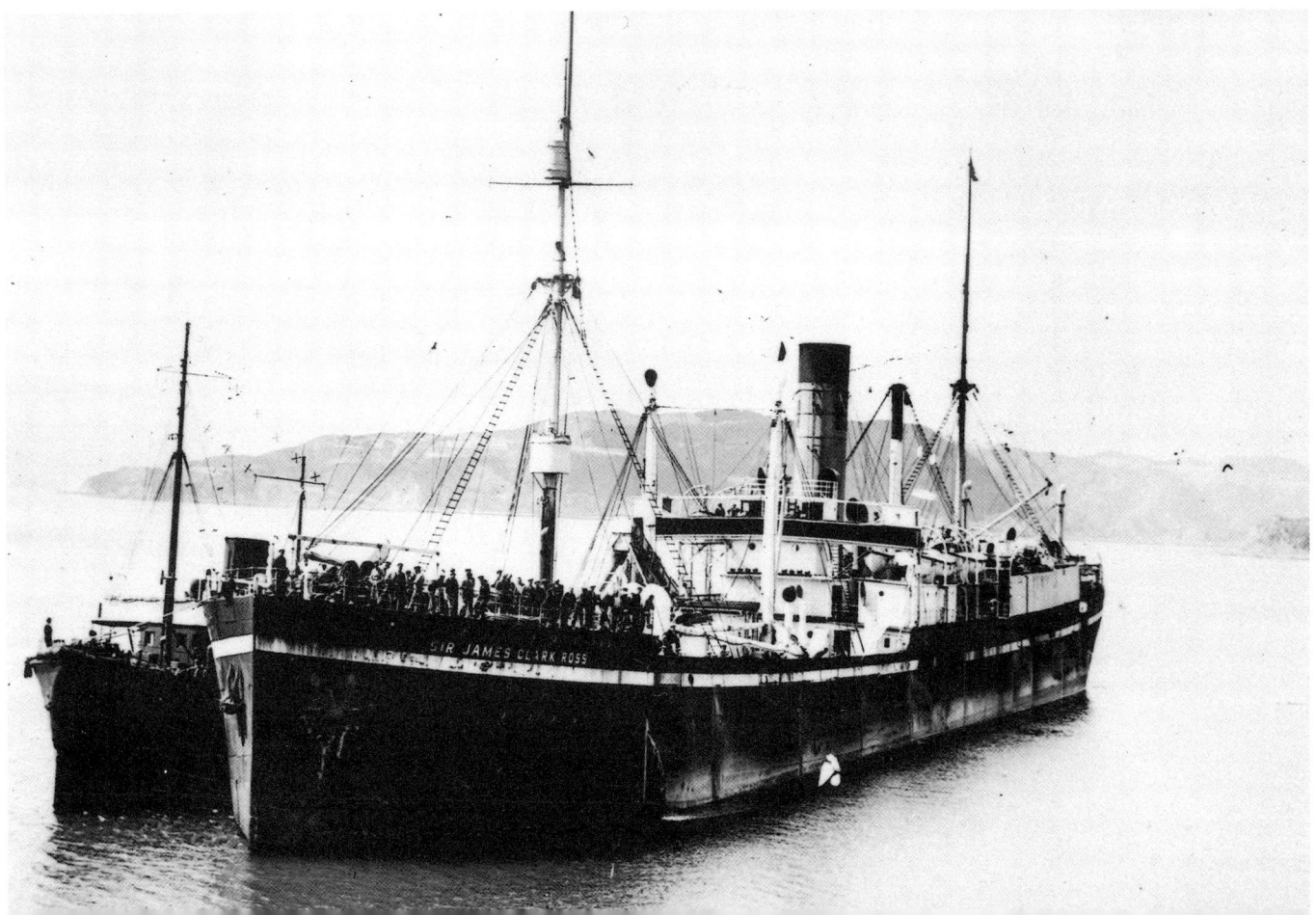

Australia to assist. Larsen applied to the British and New Zealand Governments for a whaling license early in 1923. Both governments hurriedly consulted, and, in line with furthering the earlier 1908 British claim and based on British discoveries between 1841 and 1914, laid claim to the Ross Dependency.

The New Zealand Government promptly dispatched an officer of its Marine Department as an observer with Larsen's first whaling expedition, to ensure that New Zealand whaling regulations were observed. Specific whaling regulations for the Ross Sea were introduced in 1924 and 1927, and New Zealand levied a sum on whaling licenses and whale oil. But this situation was not to last.

In the mid-1920s factory ships were developed for pelagic whaling. They could operate anywhere at sea, and needed no anchorages. Consequently, whaling companies refused to recognise national controls over whale catches. At the same time, the number of factory ships increased and over 14 100 whales were killed and processed in the Ross Sea between 1927 and 1931 (Plate 2.2). It was the usual tale of unrestricted exploitation destroying both the resource and the industry. By the late 1930s the great whales of the Ross Sea had largely disappeared.

The Byrd expeditions, 1929–30 and 1933–35

While the large mammals of the Ross Sea were facing extinction from the whaling activities in the late 1920s, a new exploring initiative was taking place. An American patrician, Richard Evelyn Byrd (Plate 2.3), captured widespread media attention in 1926 by flying over the Arctic. In an era when flying was opening up new frontiers, Byrd saw the applications which lay in aviation in Antarctica.

With the backing of family money and sponsorship from wealthy citizens such as the Rockefellers and Ford, Byrd put together a substantial expedition. In January 1929 he established a large base camp

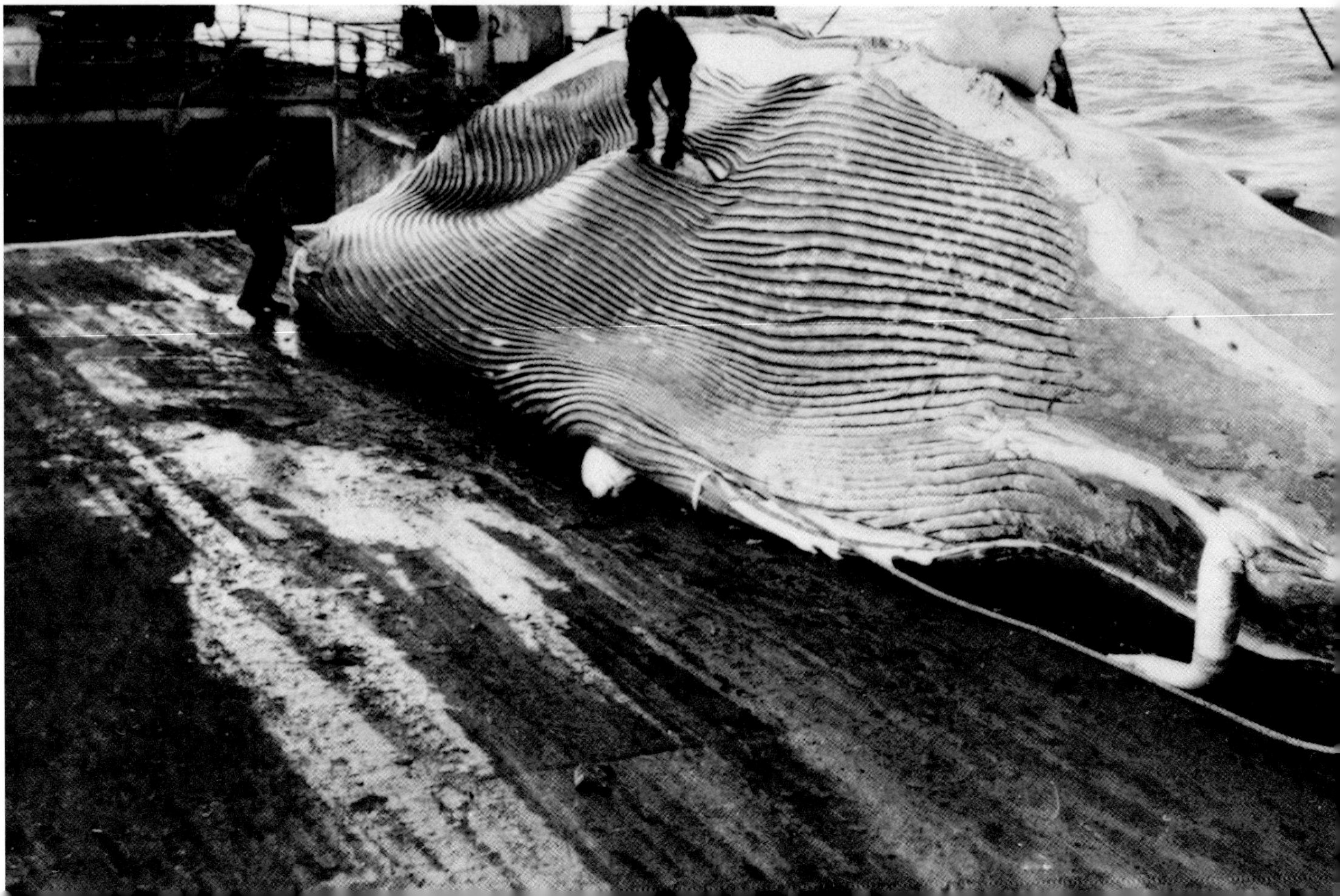

Plate 2.2. *Flensing a blue whale, 30 metres long. The whale is cut along the back and the blubber stripped off in sheets with the aid of a block and tackle or small crane. Photo: Alexander Turnbull Library, Wellington.*

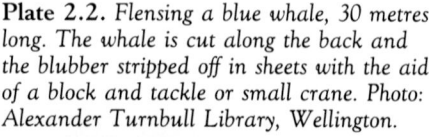

in the Bay of Whales, the site of Amundsen's Framheim base (long since disappeared with the calving of the Ross Ice Shelf). Byrd's base, Little America, served as home for 42 over-winterers. And from there, the expedition's three aircraft ranged east and south, covering, in a few hours, ground that had taken Amundsen weeks to cross. In the east, the Rockefeller Mountains were discovered. On 28 and 29 November 1929 Byrd, his pilot Bernt Balchen, and three others flew in their Ford Trimotor across the Ross Ice Shelf, up the Liv Glacier, and over the Pole (Plate 2.4). In less than 24 hours they completed Amundsen's 99 day trip and thus convincingly demonstrated the value of using aircraft in Antarctica. At the same time, a six man party led by Lawrence Gould sledged with dogs across the ice shelf and made a geological reconnaissance of the Queen Maud Mountains from the Liv Glacier, along the foothills and 200 kilometres east; a journey of over 2000 kilometres lasting 77 days.

The expedition left Little America on their support ship, the *City of New York*, in February 1930 and Byrd returned to the United States to be lionised. But he was not content to rest on his laurels; a fascination with the Antarctic had seized him. He was to return there four more times and become the personification of United States interest in Antarctica.

Within 3 years Byrd had organised a second, larger expedition to carry out further exploration and a range of research. In late January 1934 he re-established his base (Little America II) at the Bay of Whales. This time he had four aircraft (including an autogyro) and 56 men wintered-over. Byrd himself spent from March until August alone at a site 200 kilometres south of Little America II before being rescued, suffering from carbon monoxide poisoning and, perhaps, the effects of solitude.

During the summer months aerial reconnaissance covered extensive areas of Marie Byrd Land. Also during the expedition, tracked vehicles

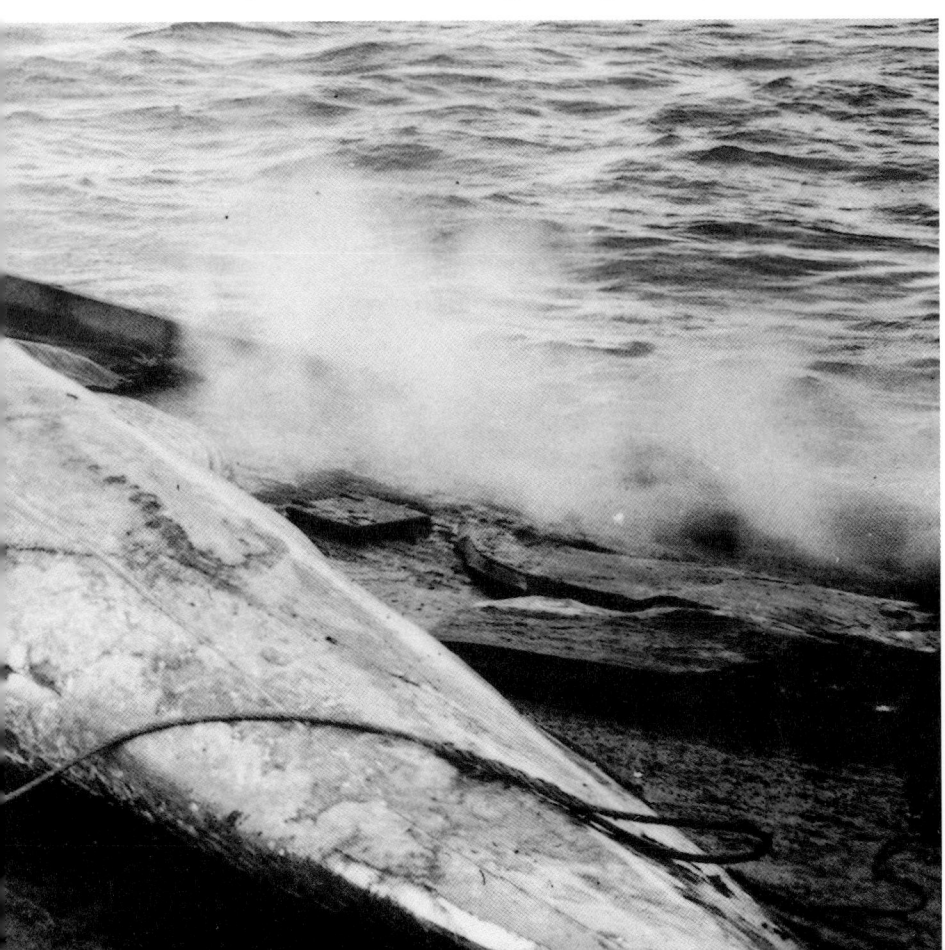

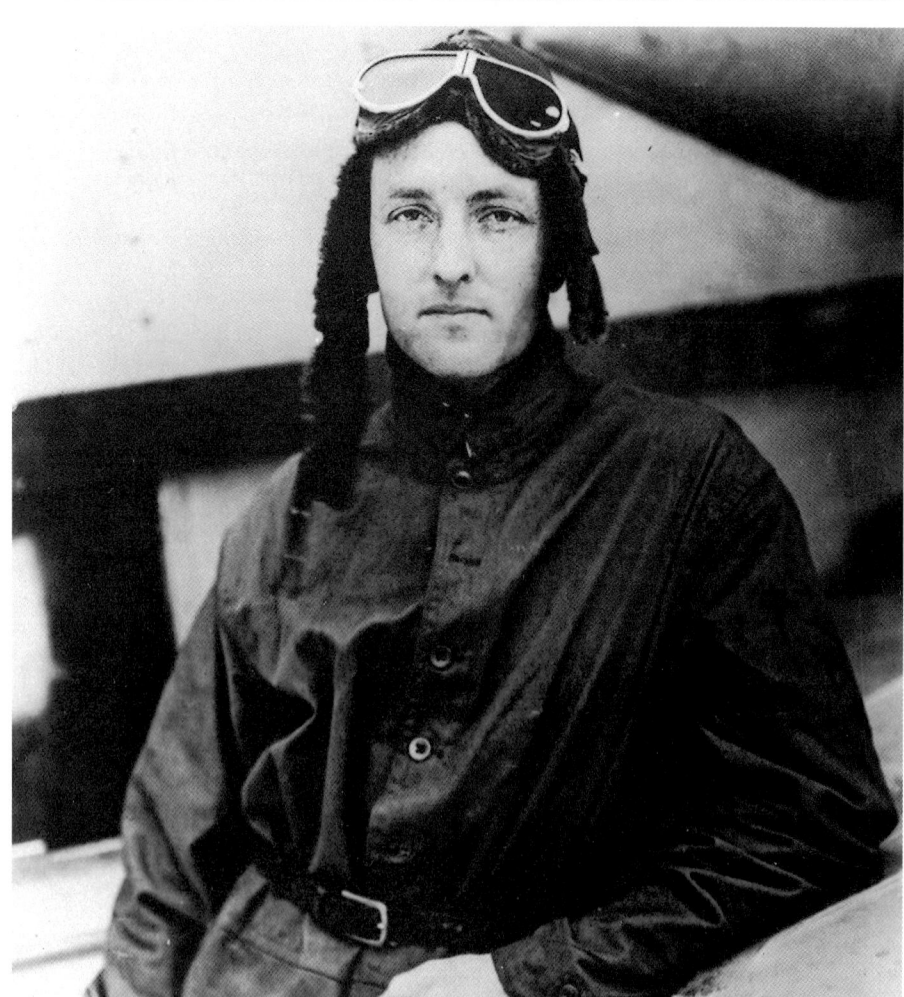

Plate 2.3. *Richard Evelyn Byrd in 1925 at the beginning of his polar career. The following year he was to fly over the North Pole with his pilot, Floyd Bennett. Photo: Canterbury Museum, Christchurch, N.Z.*

Plate 2.4. *The Ford Trimotor plane "Floyd Bennett" with which Byrd, and his pilot, Bernt Balchen, made the first flight over the South Pole. Photo: Canterbury Museum, Christchurch, N.Z.*

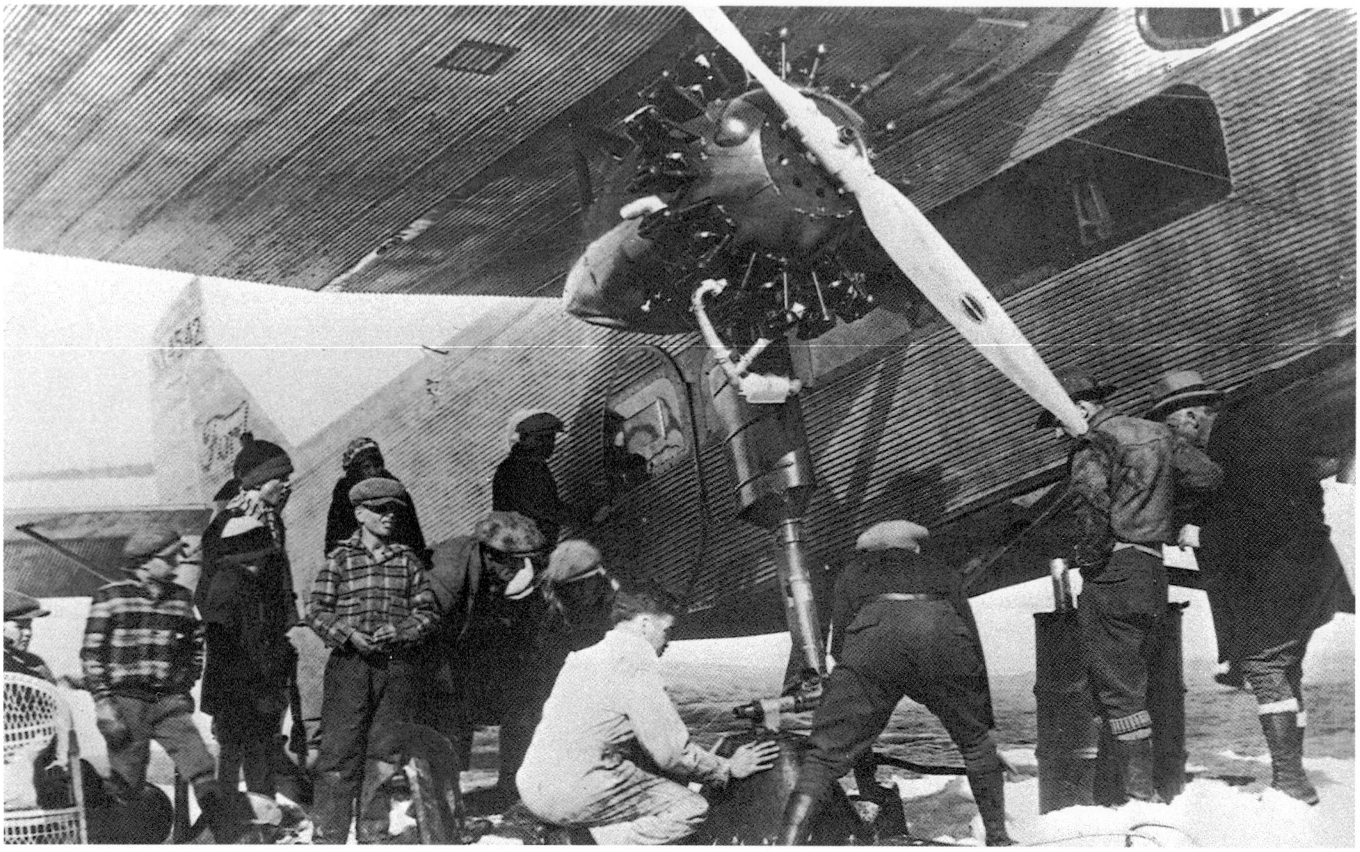

were used successfully for Antarctic travel for the first time; Citroen tractors (see Plate 1.9) covered 193 kilometres in one 16-hour-stretch. Nevertheless, the expedition also relied on traditional dog travel: 143 huskies had been taken south and these proved to be the backbone of two major overland journeys. A four man party under Paul Siple sledged to the Edsel Ford Range and a three man party led by Quin Blackburn repeated Gould's 1929–30 trip to the Queen Maud Mountains and ascended the 100-kilometre-long Thorne Glacier. They turned for home a mere 300 kilometres from the Pole and by their return to Little America II had covered 2200 kilometres in under 3 months.

Byrd's 1933–35 expedition was a major success in exploration and in the wide range of scientific research done. It was also the end of an era. The expedition showed that the days of large-scale private expeditions along the lines of Scott, Shackleton, Amundsen, and even of Byrd, with his access to large resources, were over. The costs were too great. Mounting an expedition of any size now called for government support, bringing with it government-scale administration and the mixed motives of public expenditure.

The Late Thirties—more planes and more politics

At the time Byrd was planning and running his 1933–35 expedition, another American, Lincoln Ellsworth (Plate 2.5), was attempting to make a personal dream come true by flying across Antarctica. Attempts in 1934 and 1935 failed at the Bay of Whales, and at Snow Island in the Antarctic Peninsula. Ellsworth had a personal fortune to help him sustain these setbacks though, and what he lacked in organising ability and leadership he made up for in perserverance. On 23 November 1935 Ellsworth and a Canadian pilot, H. Hollick-Kenyon, took off from Dundee Island (at the northern end of the Antarctic Peninsula) in a single-engined Northrop aircraft for what was meant to be a 16-hour flight to the Bay of Whales. In fact it took them 12 days with three

Plate 2.5. *In the front row of the crew of one of Lincoln Ellsworth's four expeditions to Antarctica are three of the pioneers of Antarctic aviation— Bernt Balchen (left), Lincoln Ellsworth (in belted jacket), and Sir Hubert Wilkins (with beard). Photo: Alexander Turnbull Library, Wellington.*

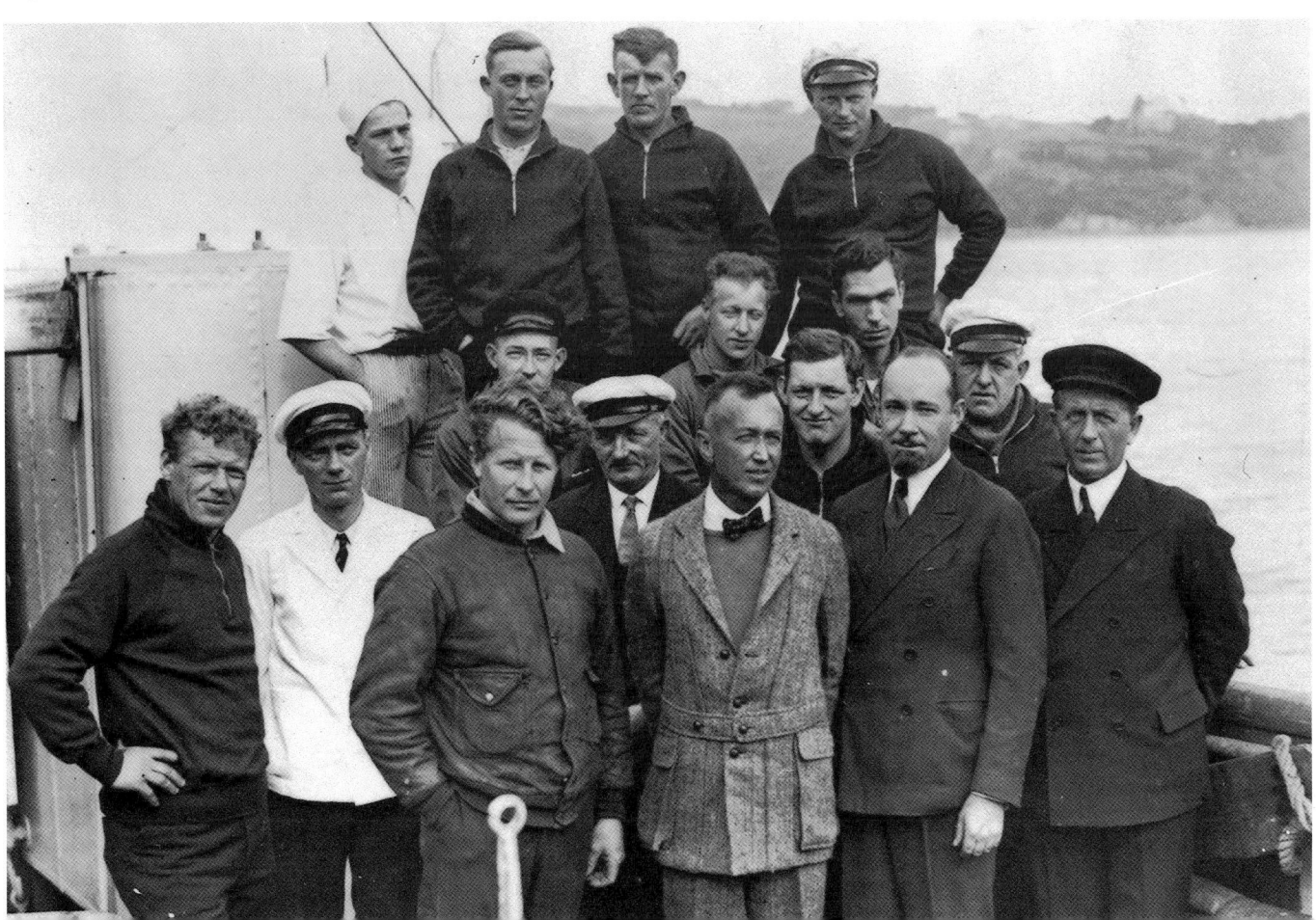

landings *en route* to get there, followed by a month-long wait until the *Discovery II* arrived to rescue them, closely followed by Ellsworth's support ship *Wyatt Earp*. These flights, with landings in unprepared terrains and made in bad weather, were a new departure in polar aviation.

The final Ross Sea expedition before World War II brought something new in terms of financial resources and material. In 1938 President Roosevelt was persuaded to support a permanent United States presence in Antarctica to consolidate that country's interests, in another version of the British 1920s "red-on-the-map" approach. Early in 1940 "West Base" under the leadership of Paul Siple was set up at the Bay of Whales ("East Base" being set up in the Antarctic Peninsula). Over the next year Curtis-Wright Condor planes flew throughout Marie Byrd Land and over the Beardmore and Scott Glaciers. A 1900-kilometre sledging trip was made to the Flood Range and a seismic station was set up in the Rockefeller Mountains. In 1941 these activities were brought to a premature end for President Roosevelt ran into funding problems with a Congress which saw little point in the million dollar plus expenditure at a time when the international situation was deteriorating. The expedition was hurriedly evacuated early in 1941.

During the war years of 1940–45 the Ross Sea sector of Antarctica was largely unvisited, except for one occasion. Early in 1941 the German merchant raider *Komet*, having mined port approaches off Australia and New Zealand, sailed south in search of the Norwegian whaling fleet. In doing so, the *Komet* ventured to the edge of the Ross Sea, and visited the Balleny Islands, before heading west to rendezvous with other raiders at the Kerguelen Islands.

"Highjump" and "Windmill" 1947–48

When the United States Antarctic Service pulled out of the Bay of Whales in 1941, there were no plans for a return. Over the previous 11

Plate 2.6. *The ice-runway at McMurdo Sound which can receive wheeled aircraft as large as this C-141 Starlifter, in the early part of the summer season.*

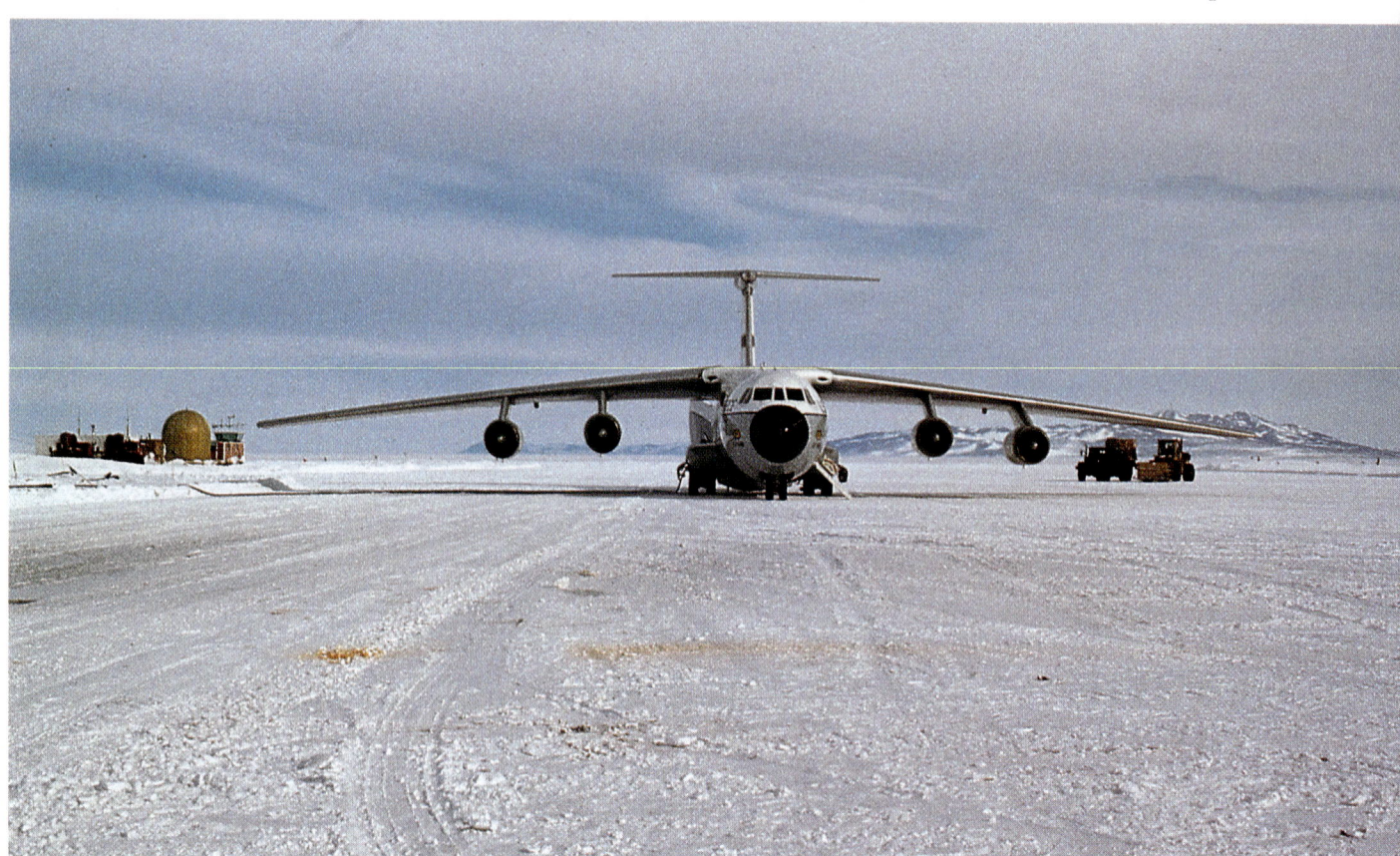

years the United States expeditions promoted by Admiral Byrd had
explored a large part of the eastern Ross Sea, eastern Ross Ice Shelf, and
Marie Byrd Land. Indeed, Byrd had acted as organiser, lobbyist, and
fundraiser for a succession of United States efforts, influencing the whole
direction of United States Antarctic activity. Byrd's influence, however,
though still strong, was about to wane.

In 1946 United States naval planners were looking at staging a
major cold-climate training exercise, primarily with their minds on the
deteriorating state of relations with the Soviet Union, but also cognisant
of a large post-war fleet on their hands in search of something to do.
There were, however, sensitivities to exercising in the Arctic. Byrd,
aware of these plans, prevailed on the Pentagon and the State
Department to use the opportunity for a major Antarctic-wide
exploration and "Operation Highjump" was born. But this time Byrd,
though nominally the leader of the exercise, was merely the figurehead.
The moving forces were a new generation of United States naval officers
who were to play a major part in United States Antarctic activities in the
next decade.

"Highjump" was a massive operation, involving over 4000 men and
13 ships operating in 3 task forces around the continent. The primary
achievement of the expedition was a sweeping coverage of the continent
by aerial photography (without, however, groundtruth and fixing; causing
the photography to be largely worthless for mapping). A small summer
camp was established, again at the Bay of Whales. Land-based activities
were limited, but 29 flights were made by 6 R-4D aircraft along the
section of the Transantarctic Mountains from the Ohio Range north to
Mackay Glacier. The aircraft and camp were abandoned at the Bay of
Whales when the expedition left. On their way north, the *Northwind*
and Admiral Cruzen paid a brief call into McMurdo Sound, visiting the
huts of Scott and Shackleton. The Admiral and his staff were thus the
first human callers for over 30 years.

*Plate 2.7. Ships drawn up against the sea
ice to offload cargo during January 1957.
The New Zealand supply ship HMNZS
Endeavour is in the foreground. Stores had
to be sledged 12 kilometres to Scott Base. By
the early 1960s the U.S. had established a
port facility at Hut Point. Photo: A. Heine.*

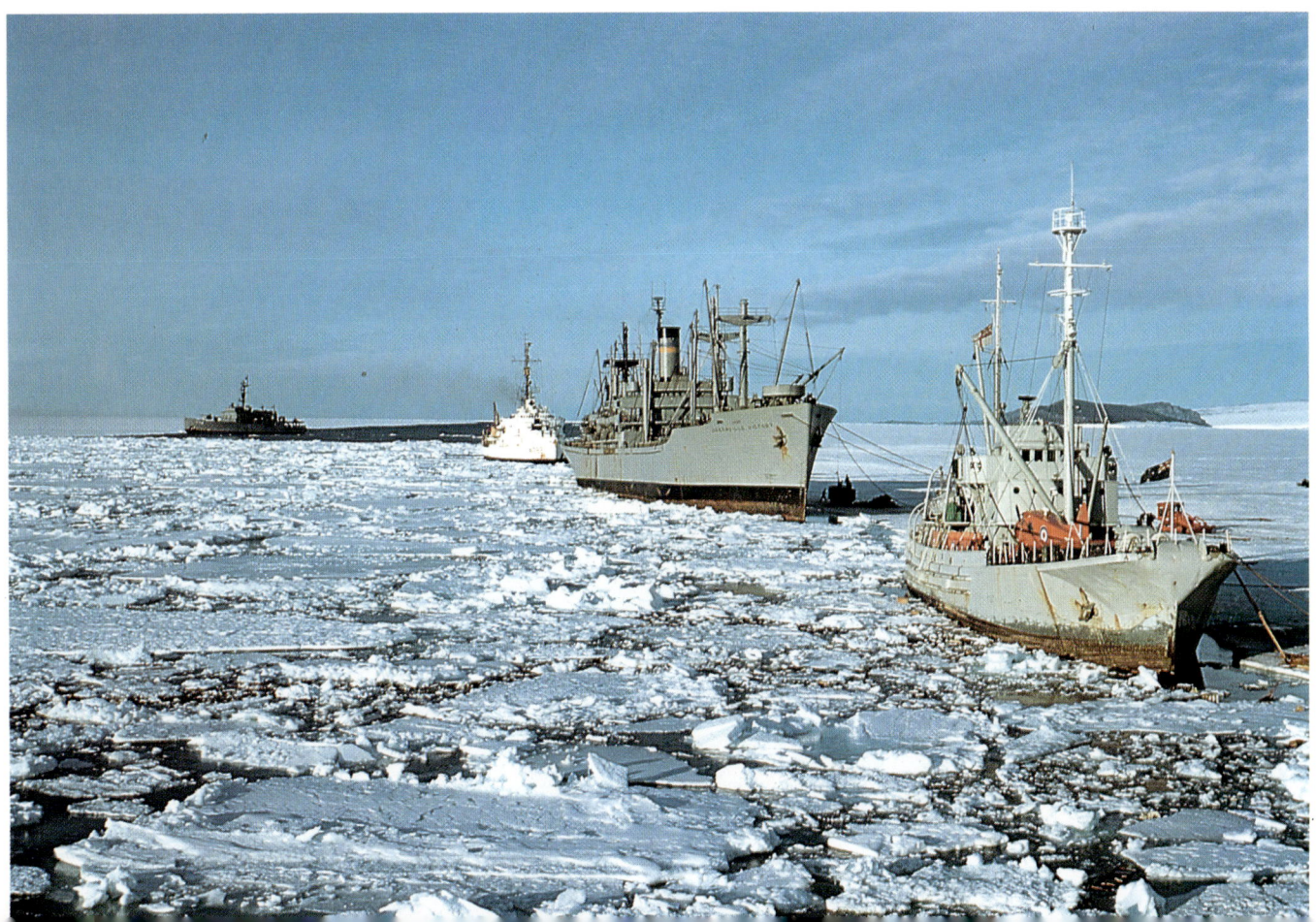

Plate 2.10. *The joint U.S.–N.Z. Hallett Station was established at the far end of a spit extending from Cape Hallett (named after Thomas Hallett, purser of Ross's ship Erebus). The Admiralty Range is in the background. Photo: B. Wood.*

▶

Plate 2.8. *McMurdo Station in 1958, viewed from Observation Hill. Note the open water in the distance and the channel cut through the sea ice by icebreaker.*

Plate 2.9. *The first Scott Base, 1958. Stores were sledged from the supply ship, using Massey Ferguson farm tractors. Photo: A. Heine.*

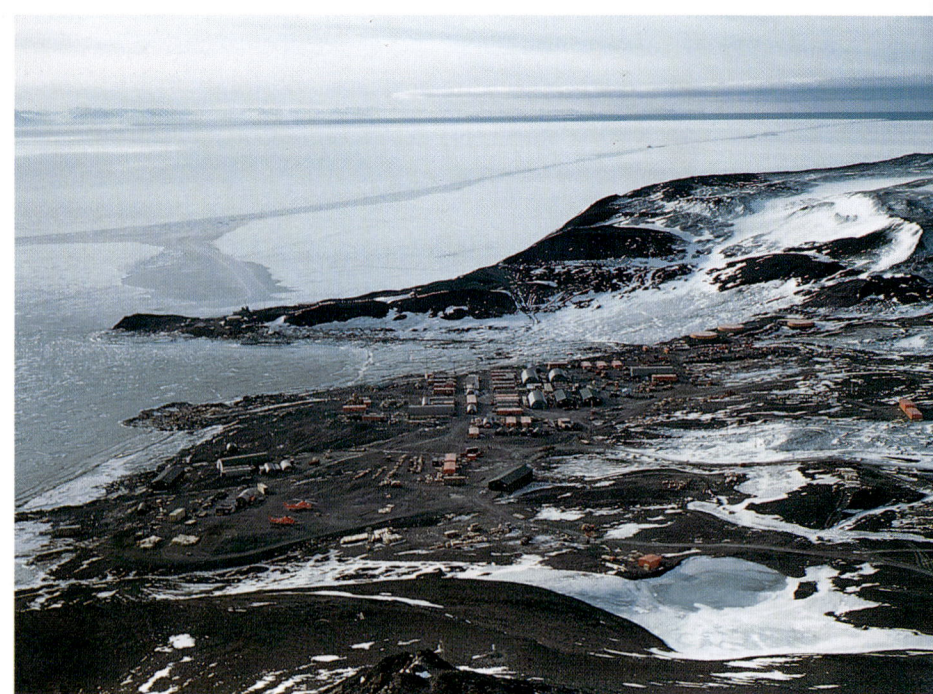

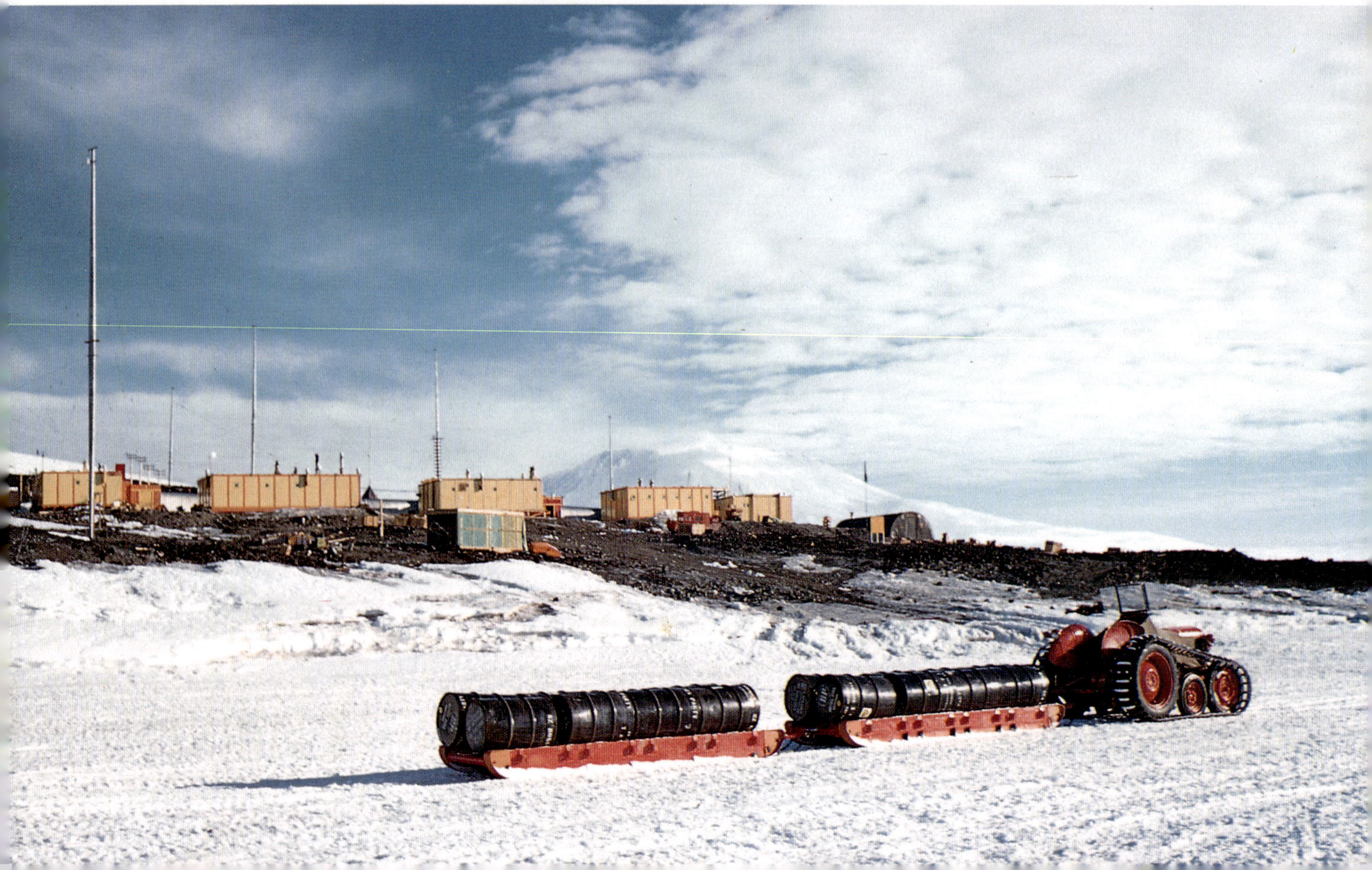

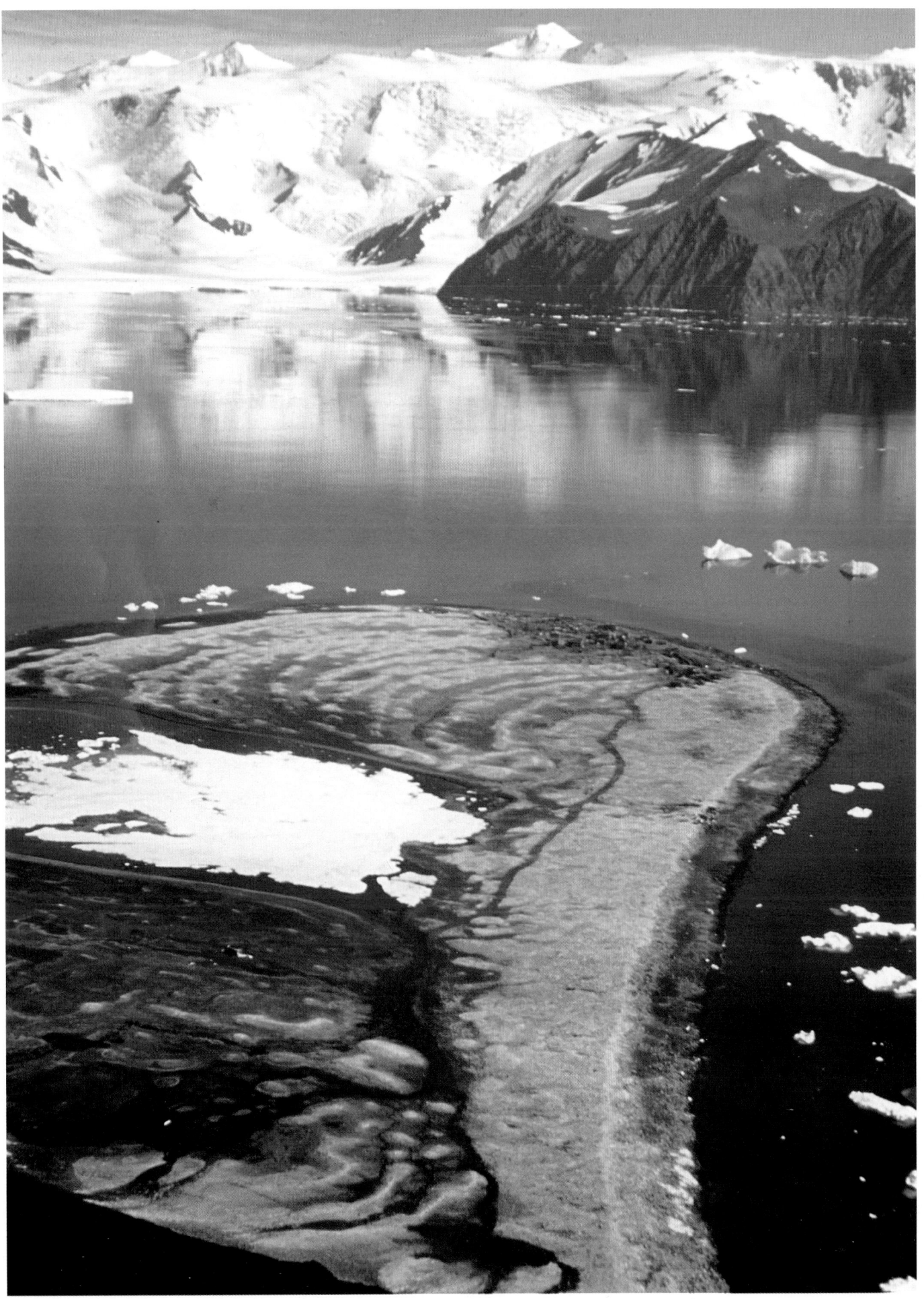

"Highjump" was followed in 1948 by the smaller "Operation Windmill", which gathered some ground fixes for the "Highjump" photography. Byrd and others planned to mount another major expedition in 1949, but the proposal was cancelled on the orders of President Truman who saw little value in further Antarctic exploration. The unsettled relationship between Truman and Byrd's powerful senator brother provided a political reason for not furthering publicity for the Byrd family. Interestingly, there were well-progressed plans in New Zealand for Antarctic expeditions in 1946 and 1949, though proposals were cancelled on the grounds of cost. Consequently, the Ross Sea region was again left to its solitude, this time for another 8 years.

The International Geophysical Year

The final lead-up to the establishment of permanent stations in the Ross Sea began in 1950. In that year the International Council of Scientific Unions (ICSU) began to formulate proposals for an International Geophysical year (IGY) in 1957–58. In 1953 ICSU identified Antarctica as a key area for geophysical research and urged that as many observing stations as possible be set up on the continent as a contribution to the IGY. Besides science though, the spirit of physical adventure was also present for in Britain Vivian Fuchs revived Shackleton's dream of a land crossing of the continent.

The two influences of science and geographical exploration were to have a major impact in the Ross Sea sector of Antarctica. The scientific community in the United States persuaded the Federal Government to fund a return to Antarctica and in late 1953 the United States announced it would set up IGY observing stations in Marie Byrd Land, at Little America, and on the Polar Plateau. In fact, the United States opted for the prime geographic site, the South Pole itself.

Meanwhile, Antarctic interests were stirring in New Zealand and science and exploration lobby groups put intense pressure on the

Plate 2.11. *The RNZAF Beaver aircraft, offloaded on the sea ice from HMNZS Endeavour, has been assembled and fuelled to support the Trans-Antarctic Expedition. Photo: A. Heine.*

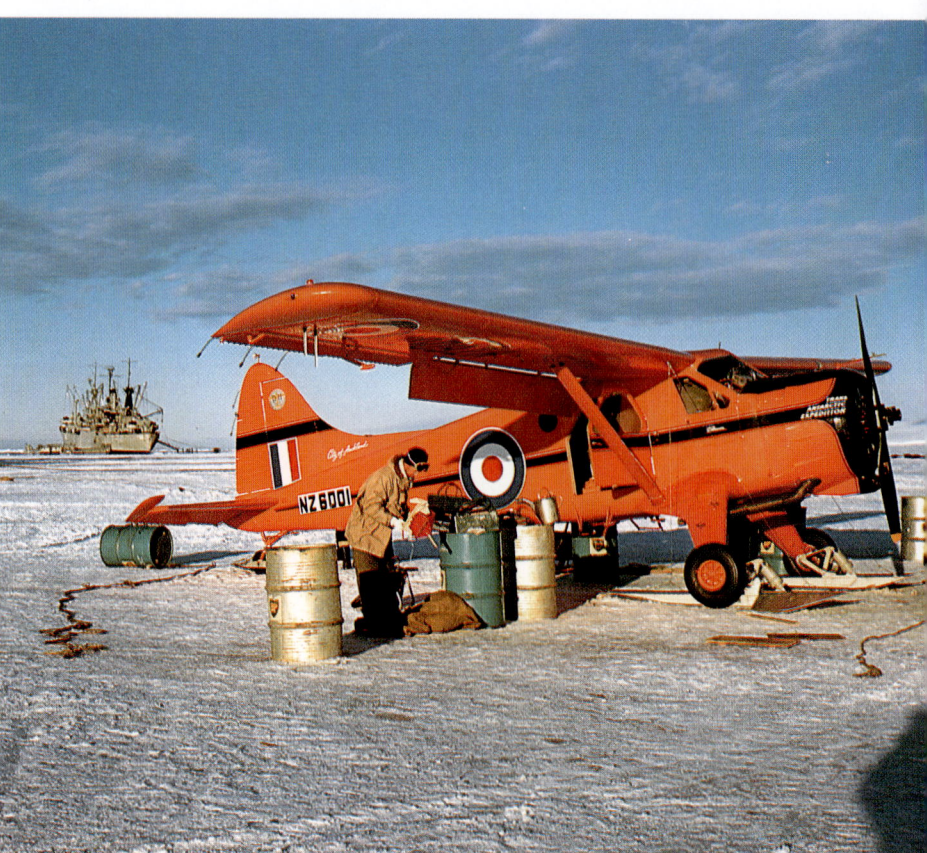

Government to set up a station on the Victoria Land coast. The main purpose of the station was to support Fuchs' Trans-Antarctic Expedition (TAE) but there was also to be an added IGY team.

In December 1955 the United States IGY effort got underway. A navy fleet of seven ships under the command of Admiral Dufek established a new Little America, this time at Kainan Bay as changes to the ice front had removed the old Bay of Whales. The same fleet also landed a party at McMurdo Sound, with the task of creating an airstrip there (Plate 2.6) (it had originally been planned to have a compacted snow runway at Little America). This gave aircraft flying out of New Zealand a staging post in establishing the inland stations. After checking Cape Royds, Cape Evans, and Marble Point, the "Naval Air Facility" was set up near Scott's 1901–03 hut at Winter Quarters Bay. From these small beginnings was to grow the main United States Antarctic base. Accompanying the United States party were three New Zealanders who examined possible base sites on the western side of McMurdo Sound as well as an access route to the Polar Plateau via the Ferrar Glacier.

The activity in the Ross Sea in the summer of 1955–56 reflected the beginning of the construction of bases across the whole continent. In addition to New Zealand, the United States and the existing occupants of Antarctica (Argentina, Australia, Chile, and United Kingdom) France, Belgium, Norway, Japan, and the USSR all planned IGY expeditions. The Japanese had been interested in a site near Cape Adare in the Ross Sea but the New Zealand Government was not keen on the idea, it being only 10 years since the war with Japan. Consequently, when the United States offered to build a jointly-manned United States–New Zealand station at Cape Hallett the offer was quickly accepted.

Preparations for both the IGY and TAE moved into full swing in October 1956. The McMurdo Naval Air Facility was by then able to receive a succession of large transport aircraft from New Zealand. Further north, Hallett Station was set up on Seabee Hook by the simple

Plate 2.12. *A ski-equipped R4D (Dakota) aircraft—the principal air support of U.S. field parties in the late 1950s and early 1960s.*

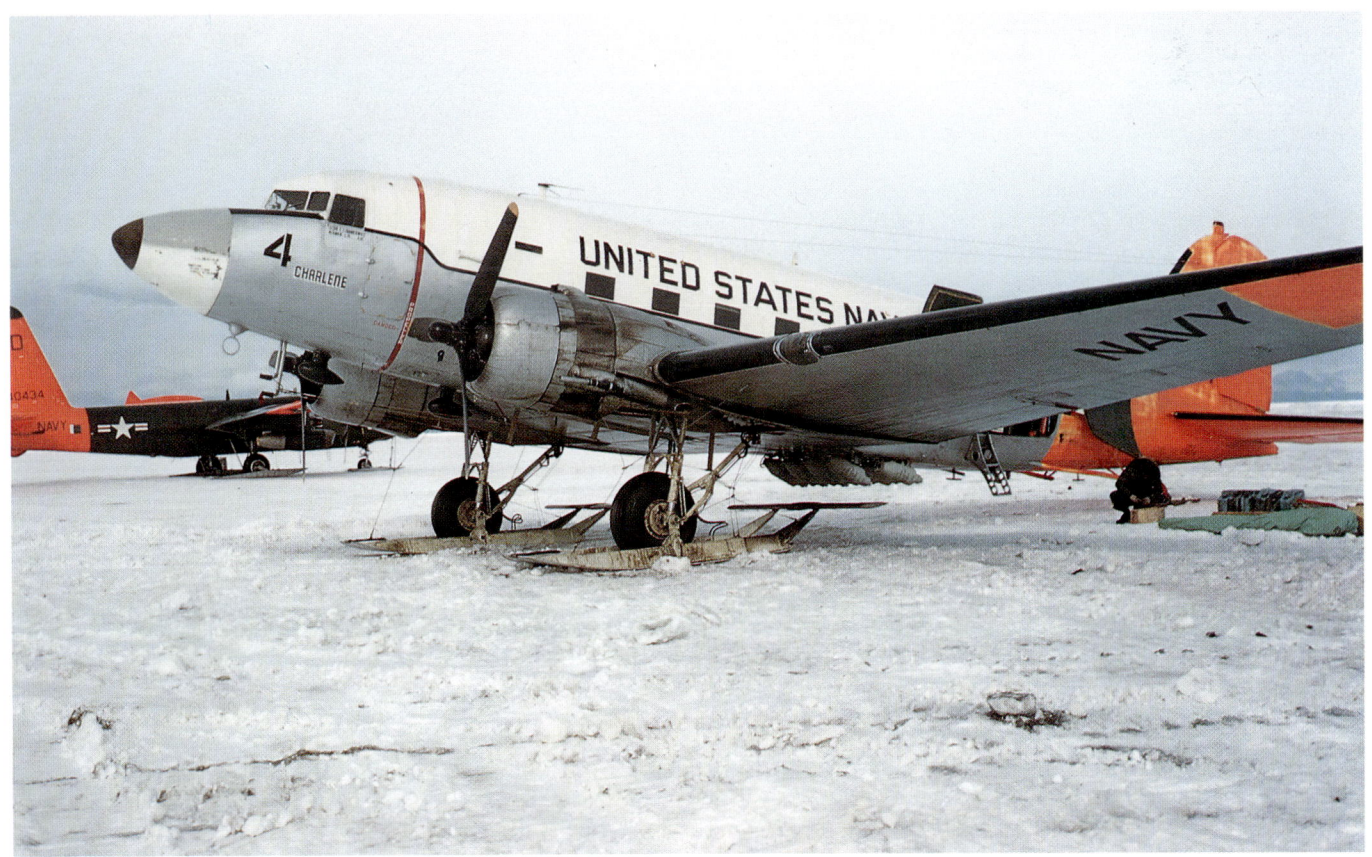

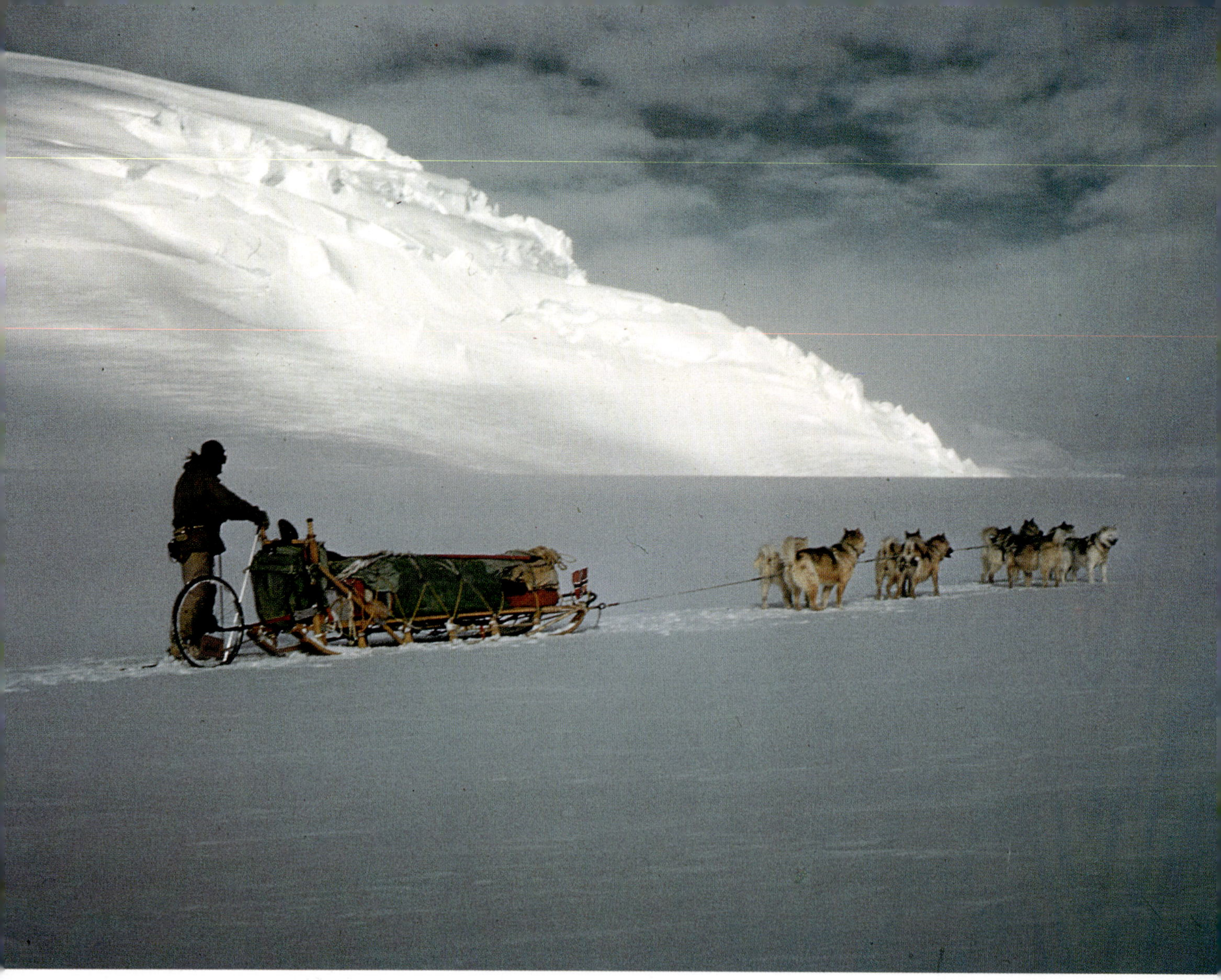

Plate 2.14. *Dog-sledging was still an integral part of the New Zealand research programme until the mid-1960s. Photo: Antarctic Division, DSIR.*

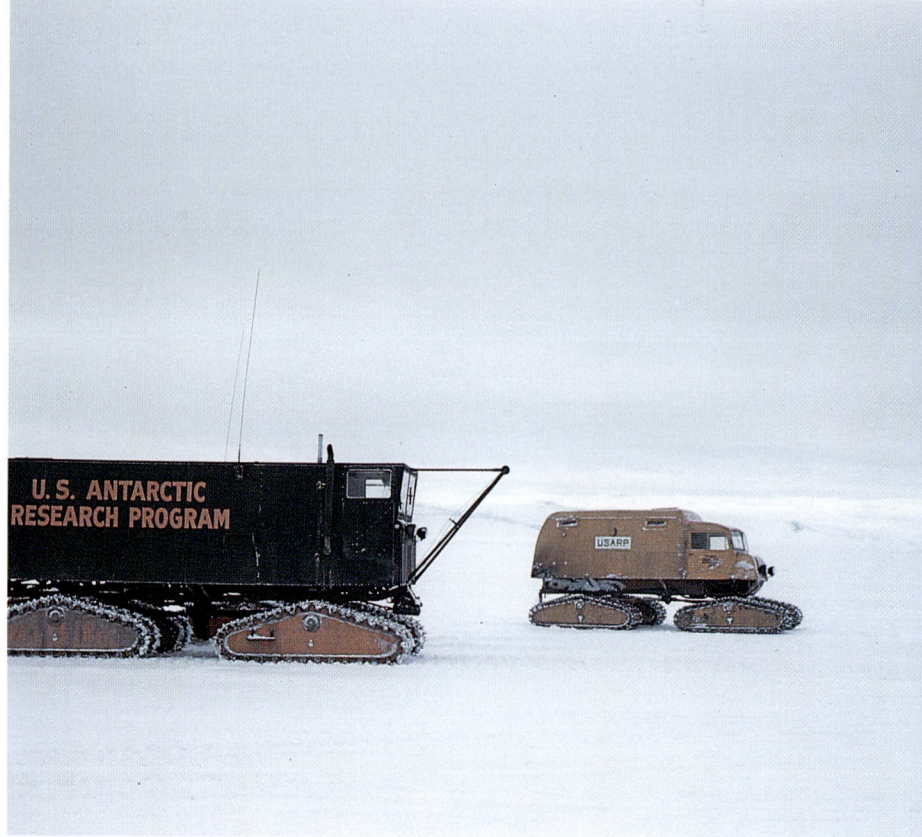

Plate 2.13. *Unlike its earlier version (in background), the larger Sno-cat was more comfortable and had a heat-absorbing black-body; it was used on U.S. traverses during the 1960s.*

expedient of forcing 8000 Adélie penguins from their nests! The most dramatic event of the year occurred at 8.34 p.m. on 31 October when United States Navy Dakota R-4D *Que Sera Sera* landed on the vast empty plain at 90°S. The first party had returned since Scott had left there in 1912, this time to build a permanent base.

Meanwhile, the New Zealand Antarctic effort was also getting underway. Using an ex-Falkland Islands Dependencies survey vessel renamed HMNZS *Endeavour* (after Captain Cook's ship), the New Zealand party reached McMurdo Sound in early January 1957 (Plate 2.7). Reconnaissance showed that their proposed base site at Butter Point was unsuitable because of access problems from sea ice to land in mid-summer. Plans were quickly changed. The base was sited at Pram Point on Ross Island only 3 kilometres from the U.S. McMurdo Station. The New Zealanders took less than 1 month to erect a sturdy prefabricated station sufficient to accommodate 23 people for 2 years. The base was occupied for 20 years before rebuilding began.

As the winter of 1957 set in and the support vessels departed, 225 men were left to winter in the Ross Sea region. There were 109 at Little America, 87 at McMurdo Naval Air Facility (Plate 2.8), 23 at Scott Base (Plate 2.9), and 18 each at Hallett Station (Plate 2.10) and South Pole Station. The IGY had brought radical changes to the region. The fruits of these endeavours were to be reaped in full in the future.

Filling in the gaps

The years from 1957 to the mid-1960s were devoted to completing the achievements of the earlier explorers. From Little America, United States traverse parties used oversnow tractor trains to explore Marie Byrd Land and investigate the Ross Ice Shelf. New Zealanders used the Skelton Glacier as a route to lay support depots for Fuchs' Commonwealth party crossing the continent from the Weddell Sea.

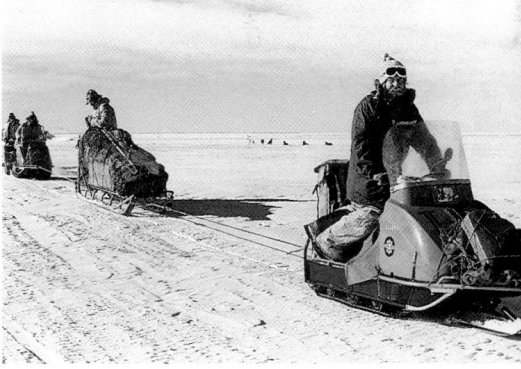

Plate 2.15. *Motor-toboggans, or "tin-dogs", now provide the transport for remote field parties. Photo: Antarctic Division, DSIR.*

The New Zealand exploits in 1957–58 were remarkable considering the limited resources. Supported by only a single-engined Beaver (Plate 2.11) and an Auster aircraft, the expedition laid the required depots to a distance of 1000 kilometres south on the Polar Plateau. From here, the expedition leader, Sir Edmund Hillary, turned a Nelsonian eye to his instructions and took three adapted "Massey Ferguson" farm tractors in a dash to the South Pole, becoming the third party to reach the Pole overland, and the first since Amundsen and Scott.

While the New Zealand tractor train headed south, three parties each with two dog teams made the first intensive topographic and geological surveys of the central section of the Transantarctic Mountains. During one of these journeys, the deputy expedition leader Bob (Holmes) Miller and the expedition doctor George Marsh ranged over the country north of the Beardmore Glacier travelling 2700 kilometres, still one of the longest sledging journeys made in Antarctica. Finally, further north, a party of geologists, led by Hillary's brother-in-law Larry Harrington, surveyed the Tucker Glacier region. This expedition was notable in being the forerunner of things to come—the summer party. Because of aircraft access between New Zealand and McMurdo Sound it was no longer necessary to winter-over to do science in Antarctica. In early March the Commonwealth Trans-Antarctic Expedition led by Vivian Fuchs arrived at Scott Base, completing the first traverse of Antarctica.

In terms of exploration 1957–58 had been a resounding success. It was also a triumph for the IGY scientific parties at the five stations in the Ross Sea area. Even before IGY was completed almost all the participating Antarctic nations had elected to continue occupying their bases and carrying out their research programmes (Plate 2.12). This was partly because of the substantial investments they had made but also for fear that they would lose political, territorial, or strategic advantages should they fail to do so. Scientists, whose frontiers are ever expanding and who realised that the surface of Antarctic knowledge had hardly been scratched, happily capitalised on this dilemma (see Chapter 16). In 1958 both the United States and New Zealand Governments announced their intention to continue indefinitely their activities in the Ross Sea region (Plate 2.13).

After IGY, the New Zealand and United States Antarctic programmes developed along complementary lines. In less than 10 years the exploration of the Ross Sea region was completed. The New Zealanders carried out summer-long surveying and broad-scale geological mapping of the entire Victoria Land mountain ranges. They worked in small teams of four to six, initially using dog teams (Plate 2.14) and, later, motor toboggans (Plate 2.15). In 1958 a party, again led by Harrington, covered the area between the Skelton and the Ferrar Glaciers. In 1959 the foothills area south of the Skelton Glacier was surveyed (the season was disrupted by a Snocat tractor falling down a crevasse and killing Ian Couzens, the driver). In 1960 the Byrd–Nimrod Glacier areas were surveyed. In 1961 and 1962 surveys were made of the areas north and south of the Beardmore Glacier and the entire region between the Davis Glaciers and the Evans Névé. In 1963–64 Bob Miller returned to lead the last, major, New Zealand dog-sledging journey of 2200 kilometres along the entire inland area of the Oates Coast. In the same year, another group surveyed the Axel Heiburg–Shackleton Glacier region, and made the first descent of the Axel Heiburg since Amundsen.

While this was going on, United States parties carried out several, mammoth, overland geophysical traverses, notably the 1959–60 Victoria Land traverse from McMurdo up the Skelton Glacier and then north to the Evans Névé. In 1960 the ubiquitous A. P. (Bert) Crary took a team with three gigantic Snocats on a geophysical traverse to the South Pole in just 65 days. During 1960–62 the United States carried out a major

Plate 2.16. *Recent U.S. field operations now use substantial remote base camps established for 1 to 3 years. Helicopters are based at these camps for local support. Photo: C. Rudge, Antarctic Division, DSIR.*

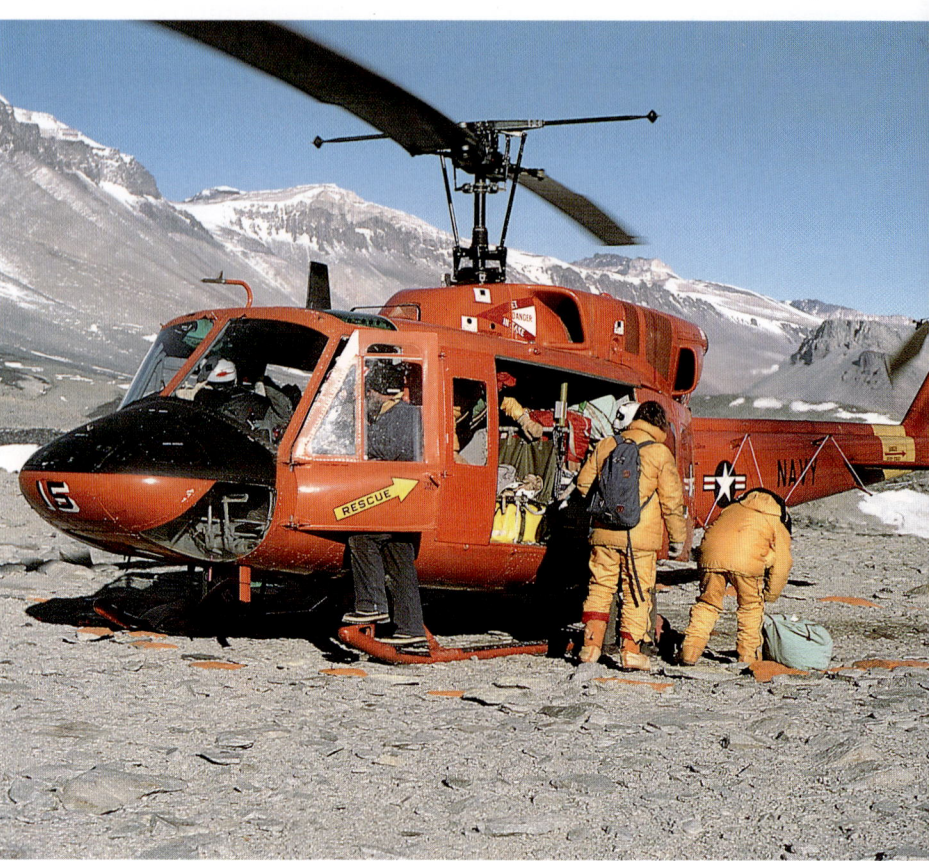

survey programme along the entire Transantarctic Mountains, and from these surveys developed the concept of aircraft-supported field camps which would operate as full base facilities for a summer to support detailed geological research (Plate 2.16).

Thus, by the late 1960s the Americans and New Zealanders had virtually completed what Sir James Clark Ross began in 1841. Although there were still major field traverses to come in the 1960s, 1970s, and on into the 1980s, and although there were still small pockets of untravelled land to cover, the Ross Sea region was no longer "unexplored" in the true sense of the word.

The close of the exploration era brought changes to the nature of United States and New Zealand Antarctic operations. In 1959 Little America was closed on the grounds that, being on a moving ice shelf, the station's life was limited and it was becoming too expensive to maintain. In any event, McMurdo Station was a more convenient headquarters for the United States Antarctic operations. In 1964 Hallett Station ceased to be occupied year-round after a fire destroyed the science laboratory and in 1971 the base was closed permanently. The New Zealanders slowly expanded their facilities at Scott Base, and in 1967 opened a second, small station at Lake Vanda in the Dry Valleys (Plate 2.17).

The exploration era had been a remarkable one. It had seen a mixture of hard work, heroism, and the influence and deeds of strong personalities. For over 100 years, the Ross Sea region had been an explorer's last frontier. But by the mid-1960s it was no longer necessary to ask what was there. By and large, this was known. The question to be asked now was why it was there.

Plate 2.17. *Vanda Station was established in 1967 to support parties in the Dry Valleys. Photo: K. Westerkov, Antarctic Division, DSIR.*

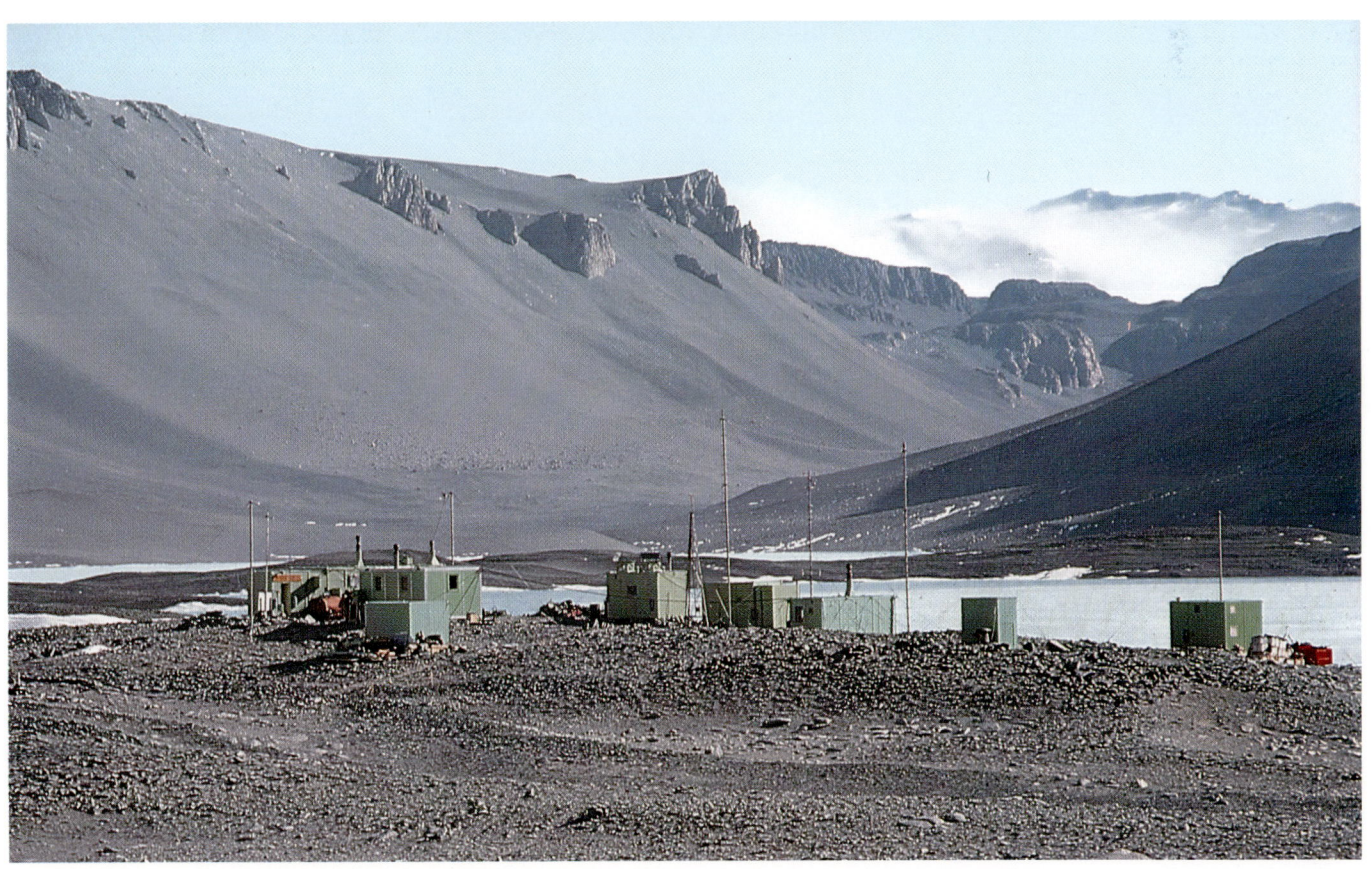

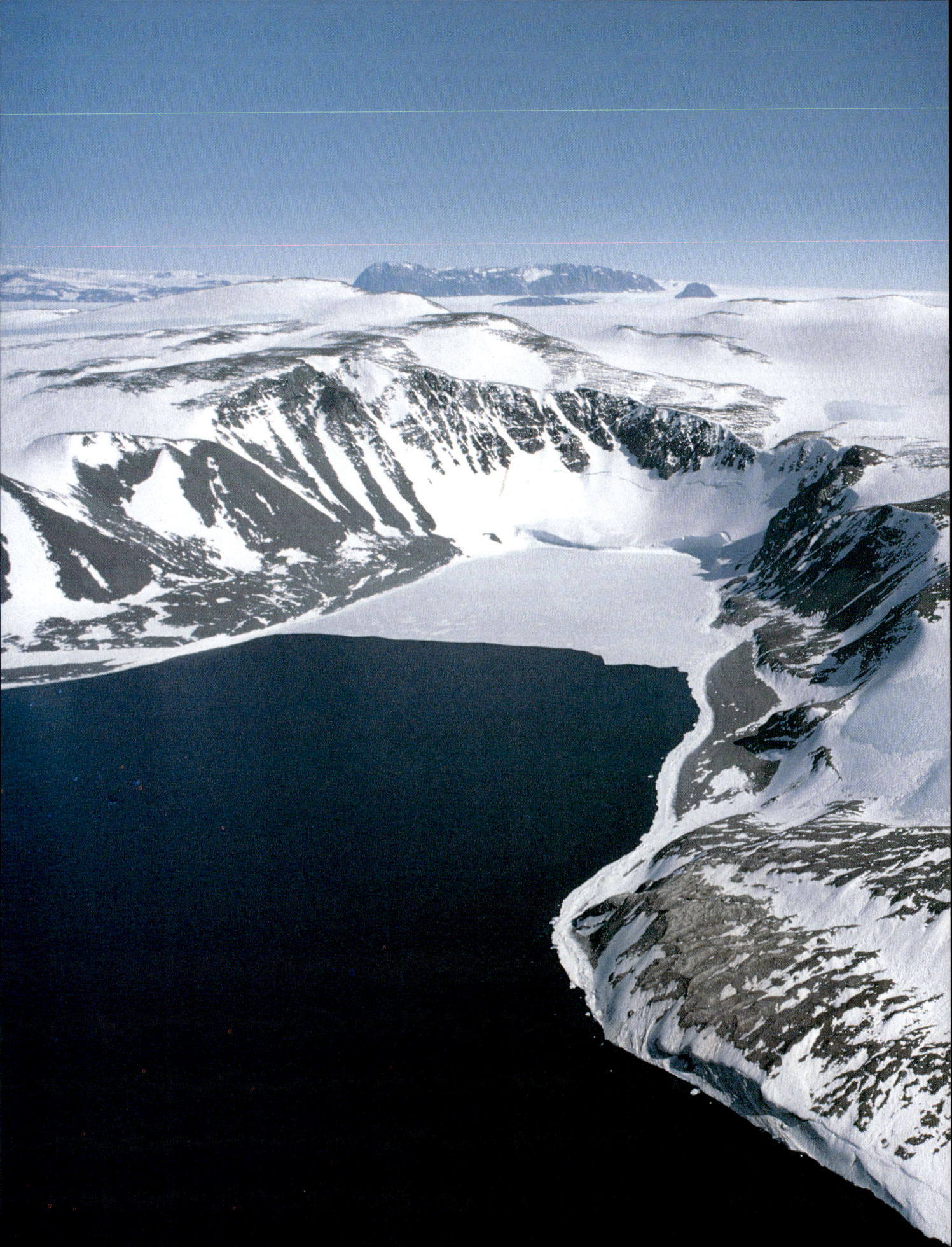

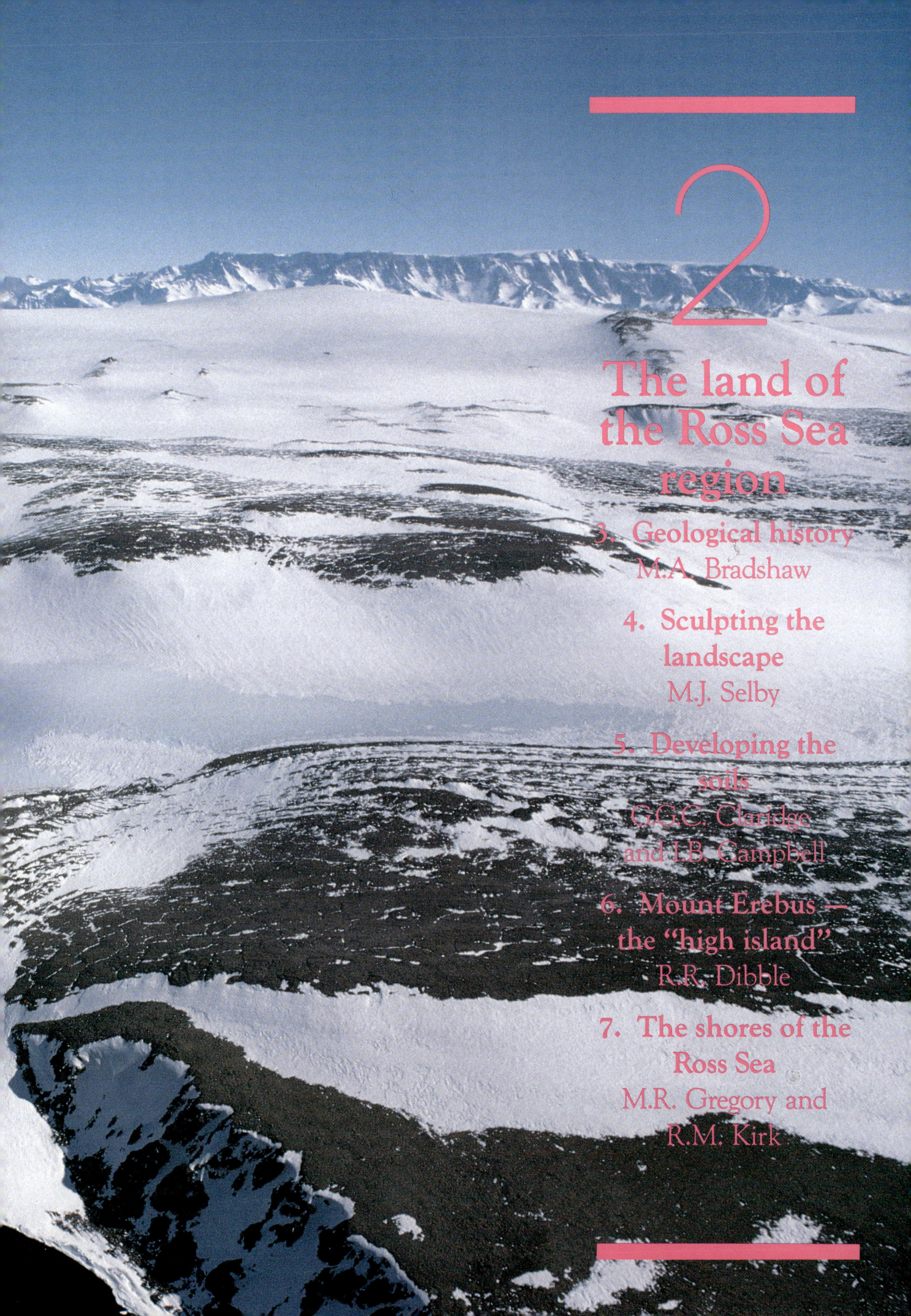

2

The land of the Ross Sea region

3. Geological history

T oday, Antarctica is the world's most inhospitable continent. But at times in the past it had a more moderate climate, sufficiently so to allow forests and land animals to flourish. This was, however, many millions of years ago, and could only be reconstructed after Antarctica's rocks have been studied in detail.

Discovering what Antarctica was like in the past by looking at its rocks seems at first an impossible task, for very few rocks appear from beneath the polar ice-cap (Figure 3.1). Much of Antarctica is a bland white expanse, a high, gently undulating plateau of ice that blankets the land below, depressing it well below the height it would normally have been without an ice cover.

Radio echo-sounding has shown that in some places large ragged mountain ranges are completely buried by the ice-cap, whereas in other areas great depressions exist that are well below sea level. The ice is so thick that in places its weight causes the underside to melt under the pressure, so that huge buried lakes lie between bedrock and the ice cover.

The same principle is used by radio echo-sounding as by echo-sounding at sea. Radio waves emitted from a plane flying over the ice rebound from both the ice surface and the denser rock surface buried below. Waves that have had to travel a long way through the ice before they are reflected by the buried land take longer to return than those that are bounced off the ice surface. From the continuous emission of signals and their recorded times of arrival after reflection, a picture emerges of what the land below the ice looks like.

The Antarctic ice-cap cloaks two very different areas. East Antarctica is a typical continental block with a crustal thickness of between 30 and 40 kilometres. West Antarctica, however, is much thinner, its crust in places is only 10 kilometres thick, and without the ice-cap it would appear as a scatter of mountainous islands in an ocean setting. The junction between the two parts of Antarctica probably lies below the Ross Sea and Ross Ice Shelf.

About 98 percent of Antarctica is covered by ice, and limited exposures of rock are found only in scattered nunataks (isolated hills or peaks that project above the ice) around the fringes of East and West Antarctica, along the Antarctic Peninsula, and throughout the rugged Transantarctic Mountains that cross the continent from one side to the other. Much of the Transantarctic Mountain belt occupies the western part of the Ross Sea region (see Figure 3.1) and provides extremely important exposures of rock that help us understand the geological history of the continent.

Searching for a continent's geological past is made doubly difficult because the rocks today may be many hundreds of kilometres distant from where they first formed. This change of position is due to later movement along fracture or fault lines. The older the rock, the more time it has had to be moved, and the more complex the problem is for geologists; also, the time involved is so great and in the past Antarctica was very different from what it is today. Seas and rivers have come and gone, and various forms of life have evolved and become extinct.

The core of East Antarctica is extremely old rock, and in this it is no different to other continents such as Australia, South America, and Africa. The ancient rocks of Antarctica are very like those of Australia and parts of India, and provide evidence that they were once joined together and have since split and drifted apart. When these southern continents were united they formed a giant land mass called Gondwana (Figure 3.2). (The name is derived from an ancient tribe in India known as the Gonds, and from Wana, meaning land.)

The oldest rocks in Antarctica are about 3000 million years old,

Figure 3.1. *Ice-free areas of the Ross Sea region.*

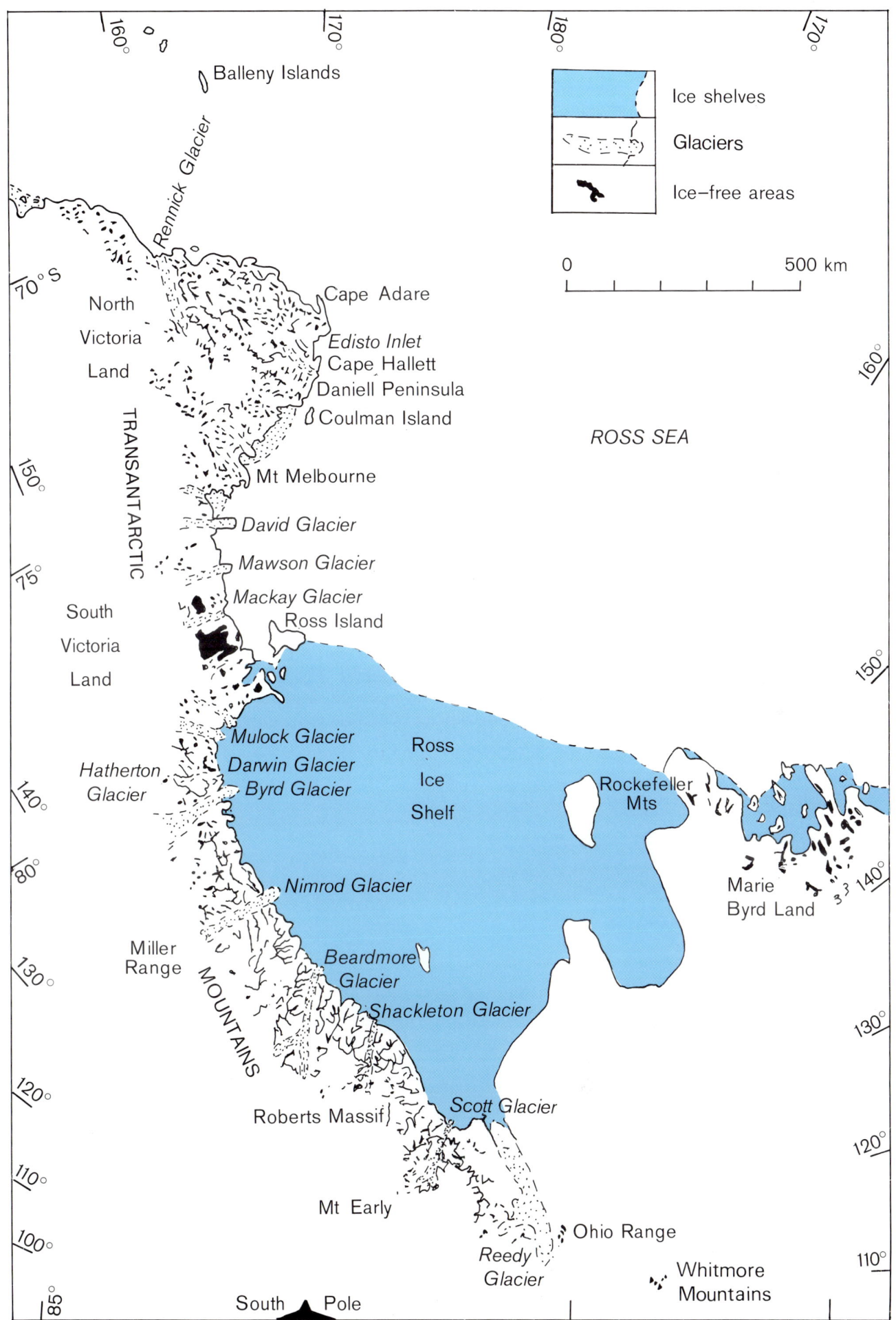

Ice shelves

Glaciers

Ice-free areas

0   500 km

Balleny Islands

Rennick Glacier

70° S

North
Victoria
Land

Cape Adare

Edisto Inlet
Cape Hallett
Daniell Peninsula
Coulman Island

ROSS SEA

TRANSANTARCTIC

Mt Melbourne

150°

David Glacier

Mawson Glacier

75°

Mackay Glacier
Ross Island

South
Victoria
Land

160°

Mulock Glacier
Darwin Glacier
Byrd Glacier

Ross
Ice
Shelf

Rockefeller
Mts

150°

Hatherton
Glacier

140°

Nimrod Glacier

Marie
Byrd Land

140°

80°

Miller
Range

Beardmore
Glacier

130°

MOUNTAINS

Shackleton Glacier

130°

120°

Roberts Massif

Scott Glacier

120°

110°

Mt Early

Ohio Range

100°

Reedy
Glacier

Whitmore
Mountains

110°

85°

South Pole

formed during the very long Precambrian era (Figure 3.3). Rocks can be dated by measuring the proportions of natural radioactive isotopes present in the rocks. Each radioactive isotope decays at a fixed rate and is unaffected by any variations of conditions. Thus, the amount of decay indicates how long a radioactive substance has been decaying, and so its age can be calculated.

Over many millions of years, younger rocks have been gradually laid down on the ancient crust, or "craton" of East Antarctica, and subsequently have become altered or "metamorphosed". By geological standards these rocks are still very old and in the Transantarctic Mountains form part of the foundation, or "basement", of this chain. We know that this basement also exists deep below at least part of the Ross Sea. Edward VII Land, which borders the eastern Ross Sea, has similar old rocks but they may have originated in a different part of the southern world to those in the Transantarctic Mountains.

Roots of the Transantarctic Mountains

The older rocks forming the basement of the Transantarctic Mountains are Precambrian in age, at least 740 million years and older—from a time when only simple bacterial, algal, and fungal life existed on earth. Sediments thought to have been deposited on the sea floor, and altered by the heat of deep burial within the earth's crust, are now uplifted and complexly folded (Figure 3.4). They are exposed only because the rocks that were above them have been removed by erosion after their uplift from the sea.

Also forming part of the basement are slightly younger rocks of early Palaeozoic age, mainly Cambrian, originally laid down as flat layers on the sea floor and which now lie on the earlier Precambrian rocks which had been bent and uplifted. The contact surface between them is called an unconformity. Usually an unconformity represents a large span of geological time during which no sediments were deposited to provide

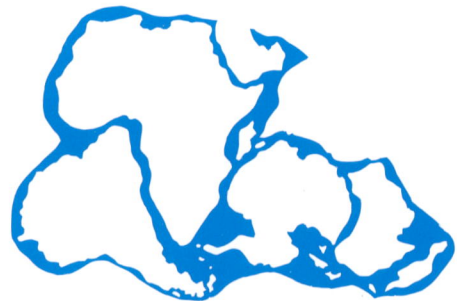

Figure 3.2. *For hundreds of millions of years, Antarctica was joined to other southern continents as a single super-continent called Gondwana.*

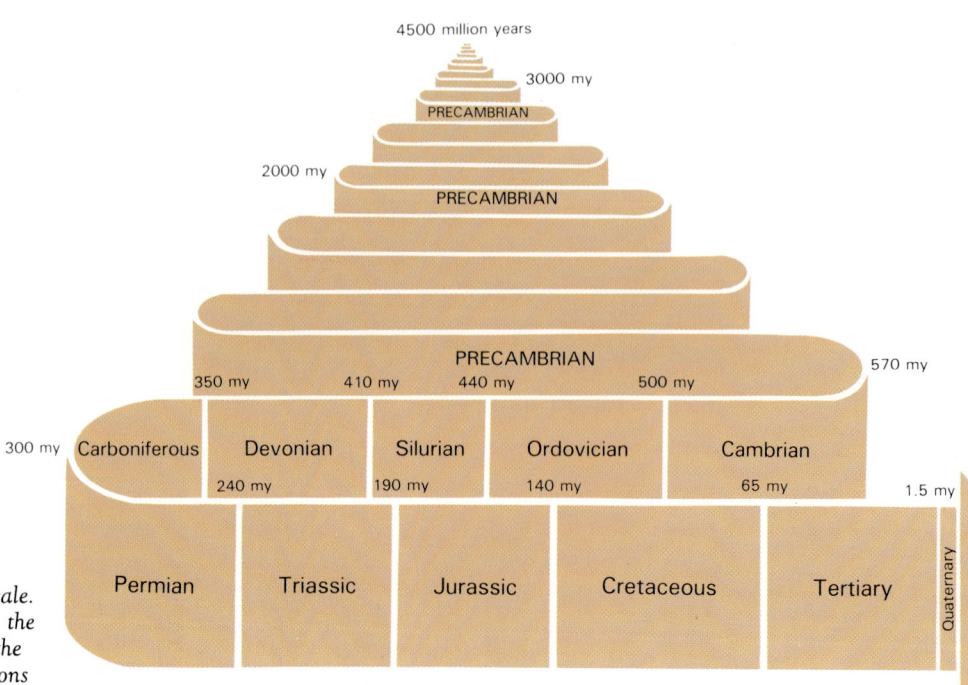

Figure 3.3. *Geological time drawn to scale. The Precambrian era, during which only the simplest forms of life existed, was by far the longest division of time. Ages are in millions of years.*

a geological record, and when erosion was important. Like the Precambrian rocks before them, the Cambrian sediments were deposited in a subsiding trough beyond the coastal margin of the East Antarctic continent. But unlike the Precambrian sediments, they contain early forms of complex animals, the skeletons of trilobites (long-extinct segmented arthropods), the shells of simple brachiopods, and the skeletons of obscure coral-like animals called archaeocyathids.

Some of these rocks are limestone, and these are particularly obvious in the Beardmore Glacier region where they reach a phenomenal thickness of over 5 kilometres (the Shackleton Limestone, Plate 3.1). Limestone is formed largely from the broken shells of animals that once lived on the sea floor and contains very little, if any, land-derived mud or sand. Several features of the Shackleton Limestone indicate it was deposited in shallow water, which means that the trough steadily subsided at the same rate that sediment accumulated. Archaeocyathids were the most common animals in the shallow seas, and are similar to ones found elsewhere in the world. Antarctica was probably in equatorial latitudes and thus had a tropical climate at this time.

Other sediments such as clays and sands were also laid down on the sea floor. After the limestone was deposited, gravels at least 1.2 kilometres thick (the Douglas Conglomerate) were spread over the trough, after increased erosion on the adjacent continent.

Although less spectacular, possible Cambrian limestones also occur in South Victoria Land near the Skelton Glacier where they are known as the Anthill Limestone. Fossils have yet to be found in them.

In North Victoria Land limited amounts of Cambrian limestone occur, commonly as large blocks which slid down the Cambrian sea floor and are now enclosed in other sediments. The blocks are particularly rich in trilobites which have close affinities with Australian fossils. The offshore Cambrian basin of North Victoria Land was probably not continuous or in line with that of the Beardmore region. In fact, North

Figure 3.4. *Geological section across the Transantarctic Mountains, showing the succession of rocks.*

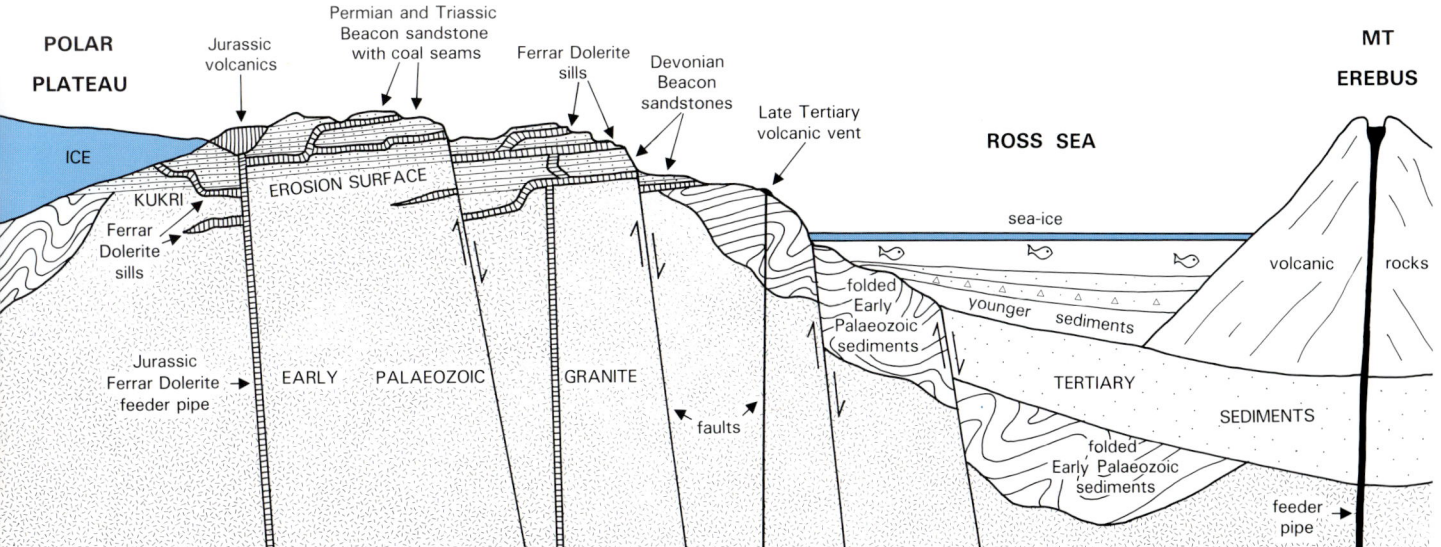

Victoria Land has proved to be an extremely complex area (Plate 3.2).
The more that is learnt about it, the more complex its geology seems. It
is now known that major fault lines obliquely cross this very wide and
rugged part of the Transantarctic Mountains, bringing together
contemporaneous rocks of quite different character which appear to have
originated in very different environments.

A mountain building phase

At the end of the Cambrian Period, the slow movements of the
Ross Orogeny (mountain building era) occurred, pushing the sediments
up from the sea floor by a few millimetres each year, and buckling them
into folds by compressive forces. At the height of the orogeny, some
sediments had been squeezed deep into the earth's crust below the new
fold mountain belt, and the heat in that region caused the sediments to
partially melt. This molten rock, with material from even deeper levels,
formed a mobile "magma" which rose and invaded the higher parts of
the mountain chain, producing enormous granitic intrusions called
batholiths. Being still well below ground level and insulated by thousands
of metres of rock above, the magma bodies took a long time to cool and
solidify, producing a coarsely crystallised rock. Radioactive isotope studies
indicate that the cooling of the granitic rocks took place between 450
and 500 million years ago, or in the early Ordovician.

These large areas of intrusive rocks are found throughout the
Transantarctic Mountains and can be easily seen in South Victoria Land
(Plate 3.3), especially in the Dry Valley region where they are referred to
as the Granite Harbour Intrusives.

Younger rocks

Mountain building phases are always followed by long periods of
erosion of the rocks above sea level. In the Transantarctic Mountains
this situation lasted for at least 60 million years, during which the Ross

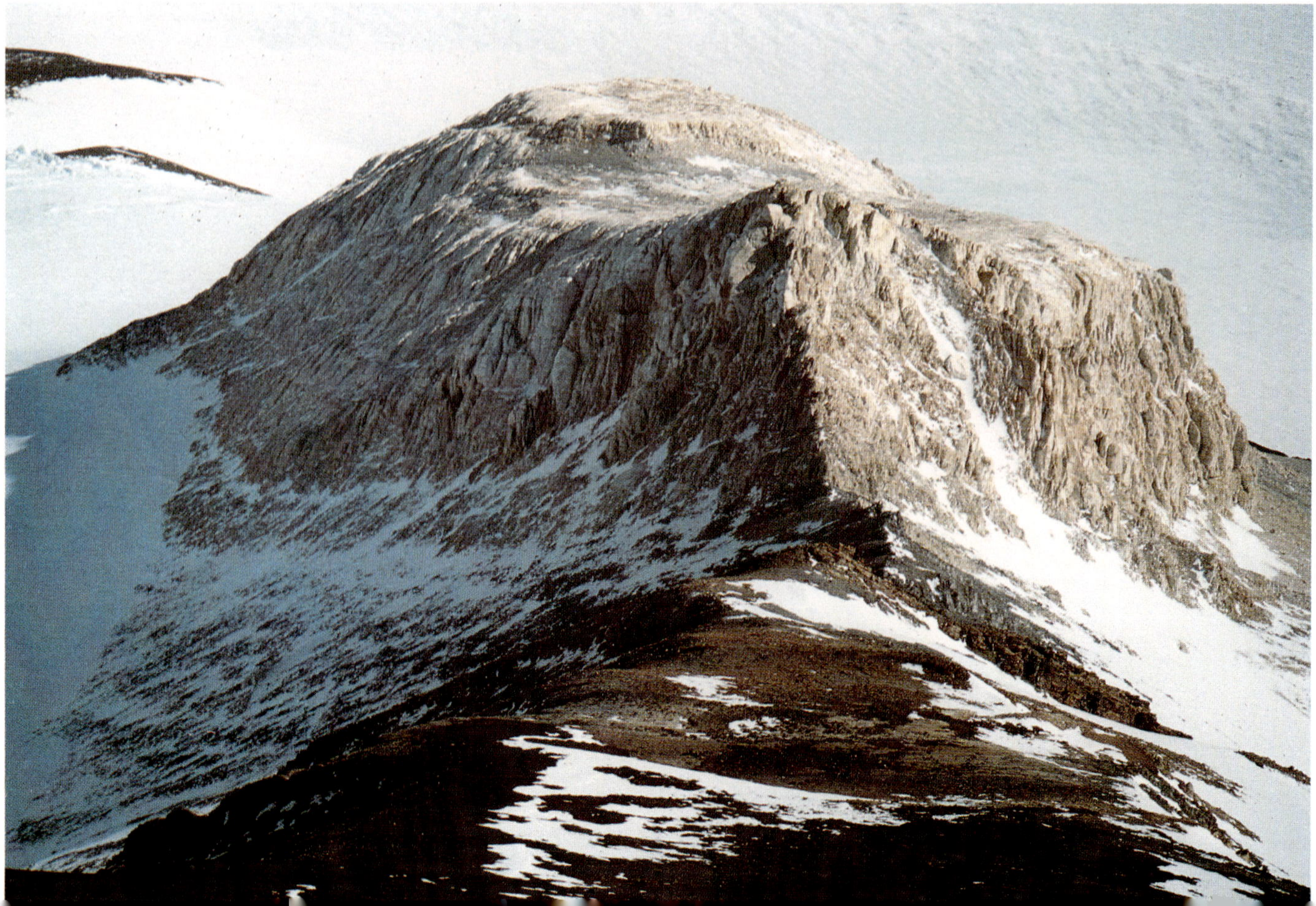

Plate 3.1. *The light-coloured Shackleton Limestone contains some of the earliest fossils in Antarctica. Photo: M. Bradshaw.*

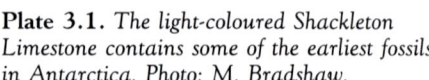

mountain chain was worn lower and lower. By early Devonian times it was low enough for shallow water seas to spread across the land, depositing sandstones and mudstones. Life had changed dramatically in the interval of over 100 million years between when Cambrian life teemed on the ocean floor in the Antarctic region, and the early Devonian. Much of it was still confined to the seas. But numerous shelled invertebrates, such as clams and brachiopods, had evolved to join the trilobites, which were themselves very different from their distant Cambrian relatives. The first large backboned animals (vertebrates), primitive fish, had also evolved. Simple, low, woodless plant species were common in coastal swamps, and some plants had already begun their successful invasion of the land.

Many of these early fossils are found in the Ohio Range of the Transantarctic Mountains. This locality is just outside the Ross Sea region, but is important because it is the only place where early Devonian sediments containing numerous shelly fossils rest unconformably on the granitic roots of the Ross Mountain chain.

In South Victoria Land, a thicker sequence of Devonian sediments rests unconformably on the Granite Harbour Intrusives of the basement, and the plane of unconformity is called the Kukri Surface because it is well exposed in the Kukri Hills (Plate 3.4). The Devonian rocks here (known as the Taylor Group) are about 1200 metres thick and are largely hard, quartz-rich sandstone with much thinner mudstone horizons. These rocks have been very little altered or disturbed since their original deposition, apart from general uplift. They are therefore seen as flat-lying, extensive and easily recognisable outcrops, forming the lowest part of a pile of yellow sandstones known throughout the Transantarctic Mountains as the Beacon Supergroup. The origin of this name dates back to H. T. Ferrar who discovered the rocks on a journey up the Ferrar Glacier during the Discovery Expedition of 1901–4. He named the rocks after Beacon Heights, a prominent land mark adjacent to the upper

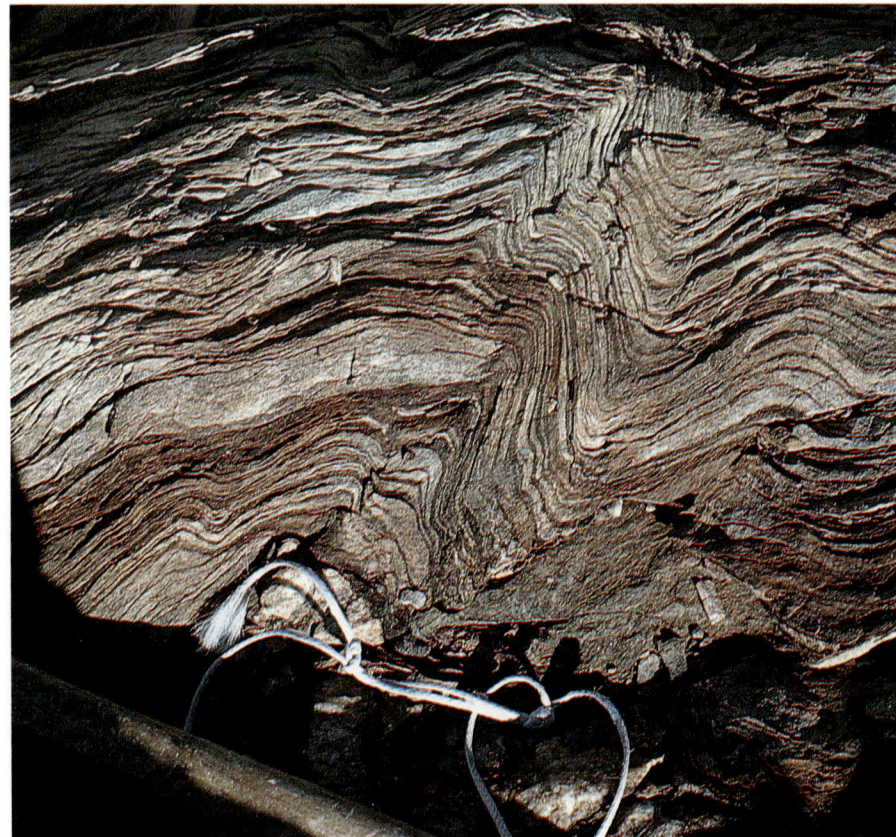

Plate 3.2. *The rocks of North Victoria Land are much more contorted than those of South Victoria Land. Photo: M. Bradshaw.*

Taylor Glacier and which he used for navigation.

The Devonian sediments, which represent the initial inundation of the old land surface by the sea, lack shelly fossils in South Victoria Land, but are especially rich in animal burrows and trackways (Plate 3.5). There is debate whether these unfossiliferous sediments were all marine, as suggested by the abundance of animal traces and certain sedimentary structures, or had been laid down by rivers. Towards the top of the sequence trace fossils become fewer and horizons containing non-marine fossil fish suggest temporary lakes and pools existed on a broad river plain. Red beds, dessication tracks, and fossil soil horizons suggest conditions were semi-arid. The fish are particularly interesting as they include a major group no longer living today, but which was highly successful during the Devonian Period. These are the "placoderm fish", which had the front part of the body covered with a rigid armour of thick bony plates. They lived alongside early sharks (known only from their teeth and the small denticles that covered their skin) and also primitive lung fish, the group which eventually gave rise to the first land animals. South Victoria Land is well known for its Devonian fish fossils. Outcrops on Mount Fleming, Mount Crean, and Mount Ritchie have yielded superb specimens, including some almost complete skeletons. Elsewhere, usually only disarticulated fish plates, teeth, and denticles are found.

At the same time that the shallow invading seas of the early Devonian were being pushed back by prograding rivers (i.e., rivers depositing their sediment loads nearshore), volcanic eruptions were occurring in other parts of Antarctica. In North Victoria Land a thick pile of lavas (mostly rhyolites) were erupted and today form the Gallipolli Heights. Between eruptions, thin plant-bearing sediments were laid down on the side of the volcanoes. Similar volcanics were erupted in Marie Byrd Land and are associated with well-preserved Devonian plant fossils in muddy sediment. In both these areas, large granitic bodies are found

Plate 3.3. *Many of the granitic rocks that form part of the basement in the Transantarctic Mountains were intruded during the late stages of the Ordovician Ross Orogeny. At this locality in the upper Taylor Valley there are two phases of granitic intrusion; an earlier grey phase containing dark inclusions, cut by a later pink (light) phase. Photo: M. Bradshaw.*

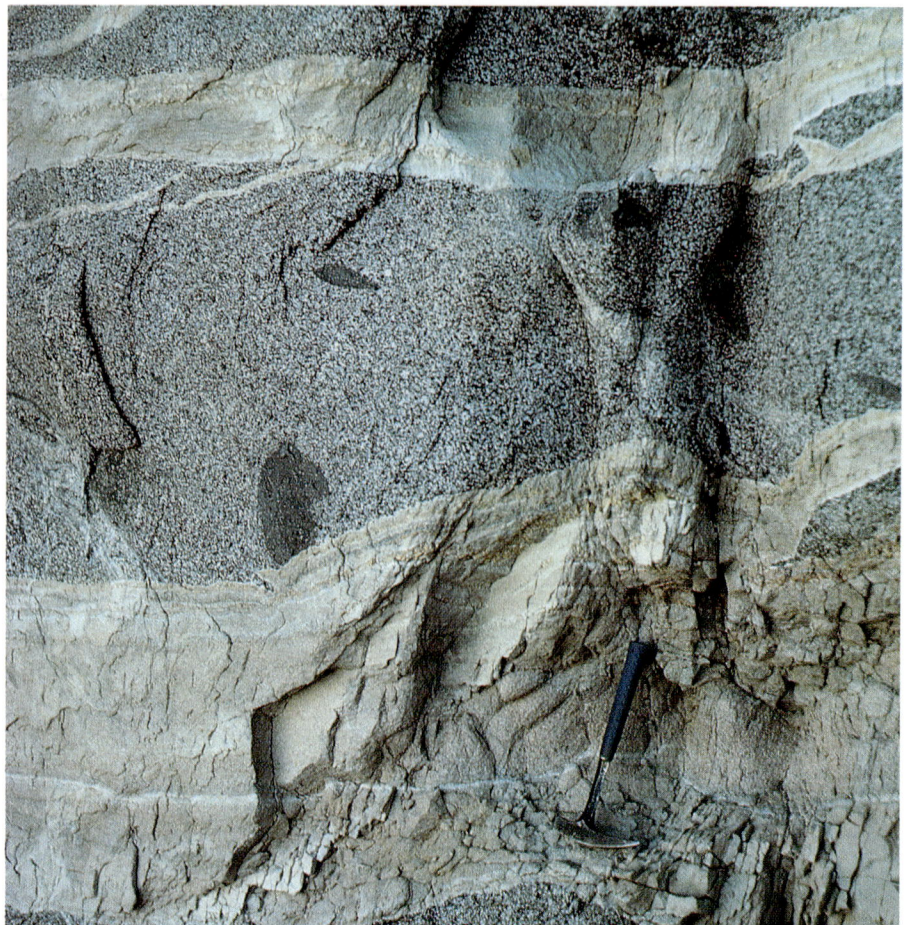

which may have been intruded into magma chambers associated with the volcanoes.

An ancient ice-cap

After the Devonian there is a break in the geological record throughout the Transantarctic Mountains. Rocks of the succeeding Carboniferous period are not known, although they are common in other parts of the world. By studying rocks in other countries, we know that at the end of the Carboniferous and the beginning of the Permian Period the world entered a time of global cooling. At this time Gondwana had a polar position, and much of the continent was affected by the Permo-Carboniferous glaciation. During the 20 million years of global cooling the land mass moved slowly across the Pole. At the onset of glaciation the Pole lay in the combined parts of Africa and South America, then progressively moved to a site in present Antarctica, before migrating into Australia during the waning phases of glaciation.

During this major glacial period very large polar ice-caps developed, as large, if not larger, than those in Antarctica today. The base of these ice caps gouged into older rocks (Plate 3.6), carrying them outwards to the ice margin as a frozen bed load of boulders, sand, and mud. Here the debris melted out and became deposited as an ill-sorted sediment known as till, which when hardened to rock is called a tillite.

Because it is rare for sediments to be deposited under a glacier that is actively eroding, most glacial sediments are preserved as the ice retreats. Those containing the largest boulders, sometimes over 7 metres across, are deposited closest to the ice front; whereas the finer gravels, sands, and muds are washed out to be deposited further away. The ice front itself may be grounded on land, or it may be floating on the water of lake basins or the sea into which it has spread. Sediments frozen into the base of such floating ice melt out and "rain" down on to the lake or sea floor below. Further thickening and advance of the ice may cause it

Plate 3.4. *Horizontally bedded Beacon sediments unconformably overlie the Kukri Erosion Surface (seen as the almost horizontal line) that had been eroded after the Ross Orogeny. Photo: M. Bradshaw.*

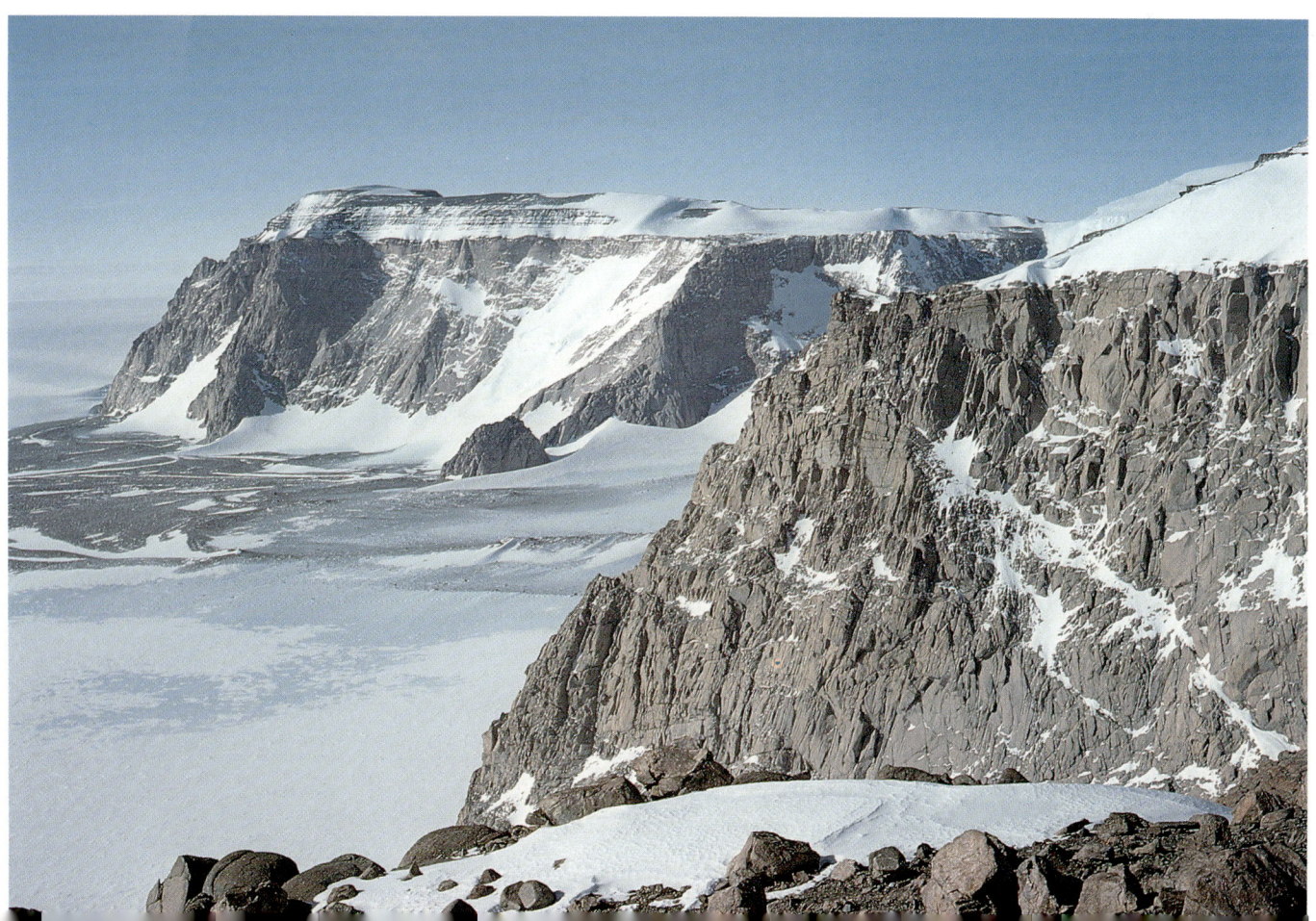

50

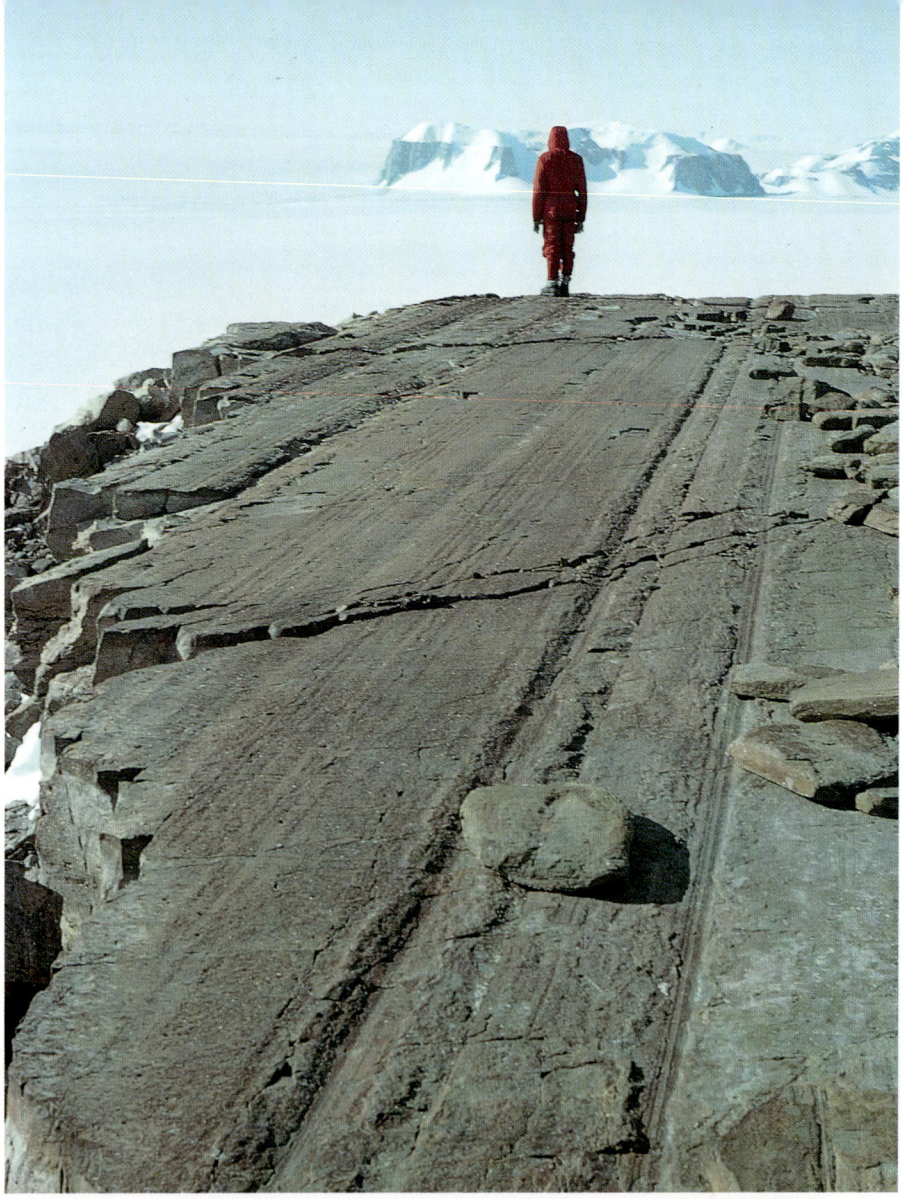

Plate 3.6. *These prominent grooves in the top of a bed of glacial sandstone in the Ohio Range were caused by the movement of a Permian glacier about 295 million years ago. Photo: M. Bradshaw.*

Plate 3.5. *Wormcasts in Beacon sediments, Asgaard Range. Photo: K. Lefever.*

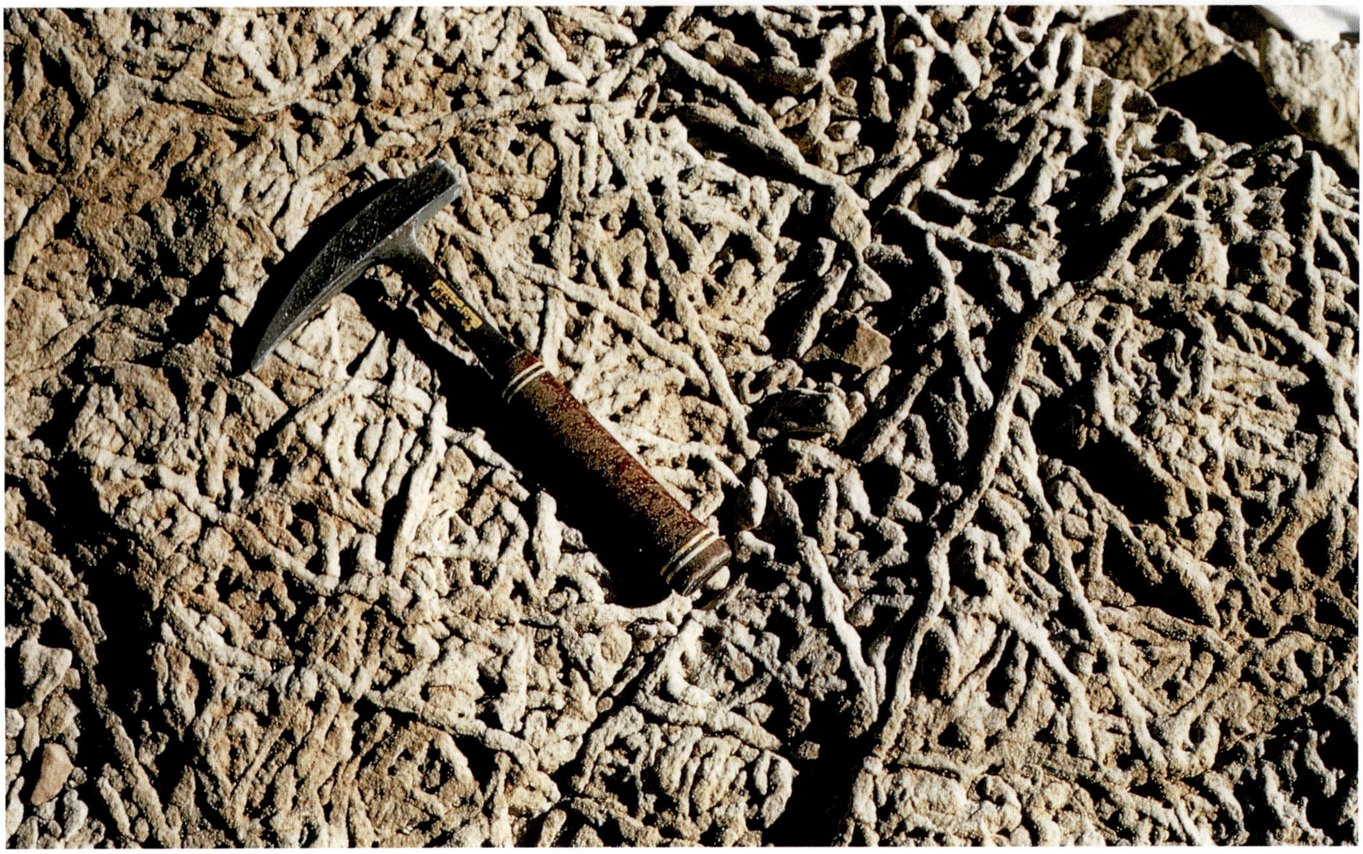

to become grounded on the lake or sea bottom, producing grooving of the earlier sediment layers. When the sediment is converted to rock, such grooves are good indicators of the direction of ice movement.

Throughout the Transantarctic Mountains there are sequences of glacial sediments which record the later stages of the glaciation during the early Permian, about 290 million years ago. The Beacon Supergroup rocks deposited after the Devonian are known as the Victoria Group, and the glacial beds were the first of these sediments to be deposited. They are similar to glacial beds in India, Australia, Africa, and South America. In different parts of the Transantarctic Mountains the glacial beds have been given different names: e.g., Pagoda Tillite in the Beardmore region, Metschel Tillite in South Victoria Land. However, they are essentially very similar and suggest a major ice-sheet that was spreading outwards over Gondwana and developed into a floating ice shelf beyond the land area.

The thickness of the glacial sequences varies considerably depending on how close the area of deposition was to the actively eroding ice. In North Victoria Land and part of South Victoria Land it is absent or thin, but in the Beardmore Glacier region it reaches nearly 400 metres in thickness. In the Darwin Mountains remnants of glacial valleys cut in older Devonian sediments are preserved, infilled with slumped tillite and river-channel outwash sands. In the Ohio Range similar valleys were gouged through Devonian sediments by ice; in some places the sediments were removed completely and the basement granite below was scoured. The Permian age of these sediments is based on characteristic spores and rare fossil leaves found in finer beds.

As the long glacial period waned and the ice caps shrank, areas that once received coarse debris from the ice front now received only very fine material that was very clearly deposited in a large area of water some distance from the ice front. In the Transantarctic Mountains post-glacial sediments are preserved in the Beardmore Glacier region and in the Ohio and Wisconsin Ranges, and the body of water in which the mud and sand accumulated was probably continuous between them and has been referred to as "Lake Mackellar". This large lake or sea is thought to have been like the present day cold, brackish and muddy Baltic Sea and Gulf of Bothnia.

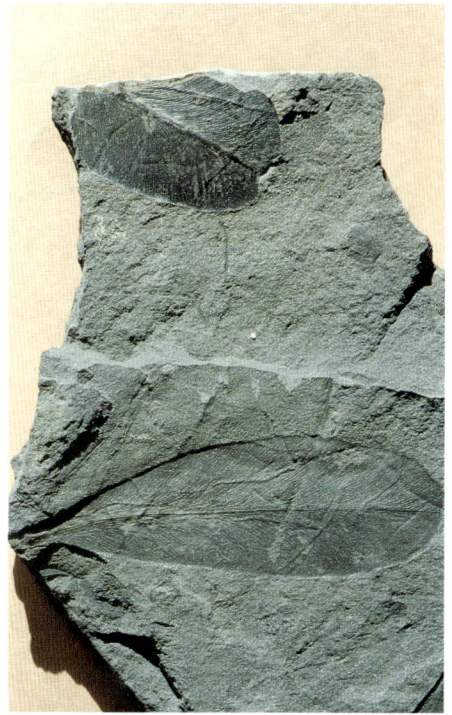

Plate 3.7. *Well preserved fossil leaves of* Glossopteris *which was widespread throughout Gondwana after the Permian glaciation. Photo: P. Barrett.*

Trees and land animals

Conditions in Gondwana changed dramatically with the demise of the glaciers. Where ice had once over-ridden the land, forests began to grow, dominated by a newly evolved and uniquely southern vegetation known as the *Glossopteris* flora. Stands of large gymnosperm trees (conifers and their allies; primitive seed plants with many fossil representatives) flourished on a land surface that was drained and built seawards by large sand- and gravel-laden rivers. These trees were deciduous, shedding their leaves every year, many to accumulate in river bed muds or in small pools. The fossil leaves are called *Glossopteris* (Plate 3.7), and occasionally fruiting bodies and seeds are preserved with them. The *Glossopteris* species of Antarctica are identical to those found in the post-glacial sediments of the other southern continents, which at that time were still firmly joined to Antarctica. Tree trunks are also found in these sediments, some in their position of growth, some obviously having been transported. They have been called *Araucarioxylon*. Roots sometimes attached to the trunks are believed to have a structure that suggests they had a "breathing" function similar to the roots of the modern mangrove, and that the *Glossopteris* trees preferred swamp habitats.

The abundance of plant life in Antarctica during the Permian was

Plate 3.8. *Coal Measures (dark rocks) are found throughout the Transantarctic Mountains, as here at Mount Glossopteris. Photo: M. Bradshaw.* ▶

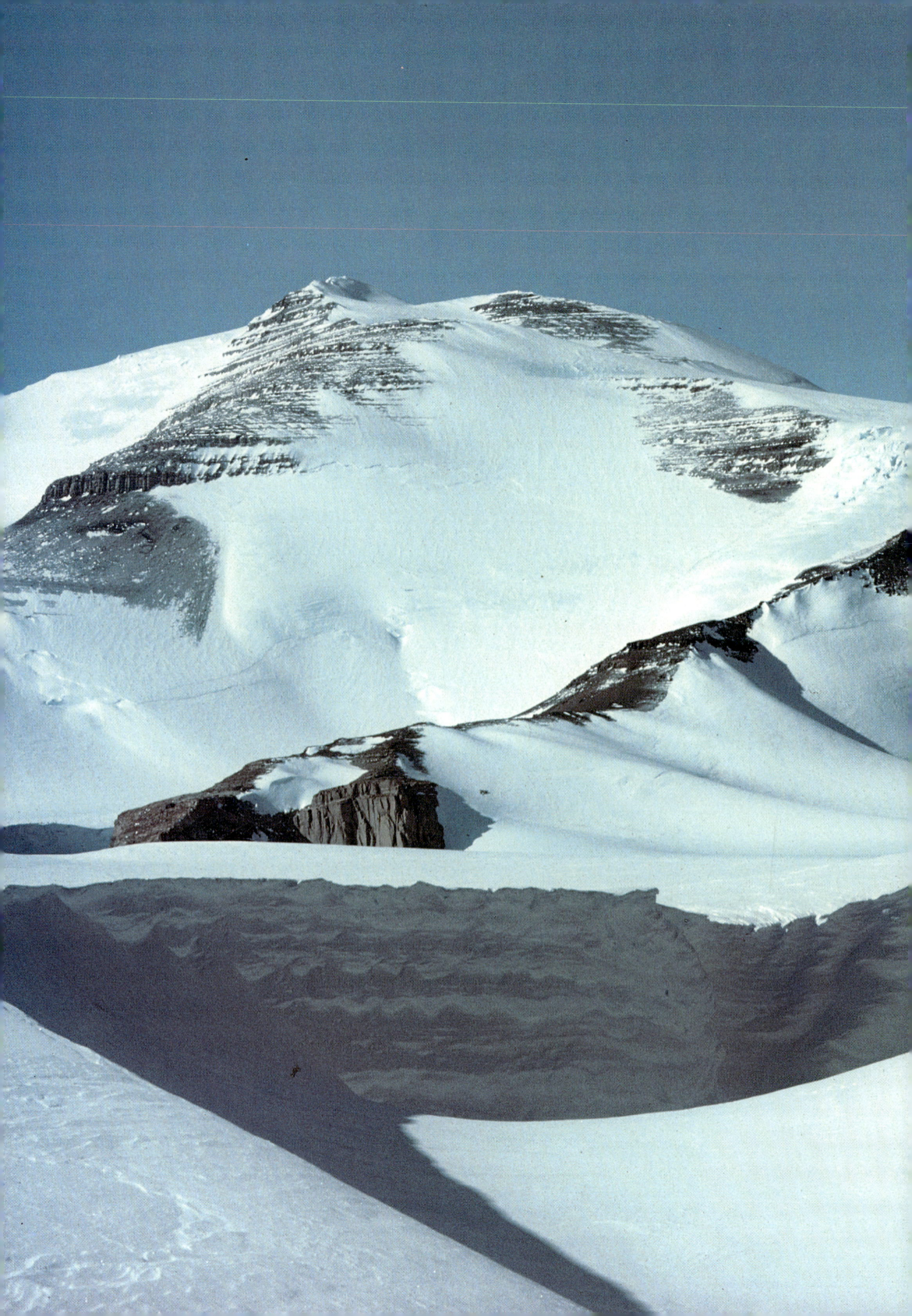

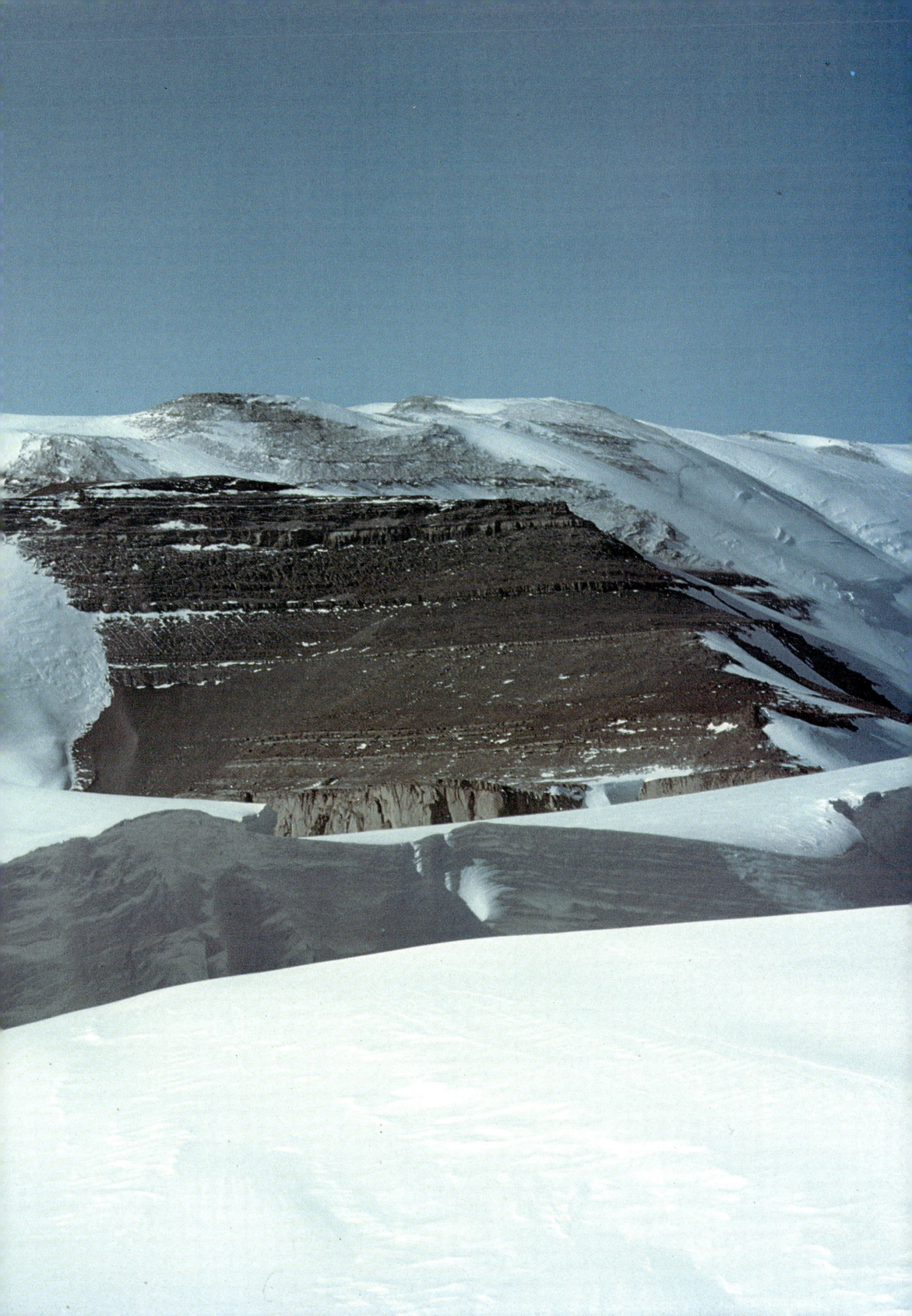

such that under suitable conditions peat accumulated. When this became strongly compressed under the pressure of piles of later sediment laid down on top, and also became heated by much later igneous intrusions, the peat became transformed into reasonable quality coal. Thus, many of the Permian sediments are referred to as "coal measures". Again, the names of the rocks are different for different parts of the Transantarctic Mountains, but they are essentially the same age. The Wellar Coal Measures of South Victoria Land, the Buckley Formation of the Beardmore Glacier region, and the Mount Glossopteris Formation (Plate 3.8) of the Ohio Range are a few of these many names.

Wood and leaves are usually the only fossils present in the Permian coal measures, apart from a few animal burrows. But occasionally insect wings or freshwater clams are found, testimony to the existence of temporary lakes on the Permian river flats. In the central Transantarctic Mountains and also the Ohio Range, sandstone containing various amounts of volcanic ash point to volcanoes erupting close to the site of sedimentation during at least part of the Permian.

Coal measure conditions continued into the following Triassic Period throughout the Transantarctic Mountains. Gravels (conglomerates), sandstone, and mudstones were laid down on the river flats by powerful rivers. Some of the muddier horizons contain plant rootlets, and many contain fossil leaves. But the Triassic flora was different to that of the preceding Permian; the dominance of *Glossopteris* was over, and the land became forested with fern-like *Dicroidium* plants, giant horsetails, and *Gingko* (a gymnosperm) trees. Fossil wood is again abundant in these sediments, and peat deposits also accumulated, later forming coal seams. One major difference compared with the Permian was the presence of important land animals in Antarctica during the Triassic.

During the late Devonian and early Carboniferous, a branch of the air-breathing "lung fish" stock, the lobe-fin fishes, had successfully attempted life on the land surface, and their descendants had slowly evolved into amphibians by developing limbs from fins. These very early amphibians were restricted to low-lying swamps during the Carboniferous but they too were evolving. By the Permian an early reptilian stock had emerged from them which had a more efficient method of quadrupedal locomotion, and the important ability to breed away from water, allowing them to colonise drier upland areas. Their descendants spread across the world. Several Triassic types are found in Antarctica, although only in the Beardmore Glacier region.

In 1967, Antarctica's first fossil land animal, part of a small amphibian, was found at the head of the Beardmore Glacier. The discovery created much interest and subsequently a large amount of terrestrial reptile bone was collected from the Beardmore and Shackleton Glacier region. One of the most important reptiles recovered was a moderate sized animal called *Lystrosaurus* (Plate 3.9). This had already been found in other Gondwana continents such as Africa and India. It was well known as a gregarious herbivore with nostrils sited high on its skull so that it could breathe when its mouth was under water, and with twin upper canines or tusks to dig out roots in the swampy areas where it preferred to browse. The animal stood about 1 metre tall, and was less than 2 metres long and is believed to have roamed the land in large herds. Because it showed no adaptations for long distance swimming, the southern land areas must have been united, for identical species of this animal are found as widely apart as Antarctica, South Africa, India, and China. Another land reptile also found in Antarctica, which probably preyed on *Lystrosaurus* herds, is called *Trinaxodon*. It was a flesh-eating animal not much larger than a small dog. Although much attention has

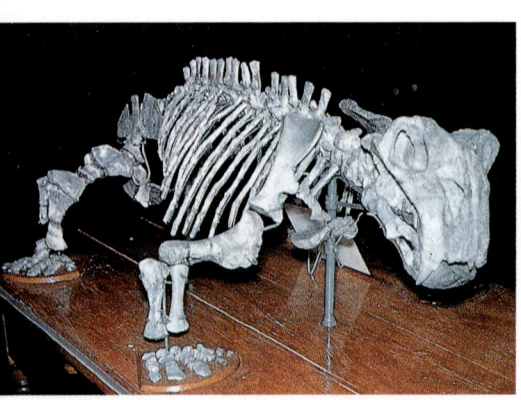

Plate 3.9. *Reconstruction of* Lystrosaurus, *a herbivorous reptile found during the Triassic Period throughout Gondwana, including Antarctica. Photo: P. Barrett.*

Plate 3.10. *These cliffs of fragmented basalt lava at Carapace Nunatak are associated with fossil-bearing lake sediments and represent the eruption of Jurassic lava into a lake-filled depression. Photo: M. Bradshaw.*

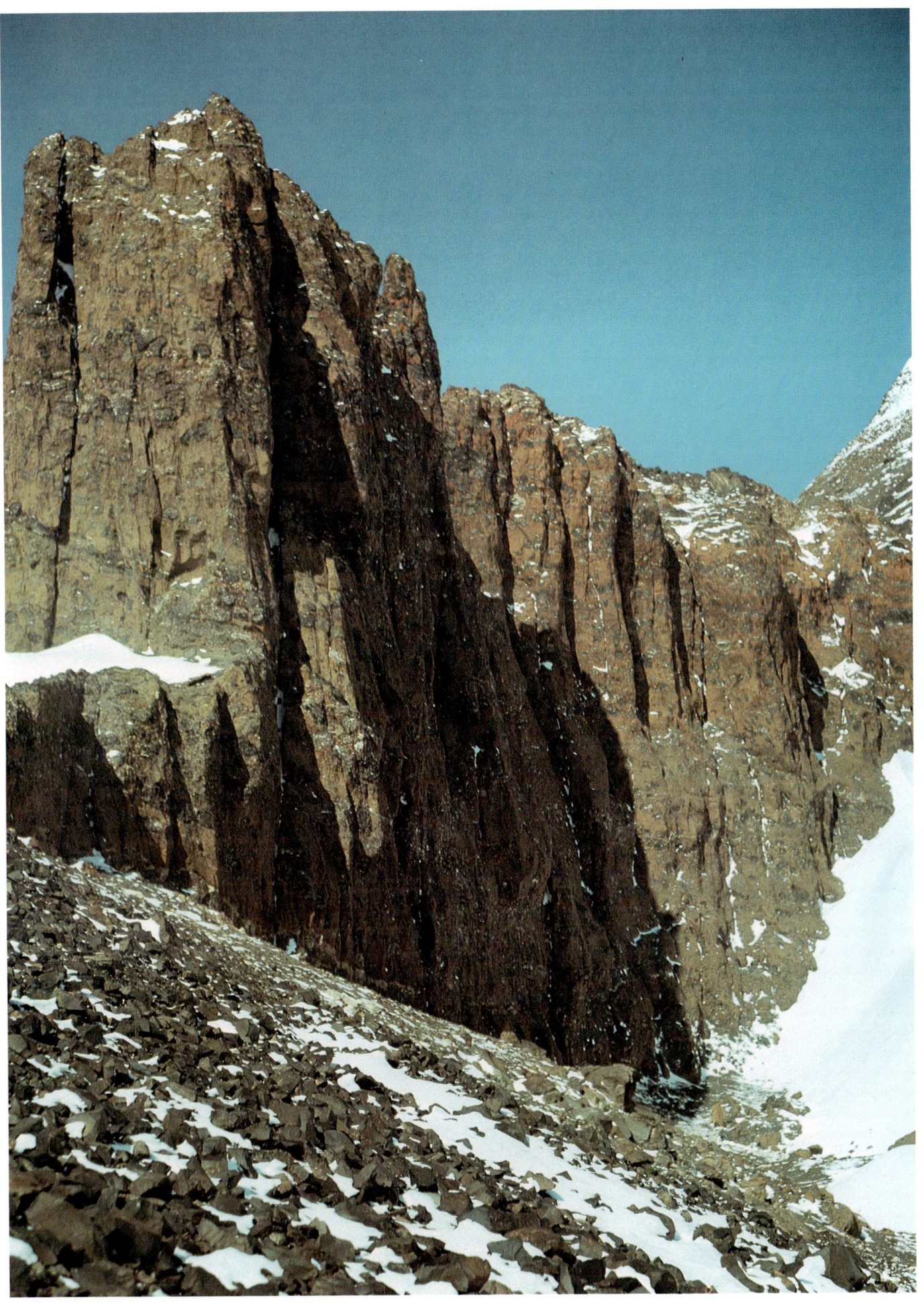

been given to the Triassic beds of South and North Victoria Land, no reptile fossils have been found in them.

For the first time in millions of years there were major signs of growing volcanic unrest and crustal instability. The Beacon rocks had been laid down on an ancient slab of continental crust that had been subsiding in a very slow and controlled manner for about 200 million years. There had been very little volcanic activity to indicate crustal rupture or a change in the tempo of its subsidence, except for a little ash in Permian and Triassic sediments. Towards the end of the Triassic, violent explosive volcanic activity increased markedly in the central Transantarctic Mountains to produce thick ash beds; and the period of stability and steady sedimentation ended for the margin of Antarctica. Changes in stress in other parts of the world began to influence the large Gondwana land mass, and during the following Jurassic Period a major fracture system was initiated within Gondwana which heralded the onset of continental drift.

The fracturing of Gondwana

Continental Drift is the theory that the continents have not always been in the positions they are today, but have drifted apart to their present positions after the breakup of an earlier, larger continent. This was proposed by Alfred Wegener in 1912 but for some time his idea was vigorously opposed and considered to be mechanically impossible. However, there is now abundant evidence that the present southern continental fragments were once closely attached to each other to form the single Southern Hemisphere land mass, Gondwana. All parts of Gondwana shared a complex late Precambrian and early Paleozoic history, with fossils and events being identical in each part. Later the Gondwana land mass shared an ice-cap followed by a *Glossopteris* flora and later still provided the home for wandering herds of early land reptiles that browsed on a *Dicroidium* vegetation. Records of these

Plate 3.11. *A Jurassic Ferrar Dolerite sill (dark rock) which has been intruded into Beacon sandstones at Finger Mountain. A similar, or the same, sill can be seen in the next range to the north. Photo: P. Barrett.*

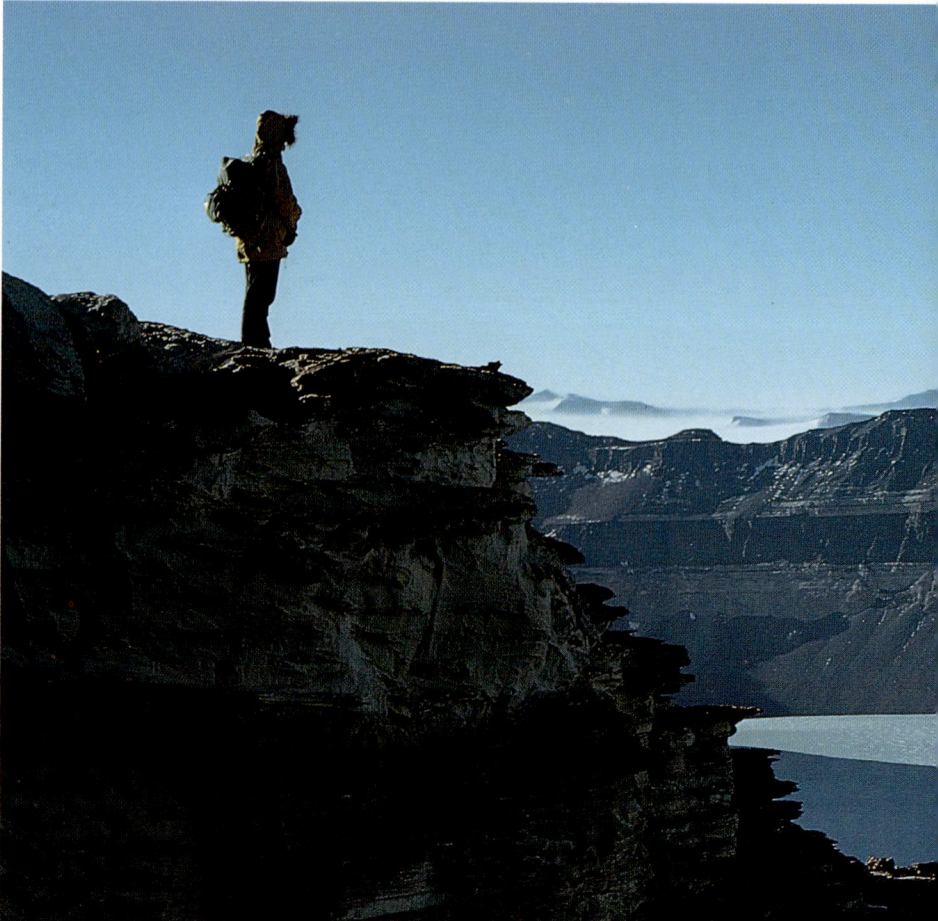

shared events are preserved in the rocks of each fragmented continental crust and are difficult to explain unless the continental fragments are fitted back together as one land mass.

The process by which continental drift has taken place is different to that originally proposed by Wegener and is known as Plate Tectonics. It can be shown that the earth's outer shell is composed of enormous plates that are moving relative to each other, and that the continental areas are carried passively within these plates. Renewal and destruction of plates, mainly involving the thinner oceanic crust, is occurring only along their margins. While new crust is formed along one edge of the plate, old crust may be being absorbed at another, so that the process is self-contained and self-perpetuating. Occasionally, the thicker continental crust of one plate collides with that of another, rumpling the thick marine sediments between so that they become welded together. This causes major changes in the stress pattern of the earth's surface, and new fractures may be propagated across continental areas. Within these fractures new oceanic crust will form to become the generative margin of a new plate.

The plate margin where new crust is forming is known as a spreading centre or ridge, and these coincide with the mid-oceanic ridges. Plate margins where oceanic crust is being absorbed by being pushed below an adjacent plate margin, cause melting at depth and generation of magma with associated volcanism, produce deep earthquakes and coincide with the deep sea trenches.

The fiery rift

A fracture system developed across Gondwana during the Jurassic, in part roughly along the line of the Transantarctic Mountains. But after a phase of dramatic volcanism, this rifting did not develop into a subsequent spreading ridge; although later fractures did eventually lead to the breakup of Gondwana.

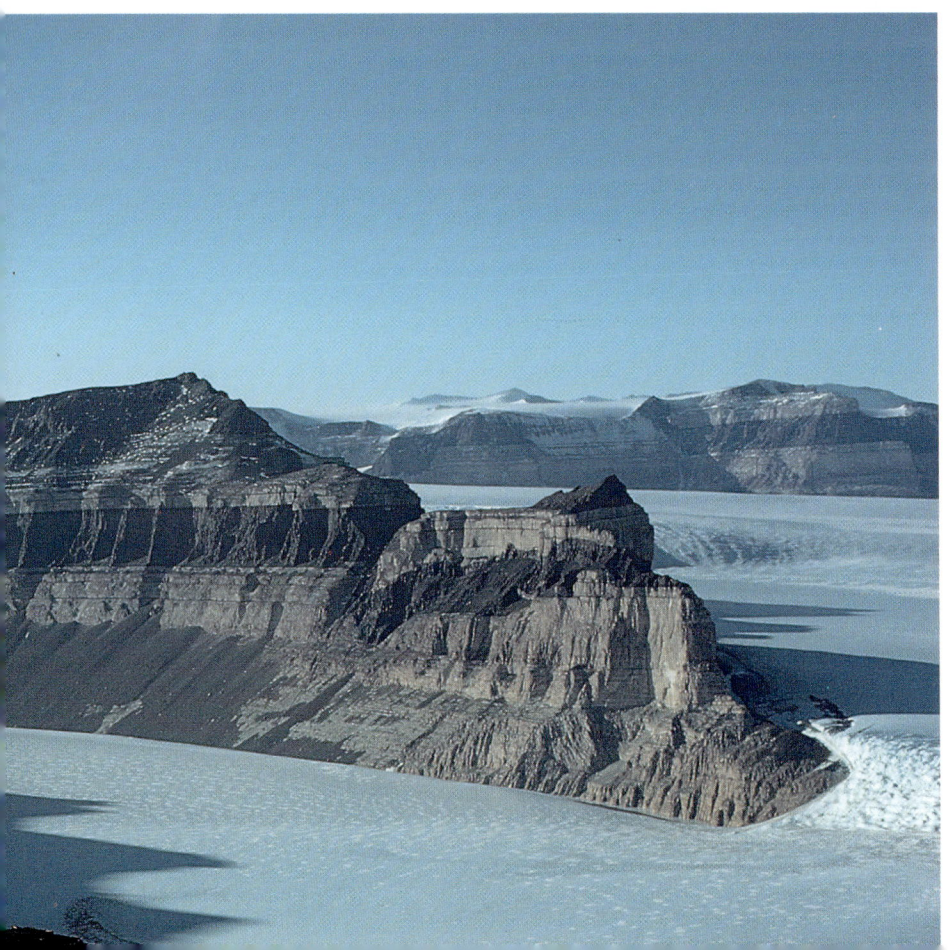

The effects of this Jurassic volcanism can be seen throughout the Transantarctic Mountains where large volcanoes erupted along a line that may have looked topographically rather like today's African Rift Valley. Initially, eruption was very violent and explosive, so that volcanic breccias and associated sediments were laid down on a surface of eroded Beacon rocks. Classic examples of these can be seen at Allan Hills and Coombs Hills in South Victoria Land. Volcanic activity later changed to relatively quiet outpouring of basalt lava from fissure vents, with local pauses in eruption during which lakes formed in the rift depression and trees grew on the volcanic slopes. Such intervals of non-eruption are represented by layers of sediment containing lake insects and crustaceans, with plant and wood fossils sandwiched between lava flows. These lava flows are known as the Kirkpatrick Basalts, named after Mount Kirkpatrick near the Beardmore Glacier where a pile of lava flows 500 metres thick has been preserved. Remnants of similar sequences can be seen at Carapace Nunatak (Plate 3.10) and Mount Brooke in South Victoria Land, and in the Mesa Range of North Victoria Land. These volcanic rocks are known only along the edge of the Transantarctic Mountains where they abut the polar ice-cap. Elsewhere they have been removed by erosion.

Deep below the ground surface, basalt magma was forced sideways into older rock as thick horizontal sheets called sills. These may be as much as 500 metres thick, and because of the insulating effects of the rock above, must have taken thousands of years to cool. Consequently, this rock has a coarser texture than the basalt erupted at the surface, and we refer to the sills as the Ferrar Dolerites (Plate 3.11). They can be seen throughout the Transantarctic Mountains, but they are especially obvious in the Dry Valleys (e.g., east side of Solitary Rocks, Taylor Glacier) where they have invaded both the granitic basement and the overlying Beacon rocks. The heat from these intrusions has caused older rock close by to become altered. Sandstones have become welded, and coal

Plate 3.12. *Greater upward movements of crustal blocks along their faulted eastern edge has produced the steep eastern face of the Transantarctic Mountains, such as the front of the Royal Society Range seen here from above the Ferrar Glacier. Photo: P. Barrett.*

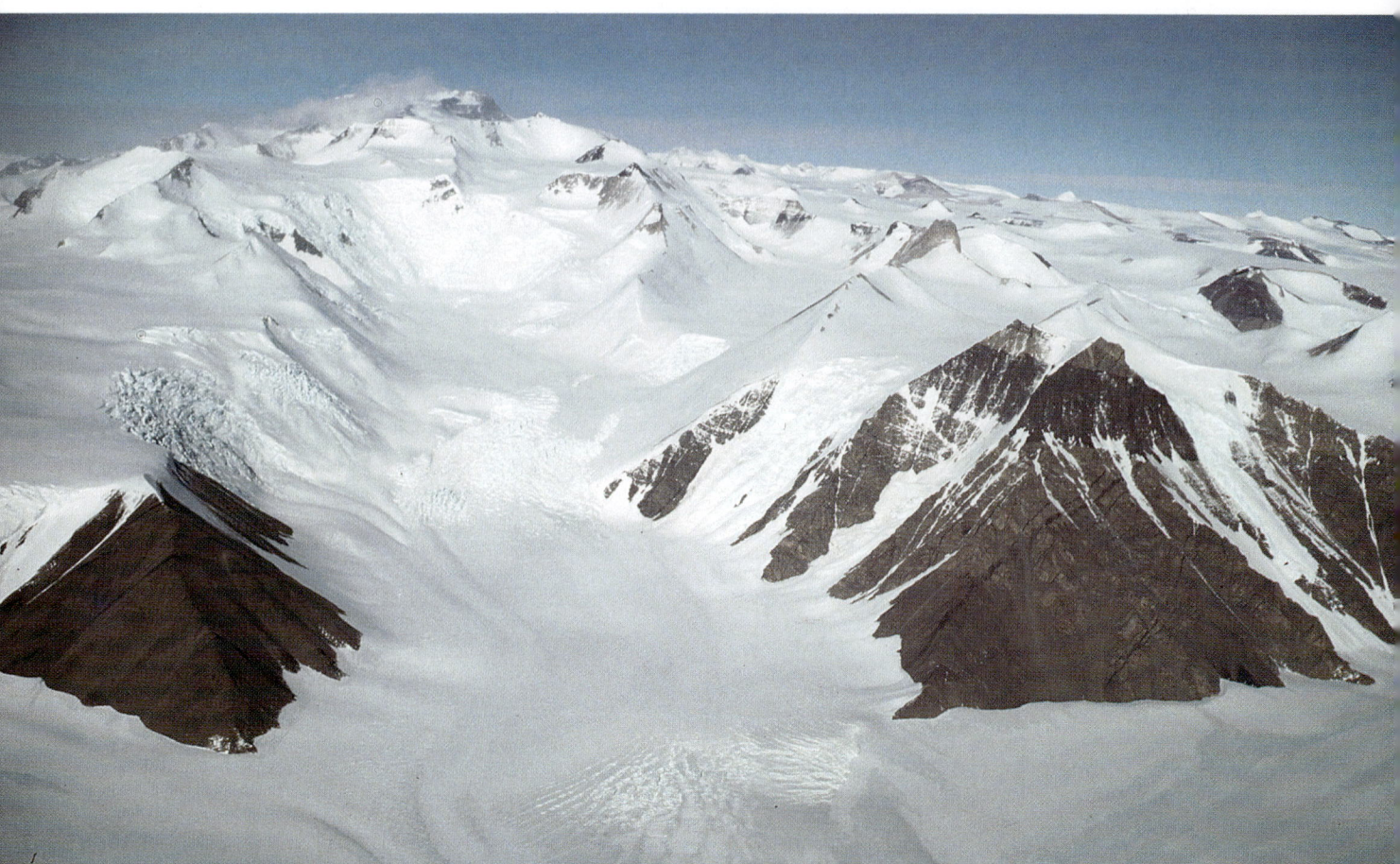

has been raised to a higher rank, in some instances reaching graphite grade.

Breakup of Gondwana

Eventually, volcanicity ceased along the early rift, and Antarctica remained otherwise unchanged. Some millions of years later a more successful rift occurred in a different part of Gondwana, steadily separating Antarctica and India. Not long afterwards, a combined Africa and South America also began drifting away from Antarctica, before they were themselves split from each other with the birth of the Atlantic Ocean between them. Later still, the New Zealand continental area drifted away from the Marie Byrd Land, Ross Sea, and Australian parts of Gondwana. And at much the same time Australia itself parted company and the Southern Ocean was born. This left Antarctica as an isolated land mass, with its Gondwana neighbours drifting further and further away from it each year.

While this fragmentation of Gondwana was taking place, no sediments were laid down in the Ross Sea region because most of it was above sea level. Consequently there is a major gap in the geological record. But towards the end of this interlude two very significant events took place, each of which had an effect on the other, and certainly changed the face of the continent. One was the growth of the Transantarctic Mountain chain, and the other was the development of the present Antarctic ice-sheet.

Growth of the Transantarctic Mountains

The Transantarctic Mountains have been formed by uplift along a system of faults that parallel the length of the chain. Blocks between the faults have been tilted more along their eastern sides than their western, to produce steep faces on the Ross Sea side, such as the front of the Royal Society Range (Plate 3.12). The most significant faults lie just seaward of the present western Ross Sea coastline. Immediately landwards of this line the crust has been vertically uplifted for approximately 4 kilometres, whereas seawards the crust has subsided more than a kilometre since the Mid-Tertiary.

Dating indicates that the Transantarctic Mountains began rising in the Early Tertiary about 45 million years ago. The initial uplift rate is estimated to have been at least 60 metres per million years, but during the Late Tertiary it may have exceeded 100 metres per million years. This increased tempo of uplift along lines of crustal weakness (faults) coincided with widespread volcanism in the Ross Sea region during the Late Tertiary. The erupted volcanic rocks are called the McMurdo Volcanics and compose many of the topographic features so familiar today in the McMurdo Sound area: landmarks such as Mount Discovery, Black and White Islands, Minna Bluff, and the entire Ross Island (see Chapter 6).

Volcanism was not confined to McMurdo Sound. In North Victoria Land, Mount Overlord and Mount Atlas are large, dormant volcanic cones, and hot steaming ground near the summit of Mount Melbourne (Plate 3.13) suggests that some cones are less dormant than others. Indeed, Ross himself had noted the apparent youthfulness of Mount Melbourne and a similarity to Mount Etna. Many of the small islands in the Ross Sea such as Beaufort, Franklin, and Coulman Islands and peninsulas such as Cape Adare owe their existence to volcanic activity in the Late Tertiary. At Cape Hallett a large thickness of volcanic rocks accumulated not in the form of a cone, but as a pile of breccia formed from eruption under ice.

Small scoria cones erupted in North Victoria Land, the Dry Valleys,

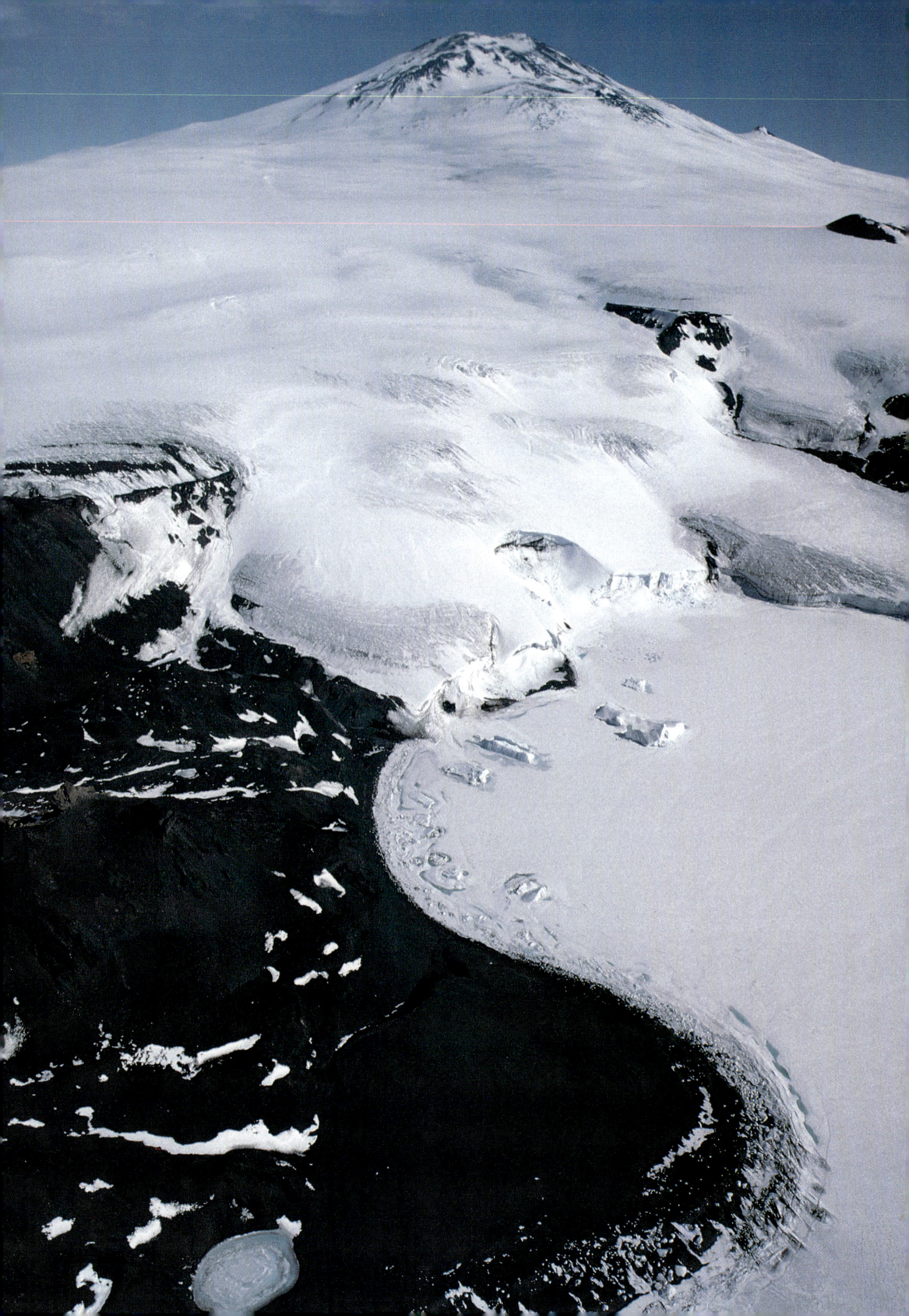

and on the sides of the Royal Society Range. The Dry Valley vents are especially significant as the volcanic material, radiometrically dated as about 4 million years old, blankets part of the exposed valley side. This shows that the main valley glaciers had retreated before the eruptions, leaving the valley sides exposed.

The most southerly volcanics are found near Mount Early at the head of the Scott Glacier. These rocks are relatively old, about 20 million years (Early Miocene), and rest on a prevolcanic erosion surface that has 900 metres of relief. The deep erosion indicated by this strong relief suggests that the Transantarctic Mountains had already begun rising. The character of the lavas and breccias, which is consistent with eruption under thick ice, point to the existence of ice this far back; this is supported by evidence in the Ross Sea.

Onset of Antarctic glaciation

Radio echo-sounding along the margins of the Ross Dependency led to the unexpected discovery of buried glacial valleys on the Polar Plateau side of the Transantarctic Mountains. This means that at some time in the past ice flowed down each side of the Transantarctic Mountains before the present ice sheet formed and ice could only flow eastwards. While glaciers flowed down the Ross Sea side of a mountain chain that was perhaps only half the height it is today, calving (breaking off and floating away as icebergs of large masses of a glacier) where they met the sea, those flowing inland coalesced to form an ice apron. This ice "piedmont" eventually merged with the growing ice fields of upland areas inland. With the accelerated growth of the Transantarctic Mountains to a maximum height of 4000 metres during the past few million years, the range began to act as a barrier to the outward, eastern flow of the ice sheet. While outlet glaciers occupied most of the earlier valleys, glaciers in the Dry Valleys became ice starved and shrank dramatically.

When the ice sheet was at its maximum, glacial debris from its base became plastered on to the valley sides. Today this bouldery-looking deposit, called the Sirius Formation, has been left stranded high above the present glacier surfaces. But apart from local glacial deposits, most of them relatively young, the story of the Antarctic Ice Sheet can only be searched for at the bottom of the Ross Sea.

We know that the Ross Sea was formed about 45 million years ago. Then, this portion of relatively thin continental crust subsided below sea level, while the relatively thickened crust further west had begun to rise into a mountain chain. Initially, the sediments laid down on the sea floor were shallow-water sandstones. But about 25 million years ago, in the Early Miocene, there was a more rapid subsidence with deposition of deeper-water muddy sediments containing plenty of evidence that debris was being rafted out to sea on icebergs before their melting allowed it to "rain down" on to the sea floor. This information has only been made available through the drilling of the Ross Sea floor from sea ice and ships. In 1973, as part of a Deep Sea Drilling Programme, four holes were drilled in the Ross Sea continental shelf from the ship *Glomar Challenger*. These holes showed for the first time that basement granite and older Cambrian marbles extended well beyond Victoria Land below the floor of the Ross Sea. The cores revealed that a thick layer of coarse, angular fragments of weathered basement rocks blanketed the basement, and was covered by Tertiary sediment 25 million years old and younger (Oligocene to Recent). The holes also revealed the presence of hydrocarbon gases.

In that same year, a drilling programme in the Dry Valleys was initiated to determine a detailed pattern of ice movement by coring

Plate 3.13. *Mount Melbourne (named by Ross in 1841 after the British Prime Minister of the time) which has steaming ground at its summit. Photo: C. Rudge, Antarctic Division, DSIR.*

sequences of glacial sediments preserved in parts of the Dry Valleys region. The deepest core penetrated to a depth of 328 metres in the Taylor Valley, and reached marine tillite about 13 million years old (Late Miocene). The tillite had been deposited by Polar Plateau ice flowing into a coastal fjord, and this fjord was much later uplifted to form the present lower Taylor Valley.

The Dry Valley Drilling Programme ended with an attempt to drill into the floor of McMurdo Sound so that the nearshore sediment could be correlated with that known on land. The difficulties in using drilling equipment under subfreezing temperatures on unstable sea ice were experienced to the full, and a depth of only 65 metres was reached.

Meanwhile, seismic surveys from ships in the McMurdo Sound region, using once again the echo-sounding principle to see what lay below the sea floor, revealed a nearshore Victoria sedimentary basin. And in 1979 a further attempt was made to drill into the edge of this basin from sea ice. This time a depth of 226 metres was successfully reached, although the preglacial beds resting on basement were not reached. In 1984 a hole was drilled near the snout of the Ferrar Glacier and encountered basement rocks at 166 metres. In 1986 another hole was drilled further seawards and successfully cored 702 metres, though still not reaching basement rock (Plate 3.14). Microscopic fossils from the overlying marine glacial sediments indicated an age of about 36 million years for the earliest sediments.

Antarctic rocks from outer space

Geologists tend to ignore rock debris on glaciers in favour of investigating rock outcrops. Thus, it is not surprising that in the first 70 years of Antarctic exploration only four meteorites had been found on the whole continent. In 1969 members of a Japanese expedition discovered meteorites lying on hard blue ice in Queen Maud Land. Soon the number retrieved in Antarctica rivalled the total collection found in

Plate 3.14. *Aerial view of the drilling complex on sea ice in McMurdo Sound by which 702 metres of core were recovered from below sea bottom, making it the deepest hole into rock in Antarctica. Photo: P. Barrett.*

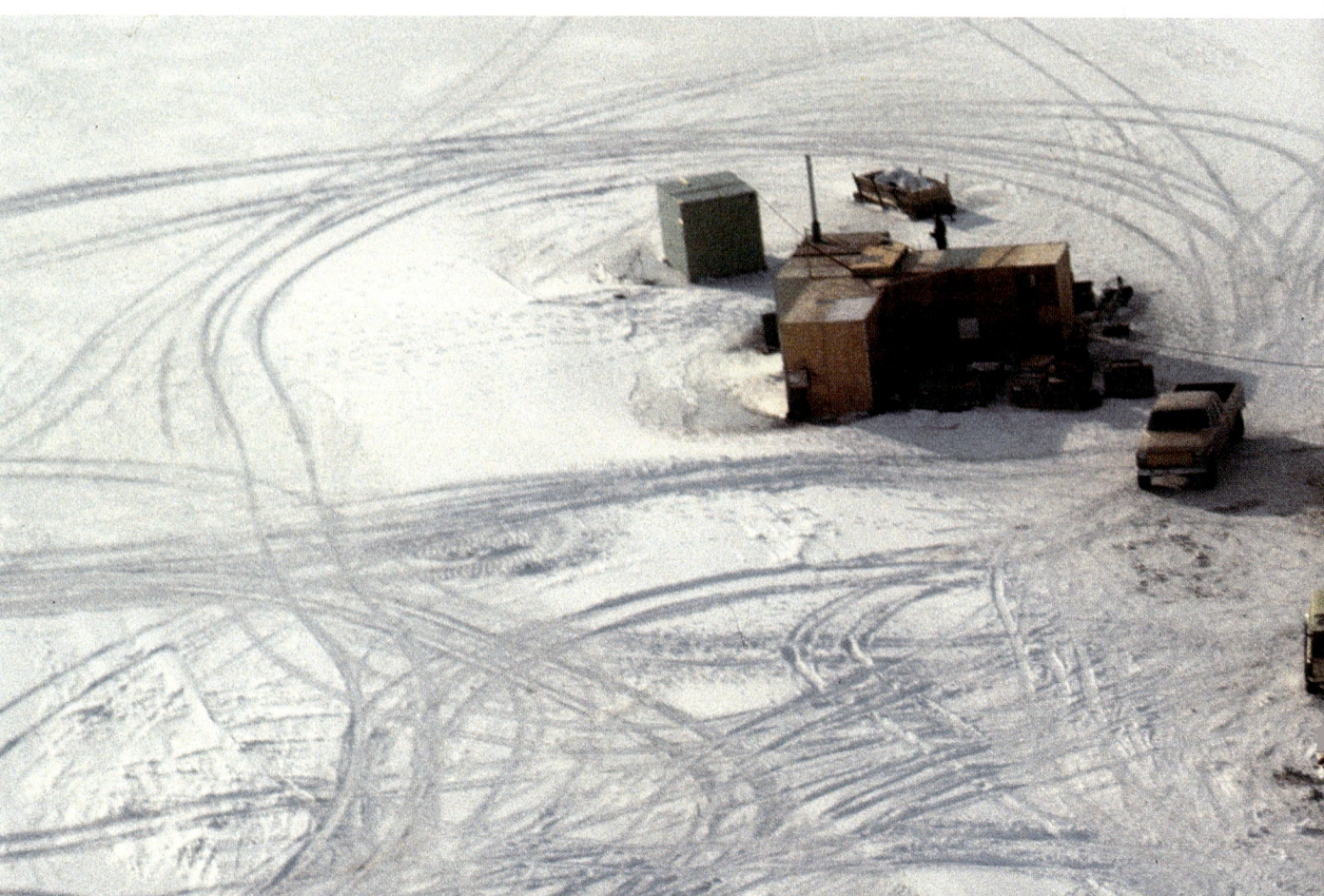

the rest of the world since meteorites were first recognised as extra-terrestrial material (not more than 200 years ago). Concentration of meteorites have been found in the Ross Sea region (Plate 3.15), and particularly around the Allen Hills, near the head of the MacKay Glacier.

The initial impact in the snow of the Polar Plateau inhibited fragmentation of these meteorites and the snow's purity meant that the organic and inorganic weathering, common in more temperate latitudes, was greatly retarded. The principal value of the meteorites still remains in determining the composition, structure, and history of their parent bodies. But their concentration in certain areas, notably the "blue-ice" fields, has great relevance to the dynamics of the Antarctic Ice Sheet. The most plausible explanation for the accumulation of so many meteorites on the small areas of blue ice is that the meteorites falling on the ice cap have been submerged and transported coastwards. In areas near mountains, upward movement of the ice combined with high rates of ablation (the combined processes by which a glacier wastes) maintains a hard surface on which meteorites have become concentrated.

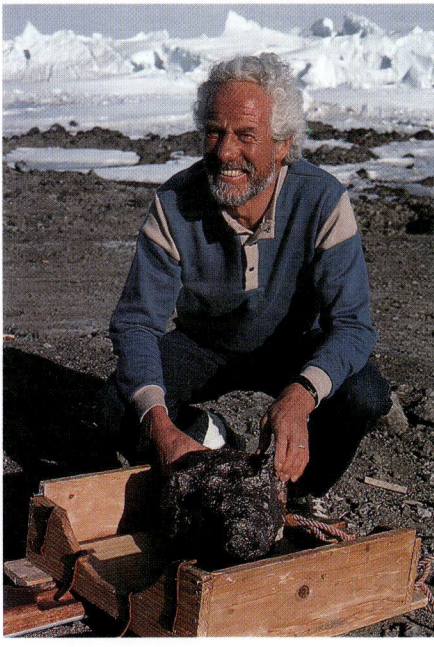

Plate 3.15. *A fragment of a large meteorite found in the Darwin Mountains. Photo: C. Rudge, Antarctic Division, DSIR.*

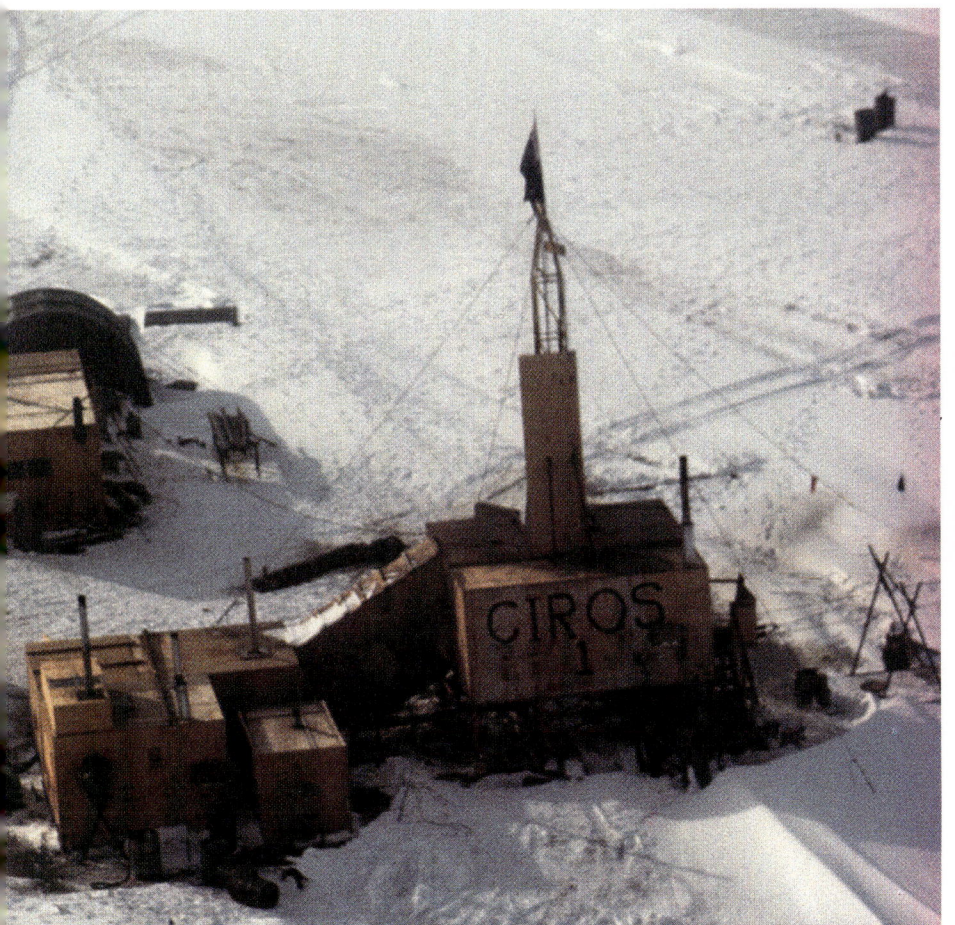

4. Sculpting the landscape

Viewed from aircraft along the western side of the Ross Sea the Transantarctic Mountains appear as a long chain of rugged peaks stretching southwards. And the Polar Plateau appears as a great white ice-blanket extending westwards into a hazy shimmering distance where the icy dazzle merges with the deep blue sky. From the Polar Plateau, ice streams cut through the mountains toward the Ross Sea where icebergs, up to 20 kilometres wide, calve from the glacier fronts. The mountains are covered with a thick mantle of snow and ice which obscures nearly all of the underlying rock. The ice stream surfaces have flow lines which show that ice is flowing from the mountain basins into the wide straight valleys. From the sea the faulted edge of the mountains shows as a steep face deeply fretted by the glacial valleys, above which rise both peaks and plateau tops crowned by thick, local ice caps.

The triple volcanic cones of Ross Island break this pattern at McMurdo Sound but southwards the form of a steep, straight mountain front is continued facing, not the sea, but the Ross Ice Shelf.

The snow and ice cover on the land is interrupted at few places except in the "Dry Valleys", west of McMurdo Sound, although in some other parts of the Transantarctic Mountains there are smaller ice-free valleys and ridges alongside the ice streams which drain the ice sheet of the Polar Plateau through the mountains. These ice streams, usually called valley glaciers, cut the mountains into separate ridges or blocks, and their development has had a major influence on the large-scale landforms of the entire mountain system (Plate 4.1).

Valleys

The major valleys developed as the Transantarctic Mountains were uplifted. Much of the uplift occurred when the East Antarctic Ice Sheet was forming. We assume that mountain glaciers formed in depressions and flowed both inland and towards the sea, progressively deepening and

Plate 4.1. *The Hatherton Glacier, flowing from the distant Polar Plateau, in a trough which divides the Britannia Range from the Darwin Mountains on the right. Photo: M. Selby.*

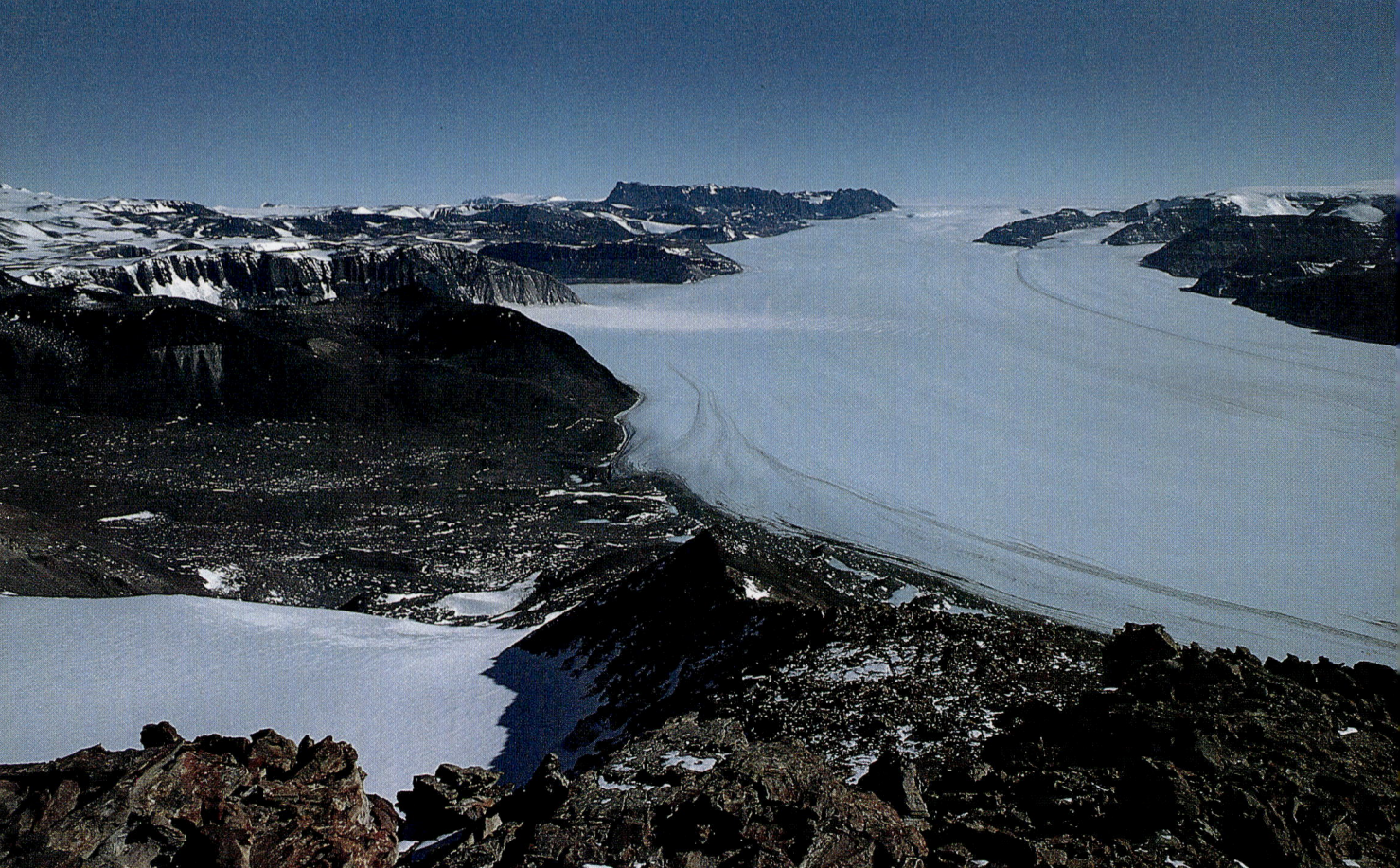

widening the valleys they occupied. There is no trace of an ancient river valley system. But it is probable that rivers drained the rising mountains and that, as the ice volume increased, the glaciers would have followed the older river-cut valleys. Some of the valleys were formed along the crushed and weakened rock of fault lines, others were cut down the slopes of the rising mountain flanks.

Over several million years the ice draining east and west from the Transantarctic Mountains so enlarged and cut back the valleys that a breach developed in many places along the crest-line of the ranges. And as the Polar Plateau ice grew to over 2 kilometres thick, it flowed through the breaches and formed outlet ice streams. One such stream is the Byrd Glacier which, moving at 2 metres a day, now drains a substantial part of the East Antarctic Ice Sheet through a valley which is 20–30 kilometres wide and 2 kilometres deep (Plate 4.2). If it became ice-free the Byrd Valley would be the largest fjord in the world.

The Polar Plateau ice has overwhelmed the inland side of the Transantarctic Mountains, burying the old valley systems and leaving only a few peaks which stand as nunataks above the ice surface. The ridge which formed the drainage divide between the east- and west-flowing ice has been cut away by the bigger outlet glaciers, but it has survived in a few valleys. As the mountains have risen further, and the Polar Plateau ice level dropped a little, some of the divides have emerged from beneath the ice to form either a high cliff and ice-fall over which some Polar Plateau ice could continue to flow towards the sea, or a narrow gorge through which the restricted ice flows relatively rapidly. These situations are recognisable at present in the heads of the Taylor (Plate 4.3), Wright, and Victoria Valleys.

Ice-free troughs

Climatic changes have caused the ice sheet of the Polar Plateau to alternately thicken and thin and the amount of ice in the high mountain

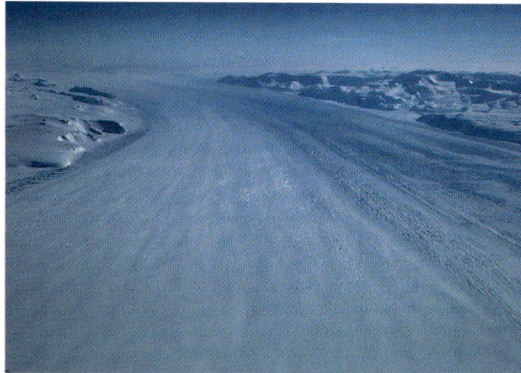

Plate 4.2. *The Byrd Glacier cutting its trough as it drains from the Polar Plateau. The valley walls are very steep as a result of vigorous erosion. Photo: Antarctic Division, DSIR.*

Plate 4.3. *Many ice streams draining the Polar Plateau have flows restricted by bars forming the valley floors. In the Dry Valleys the ice volume crossing the bars may be insufficient to maintain a glacier throughout the valley, as is shown here in the Taylor Valley. The glacier loses ice by evaporation and melting, forming streams during the summer warmth and maintaining lakes such as Lake Bonney. Photo: M. Selby.*

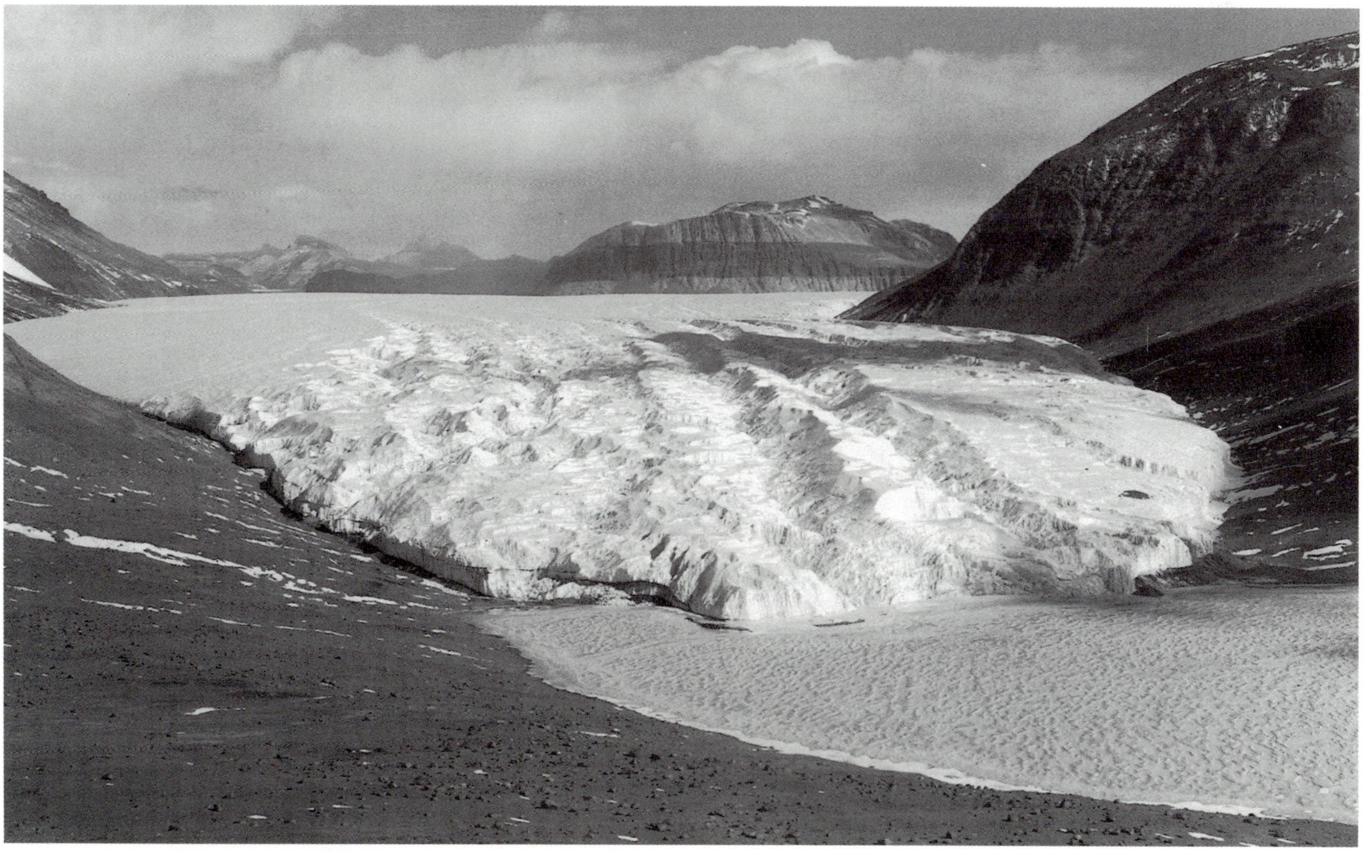

Plate 4.6. *The top of the Asgaard Range as seen from the Olympus Range. In the middleground outcrops of Ferrar Dolerite form cliffs above the trough of the Wright Valley. In the distance large cirque basins have merged to form through valleys with flat floors and steep cliffs formed in Beacon Supergroup sedimentary rocks. Photo: M. Selby.*

basins has also fluctuated. Alongside many of the outlet glaciers through the mountains, and especially in the McMurdo Sound area and farther south, shrinking of the ice has exposed bare rock. In the McMurdo region several valleys have been left ice-free and these form the largest, dry desert area of Antarctica.

The McMurdo Dry Valleys are broad, open, nearly straight valleys which were cut by former outlet glaciers. Victoria and Wright Valleys have been ice-free in their central sections for about 4 million years, but the Taylor and Koettlitz Valleys, a little to the south, have been invaded on many occasions by ice from a thickened Ross Ice Shelf (see Chapter 10).

It has been possible to date some of the glacier fluctuations, because moving ice carries rock debris which has fallen from the valleys walls, or been eroded from them, and leaves it as irregular sheets of broken rock fragments and sandy soil. This glacial till may later be flooded by a freshwater lake, or by a trapped seawater lake, where basins occur in the valley floor. The remains of marine plankton, shellfish, or thin layers of precipitated calcium carbonate accumulate on the lake floors. As the lakes evaporate, or as a fresh flow of glacial ice buries them with new till, these fossils and lake muds are left as a record of the existence of the lakes. Geochemists can analyse the muds, and geologists can estimate the age of many of the fossils which are then used to provide a dated record of the fluctuations of glaciers as indicated by the surviving tills.

In the Wright, Taylor, and Koettlitz Valleys, young volcanic cones have erupted on many of the valley-side slopes and their ages can be determined by radio-isotope dating (Plate 4.4). Where the volcanic material lies on a sheet of till, the volcano is clearly younger than the till below it and also older than any till which may overlie it. Thus, in the Koettlitz Valley, where the tills are inter-layered with lavas, a giant sandwich has been formed preserving a record of ice advances shown by till, and ice retreats shown by lava or scoria deposited on till. The time-span in which each layer of till was formed is determined from the volcanic rocks. From this evidence we not only know something of the age of the landforms in the ice-free valleys, but also much about the variations in the volume of the ice sheet which is drained by the glaciers.

The ice-free valleys now have broad floors and steep sides and also an irregular slope down the valley floor of basins, often occupied by ice-covered lakes, and rock ridges—all inherited from periods of erosion by the outlet glaciers.

Ridges between the valleys

Broad mountain ridges separate the major trough valleys. The form of these ridges is controlled partly by the nature of the rocks and partly by the high-level armchair-shaped basins, called cirques, originally cut by small glaciers (Plate 4.5).

In North Victoria Land many of the ridges are formed on basement igneous and metamorphic rocks that have rather irregular and relatively closely spaced joints. Many of these ridges have uniform slopes leading up to narrow, sharp ridge crests. Around the McMurdo area, and south of it, ridges are formed on the nearly horizontally bedded sedimentary rocks of the Beacon Supergroup, and the thick sills of Ferrar Dolerite which usually have a regular, and nearly vertical, dominant joint pattern. These two types of rocks have a major effect on the form of the ridges developed on them. The glacial basins tend to be rather flat-floored where major joints have influenced the landforms, and the cliffs are influenced by the major vertical joints which have their most obvious expression in the great columnar masses of exposed dolerite. Rock debris from slow erosion of the cliffs has produced long slopes, with a uniform

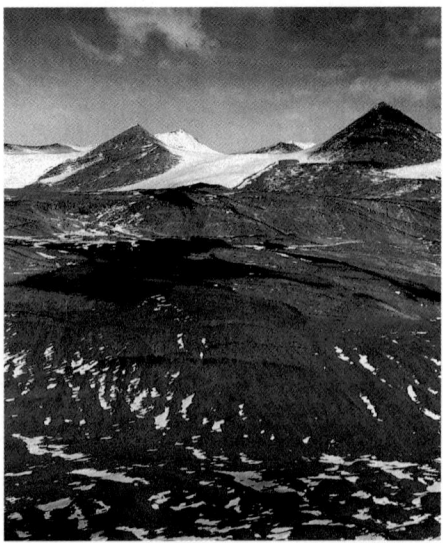

Plate 4.4. *The dark patches on the slopes of the Taylor Valley are remnants of small volcanic cones which post-date the surfaces on which they lie. Photo: G. Claridge.*

Plate 4.5. *Glacially smoothed topography forming the crest of part of the Britannia Range and a large cirque formed as a lee feature in the middleground. The Hatherton Glacier is in the foreground. Photo: M. Selby.*

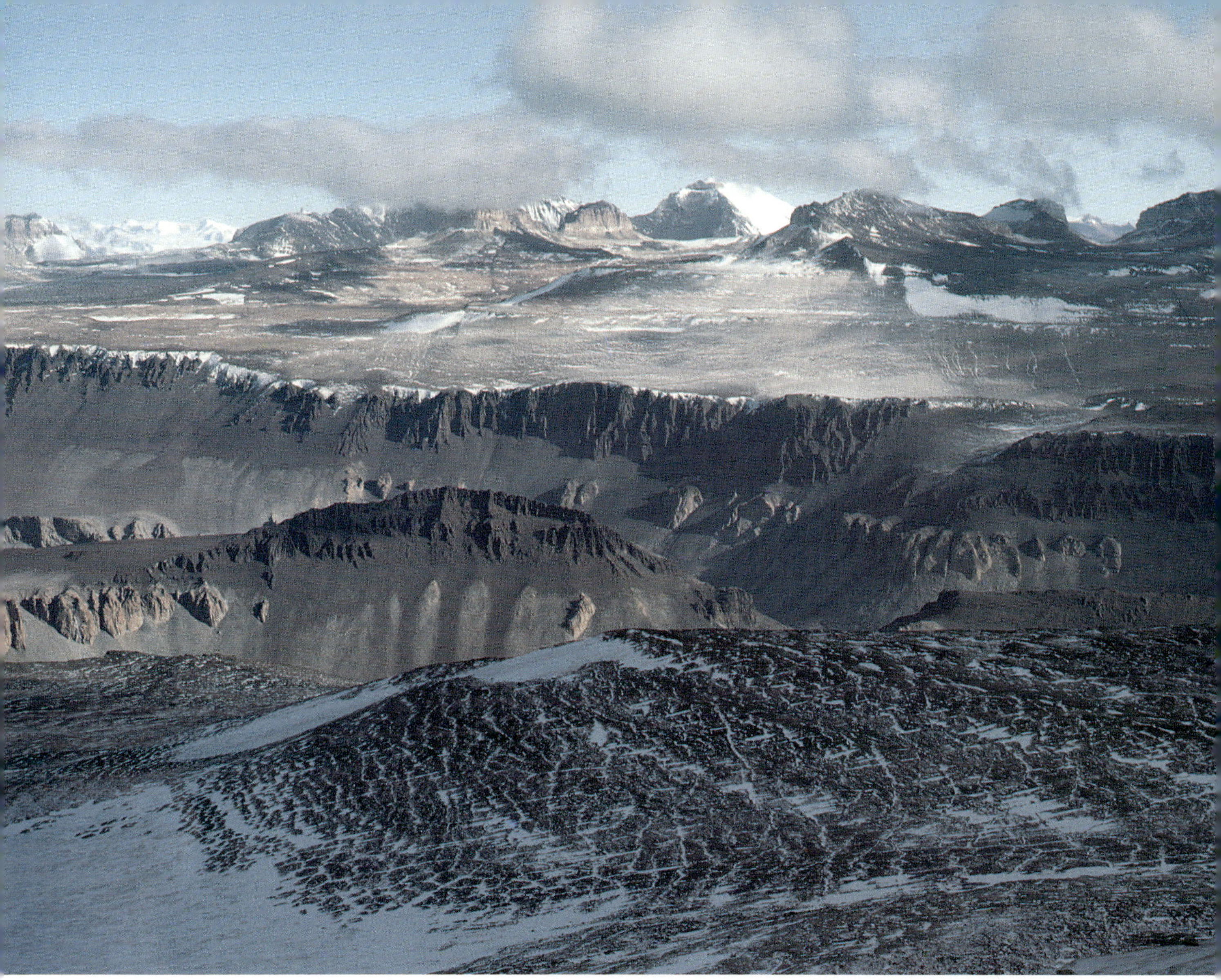

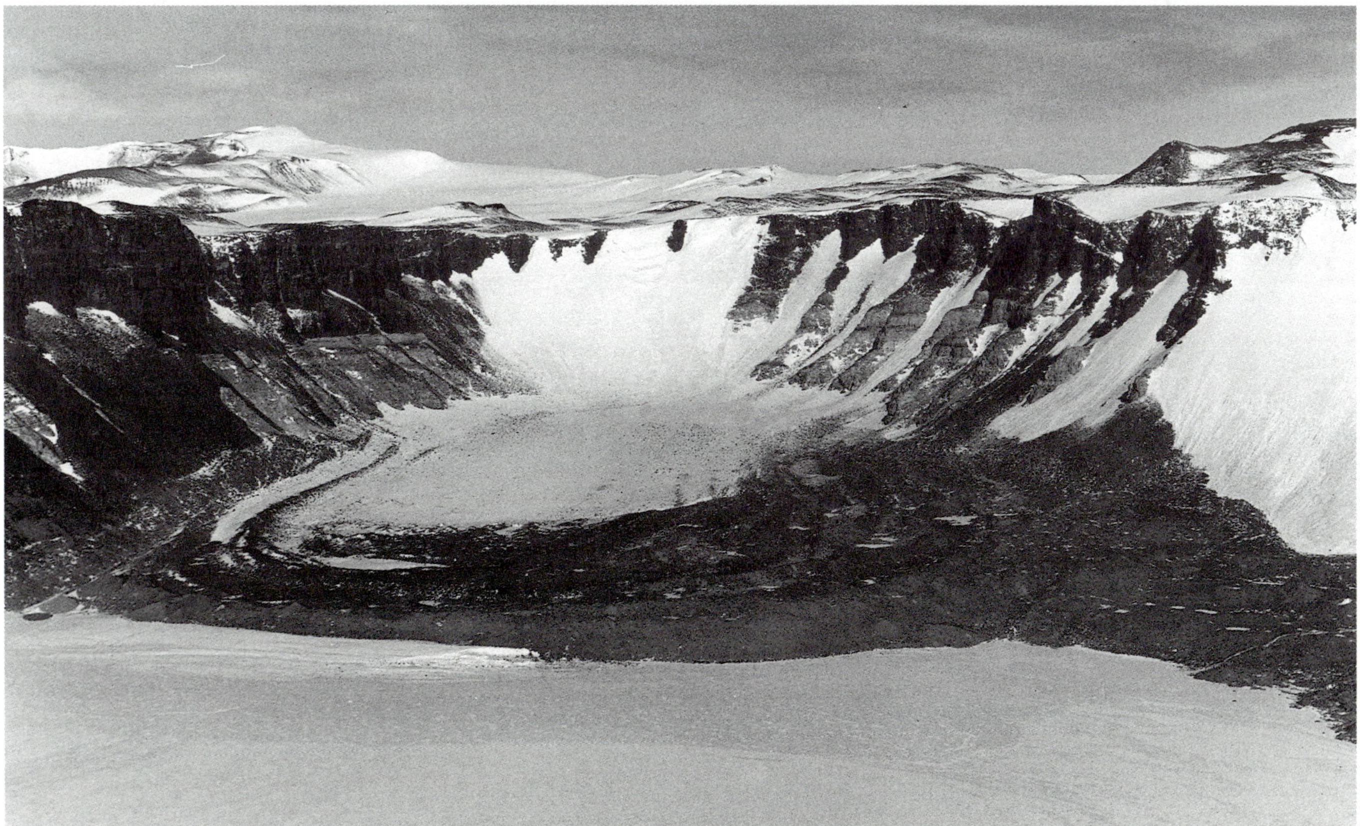

litter of fallen joint blocks and broken fragments. The erosion which has occurred since ice ceased to occupy the cirques has greatly altered the detailed landforms of the basin floors and of the cliffs.

The altitude of the cirque floors is remarkably uniform along each major ridge. Progressive widening of the cirques by glaciers has removed many parts of the cliffs and ridge tops to form a relatively uniform high-level platform, as on the top of the Asgaard and Olympus Ranges, which separate the main Dry Valleys (Plate 4.6). On this platform stand the massive bastions which are the relicts of the old glacial cirque walls.

Valley floors

Large areas of the ice-free trough valleys and the high-level cirques are covered by tills or, more rarely, consist of expanses of bare rock. Since the glaciers disappeared the tills have been considerably modified by the harsh climate. Moisture and salts, derived both from the sea and the rock, have split fragments from the rock surfaces, often leaving them deeply pitted by various sized hollows called "tafoni". In Antarctica this process of weathering (see Plate 10.2) is particularly effective where moisture is available from melting snow, for as water turns to ice in the cracks the consequent expansion breaks the rock.

Melting is greatest where dark rock is exposed to sunlight. This is because, although the air temperature may be below 0°C, the dark rock absorbs heat and may be several degrees above 0°C, so that snow in contact with it will melt. Freezing occurs again as shadows fall across the rock or as air temperature falls. Freeze–thaw and crystallisation of salts in meltwater together are potent weathering processes; they slowly break down large boulders into fragments and allow soils to develop (see Chapter 5).

Two other processes act to modify till sheets—ground freezing and wind. The very low temperatures of Antarctic winters (−40° to −70°C) cause the ground, even the bedrock, to contract. The contraction is

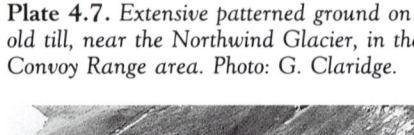
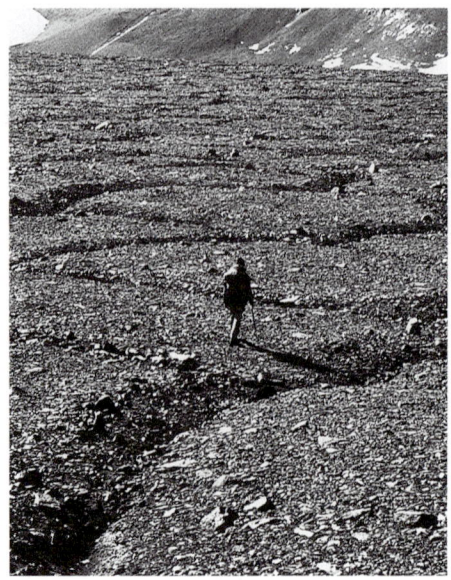

Plate 4.7. *Extensive patterned ground on old till, near the Northwind Glacier, in the Convoy Range area. Photo: G. Claridge.*

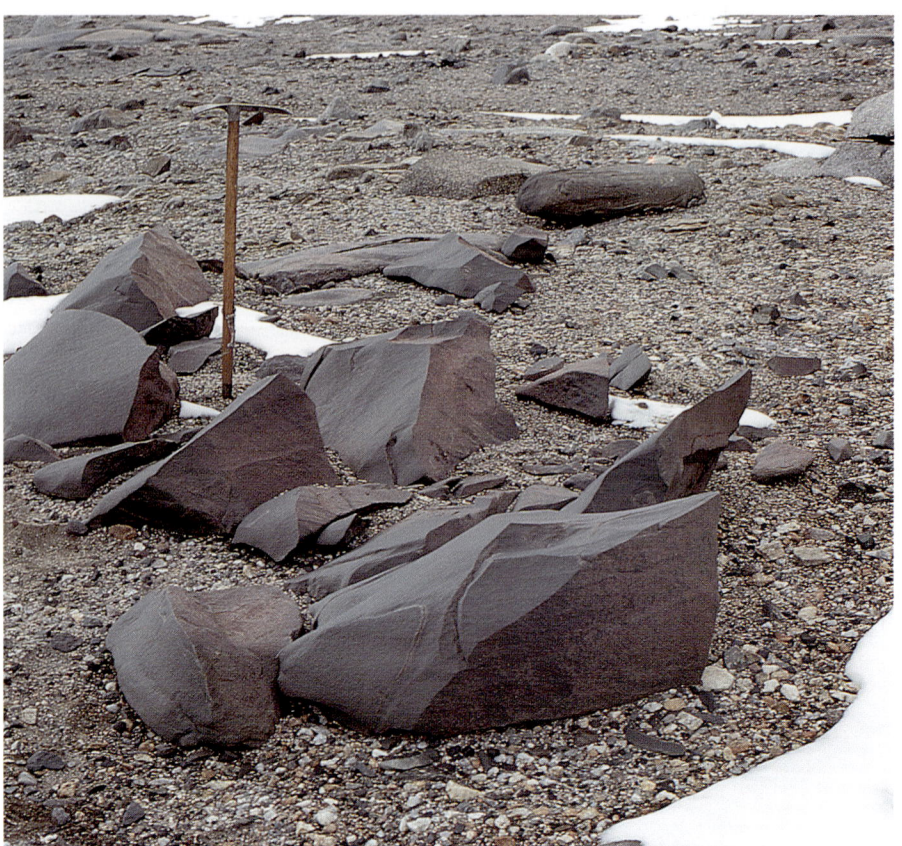

Plate 4.8. *Sand-blasted boulders with polished faces toward the dominant wind. Such features are called "ventifacts". Photo: M. Selby.*

evident from the wide cracks which divide the surfaces of till sheets into a mosaic of 4- to 6-sided polygons with a length along one side of up to 20 metres (Plate 4.7). Permanently frozen ground underlies most of the surfaces of the Dry Valleys, but it is particularly obvious where there are till sheets. Where meltwater is produced in summer it cannot penetrate into the rock and till because they are frozen. Consequently, some flat areas become saturated and form soft ground, which on low-angled slopes allows rock and soil to "flow" as a thin sheet at rates of several centimetres a year.

Very strong winds occur in Antarctica. Cold air over the Polar Plateau is very dense and flows downslope towards the edges of the continent, picking up speed as it falls down the steeper gradients. Wind speeds over 150 kilometres per hour (80 knots) occur in many valleys every winter and occasionally in summer. The cold dense air can pick up sand and even pebbles as large as billiard balls, carrying them to heights of a metre or more. Consequently, boulders exposed to the wind-driven sand have smooth faces cut across them (Plate 4.8). On many old till sheets nearly every small stone exposed has a cut-face at right angles to the dominant wind direction.

Much of the wind-blown sand is carried beyond the Dry Valleys on to the sea ice or into McMurdo Sound. But in the eastern end of Victoria Valley it falls on to the Lower Victoria Glacier from where it is washed by meltwater back into the Victoria Valley. There it is picked up again by wind and blown into extensive sand dunes (Plate 4.9). These dunes are similar in shape to those found in the hot deserts of the subtropics but there is one major difference. In Antarctica snow is often mixed with the sand; the snow may melt in midsummer and subsequently become ice, so freezing the dunes solid. Only the few centimetres-thick layer of sand at the surface is truly mobile.

Meltwater from the glaciers forms streams which flow for a few days of the year. These streams carry sand but are usually incapable of

Plate 4.9. *A dune field in the lower Victoria Valley. Photo: M. Selby.*

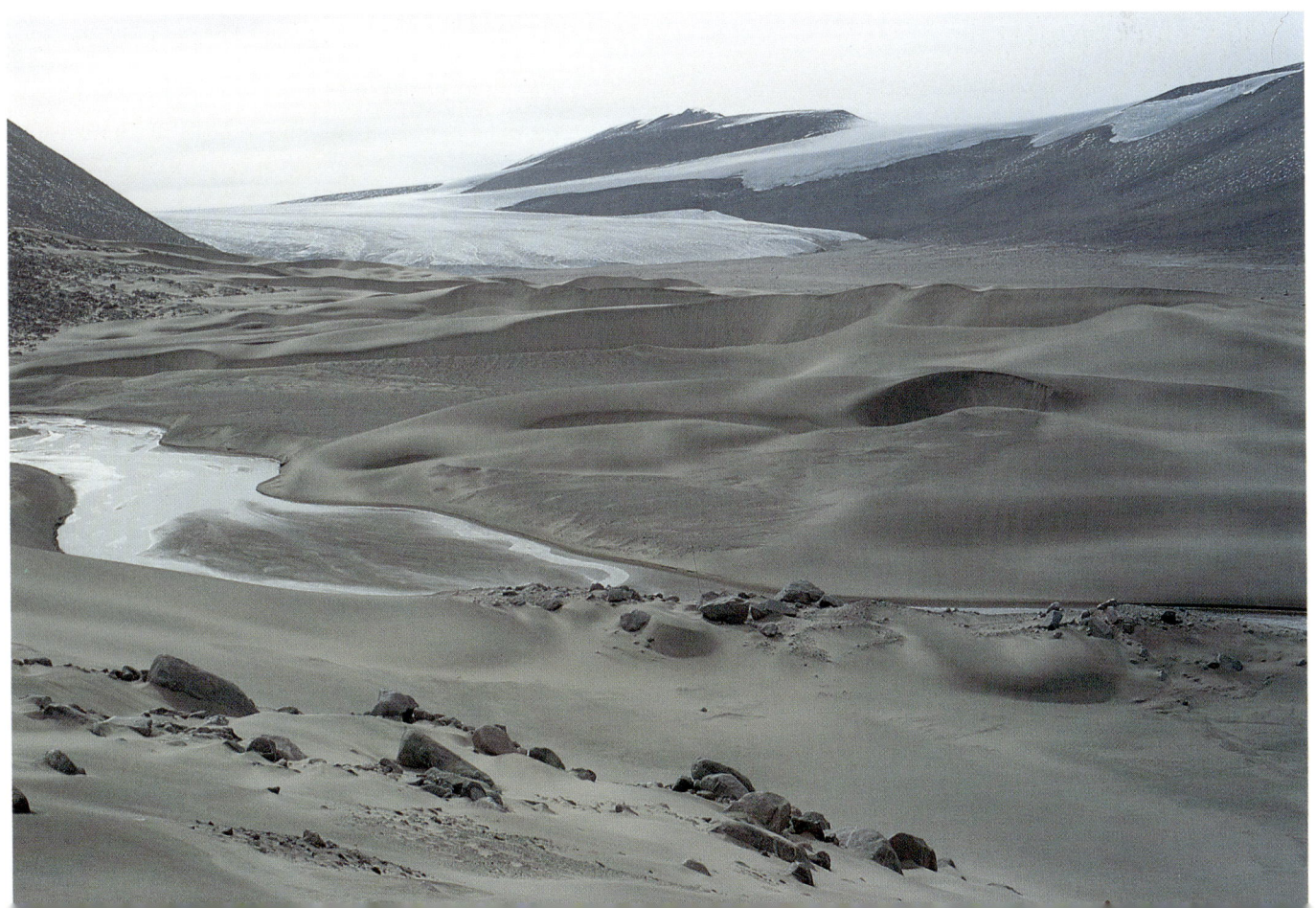

eroding into bedrock. This is because they are shallow and so do not have enough energy or "stream power" to carry the large rock fragments needed to erode bedrock, nor are they active for long enough or have a large enough flow of water to be effective against rock. But they do cut into ice on glacier margins and surfaces.

The most spectacular feature of the Dry Valleys is The Labyrinth. It consists of a series of canyons and box-shaped cavities developed on a dolerite sill at the head of the Wright Valley (Plate 4.10). It is not known how the canyons formed but a favoured hypothesis is that a massive flow of meltwater from the Polar Plateau carved them out. What caused the flow is not known—was it the result of volcanic activity beneath the ice? Another hypothesis is that salt crystallising in the joints of the rock broke off individual grains which were then blown away to settle on the floor of the Ross Sea. We may never know the true reason.

Plate 4.10. *Looking over the canyons of The Labyrinth down the Wright Valley to Lake Vanda. Photo: Antarctic Division, DSIR.*

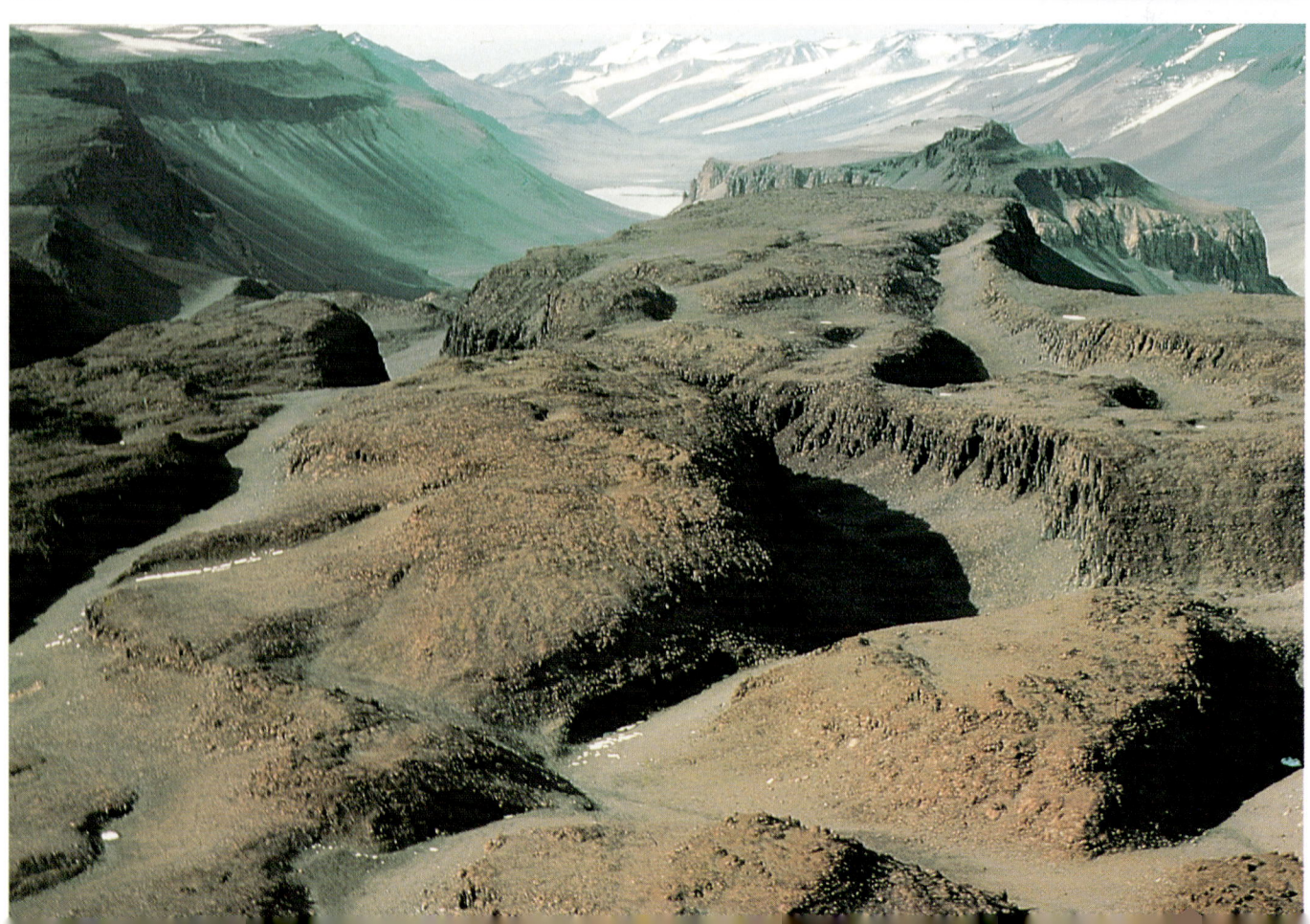

The soil cover on the ice-free areas of Antarctica is produced by atmospheric and biological processes acting on the rocks and other surface deposits. The soils result from the most recent processes forming the Antarctic landscape but, to some extent, the influences of past events can still be distinguished.

Antarctic soils differ from soils of other continents in that plant and animal life are almost absent and, therefore, nutrient cycling and biological weathering play an almost negligible part in soil development. The rate of soil development is also very slow because the extreme cold restricts the availability of water for chemical weathering, as well as for biological processes.

Most Antarctic soils are formed from till, deposited by the glaciers that have shaped the continent in the last few million years (Plate 5.1). Soil properties largely depend on the rock type from which the till was formed and the extent to which the rocks have been broken into smaller fragments during the till's deposition. Throughout much of the Transantarctic Mountains tills are formed largely from fragments of Beacon Sandstone and harder, more resistant Ferrar Dolerites. The sandstone crumbles readily into its component grains and the proportion of fine material, largely quartz sand, can be relatively high in the soil parent material. On the other hand, tills formed mainly from granite, metamorphic rocks, or very hard sediments tend to be much coarser in texture.

Where tills are not present and the bare rock predominates, soils are limited to thin veneers or accumulations of fine material, within cracks or crevices. Most of this material consists of mineral grains fretted from the surrounding bedrock, although some wind-blown material may be present.

Because of the stability of the landscape and the slow rates of erosion and transportation of debris, soils are forming on surfaces that have been weathering for millions of years. Thus, although weathering rates are slow (for the reasons mentioned above), soil-forming processes have had a long time in which to operate, sufficient for characteristics such as redness, soluble salt content, and rock breakdown to develop.

Although temperatures are generally low and the atmosphere is extremely arid, the climate varies sufficiently from place to place to be reflected in the properties of the soil. At one end of the climatic range is an extremely cold, arid upland zone on the edge of the East Antarctic Ice Sheet, where temperatures rarely rise above freezing point and liquid water is almost never available. Water is sometimes available in the soils in a climatic transition zone. And at the other end of the range are relatively warm regions at low altitude, along the coast of the Ross Sea and Ross Ice Shelf, where soils may become wet more frequently.

Most of the soils contain no visible plants and life is confined to a few micro-organisms; in the most extreme situations soils may be sterile. In sheltered spots, where moisture is available, some soil may be biologically active although even here the organic cycle contributes little or nothing to soil development. These soils, marked by the presence of mosses, lichens, and a diverse microflora and microfauna, are usually very young, disturbed by freezing and thawing, and are generally very weakly developed.

Soils of the Transantarctic Mountains

A well-developed soil which has been weathering for a considerable period shows several distinctive features (Plate 5.2). The soil surface is usually covered by a stone pavement or lag gravel of the more resistant rocks (and in older soils these are often stained and polished). The size of the boulders protruding through the stone pavement gives some indication of the soil's age, because in old soils all but the most resistant rocks may be eroded to ground level. Below the stone pavement there is often a thin (1–4 centimetres) horizon of fine gravel and sand which accumulated through fretting of the surface rocks.

The soil proper, especially if formed from till, generally consists of fine textured, oxidised and stained material (containing up to 4% clay) with horizons or scattered accumulations of salts. The most strongly weathered soils are reddish brown or brownish yellow. The upper soil horizon is the most strongly oxidised and staining diminishes downward, but may be detected to depths of a metre or more in the oldest soils.

Throughout the soil, but more particularly near the surface, "ghosts" or the crumbled remains of more easily decomposed rocks may be found. These are often sandstone boulders which have disintegrated to form a mass of loose sand. Coarse-grained granite also crumbles readily into its constituent minerals due to the weathering of the less stable components, the disintegrating effect of salt films in intergranular spaces, and even the freezing of thin films of moisture. Resistant rocks such as basalt and dolerite may be fresh and angular within the soil, whereas similar rocks on the soil surface are almost completely destroyed.

Most soils are underlain by ice-cemented material termed frozen ground or permafrost. In young soils frozen ground is near the surface and some melting may occur when soil temperature rises in the summer, but in older and drier soils the ice-cemented layer is much deeper. The loose material above frozen ground is permanently frozen but not ice-cemented. Where this dry material is thin, patterned ground may occur (see Plate 4.7). In younger soils the polygon edges are small cracks, in older soils they may be trenches up to a metre in width and 50 centimetres deep.

Plate 5.1. *An ice-free valley in the Asgaard Range showing tills of various ages on the valley floor and on some slopes, with bedrock outcrops on the ridges. Photo: G. Claridge.*

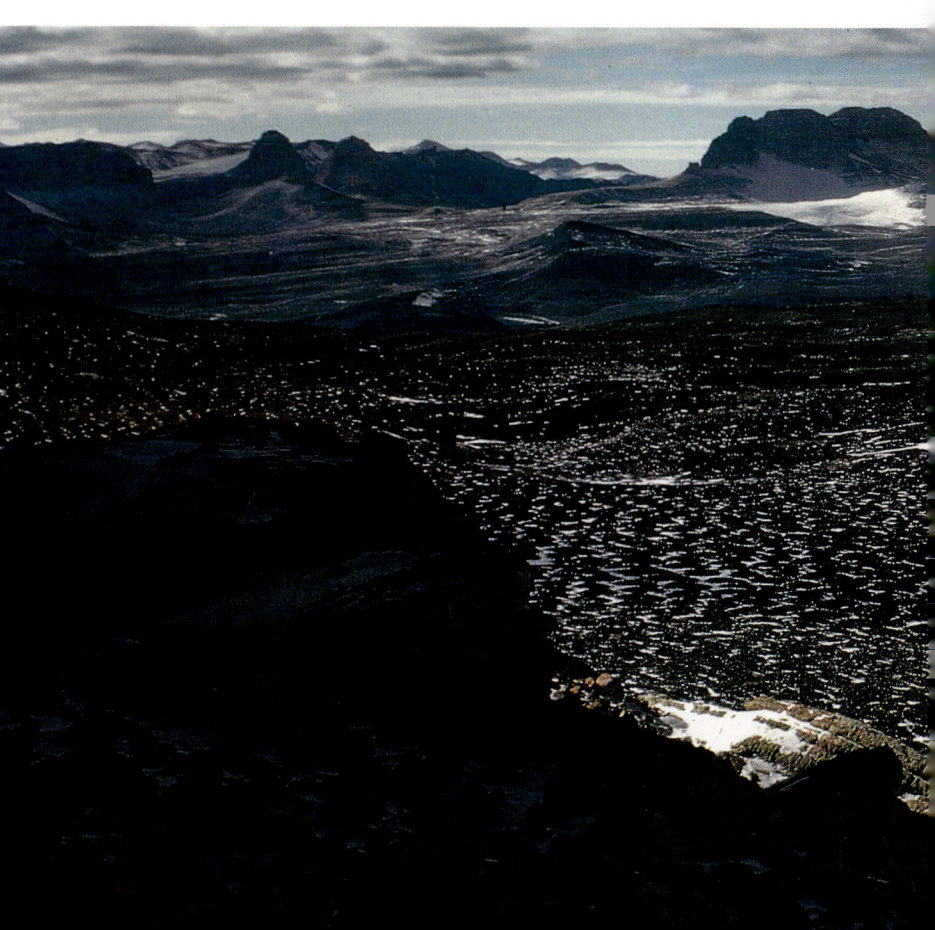

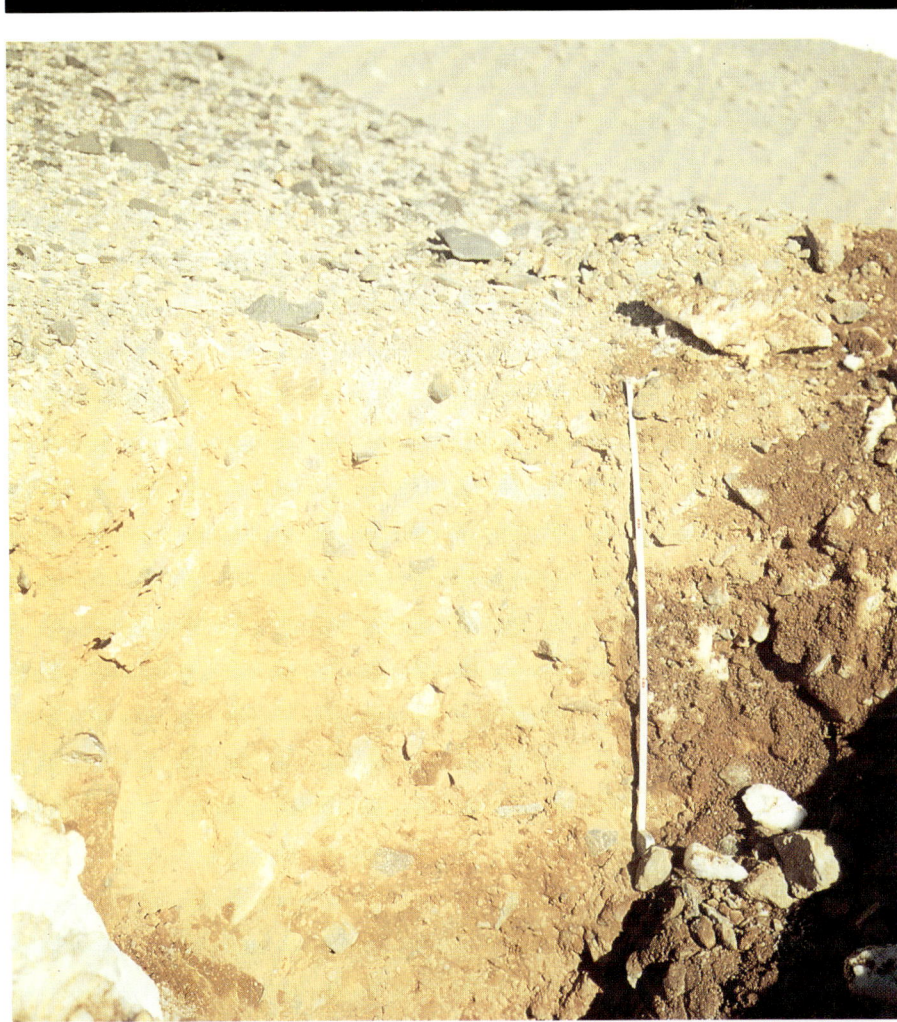

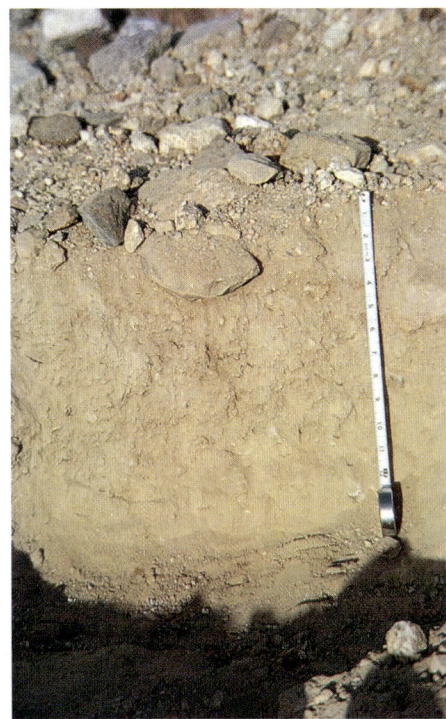

Plate 5.2. *A soil formed on till deposited by the Meserve Glacier, Wright Valley. Note the stone pavement, the sandy material below, the thick salt horizon, and the compact silt-cemented till below. The soil is about 3 million years old. Photo: I. Campbell.*

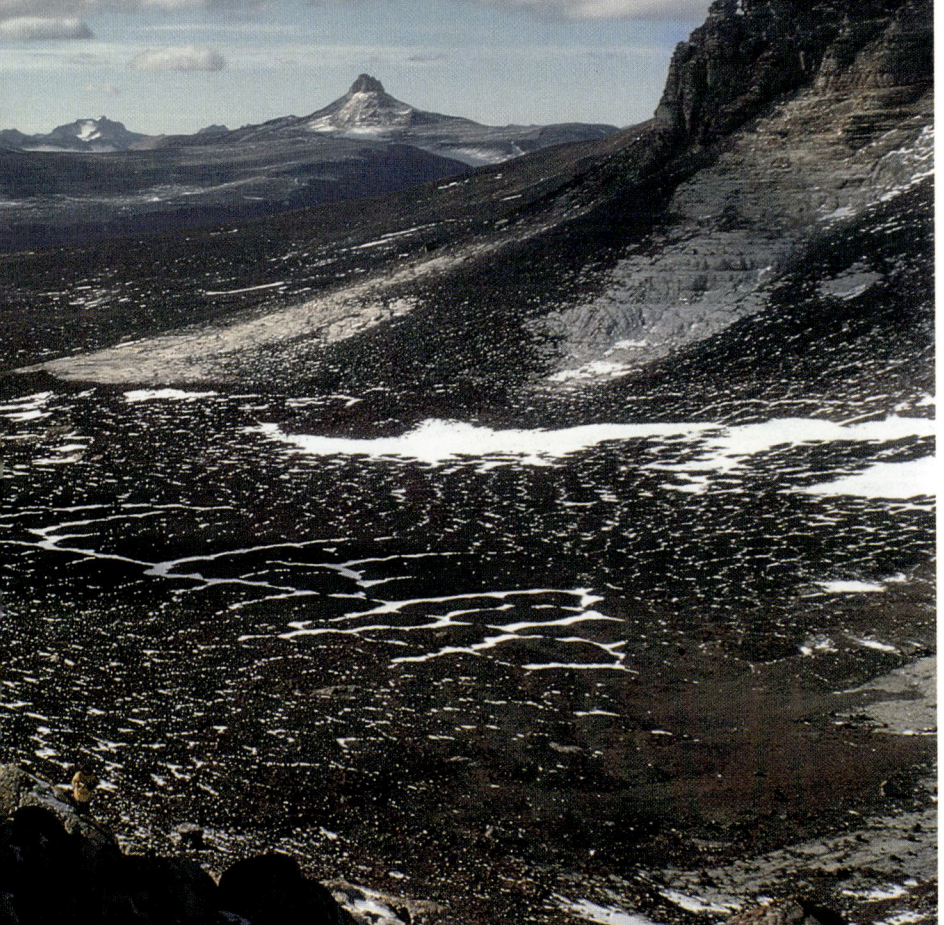

Plate 5.3. *A soil on granitic till from the Brown Hills region, beside the Darwin Glacier. The till on which the soil is formed was deposited by the Darwin Glacier within the last 50 000 years. Photo: G. Claridge.*

Soils and glacial history of Antarctica

When soil studies first began in Antarctica it was thought that most of the features seen were related to events of the last glaciation which, by analogy with mountainous regions in the rest of the world, was considered to have been at its maximum between 50 000 and 20 000 years ago. However, it is now realised that because of the slow rate of weathering, some of the more strongly weathered Antarctic soils must be extremely old. It was only after the ages of some soils had been determined by radio-isotope dating of their associated volcanic debris that it was possible to show that certain soils had been exposed to weathering for at least 2–4 million years. With the ages for some soils known, the ages of others could be interpolated using differences in degree of weathering and soil formation.

By these means it has been possible to recognise several distinct surfaces throughout the Transantarctic Mountains. The youngest of these surfaces is probably less than 50 000 years old (Plate 5.3), whereas the oldest is considered to be 10–15 million years old (Plate 5.4). By making allowances for variations in soil development it is possible to correlate weathering stages in different parts of Antarctica. Thus, a detailed chronology can be established where surfaces have not been dated by the radio-isotope method.

Significance of salts in Antarctic soils

Because Antarctic soils are forming in an arid climate with little moisture available for flushing, salts accumulate either as encrustations, surface coatings under stones, or concentrated as horizons within the soil profile. A thick salt horizon can be seen in Plate 5.2, and a small undrained hollow with salt efflorescences is illustrated in Plate 5.5. The salts differ greatly in composition from place to place. Some are derived from weathering of the rocks from which the soils are formed but the

Plate 5.4. *Old landscape in the Roberts Massif, Shackleton Glacier area. The terraces above the ice tongue have been exposed to weathering for about 15 million years. Photo: G. Claridge.*

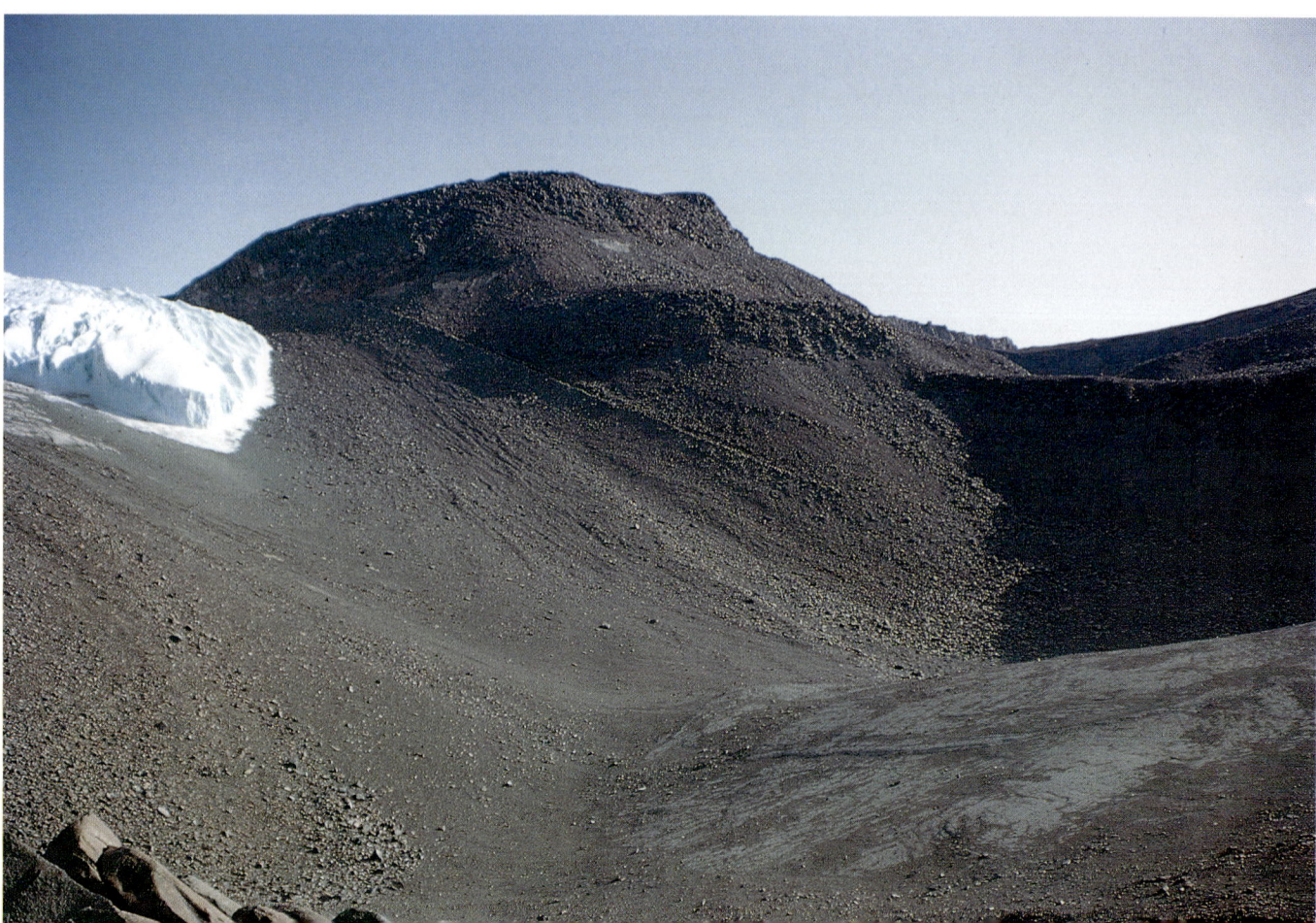

dominant source of the salts is precipitation.

In coastal areas the salts in the soils consist largely of sodium chloride and sodium sulphate, derived from sea water caught up in winds and blown inland. Further inland sulphates of sodium and magnesium are deposited by windblown snow from the inland ice sheet. These compounds originate in the protein-rich froth on the surface of tropical oceans and are transported to the polar regions by circulation in the upper atmosphere. During transport, the proteins are oxidised to nitric and sulphuric acids while the sodium chloride absorbs moisture and is precipitated before the air masses have travelled very far. The acids combine with cations (positive ions) released from the rock to form the salts found in the soils. Nitrate salts accumulate in Antarctic soils because there is no water to leach them out nor organisms to use them. Detailed studies of the salts may reveal major changes in accumulation rates or distribution patterns which will reflect past variations of climate or, with nitrates, be useful in detecting past changes in global circulation patterns. The stability of salts in Antarctic soils is such that the products of widespread global pollution could accumulate and persist in these soils.

Relationship with soils of the hot deserts

Antarctic soils have formed in a cold desert but closely resemble soils of the hot deserts. There are parallels in the occurrence, origin, and composition of salts (including nitrates), the disintegration of surface rocks, the types of lichens and algae found on the rocks, the formation of desert pavements and surface soil crusts, and the presence of pale horizons and red colours in old soils. The major differences between cold and hot desert soils are due to the almost complete absence of water in most Antarctic soils.

Influence of man on Antarctic soils

The very simplicity of Antarctic soils makes them more susceptible

Plate 5.5. *An undrained basin in shallow till in the Alatna Valley, showing salt efflorescences in the centre and a zone of moist and highly saline soil surrounding the basin. Photo: I. Campbell.*

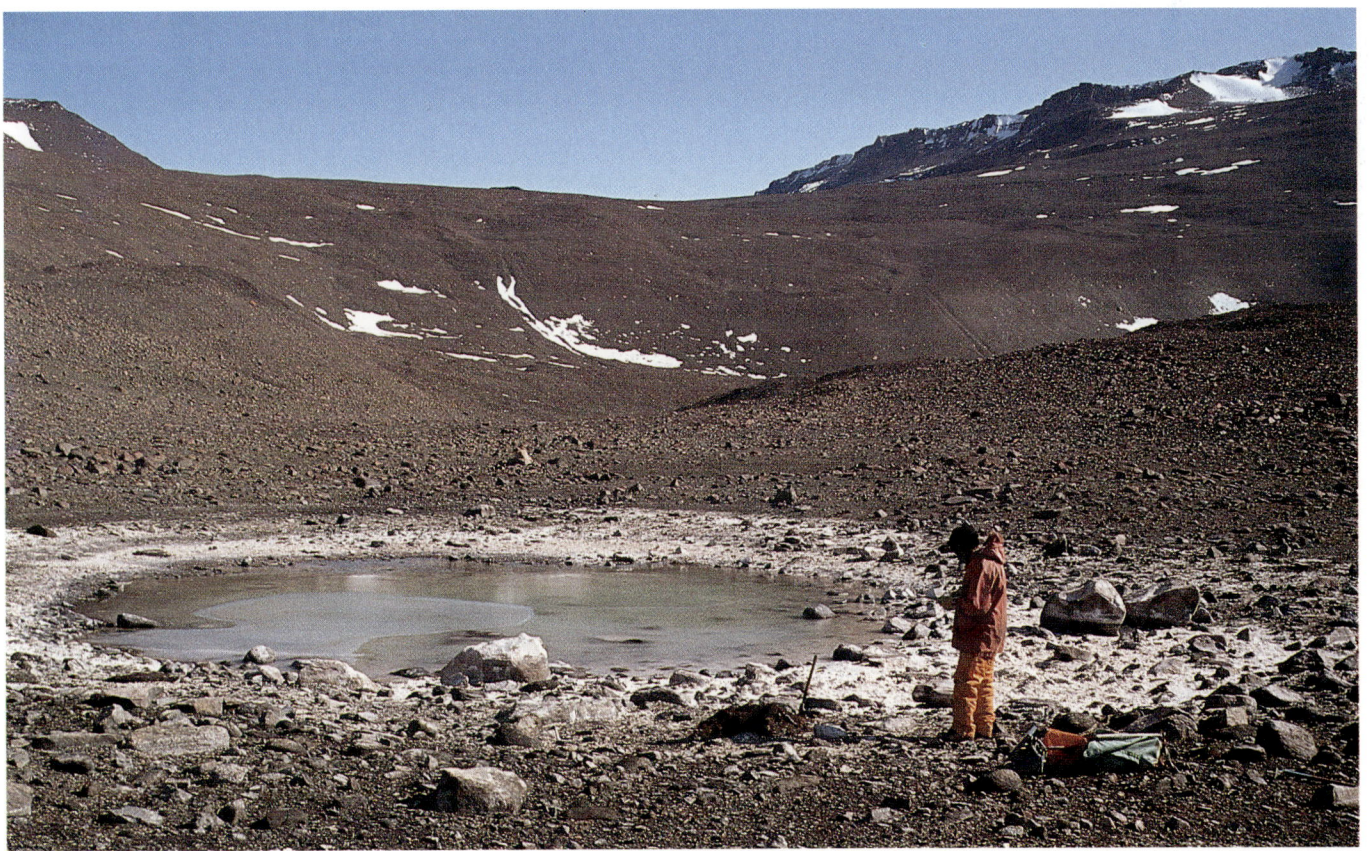

to damage than the soils of temperate lands. Introduced organic materials and the products of combustion have an exceedingly long life and remain in the soil because of the slow rate of decomposition and the absence of effective leaching. The scraping of snow for water, or the smoothing of the ground surface, causes less snow to be trapped and results in increased local warming and thawing through greater exposure of soil and rock surfaces. Dust from mechanical operations also accumulates on snow surfaces and accelerates thawing; the resulting increase in available moisture causes soil activation and patterned ground movements. The effects of direct mechanical disturbance of the soil by digging or scraping are almost permanent (Plate 5.6) and recovery of the soil to its natural undisturbed state is virtually impossible because of the exceedingly slow rate of soil development. Thus, the soils form one more testimony to the fragility of the Antarctic environment.

Plate 5.6. *Part of McMurdo Station, showing extensive ground disturbance by earthmoving work. Photo: I. Campbell.*

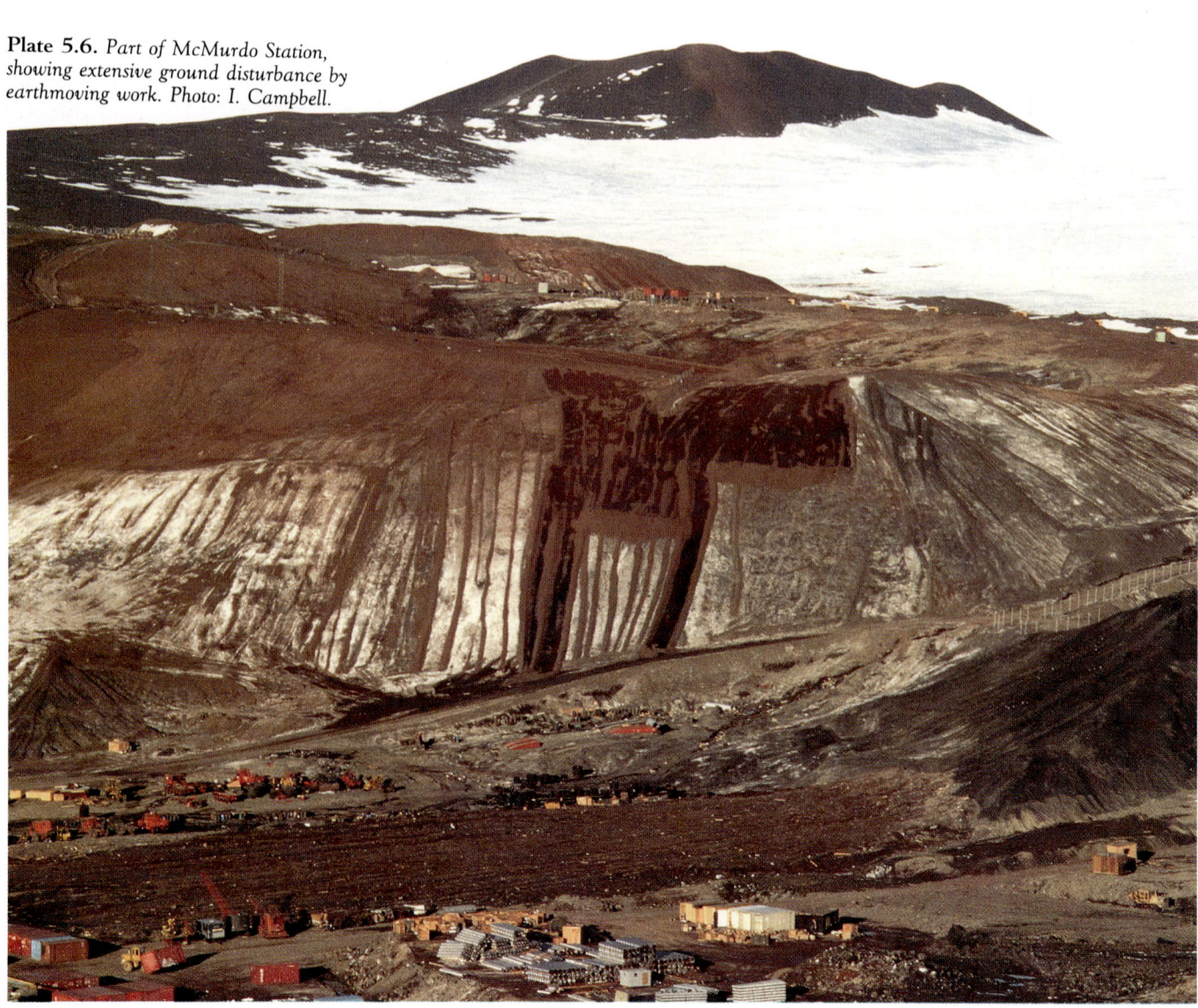

"Erebus, a deity of hell, son of Chaos and Darkness. He married Night by whom he had the light and the day. The poets often used the word Erebus to signify hell itself and particularly that part where dwelt the souls of those who had lived a virtuous life, from whence they passed into the Elysian fields." *Lempriere's Classical Dictionary.*

Discovery

The dominant and most universally recognised topographic feature in the Ross Sea region is undoubtedly Mount Erebus. As Ross's ships sailed southwards along the coast of Victoria Land they approached, on 27 January 1841, "some land which had been in sight since the preceding noon, and which we called the "High Island"; it proved to be a mountain twelve thousand four hundred feet in elevation above the level of the sea emitting flame and smoke in great profusion". Ross named this volcano Mt Erebus after his own vessel and the inactive volcano to the eastward, some 3345 metres (10 900 feet) high, was named Mt Terror after his other ship (Plates 6.1, 6.2).

The assistant surgeon, Joseph Dalton Hooker, then just 21 years old and destined to be one of the greatest of nineteenth century botanists, wrote: "This was a sight so surpassing everything that can be imagined and so heightened by the consciousness that we have penetrated, under the guidance of our commander, into regions far beyond what was deemed practicable, that it really caused a feeling of awe to steal over us . . .".

The blacksmith, Cornelius Sullivan, brought his imagination to bear: "On the 28th we discovered Mount Erebus this splendid Burning Mountain was truly an imposing sight".

The easternmost point of what was later to be called Ross Island was named Cape Crozier by Ross after the Commander of the *Terror*, Francis Crozier, a friend and fellow officer of 20 years standing; and the

Plate 6.1. *An artist's impression of Mt Erebus and Mt Terror drawn during Ross's expedition, 1841. Photo: Alexander Turnbull Library, Wellington.*

Plate 6.2. *Mt Erebus reflected in the waters of McMurdo Sound; a view from the WSW. The large snow-free area near sea level on the left is the Cape Royds–Cape Barne area, and the smaller snow-free area on the right is Cape Evans; the Barne Glacier flows between them. Abbott Peak is visible on the skyline on the left slope, Hooper's Shoulder just below the steep ascent to the cone in the centre, and Williams Cliff below the cone to the right.*

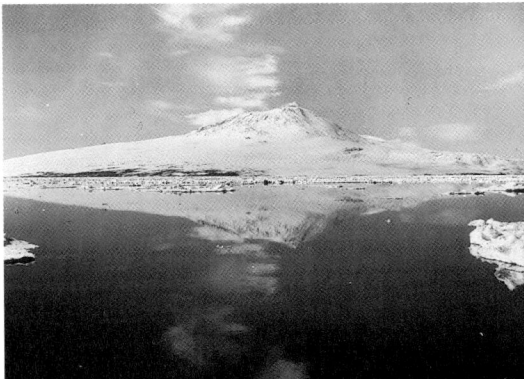

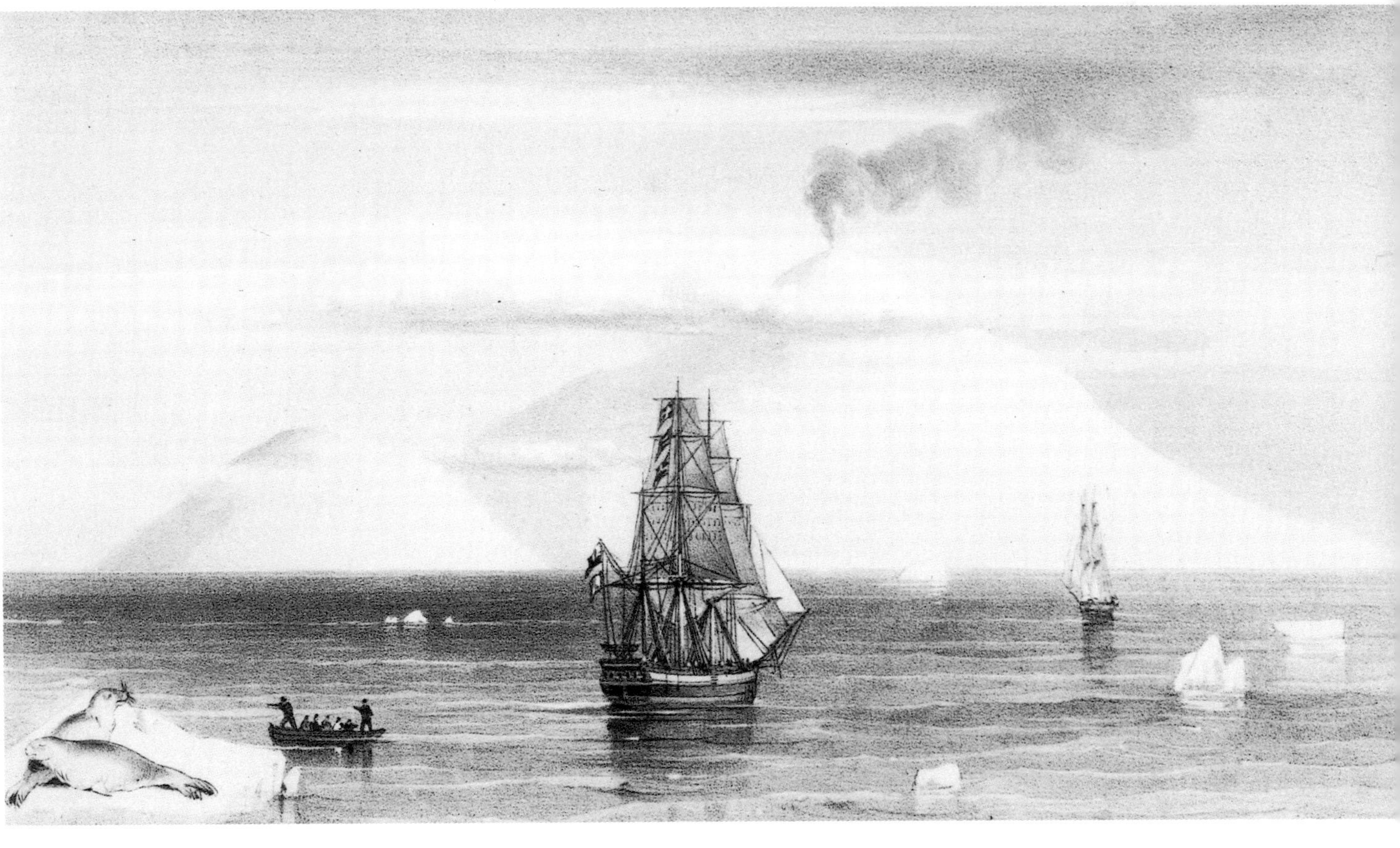

western promontory at the foot of Mt Erebus he named Cape Bird after the senior lieutenant of the *Erebus* "whose worth was so well known to me and who . . . had ever shown so much prudence during the arduous voyages to the arctic regions, in which we sailed as messmates".

Ross Island's volcanic features

Ross Island has a 120° pattern of radial symmetry centred on Erebus (Figure 6.1). The eastern arm is Mt Terror, a simple conical volcano of 3230 metres elevation and nearly equalling Erebus in volume (1670 cubic kilometres). The NNW arm, Mt Bird, is a much smaller, flat-topped dome of 1800 metres elevation. The SSW arm is the narrow Hut Point Peninsula of only 200–300 metres elevation, except for Castle Rock of 413 metres. Mts Erebus, Terror, and Bird are three large volcanic cones which developed around the sea floor and eventually coalesced to form Ross Island. These cones are part of a series of "McMurdo Volcanics" which stretch over 2500 kilometres along the edge of the Transantarctic Mountains from Mt Weaver in the south to the Balleny Islands in the north.

Mt Bird appears to have been the first cone on Ross Island to develop, followed by Mt Terror and finally Mt Erebus, the youngest and still active cone. A line of small craters extends from the flanks of Mt Erebus to form Hut Point Peninsula. Castle Rock is one of the oldest of these small vents and is composed of volcanic breccia that formed from eruptions below the ice when the Ross Ice Shelf was thicker and higher than at present. The Peninsula ends at Cape Armitage with the extinct cone of Observation Hill, which now has at its top the memorial cross erected to Scott's doomed polar party by the survivors of the expedition.

The lava of Mt Erebus is a special type of basalt called Kenyte, found at only a few localities in the world. Anyone visiting Cape Royds or Cape Evans will be able to see the characteristic lozenge-shaped crystals of anorthoclase feldspar that are found in this rock. Those

Figure 6.1. *Ross Island showing the major features. Expedition bases and huts are shown by circles, seismometer stations by solid squares (see Plate 6.3 for details near the summit of Mt Erebus). The site of the DC-10 plane crash in November 1979 is by the seismometer station CRA, south of Lewis Bay. (BOM and TER are other seismometer stations).*

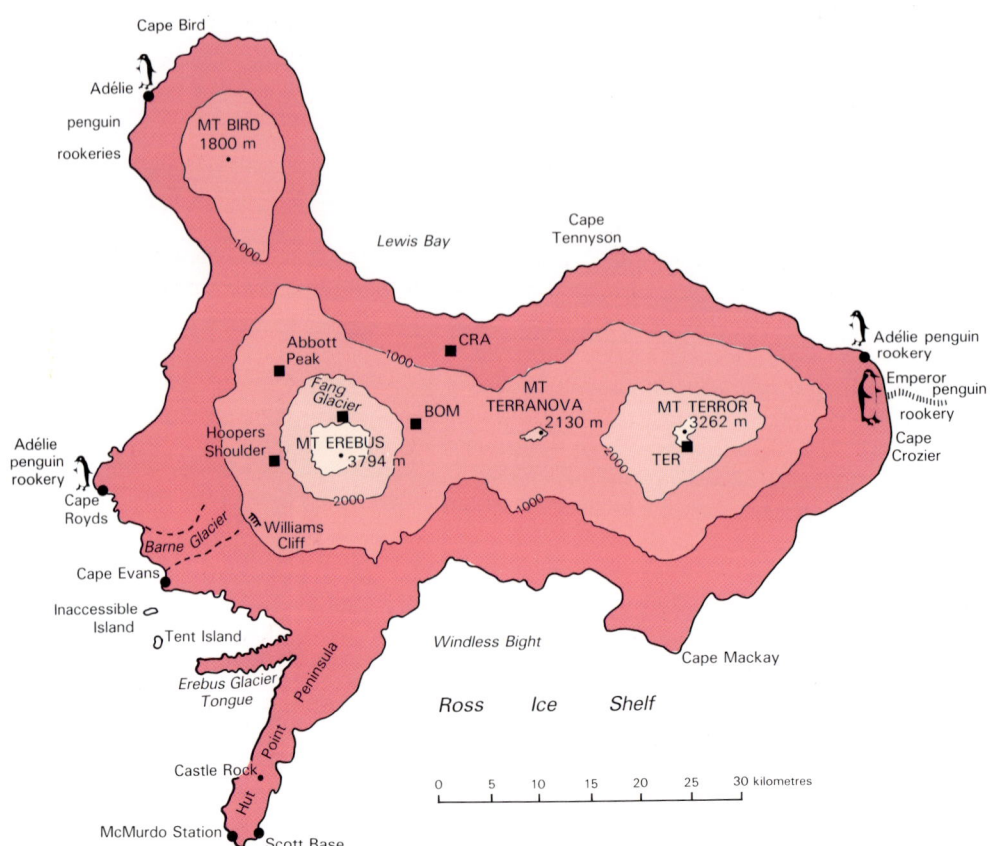

fortunate to visit the summit crater will see perfect crystals of this
mineral littering the crater rim, blown out from the volcanic neck during
explosive eruptions.

Mt Erebus and its ascents

From sea level Mt Erebus slopes gently at 1 in 7 to an elevation
of 2000 metres but then steepens to a grade of 1 in 2.5. The approach
from the NNE side leads to Fang Ridge (3159 metres) which is an earlier
volcanic remnant, breached on the west side before being buried by later
eruptions. The glacier between Fang Ridge and Erebus crater is the
easiest route by which to ascend the uppermost part of the mountain
(Figure 6.1). On the WSW side the steep slope continues to 3200 metres
where it decreases abruptly at a second old crater rim or "somma", 4
kilometres in diameter (Figure 6.1). The somma rises in a spiral pattern
clockwise round the mountain to 3450 metres. It then disappears under
more recent lavas on the ENE side of the summit only to emerge again
as the Truncated Cones on the south-west side. The plateau-like old
crater floor above the somma ring has typical slopes of only 1 in 8,
allowing easy, safe access by motor toboggan. Final access to the summit
cone is again steep before the Main Crater is reached at 3795 metres.

The first ascent of Mt Erebus was made from Shackleton's base at
Cape Royds. A party of six led by Edgeworth David, all of whom were
novice climbers with makeshift equipment, manhauled a 3.6-metre sledge
to 1700 metres. Leaving the sledge, they carried 20-kilogram loads to the
western somma rim and camped overnight. One of the party,
Brocklehurst, was found to have frostbitten feet (and eventually had his
big toes amputated) and stayed in his sleeping bag while the others
crossed the plateau and climbed up to the rim of the active crater on 10
March 1908. They reported that "after a continuous loud hissing sound,
lasting for some minutes, there was a big dull boom and immediately
great globular masses of steam would rush upwards . . . There were at
least three well defined openings at the bottom of the cauldron, and it
was from these that the steam explosions proceeded." Marshall, a
surgeon and cartographer, measured the altitude as 4120 metres using a
hypsometer, a value much greater than Ross's earlier and much more
accurate estimation. The round trip from Cape Royds took 6 days; 5 up
including 1 day tent-bound by a blizzard, and 1 down. Subsequently,
Edgeworth David wrote an account of the ascent of "one of the fairest
and most majestic sights the earth can show" in the expedition's book
"Aurora Australis". Among the many dramatic sights at the top were the
"extraordinary structures which rose every here and there above the
surface of this snowfield" (Plate 6.3).

The second ascent, by a similar route, was made by a party from
Scott's *Terra Nova* expedition led by Raymond Priestley. They were
much better equipped but, due to bad weather, took 9 days for the
round trip. At the rim, which they reached on 12 December 1912, they
heard a loud explosion and, among the smoke, saw large blocks of
pumice being hurled aloft. After starting down, one of the party, Gran,
went back to retrieve a can of exposed film left by mistake. He wrote
"the volcano suddenly began to make a noise—the stones under me
began to move—everything became dark . . . ".

Because of the 40-year gap in exploration of the McMurdo Sound
area, Mt Erebus was not climbed again until the 1958-59 summer when
the summit was reached by two parties. A New Zealand team of three
crossed Windless Bight by "Sno-cat" on 30 December 1958 and
manhauled a sledge to 1000 metres. Climbing on skis and then crampons
they reached the summit on 4 January 1959 and sketched and
photographed the summit crater, recognising the Main, Inner, and Side

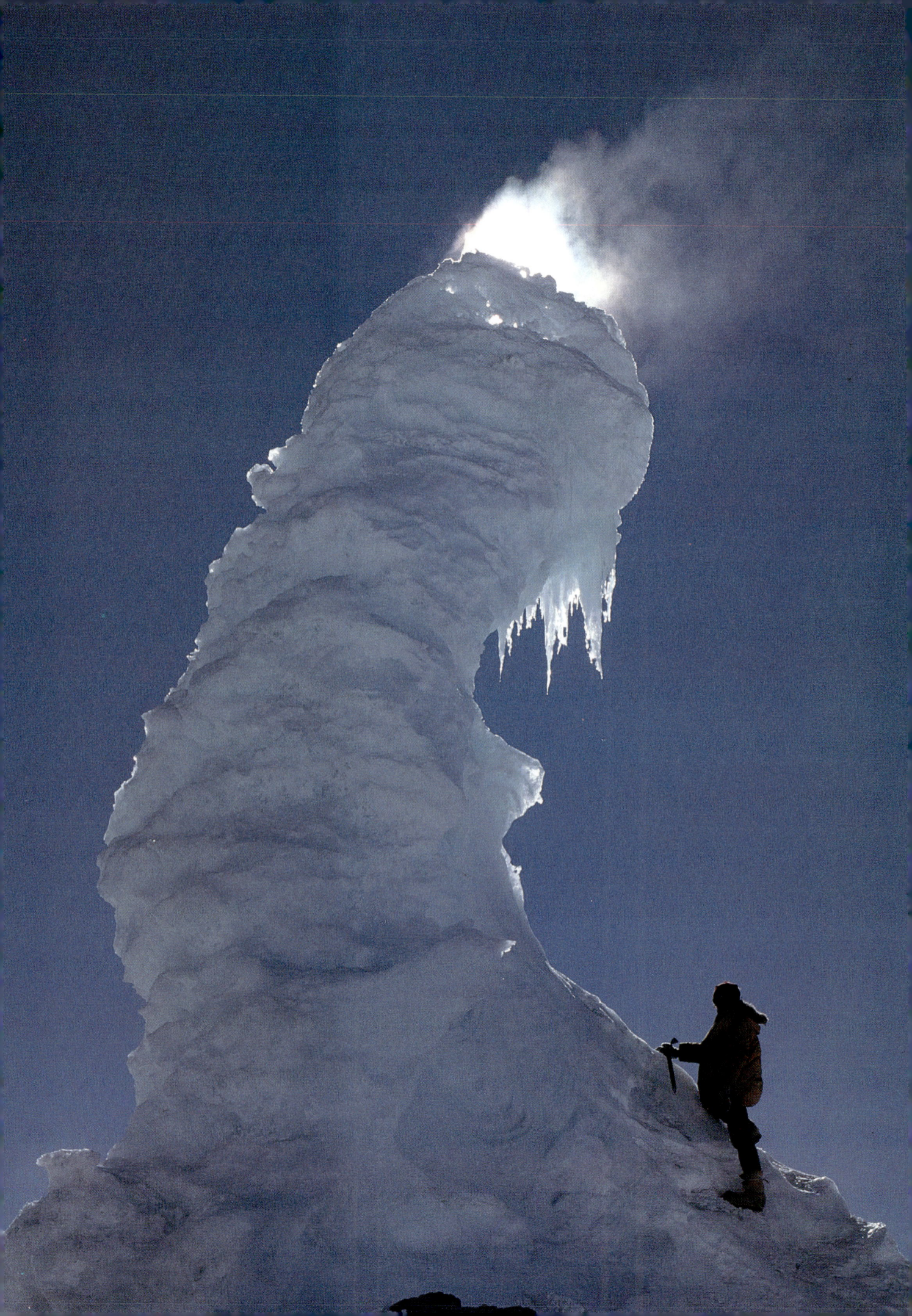

Craters for the first time (Plate 6.4). They skied most of the way down, returning along the top of the Hut Point Peninsula and finally tobogganned down to Scott Base on 6 January. The round trip took 8 days but the distance was three times that of the previous journeys.

As the New Zealand party was returning to Scott Base the first American party was reaching the summit. A four man team had set up a base camp 8 kilometres from the summit and two of the party went to the crater in cold windy conditions with poor visibility. Above 3500 metres the only food they carried was in their pockets and the pair spent only 10 minutes on the rim. Back at their base camp, they radioed for a helicopter to evacuate them and so set an important precedent.

Several subsequent surface ascents were made and on the seventh ascent in 1966 motor toboggans were used for the first time, reducing the round trip to 2 days. Since 1972, helicopters have become the principal mode of transport for those working on the mountain, although in 1985 a party of three from the "In the Footsteps of Scott" Expedition traversed Ross Island via Mts Erebus, Terra Nova, and Terror to Cape Crozier, returning by the Ross Ice Shelf.

Exploring the crater

For almost 70 years the purpose of the ascents had been to reach the crater. The possibility of descending into the crater had been discussed by a French volcanologist, Haroun Tazieff, and Sir Edmund Hillary as early as 1959. But it was not until late 1972 that the preliminary to such a descent occurred, when two New Zealanders were lowered 60 metres by winch from the rim to the floor of the Main Crater. The Inner Crater was obscured by fumes at the time but some days later incandescence was seen in four areas, the largest two being pools of molten lava 20–30 metres across.

The sighting of the lava rekindled enthusiasm to descend into the Inner Crater and an international expedition of experienced United

Plate 6.3. An "ice-fumarole" on Mt Erebus, caused by the freezing of the water vapour in the volcanic gases escaping from the hole. Photo: H. Keys.

Plate 6.4. Vertical aerial photograph of summit area of Mt Erebus showing main natural features and equipment installations. The solid squares are seismometer installations. Photo: U.S. National Science Foundation.

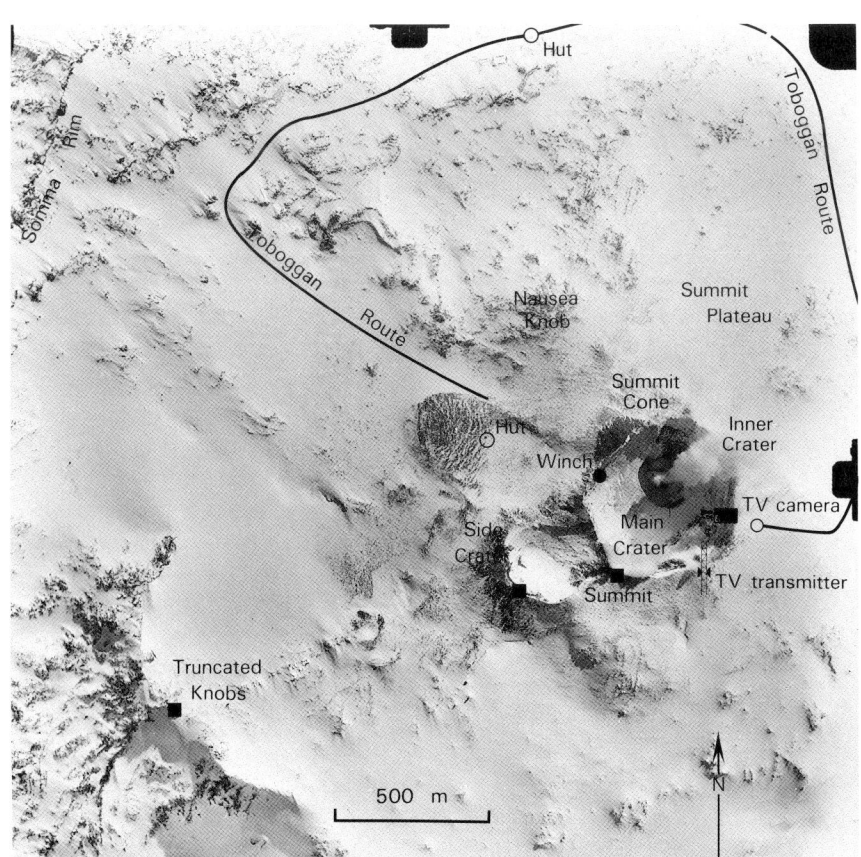

States, New Zealand, and French volcanologists was gathered for
the 1974–75 season. Climbing ropes, ladders descending from the winch
site to the Main Crater floor, and a "flying fox" were installed to
transport personnel and equipment into the crater. Then the troubles
associated with a dangerous volcano at high altitudes in cold windy
conditions began. Altitude sickness affected some of the scientists,
blizzards pinned them in their tents at times, an eruption broke the
flying fox cable 2 days after it was installed and, throughout, the intense
volcanic activity prevented descent into the crater.

Expeditions in subsequent years suffered from similar troubles but
in spite of the hazards the United States provided a hut at the Summit
Camp during the 1978–79 season. Two New Zealanders were lowered 65
metres down the overhanging walls to the Inner Crater using a "Z-
pulley" system (Plate 6.5). The geochemist, Werner Giggenbach,
intended to take gas-sampling equipment to the lava lake and needed to
climb down a steep slope at the end of the hauling rope. As he prepared
to climb down, a large explosion occurred in the active vent. A warning
shouted over the radio from an observer above enabled Giggenbach to
swing his ropes clear of the hurtling bombs, though one hit him on the
leg, singeing his trousers. Lava bombs also landed close to the ropes laid
out on the Main Crater floor and an ash cloud obscured the scene. The
watchers were relieved to see Giggenbach being hauled out to safety
(though without the much sought after gas samples) after the cloud
dispersed.

Although in almost a decade of effort at least 70 descents had been
made into the Main Crater, only two partial descents into the Inner
Crater had been made. In 1983 the descent phase was decreed to have
ended, at least temporarily, and attention turned to "remote sensing" of
the volcano's activity.

Understanding Mt Erebus

Most volcanoes occur in association with large earthquakes at the
junctions of the tectonic plates which make up the earth's outer layer
and which are continually in motion relative to each other. The
volcanoes of the Pacific Ocean perimeter are generally of this type. Mt
Erebus is one of the few volcanoes, however, which occur within a plate
and not at its boundary. Its position, within a polar continent devoid of
large earthquakes, its lava lake and its high altitude all combine to make
Mt Erebus well worth studying.

The simplest way of studying a volcano, and the one almost
exclusively used until recent years, is simply to observe it. Visual
observations of Mt Erebus's activity date from 1841 and were relatively
frequent between 1900 and 1915. Shackleton wrote in 1909 that "it
became quite an ordinary thing to hear reports from men who had been
outside in the winter that there was a strong glow on Erebus . . . and at
other times we have seen great bursts of flame crowning the crater".
Since 1956, when continuous occupation of the bases in McMurdo
Sound began, observations of incandescence have been infrequent and it
may be that the earlier periods were more active—or the observers more
imaginative.

Intermittent visual observations, hampered also as they are by
variable conditions of light and cloud, are inadequate to give a
continuous record of the fluctuating activity of a volcano, much less to
enable this activity to be understood. Marvellous though the eye is, for
continuous surveillance the assistance of the constant presence of
"untiring" instruments is needed.

The first requirement of volcanic study is still to know when
eruptions are taking place. The "infrasonic" recorder has been found to

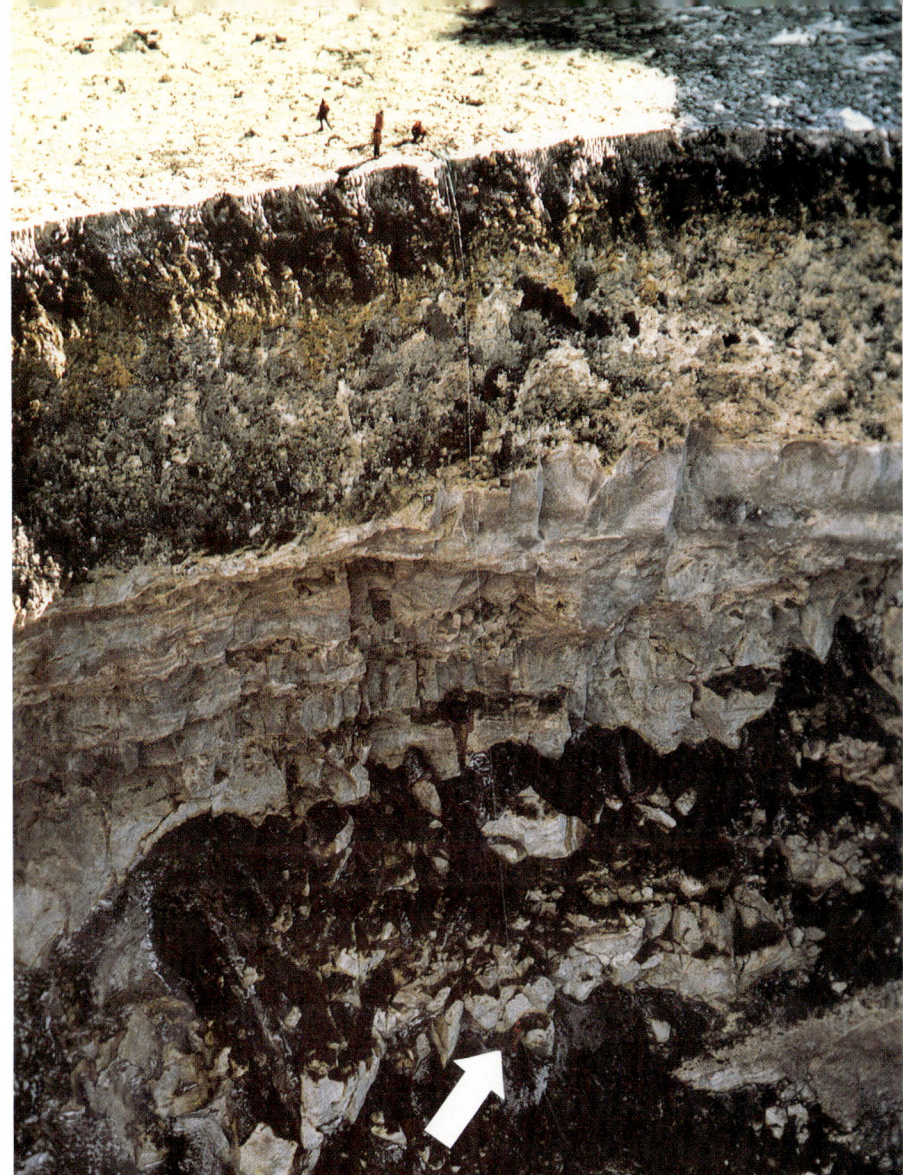

Plate 6.5. *The side of the Inner Crater showing a scientist (arrowed) being lowered on a rope to take gas samples. Photo: R. Dibble.*

Figure 6.2. *Seismograms of an earthquake originating near the summit of Mt Erebus. The Loop recording is due to magnetic induction in a wire encircling the summit. The Infrasonic recording is from a microphone at the summit. Note that in this instance the earthquake is too small to be recorded at Mt Terror and Scott Base.*

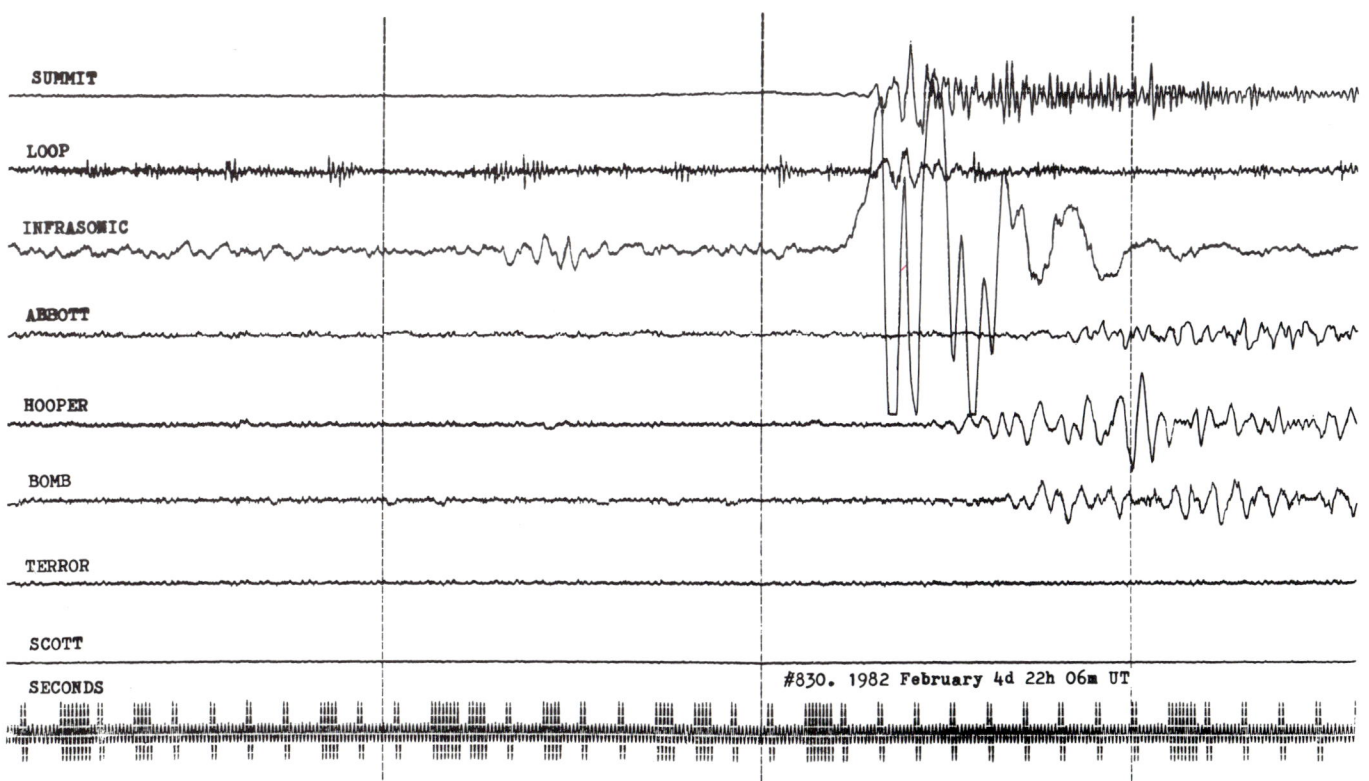

SUMMIT

LOOP

INFRASONIC

ABBOTT

HOOPER

BOMB

TERROR

SCOTT

SECONDS

#830. 1982 February 4d 22h 06m UT

be a useful, though not complete, substitute for the eye in this respect. Sound is produced by motion of one kind or another, and the sound of a violent eruption of Mt Erebus has been heard at Scott Base and McMurdo Station during the most intensive eruptions of modern times in late 1984. Sound below the threshold of human frequency, however, travels much more efficiently than the higher frequencies and recording "infrasound" enables much smaller eruptions to be detected. Most recently, in 1986, continuous "visual" observations of the crater were made using a television camera mounted at the crater rim from which continuous pictures were transmitted to a videotape recorder at Scott Base. The details of the accurately timed eruptions can thus be correlated with observations on other instruments, such as seismographs.

Earthquakes on Mt Erebus

The study of earthquakes and their characteristics are a popular and effective method of investigating the interior "plumbing" of volcanoes in other parts of the world. No one has yet convincingly reported feeling an earthquake on the Antarctic Continent. Apart from the deep ocean basins, Antarctica is seismically the quietest place on earth. The only earthquakes which can be definitely located within the Ross Sea region are associated with the movement of ice or with volcanic activity on Mt Erebus.

Small earthquakes suspected to originate in the Mt Erebus volcano were first recorded at Scott Base in 1957, but it is impossible to locate an earthquake from readings at just one point. With the advent of intensified investigation of Mt Erebus in 1973, seismographs on the volcano were used to confirm that shocks did originate from Mt Erebus and to locate them more precisely. In 1980, the United States, Japan, and New Zealand inaugurated a co-operative programme of year-round seismic studies. By 1985 the network on the mountain had grown to 10 stations, with the results being telemetered to recorders at Scott Base by means of transmitters powered by batteries and solar panels (Figure 6.2). These panels, of course, fail to provide power in the winter darkness, but are reactivated on the return of the sun. The top of Mt Erebus is just about the most hostile place on earth for the operation of remote automated equipment, but considerable success has been achieved in using seismographs to locate earthquakes on the mountain and delineate the "plumbing system" of the volcano. The accurately timed TV-surveillance pictures of eruptions, begun in 1986, have added a new dimension to the seismic observations for they allow the relationship of the earthquakes and the eruptions to be studied.

Other science programmes aimed at understanding Mt Erebus have been the measurement of deformation of the crater as it responds to the forces of the magma moving upwards in the volcano; the use of a wire loop around the crater to record, by magnetic induction, the movement of material within the loop; the measurement of the flux of sulphur dioxide gas in the plume by correlation spectroscopy; and the measurement of temperature by infrared thermometry.

Tragedy on Mt Erebus

Public awareness of Mt Erebus became worldwide when, on 28 November 1979, an Air New Zealand DC-10 airliner on a sightseeing flight to the Ross Sea region crashed into the slopes of the mountain at a height of about 500 metres, south of Lewis Bay (see Figure 6.1). All 20 crew and 237 passengers were killed in the crash which is New Zealand's greatest disaster, the death toll being even higher than that of the Hawke's Bay earthquake of 1931. Thus, in one blow, several times as many people lost their lives than had perished in the 80 years of arduous exploration and scientific study of the continent (Plate 6.6).

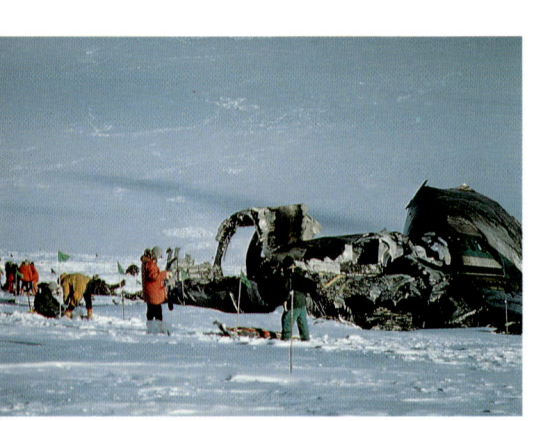

Plate 6.6. *Aftermath of the crash of the Air New Zealand DC-10 airliner on the lower slopes of Mt Erebus, 28 November 1979. Photo: R. Thomson, Antarctic Division, DSIR.*

The shape of the Antarctic Continent comprises the broadly circular, little-indented coastline of East Antarctica, the two large, ice-shelf covered identations of the Weddell Sea and the Ross Sea, and the long projection of the Antarctic Peninsula. East Antarctica occupies three-quarters of the continent's area and is a vast, domed ice-sheet rising to over 4000 metres, with only small ice-free areas at the coast. A major exception to this is the coast of Victoria Land where the Transantarctic Mountains confine the ice to valley glaciers and where smaller ice fields and shelves occur on the gentler slopes bordering the mountains. McMurdo Sound, shaped by the constriction of the northward-flowing Ross Ice Shelf against Ross Island and in the strait between the island and the Victoria Land coast, forms the southern boundary of the shore. Thus, the extensive ice-free shores of Victoria Land contrast with those of most of East Antarctica. Similarly, there are contrasts between West and East Antarctica. West Antarctica has a much smaller area and can be thought of as a large archipelago submerged in ice. Much of its ice-free coastal area is mountainous and occurs in the Antarctic Peninsula, so that its shores have intricate outlines and there is much variation in exposure to wave action.

The continent has a coastline of 29 000 kilometres. About 4000 kilometres, or 14 percent, of this occurs in the Ross Sea where by far the largest feature is the cliffed edge of the Ross Ice-Shelf. There are wide ranges of forms and origins of coastal landforms, affected by such factors as the extent, thickness, and history of ice cover on both land and sea, the duration of open water in summer, and the rate and type of rock weathering and erosion.

The coastal landforms are important for three reasons. Firstly, they are formed by distinctive cold climate processes. Secondly, study of the coast, especially of ancient beaches (many of which are now raised high above sea level), helps us unravel the history of Antarctica and its changing relationships with world climates and sea levels. This study of the past has important applications in the debates about global warming to be expected from the increasing content of atmospheric carbon dioxide and the "greenhouse effect". The work helps us predict future higher sea levels, possible inundation of populous coastal lowlands around the world, and greater severity and extent of coastal erosion. The most significant effects of any warming will occur at high latitudes, and this could lead to appreciable melting of ice. Thirdly, we need to clearly understand the environmental functioning of Antarctic coasts and inshore waters, not least their sensitivities to such activities as hydrocarbon exploration and exploitation already carried out on very shallow shelves in the Arctic with prospects for similar activity on deeper shelves in Antarctica. The Antarctic continental shelf covers about 4 million square kilometres, some 60 percent (2.4 million square kilometres) of which is ice-free. The shelf has a mean depth of 350 metres (almost twice that of other shelves), a mean width of 128 kilometres and a minimum width of 48 kilometres. Maximum widths occur in the Weddell Sea (611 kilometres) and the Ross Sea (1127 kilometres). These characteristics together with the problems posed by stormy seas, drifting pack ice, and icebergs impose important constraints on the intensifying search for exploitable Antarctic resources. But the challenges, at least to exploration, will not necessarily be a deterrent or be insuperable.

The Victoria Land coast

The Victoria Land coast is ice-free to a much greater extent than most of the coast of East Antarctica, and thus shows a much wider range of coastal landforms. Nevertheless, it is still dominated by the influence of ice in all its various forms, both past and present. Even in the ice-free coastlines, seasonal development of a persistent ice foot ensures that the influence of ice in shaping the shore remains strong. During winter the continent is encircled by a belt of sea ice over 1500 kilometres across, so that wave action does not then affect the shore. Although this pack breaks up in summer, large areas of landfast sea ice can persist in some coastal embayments for several seasons. Even where the ice breaks out, large fields of pack ice moving with winds and currents damp the energies of oceanic storm and swell waves penetrating to the coast.

There are five broad categories of coastal landforms. In order of areal significance they are: high ice cliffs; high rocky cliffs; low rocky shores (partially ice and snow covered); beaches; and landfast sea ice (but its significance varies both from place to place and from year to year).

The distribution of each coastal type has been mapped along most of the Victoria Land coast and a typical section (Cape Hallett) is illustrated in Figure 7.1. About 28 percent of the Victoria Land coast is free of glacial ice, a much higher proportion than the continent as a whole (5 percent). Beaches occupy less than 5 percent and low rocky shores about 6 percent of the Victoria Land coast. These two shoreline types are the principal wildlife habitats (especially for penguin colonies) and are the preferred sites for human activities because of ready access and the relatively flat land. They are the region's "prime real estate" and the conflict between wildlife and human requirements is already one of the more severe conservation problems in Antarctica, illustrated by the controversy over the proposed construction of an airstrip which would displace a penguin colony in Terre Adélie.

There has already been some human interference with many beaches (e.g., Cape Adare, Seabee Hook, Marble Point, Cape Bird, Cape Evans). Others have been designated Specially Protected Areas or Sites of Special Scientific Interest to which access is controlled or restricted (e.g.,

Figure 7.1. *A typical section of Victoria Land coast, at Cape Hallett, showing mapping of coastal types.*

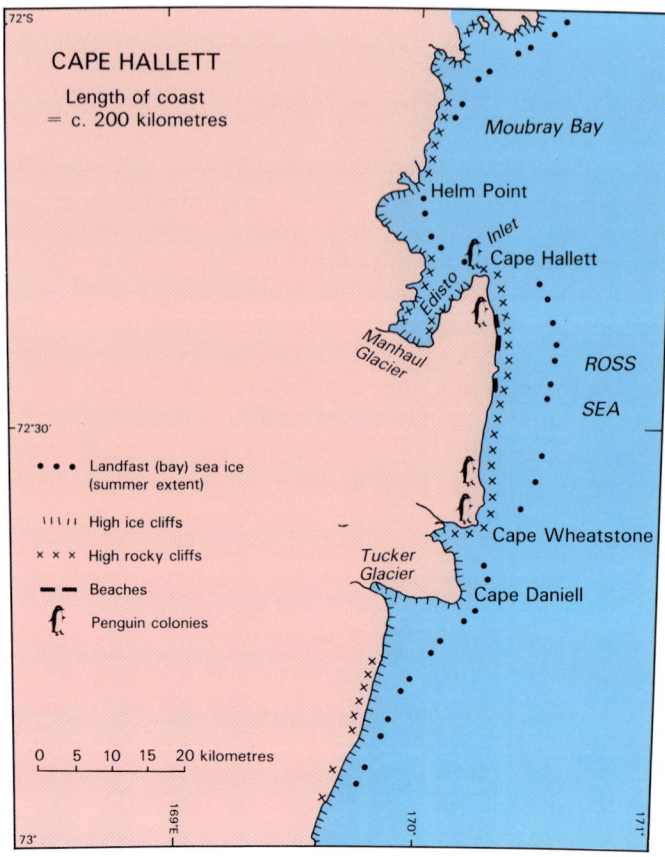

Cape Crozier, Cape Royds, Beaufort Island, Cape Hallett). A few have historic significance (e.g., Cape Adare, Inexpressible Island, Cape Evans, and Cape Royds).

The shore types are also closely related to the geological structure of the coast. High cliffs are consistently developed in Cenozoic volcanic complexes, and also where open water persists for some time. Such complexes are more common in the northern part of the coast and form most of the offshore islands, including Ross Island. There would be a larger proportion of coast in the high cliff category if the shores of Coulman Island and Cape Daniel were not more or less continuously protected by ice piedmonts.

In contrast, low rocky shores are typical of hardened Paleozoic basement rocks that predominate in the south and where sea ice cover is generally more persistent.

Beaches develop where there are adequate supplies of detritus and coastal processes which lead to its accumulation, principally in coastal embayments and in the lee of peninsulas and islands. Consequently, the most extensive beaches of the Victoria Land coast are associated with high cliffs and volcanic terrains.

High ice cliffs (glacial shores)

The most common of Victoria Land shores are high ice cliffs, 50 metres or more in height, formed at the seaward margin of ice shelves, ice piedmonts, and glaciers (Plates 7.1, 7.2). Most of this ice front is floating (e.g., Ross, McMurdo, Nansen, and Lady Newnes Ice Shelves; and also Erebus, Nordenskjold, Drygalski, Borchgrevink, and Mariner Glacier Tongues) and is maintained through calving of tabular icebergs and other wastage. This is generally a dynamic landform of constantly changing promontories and inlets. The intricately sculptured ice cliffs are generally notched at the base, like icebergs, and are often protected by a seaward submerged ramp. In McMurdo Sound, some glaciers and the ice

Plate 7.1. *High ice cliffs and intricately sculptured coastline of the Ross Ice Shelf at Cape Crozier. Photo: M. Gregory.*

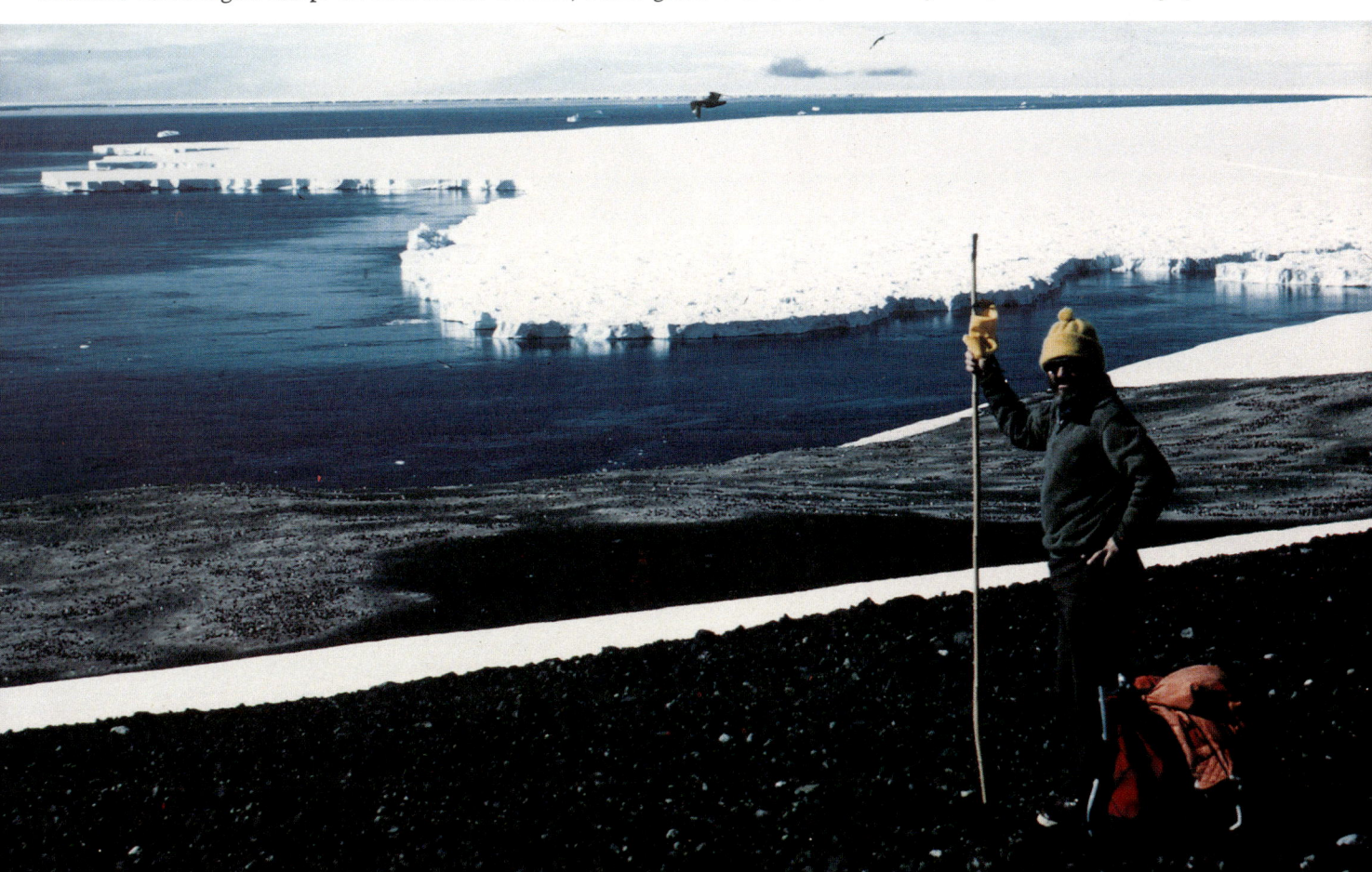

shelf are locally and heavily mantled with glacial debris and submerged ice cliffs have been observed beneath these deposits. However, this material remains only briefly in the coastal zone before calving and melting delivers it directly to deeper water.

Ice cliffs also form where the grounded ice piedmonts and glaciers spill directly over bedrock and into the sea (e.g., Wilson Piedmont Glacier).

Landfast (bay) sea ice

By mid summer, pack ice has broken and dispersed and the Ross Sea is largely open water. Landfast ice may persist in bays and inlets over some summers (see Plate 7.2) but the North Victoria Land coast is largely ice free by late January. Extensive areas of thick multi-year ice, as typically develop in the Arctic and have been reported from the Amundsen, Bellingshausen, and Weddell Seas, are unusual. Except where rafting of flows has occurred or pressure ridges have formed, the thickness of this sea ice cover seldom exceeds 2–2.5 metres. Sea ice begins to form in the Ross Sea by late March or early April, although significant areas of nearshore open water (polynyas) persisting throughout winter have been reported from Terra Nova Bay and elsewhere along the North Victoria Land coast and off the Ross Ice Shelf (see Plate 9.6).

Low rocky shores (partially ice and snow covered)

These are typically low (under 30 metres), muted, hard rock shores, still carrying a glacial imprint and have been little modified by modern marine processes (Plate 7.3). In many situations, cliffs are more actively attacked by frost fissuring as seasonal snow and ice accumulations enlarge in cracks and joints (Plate 7.4). Wave-washed, bedrock surfaces extending from contemporary sea level to the marine limit are conspicuous in McMurdo Sound between Granite and New Harbours. Elsewhere, low rocky shores may be made up of coarse morainal talus or other rubble.

Plate 7.2. *High rocky cliffs protected by landfast sea ice at the southern end of Beaufort Island. Note the ramped pressure ridges in sea ice and talus cone built out at the cliff foot. Photographed in January 1982. Photo: M. Gregory.*

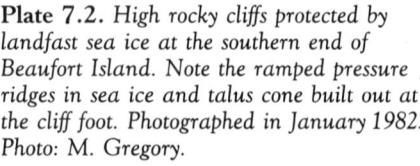

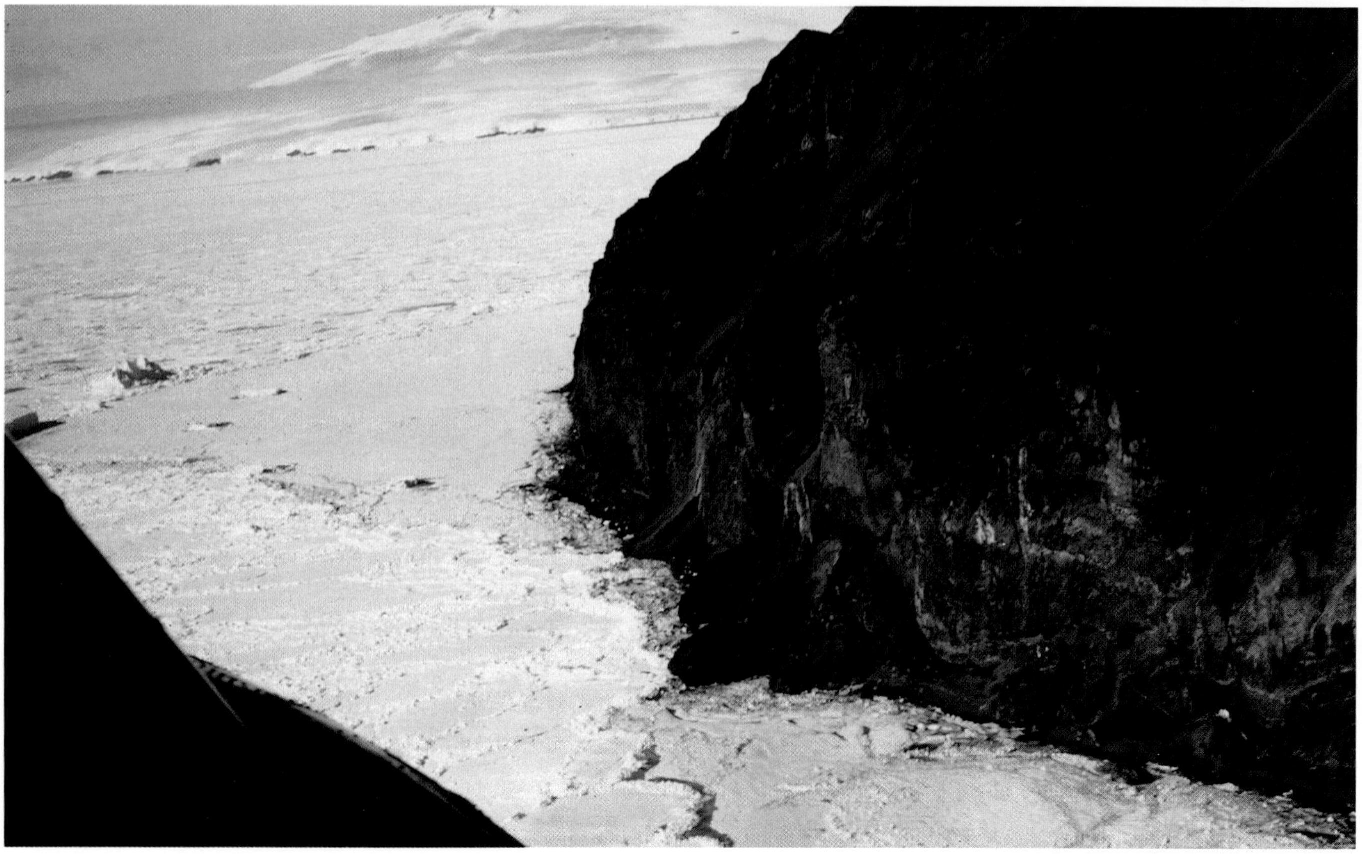

Snowdrift-ice slabs, often containing accompanying talus and beach deposits, are common on many shores of this type.

High rocky cliffs

Along the North Victoria Land coast often spectacular cliffs develop, with precipitous drops into deep water (Plate 7.5). In the volcanic breccia and lava complexes of the Adare and Hallett Peninsulas, cliffs reach heights of 2000 metres. Here, as in similar rock types elsewhere around the Ross Dependency, stream and slope processes are seasonally important, and narrow ice-cored beaches fed by scree and talus are temporarily present at the foot of the cliffs (Plate 7.5).

The Downshire Cliffs of Adare Peninsula and the Cotter Cliffs of Hallett Peninsula (Figure 7.1), as well as the similarly high east-facing cliffs of Franklin and Beaufort Islands, are more or less continuously fronted by a shallow subtidal platform up to 200 metres across. From the Possession Islands there is evidence that this platform extends to depths exceeding 10 metres, and that it is locally veneered by rounded boulders up to one metre or so across, which in places may be tightly packed. Marine boulder pavements near Gneiss Point, McMurdo Sound, may be elevated equivalents of these subtidal or intertidal platforms.

Beaches

Beaches of high latitudes differ from those of tropical and temperate latitudes in many ways. Perhaps most obvious are the seasonal effects of ice on land and sea in the polar regions. Other important differences relate to changing climate and to the role of stream action.

The processes that create coastal landforms in the Ross Sea today are little different (apart from the extent of ice cover and the relative levels of the land and sea) to those that occurred throughout the waxing and waning of the Quaternary Ice Ages. In temperate latitudes, coasts have undergone repeated, wider ranging fluctuations of environmental conditions. Particularly in semi-arid areas, river and stream action is the principal supplier of beach sediments. In contrast, on the Victoria Land coast, streams occur only sporadically for short periods in summer and are much less important as suppliers of coastal sediment. However, the few streams that do occur for short periods can be locally very effective agents of erosion and sediment transport because there is no vegetation to bind what passes for soil, no water loss by transpiration of plants, little evaporation, and no loss of water to ground water tables owing to the presence of permafrost. In winter the ground is frozen and in summer a shallow thaw layer (up to 30 centimetres thick) develops over the permafrost, particularly in dark sediments that absorb the sun's heat.

Polar beaches are characterised by ice-push ridges that develop from the "bulldozing" action of pack ice driven ashore by winds and currents, by ice melt features, and a range of ice-contact and soil creep features.

Two kinds of polar beach have been distinguished on the Victoria Land coast: strandlines, and cuspate forelands and spits. Strandlines are accumulations of beach sediments formed by marine reworking of in situ glacial debris, slope deposits, or bedrock material. There is little external input of sediment to these beaches while they are forming. This type of beach is very common in the southern Victoria Land coast, especially on the western shores of McMurdo Sound around Marble Point. The strandline beach type is, therefore, also associated with low rocky shores composed of hardened Paleozoic basement rocks and they commonly show evidence of much relative change in sea level.

Cuspate forelands and spits are the main and most spectacular coastal depositional landforms along the Victoria Land coast (Plate 7.6). Named after the geometrical form (cusp) where two branches of curves meet, they consist of suites of elevated beach ridges composed mostly of material derived from late Cenozoic volcanic terrains. They are usually formed at the downdrift ends of coastal-sediment transport paths fed by eroding, high rocky cliffs; or at the leeward ends of islands. Such cliff–beach relationships are best developed at Cape Adare where the sediment comes from the Downshire Cliffs (see Plate 1.2), at Cape Hallett where it is from the Cotter Cliffs, and on Possession (Plate 7.6), Franklin, and Beaufort Islands. Formation of the beach ridges requires a long open-water season, significant wave action, adequate sediment supply, and a shallow nearshore.

The Victoria Land beaches arise from landform development processes interacting from almost extremes of the erosion–sedimentation spectrum. The ice action is completely unselective in eroding, transporting, and depositing sediments of a wide variety of sizes and shapes, whereas the winds (that can transport both sand and pebbles), waves, and currents are very selective.

This combination is unique to very cold climates because wave action is seasonally excluded from the beaches by the formation of sea ice. Ice action and wave action thus follow each other in a seasonal rhythm of derangement and rearrangement of the sediments and the persistent permafrost acts as a "glue" to consolidate and hold the resulting deposits.

Most of the beaches are evidently ice-cored and even in summer the permafrost level effectively limits the depths within beaches to which waves and other processes actively move or otherwise disturb material. These depths vary greatly from less than 1 centimetre (as in a veneer of single sand-grain thickness on an ice platform) to over 1 metre on steep, high-energy beaches such as Ridley Beach, Cape Adare.

It is not surprising that under these conditions the sediments of the beaches are mostly coarse-grained and contain a wide range of sizes. Sediment particles are also commonly angular, though with a tendency

Plate 7.3. *West-facing boulder-strewn, low rocky coast of Foyn Island, with a discontinuous ice foot developed. Photographed in January. Photo: M. Mabin.*

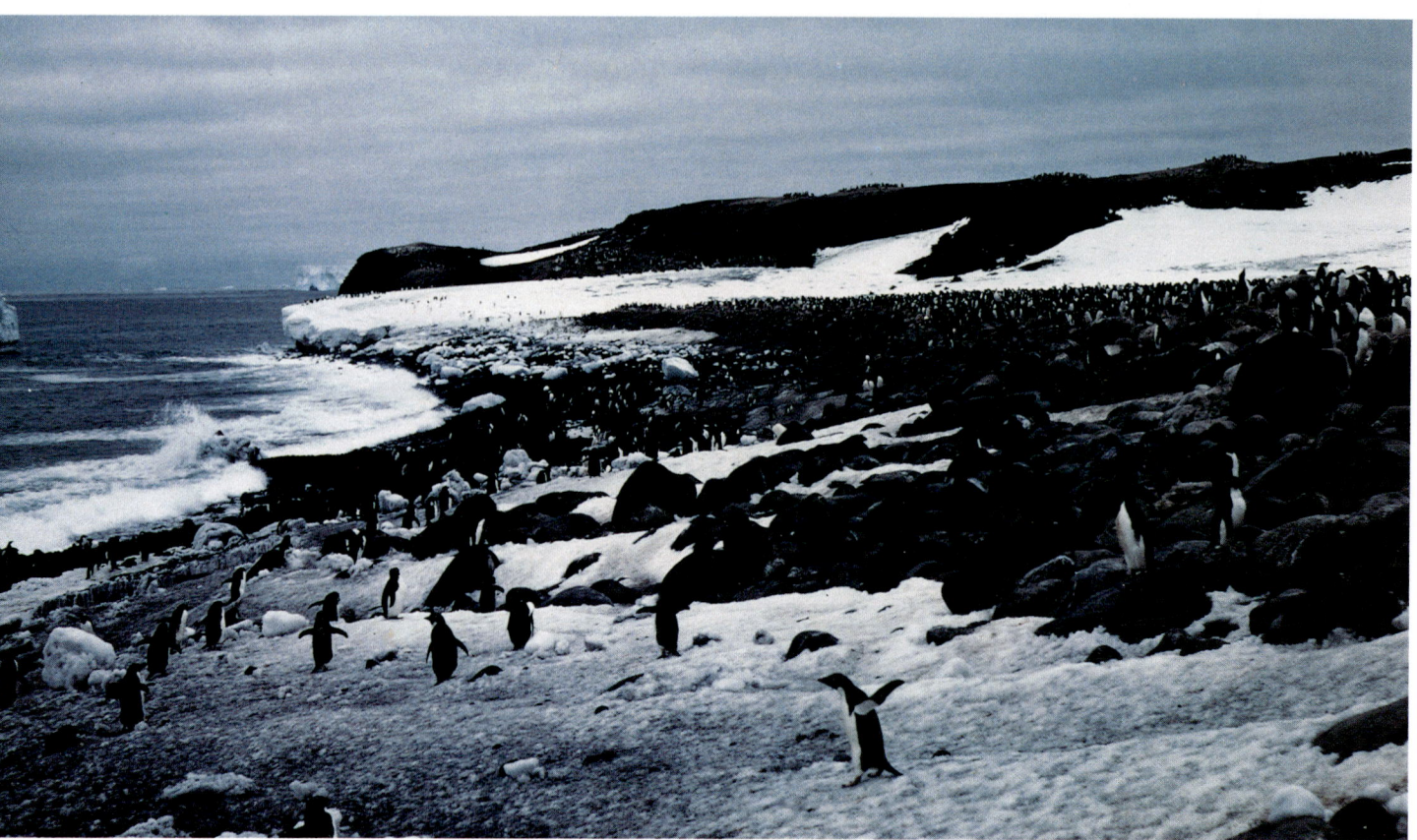

for the larger grains to be better rounded. There are only a few examples of true sandy beaches (e.g., New Harbour and Blacksand Beach, Cape Royds), these being supplied by ephemeral streams or by sand blown into winter snow banks that melt and release sand in summer.

Finally, since ice seals most of these beaches from wave action for all but 2 months of the year, it is remarkable that they show marked changes of form in summer and such a strong imprint of the sea.

A very distinctive feature of many Victoria Land beaches, especially of the forelands and spits, is the influence of penguins on the development of beach ridges. Because Adélie penguins prefer the higher, drier ridge crests as nesting sites and use pebbles to form their nests, the depressions between ridges become almost devoid of pebbles smaller than about 4 centimetres in diameter. The ridge crests are heightened by generations of pebble gathering and are also both added to and partly cemented by large quantities of guano. Such "avian ridges" are rare on the strandline beaches.

The ice foot, of which there are several types, is important in the evolution of high latitude shores (see Plates 7.3, 7.5, 7.6). On Ross Sea beaches, once the ice foot has disappeared a true beach-foreshore profile develops rapidly. Here it is clear the ice foot has played a protective role, reducing wave energy at the shore line even after most of the sea ice has disappeared. However, through incorporation of beach material the ice foot can also be an agent of erosion and transport.

The past: raised beaches

Beach deposits occurring above the limit of present marine action (including that of ice-push) indicate that there have been changes in the relative position of sea level with respect to the land. Perhaps the greatest attention given to polar coastal landforms in both hemispheres has been the study of raised beaches and platforms because of their significance in unravelling Quaternary glacial and interglacial histories

Plate 7.4. *Low rocky cliff at Cape Royds, Ross Island. This cliff is developed in Cenozoic volcanic rocks and is subject to frost riving and wedging rather than attack by marine processes. Photo: M. Gregory.*

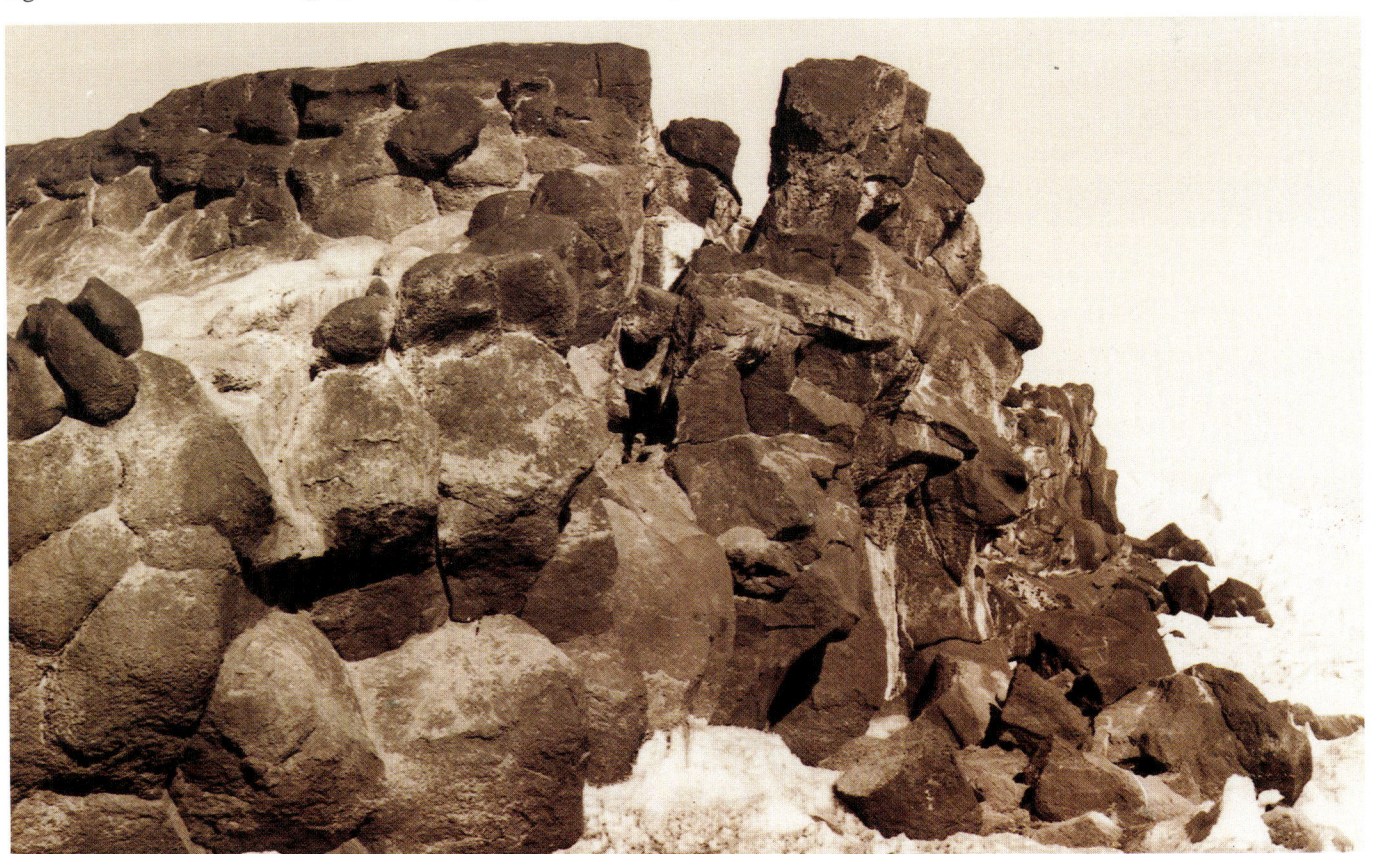

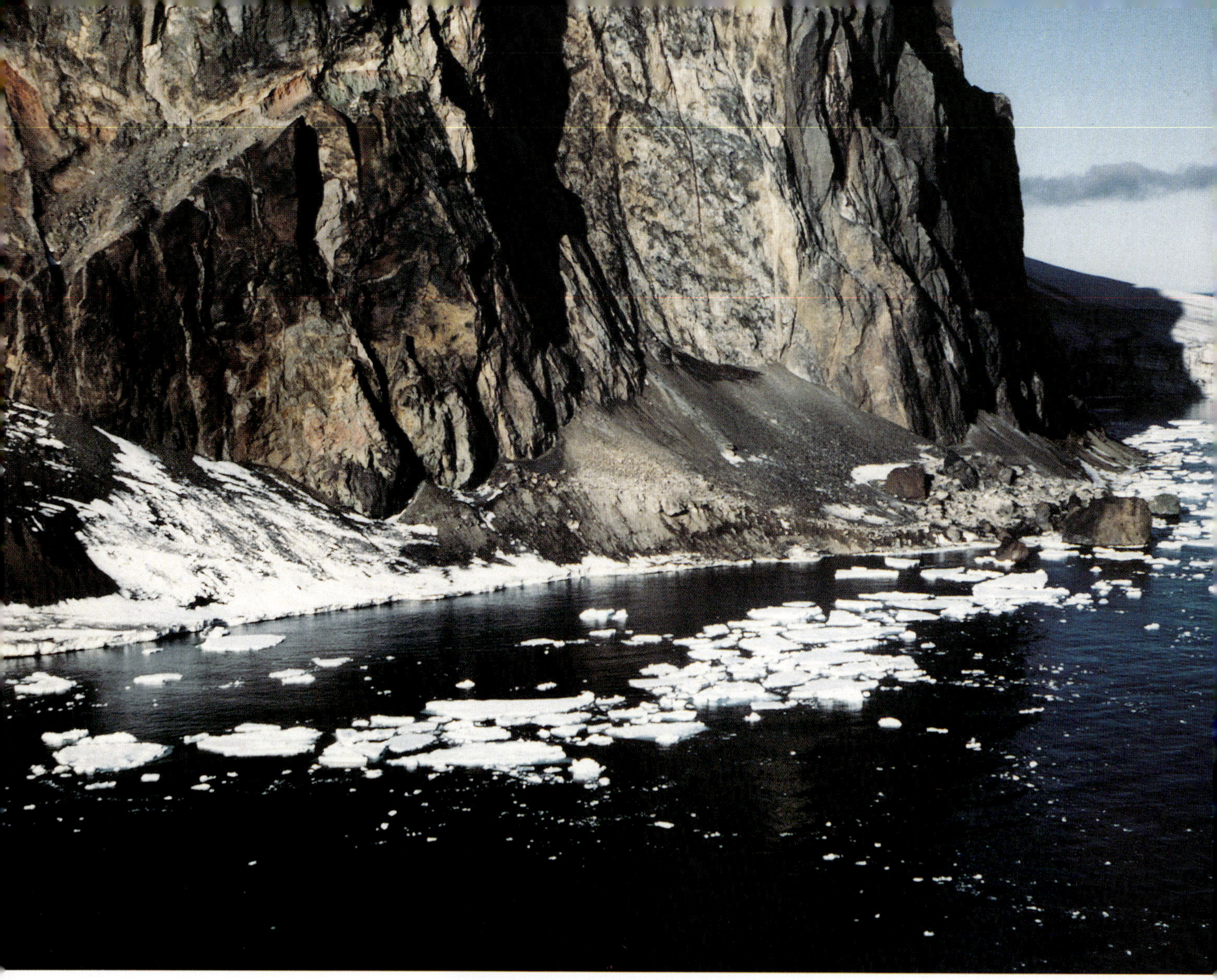

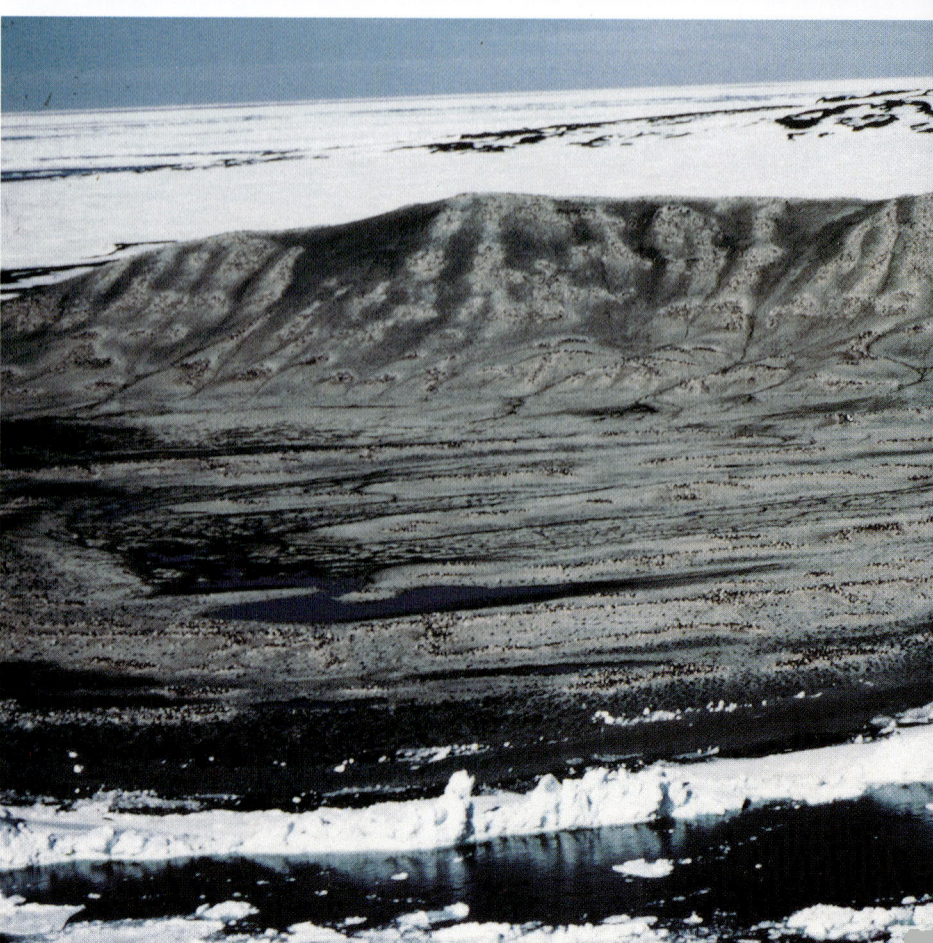

Plate 7.5. *High rocky cliffs on Cenozoic volcanics at Cape Wheatstone. Note the extensive talus cones and protective ice foot. Broken pack ice dampens wave attack on this shore. Photo: M. Mabin.*

and their relationships with changing world climates and sea levels. Raised beaches have been found on all sides of the Antarctic Continent and some are thought to predate the most recent glacial episode, but most are believed to be younger. Commonly, these deposits occur at less than 60 metres above sea level, though higher ones exist.

On the Victoria Land coast, raised beach ridges locally extending to heights of over 30 metres above sea level are well known. The distribution, heights, and patterns of these ridges, especially when determined over wide regions, are an important element of the glacial history of the Ross Sea Region. They can yield much information for studies of both change of land level due to changing ice cover and change in ocean levels due to changing ice–water volumes.

Strandlines (principally along the southern parts of the coast and in hardened Paleozoic rocks and related glacial deposits) commonly show evidence of large changes in sea level. These sites are potentially useful indicators of the greatest elevation of the land since unloading of the glacial ice, because the marine limit is often clearly defined. However, such sites rarely contain material suitable for dating by the carbon 14 method.

Cuspate forelands and spits are commonly sites of extensive penguin rookeries and so they contain bird and other organic materials suitable for carbon dating. However, because of the conditions necessary for this type of beach to develop, the ridges present may not be a complete record of sea levels and other coastal changes since deglaciation. Thus, the oldest (first deposited) ridge in a foreland complex may not represent the marine limit. Clearly, it is essential to discover sites where the marine limit can both be reliably defined and dated.

Along the northern part of the coast, from Wood Bay to Cape Adare, raised beaches indicate relative sea level variations to be generally less than 5 metres. In the central part from Terra Nova Bay southwards to Cape Ross variations exceed 25 metres, with a maximum of 32 metres

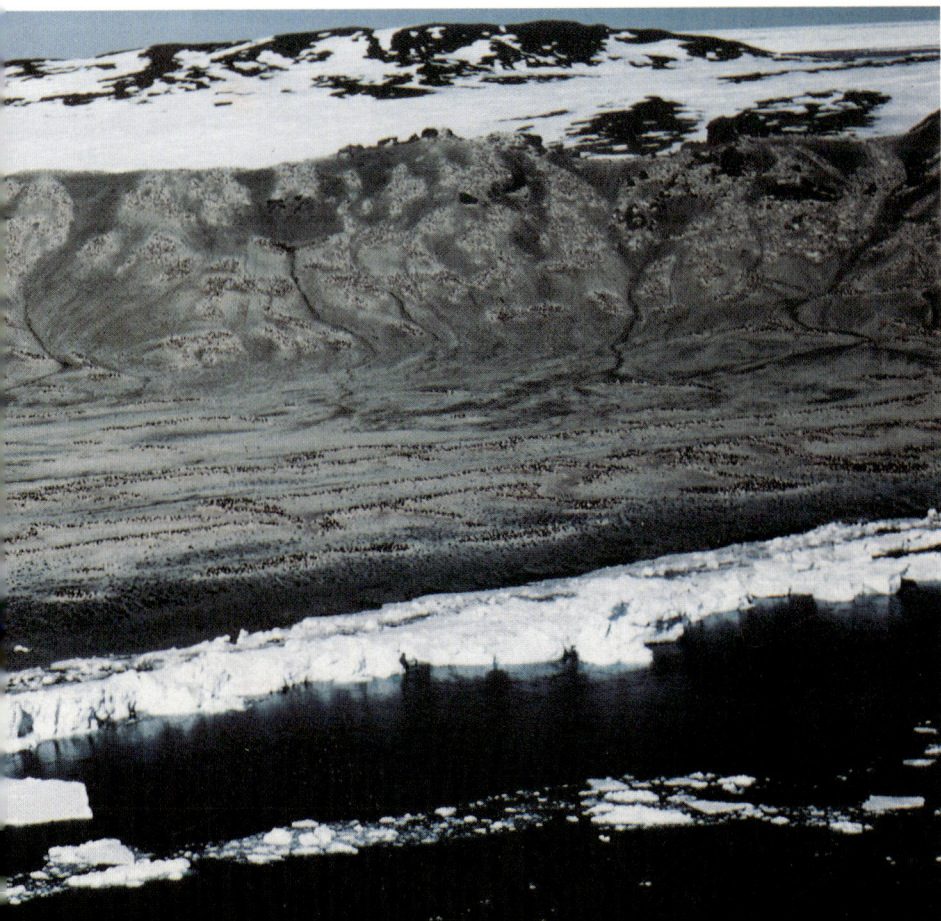

Plate 7.6. *The cuspate foreland with raised beach ridges at Possession Island and with its beach continuously protected by an ice foot and high ice-push ridges. Photo: M. Mabin.*

at Cape Ross. South from Granite Harbour sea level variations are less than 20 metres and show a general southward decline to zero in southern McMurdo Sound. Offshore islands in the southern Ross Sea show relative sea level variations of 5–10 metres, considerably less than the mainland sites at the same latitudes. Raised ridge heights on the islands also show a southward decline from Franklin Island to Beaufort Island to Cape Bird.

The future

Over recent years there has been much speculation that liquid and gaseous hydrocarbon resources may be hidden beneath the continental shelf and slope of Antarctica. The most prospective region, on present knowledge, is the Ross Sea beneath which at least three basins with appropriately thick sedimentary sequences have been located. Asphaltic residues identified near the base of a drillhole in McMurdo Sound will only serve to further tantalise some prospectors, perhaps until drilling for a seismically identified target occurs. Despite the harsh climate, and operational and environmental hazards, should a discovery be made, exploitation could shortly follow using existing concepts, materials, and technologies. Of economic necessity, only fields of the "super giant" category would be contemplated for development. It follows that the potential would exist for a major oil spill after a tanker accident or wellhead blowout. The short- and long-term environmental consequences of a large hydrocarbon spill in Antarctic waters are currently impossible to evaluate, but they are likely to be grave. The strong, clockwise gyral circulation of the Ross Sea suggests that much of the shoreline of North Victoria Land and Ross Island could be at risk in an offshore spill. Impact potential would be greatest during relatively ice-free periods of the short summer (December–March). The environment's response would be determined by hydrocarbon type and weather conditions, as well as by coastal character.

It is likely that oil deposited on high ice cliffs and other permanent and seasonal ice shores would rapidly self-cleanse through vigorous wave action and abrasion by drifting ice. There would also be little retention of oil on rocky shores, whether low, glacially muted bluffs, or high precipitous cliffs, for these would be similarly self-cleansed. In addition, landfast ice or a protective ice foot would further shield these rocky shores from impacting oil slicks.

Once the protective ice foot disintegrates and disappears in summer the beaches would be exposed to oil spill contamination. In the porous, coarse gravels of the intertidal zone and wave-swept active beach face, stranded oil could rapidly penetrate, probably to the depth of the impervious ice or permafrost core. It could also be buried by rapid changes in beach morphology. The irregularly hummocky relief and large hollows left by the melting of stranded blocks of ice (Plate 7.7) would also provide important sumps for stranded oil.

Storm waves would probably throw oil on to and across the most seaward ridges of cuspate forelands, gradually immobilising sediment and ashphaltic pavement. The result could be to enhance the erosion that is already evident at the southern ends of most of the beaches, and also to reduce the availability of pebbles for penguins' nest-building.

Should serious exploitation of offshore hydrocarbons beneath the Ross Sea ever eventuate, there would probably be a demand for moderate to large-scale adjacent onshore construction. This would possibly include accommodation, pipeline landfall and tank farm storage, and facilities for processing, fabricating, maintenance, and ship berthage. Geotechnical and economic factors suggest that beaches would be the preferred sites for these facilities, although some low rocky shores could also be suitable. Beaches and low rocky coasts comprise no more than 10 percent of the North Victoria Land coast, and as most are already occupied by penguin rookeries there is considerable potential for discord and conflict arising from contentious resource exploitation and land-use.

Conservationists though will take considerable solace from the current view of many oil industry representatives that economic and operational constraints will preclude exploitation for decades, if not forever.

Shorelines adjacent to populated regions are generally polluted by unsightly plastics, persistent synthetic debris, and other man-made materials. Distant fishing activities have already led to some distasteful contamination of the Subantarctic and South Shetland Islands, as well as South Georgia. But the litter stranded on shores of the Ross Sea is minor and at present is environmentally insignificant. The only persistent and conspicuous material is dressed lumber of local origin, most of which comes from the disintegration of packing crates, boxes, and cargo pallets.

Nevertheless, there are conceivable commercial developments which could environmentally jeopardise these shores. Perceived public sensitivities to change and interference suggest that a "sympathetic" management plan is necessary for the coastal zone.

Plate 7.7. Large blocks of ice stranded some distance inland and to a considerable height above sea level on Ridley Beach, Cape Adare. Note the irregularly hummocky surface developed as a consequence of ice-push followed by later decay and ablation of the stranded ice. Photo: M. Gregory.

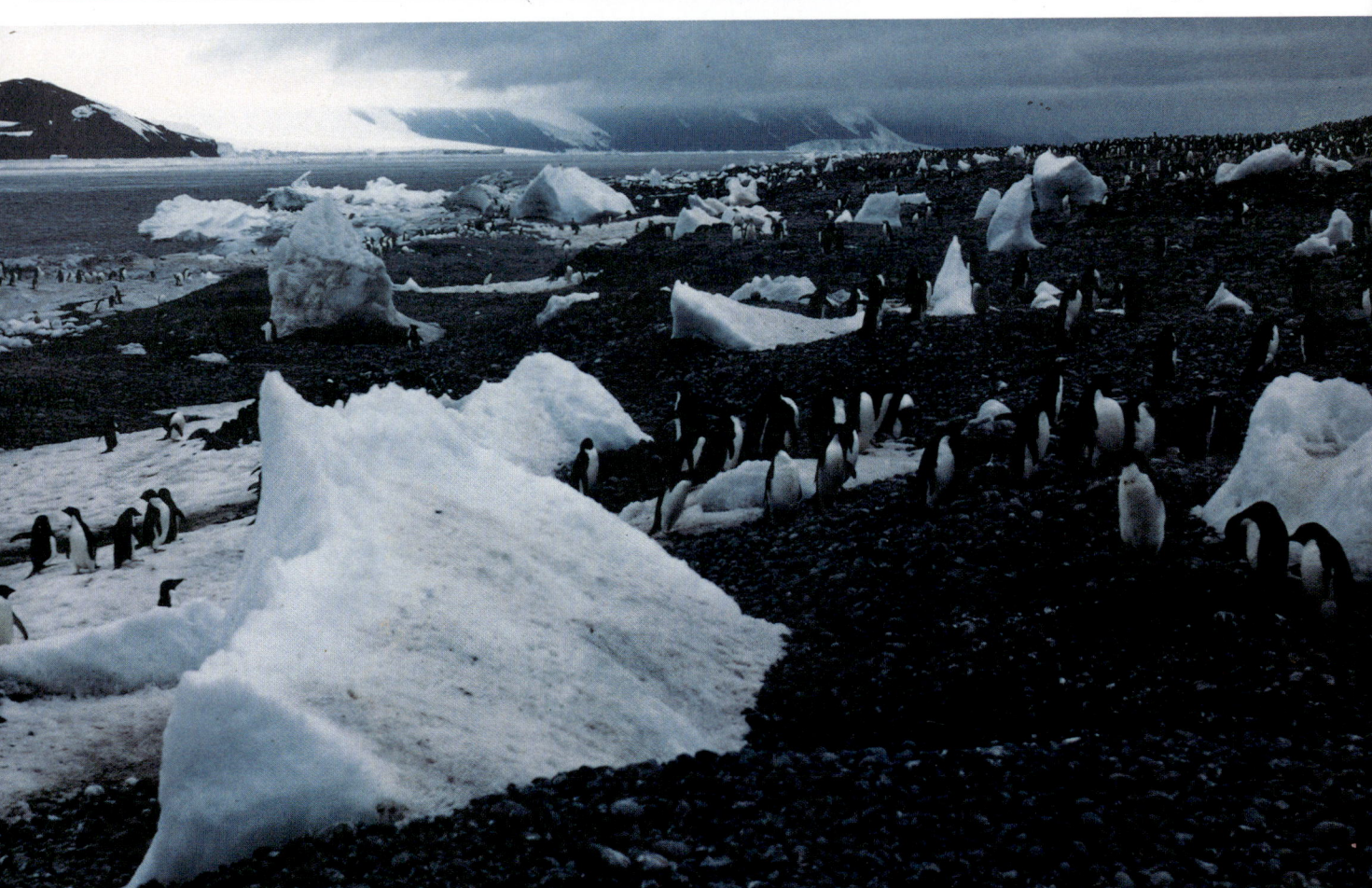

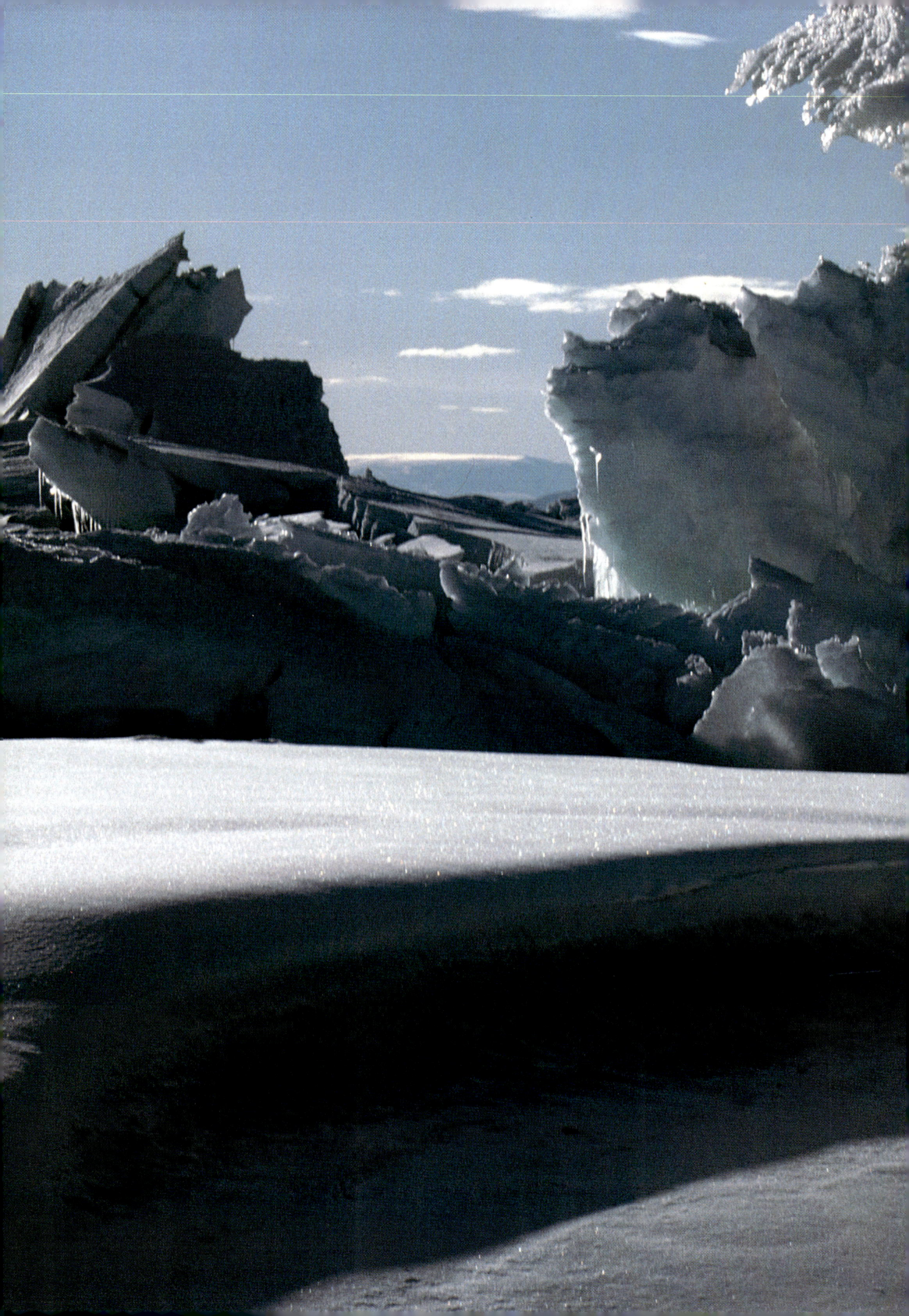

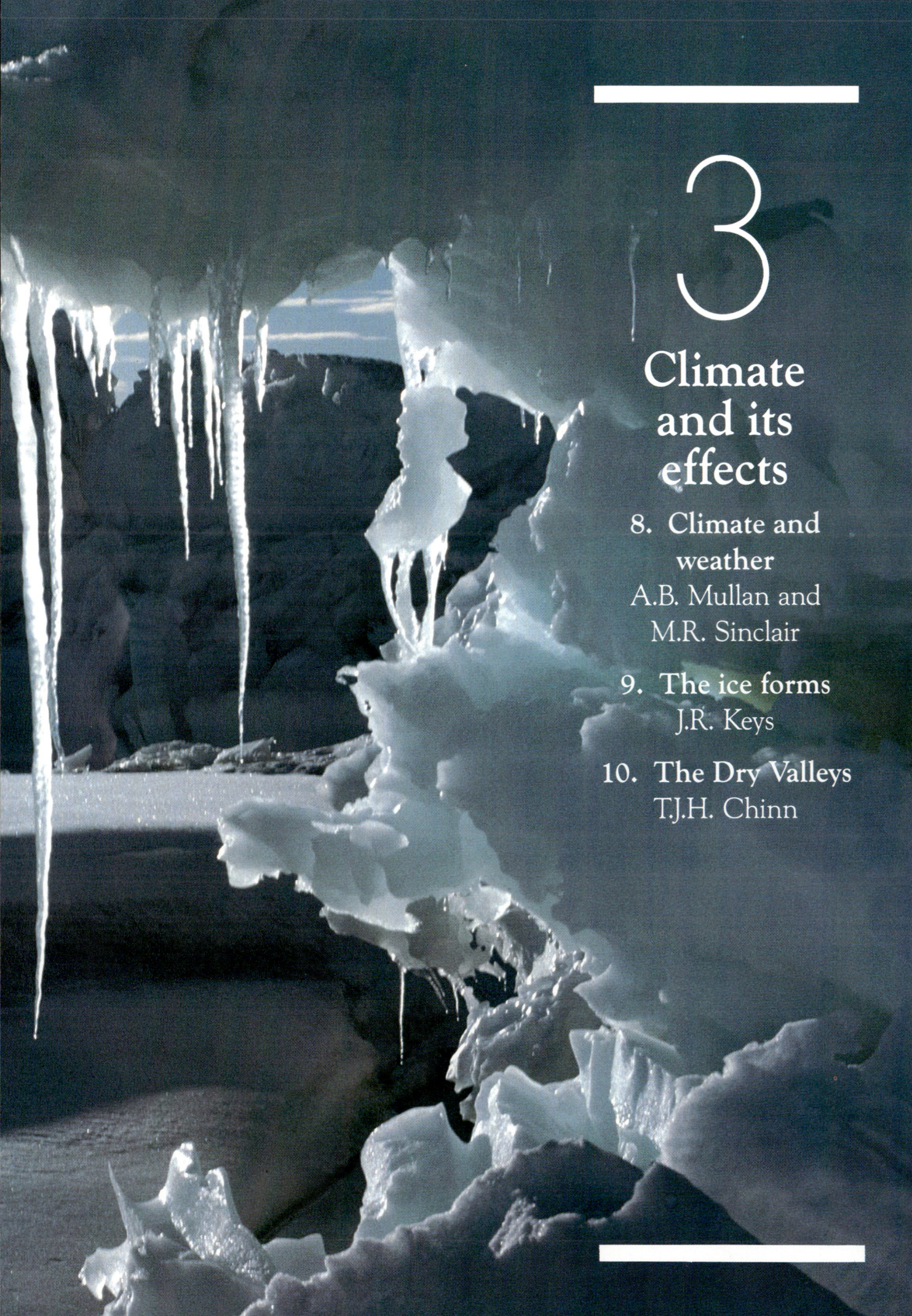

3

Climate and its effects

8. Climate and weather

Antarctica is the coldest, windiest, and driest continent on earth; it is also the highest, with a mean altitude of over 2 kilometres. The high altitude of the continent, together with its polar latitude, largely accounts for its harsh climate. Biting cold, strong winds and blinding snow often make life both uncomfortable and dangerous, and early explorers suffered many tragic losses because of bad weather. Today, despite improved technology, comfort and activity are still limited by the inhospitable environment. "Weather" is usually thought of as the short-term variations in the state of the atmosphere that affect human activities. Thus, the weather of a particular locality involves a knowledge of elements such as temperature, wind, humidity, precipitation, cloudiness, and visibility, as well as how these vary with, for example, the passage of a cyclone. "Climate", on the other hand, is the statistical summary of weather conditions over a much longer period, typically several decades.

Climatic variations occur too, and there have been epochs when the poles were ice-free. About 50 million years ago forests of *Araucaria* (the Southern Hemisphere conifer) and *Nothofagus* (southern beech) grew in Antarctica, as evidenced by fossil remains from the northern tip of the Antarctic Peninsula. As these trees require a temperate climate, Antarctica must have been largely free of ice, even though the continent was in essentially the same geographic position as at present. Around 35 million years ago, there was a sharp drop (4 – 5°C) in ocean surface temperatures, and by 25 million years ago glaciers had reached sea level in the Ross Sea region. The ice sheet attained approximately its present volume 5 million years ago, although further climatic fluctuations have followed. The last worldwide ice age, the Quaternary Ice Age, began about 1.8 million years ago and may not yet have ended. Ice ages are made up of alternating glacial (colder) and interglacial (warmer) periods, and it is generally accepted that the most recent glacial period reached its peak about 18 000 years ago, when the sea ice around Antarctica extended as far north as 45°S in winter (i.e., almost to the latitude of Christchurch). For the past 10 000 years, the world has been experiencing the warmer conditions of an interglacial period, although glaciologists are uncertain whether the Antarctic Ice Sheet is currently growing or shrinking. The following discussion of weather and climate covers the last 30 years of instrumental records, with occasional reference to data collected by the early explorers.

Radiation balance and temperature

The one abiding characteristic of the Antarctic climate is the extreme cold and in many inland areas of the high Polar Plateau the temperature seldom rises above –50°C for 6 months of the year. At these temperatures, exposed parts of the body can be frostbitten in seconds and special precautions are necessary to avoid freezing of the air passages. Normally-flexible materials such as rubbers and plastics can crack or shatter like glass, and fuel oil freezes to a jelly.

The key to understanding Antarctica's low temperatures and how they vary seasonally, and with location lies in the radiation balance. This is the difference between the incoming solar radiation (or "insolation") that is absorbed at the ground, and the longer-wavelength "thermal" radiation emitted back to space by the ground. The amount of sunlight entering the atmosphere at a particular latitude varies with the time of year because the earth's spin axis is tilted at an angle (66°33′S) to the plane of its orbit. At the latitude of the Antarctic Circle (66°30′S), there is 1 day each year of 24 hours of daylight (21 December) and 1 day of 24 hours of darkness (21 June). Further south, the periods of continuous sunlight in summer and continuous darkness in winter lengthen, until at the South Pole each period lasts for 6 months. The solar radiation also strikes the atmosphere at a more oblique angle at higher latitudes, thus reducing the incident energy per unit of surface area. The effect of this oblique angle is more than compensated for during summer months by the increased number of hours the sun stays above the horizon. The net effect is that in the height of summer the polar region actually receives more solar radiation than the Equator. Much of this solar radiation, however, is reflected back to space either by clouds or by the icy surface, leaving only a small percentage to be absorbed. At the South Pole, for example, about 80 percent of the solar radiation at the top of the atmosphere reaches the plateau surface and about three-quarters of this is immediately reflected, so only about 20 percent of the original radiation remains to warm the surface. The reflectivity (or "albedo") of the surface is clearly a crucial factor in determining the surface radiation balance, particularly as it varies so considerably with the nature of the surface. A fresh snow surface has an albedo of 0.75-0.85, whereas over the dark exposed rocks in the Dry Valley region a much lower value of about 0.20 has been measured. In addition, the surface itself, like any other body not at absolute zero temperature, radiates heat continually to space. The strength of this "thermal" radiation depends only on the temperature, varying as the fourth power of the temperature in degrees Kelvin. Hence, if the surface temperature falls from 273° Kelvin (0°C) to 233° Kelvin (-40°C), the emitted thermal radiation will decrease by a factor of almost 2. The outgoing radiation can also be absorbed by the air and clouds, and re-emitted both outward into space and back to the surface. However, because of the great clarity of the air, much of this outgoing radiation is lost to space, especially over the elevated plateau where the blanket of air is thinner.

Radiation measurements have been made in Antarctica for many years. Of particular interest are measurements from the Dry Valleys, and Figure 8.1 shows readings at Vanda Station from April 1969 to

Figure 8.1. *Incoming solar radiation and net radiation balance at Vanda Station. Negative values during the winter seasons indicate a loss of radiated energy from the surface.*

Figure 8.2. *Annual variation of temperature at Scott Base and the South Pole. The "coreless" phenomenon of almost constant winter temperatures can be seen at both sites but particularly at the South Pole.*

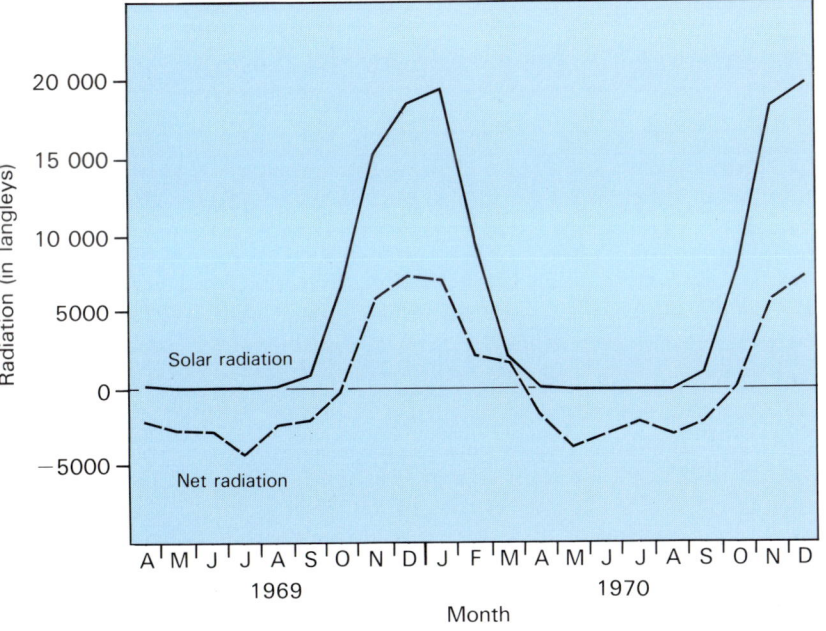
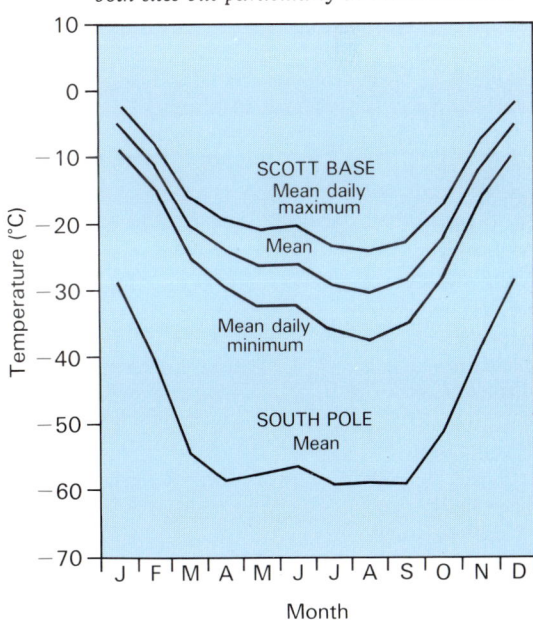

December 1970 inclusive. The incoming solar radiation is at a maximum during summer, and drops to zero during the period of continual darkness. The net radiation balance, or the net radiative energy input after accounting for reflected solar radiation and thermal emission, is negative during winter—implying that there is a net radiational deficit during that season. Although the Vanda site loses energy for about 6 months of the year, there is sufficient gain during summer to produce a net surplus over the year. Scott Base has an annual surplus too. This is primarily because the local albedo drops during summer as bare rock is exposed, allowing more solar energy to be absorbed, though in addition it receives more insolation than Vanda Station where mountains elevate the horizon. For most of the Antarctic Continent, however, which has continual snow cover, there is a net annual deficit.

Thus, except for a few weeks in mid-summer, Antarctica as a whole is continually losing heat by radiation. Because the temperature doesn't fall year after year, this radiational deficit must be balanced by an input of energy from elsewhere. One possible source is heat conduction through the surface from below. This is indeed an important source in the Arctic and over the ice shelves of the Antarctic, but is negligible over the continent itself. The main input of heat to Antarctica is by the transport (or "advection") of warm air from lower latitudes. This energy input prevents the polar temperatures from continuing to fall during the long winter night. Unlike middle latitudes where temperatures show a well-defined winter minimum, Antarctic temperatures decrease little during the long winter (April to August), showing the so-called "kernlose" (or "coreless") structure (Figure 8.2)—a type of temperature curve first noted over the Arctic pack ice and in Greenland. The temperature reversal can occur in any winter month anywhere over Antarctica, but is more common in the Ross Sea region than in other parts of the continent. The largest temperature reversals usually occur in June or July. Before mid-winter, the heat lost by radiation is offset by

Table 8.1. Mean temperatures (a) and extreme minimum temperatures (b) at Scott Base, McMurdo, and Cape Hallett, and other meteorological factors (c) at McMurdo.

	Jan.	Feb.	Mar.	Apr.	May	Jun.	Jul.	Aug.	Sep.	Oct.	Nov.	Dec.	Year
(a) °C													
Scott Base	-4.9	-10.8	-20.4	-24.2	-27.4	-26.2	-29.4	-31.0	-28.1	-22.7	-12.0	-5.4	-20.2
McMurdo	-2.8	-8.4	-17.6	-21.3	-23.8	-23.2	-26.1	-27.8	-24.8	-19.9	-9.5	-3.4	-17.4
Cape Hallett	-1.1	-3.2	-10.5	-17.8	-22.6	-23.0	-26.4	-26.6	-24.5	-18.3	-8.0	-1.7	-15.3
(b) °C													
Scott Base	-19.7	-30.2	-44.6	-50.4	-53.2	-52.2	-54.2	-56.6	-57.0	-52.0	-37.2	-22.8	-57.0
McMurdo	-15.6	-23.9	-43.3	-39.4	-44.4	-41.1	-50.6	-49.4	-43.9	-40.0	-28.3	-16.7	-50.6
Cape Hallett	-9.4	-8.9	-22.8	-31.1	-35.0	-36.7	-40.6	-47.8	-40.0	-37.2	-24.4	-11.7	-47.8
(c)													
Cloud cover (octas)	5.0	6.1	5.9	4.4	4.7	4.1	4.0	3.9	4.6	4.6	4.6	5.0	4.7
Days with blowing snow	2	3	11	14	12	14	13	11	16	10	4	3	9.4
Frequency of snow (%)	10	19	17	19	18	19	16	19	15	13	11	13	16
Frequency of fog (%)	2	1	4	4	3	2	2	3	2	3	1	2	2
Freq. visibility <4 kilometres (%)	3	4	12	14	13	16	13	12	17	11	4	5	10
Precipitation (millimetres water equiv.)	19	28	16	13	14	23	10	16	12	12	11	15	188

warm air advected from the surrounding ocean. Once the winter solstice has passed, any warming is delayed until late August by the increased area of sea ice that surrounds Antarctica.

The temperature "inversion" is another important feature of the Antarctic temperature regime. Because the snow surface radiates heat more efficiently than the air, the atmosphere tends to be cooled from below by the colder snow surface. This results in a surface layer of cold air in which the temperature actually rises with increasing altitude, instead of falling as it normally does (hence the term "inversion"). The steepest temperature rises occur in the lowest 20 metres or so above the surface. Over inland areas of the Polar Plateau, the air temperature 1000 metres above the plateau can be over 30°C warmer than at the surface. Inversion formation is favoured by clear skies, which maximise the net loss of thermal radiation, and by light winds. Strong winds, on the other hand, mix down warmer air from above and destroy the inversion. The temperature inversion is most pronounced during winter when the solar input is minimal, but it can be found at any time of the year, even near the coasts. In the McMurdo area, the inversion is seldom over 10°C.

Thus, temperature experienced at a particular location is the result of the interplay between the various radiative and advective heat fluxes. Because the layer of very cold air at the surface is so shallow, small changes in this delicate balance can produce large temperature changes. Rapid temperature fluctuations are a well-known feature of the Antarctic climate. It is not uncommon for the temperature to rise or fall as much as 30°C over a few hours. For example, in winter the appearance of low cloud is typically accompanied by a rise in the surface temperature. This warming results because cloud is a much more efficient radiator of heat than clear air. The increased radiation back to earth ("back-radiation") from the warm cloud layer increases the heat input at the surface. When the sky clears, this back-radiation decreases markedly, the surface cools, and the inversion becomes re-established. As the surface temperature falls, the amount of heat radiated to space decreases. After a few hours, a stage is reached where the heat lost at the surface is just balanced by the reduced back-radiation from the warmer air at the top of the inversion layer. It is this equilibrium that limits the strength of the surface inversion to not much more than about 30°C during the long winter night.

Near the coasts where the ocean exerts a moderating effect, winter minimum temperatures seldom drop below –40°C. Summer temperatures usually remain a few degrees below 0°C, although summer maxima as high as 8°C sometimes occur. Such conditions are typical of the McMurdo Sound area, summarised in Table 1. However, the warmest summer temperatures are found in the Dry Valley area, where the ice-free surface absorbs most of the incoming solar energy; at Vanda, a high of 15°C has been recorded. Although Vanda summer temperatures are some 6–8°C higher than at Scott Base, in winter Scott Base is warmer than Vanda because heat flow through the adjacent sea ice maintains the Scott Base temperatures at a higher level. Cape Hallett is warmer than Scott Base by an average of 5°C, reflecting the lower latitude of Cape Hallett which lies 600 kilometres to the north of Ross Island. Furthermore, Ross Island is embedded within a northward-moving current of air that flows up the western edge of the Ross Ice Shelf. Thus, temperatures near Ross Island reflect the continental influence of the Ice Shelf. By contrast, Cape Hallett enjoys the moderating effect of the Ross Sea. The differences between Cape Hallett and Ross Island are largest in the autumn months when the Ross Sea is mostly ice free.

Although Scott Base and McMurdo are only 3 kilometres apart, Scott Base is nearly 3°C colder. This difference is larger when only minimum temperatures are considered. The air reaching Scott Base originates in the Windless Bight area to the north-east. Because of the light winds in this area, the cold layer is well-established. The coldest temperature encountered at any time by Scott's 1910–1913 expedition was –60°C in the Windless Bight. Average temperatures in this area are 8°C lower than at Cape Evans and Hut Point. Scott Base is situated on a low (16 metres) promontory which projects into the cold current of air that flows down the eastern side of Hut Point Peninsula. McMurdo on the other hand, located leeward of the peninsula at a slightly higher elevation (40 metres), is less affected by this shallow cold layer. Such large local variations in temperature are a feature of the Antarctic climate.

Inland, average temperatures decrease by 1°C for every 100 metres increase in elevation (Figure 8.3a, b). Hence, the lowest temperatures on the continent are on the high plateau of East Antarctica, where the Russian station Vostok is situated, rather than at the Pole. The annual average temperature at the South Pole (2800 metres above sea level) is –51°C, whereas that at Vostok (3400 metres a.s.l.) is about 5°C colder. The lowest temperatures occur at the end of winter; on 24 August 1960 Vostok recorded –88.3°C, making this the coldest place on earth.

Atmospheric circulation over Antarctica

The energy which drives the atmospheric circulation is derived from the temperature difference between the Equator and the poles. Antarctica, as we have seen, is a global heat sink, and the atmospheric circulation transports heat from the equatorial region, where there is a net radiational surplus over the year, to the poles where there is a net deficit. However, this transport is not accomplished by a simple north-south ("meridional") air movement (except in the tropics), because atmospheric motions are affected by various forces such as the rotation of the earth and pressure differences. The result is that in middle and

Figure 8.3. (a) Topographic elevation (in metres); (b) mean annual surface temperatures (in °C); (c) streamlines of prevailing surface-wind flow; and (d) mean annual snow accumulation (in centimetres water equivalent) over the Antarctic Continent. The dominating influence of topography on Antarctic climate is obvious.

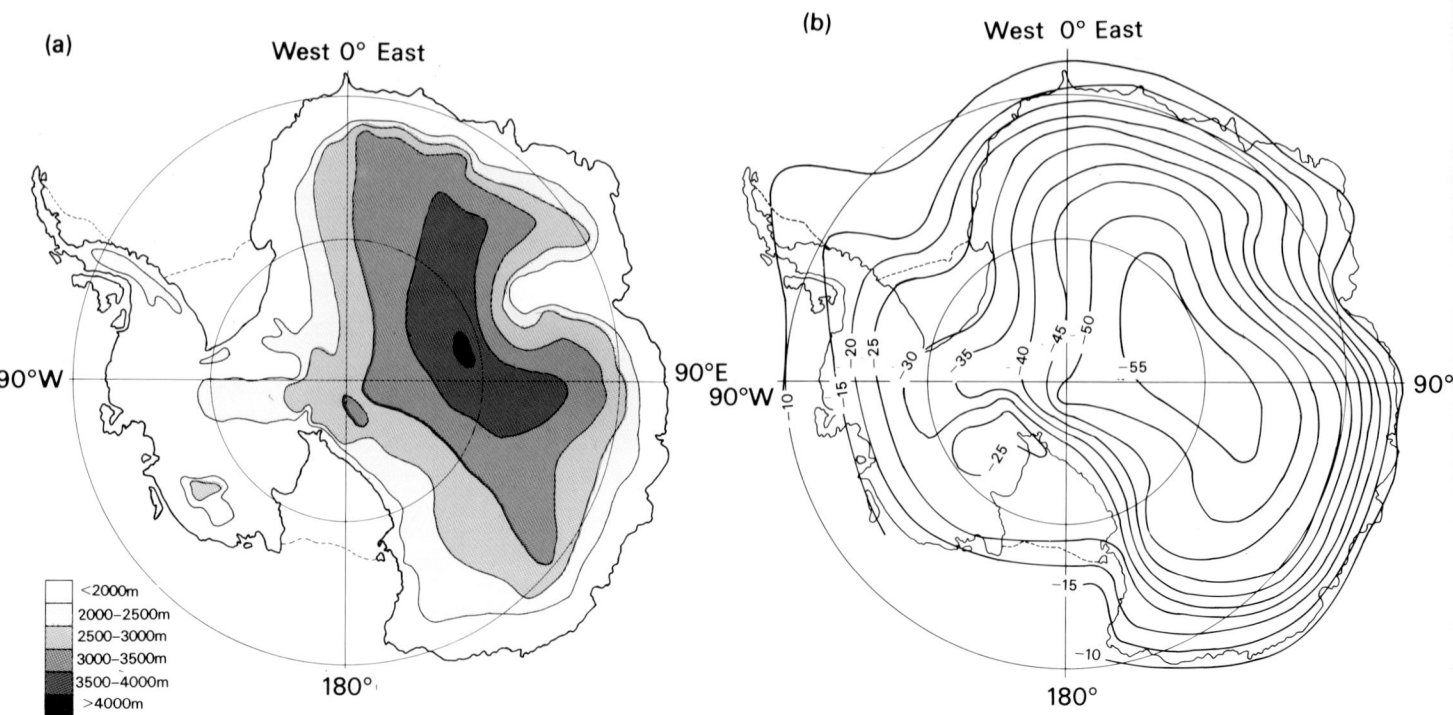

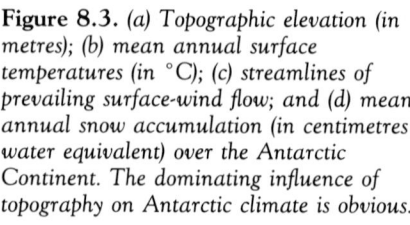

high latitudes the time-averaged flow (Figure 8.4a) appears to be largely
"zonal": i.e., easterly or westerly winds, paralleling lines of latitude. Over
the Southern Oceans there is a wide belt of westerly winds, that extends
from the subtropical high-pressure belt near 30°S to a band of minimum
pressure (called the "Antarctic Circumpolar Trough") encircling the
South Pole at about 65°S. This meridional pressure gradient is much
stronger than that encountered in the Northern Hemisphere, and
consequently the westerlies are also stronger and more persistent in the
South—hence the reputation of the "roaring forties", "furious fifties",
and "screaming sixties". The rapid drop in pressure as one moves
southward has indeed been known for a long time, for James Clark Ross
commented on the "apparent deficiency of atmosphere" over the middle
and high latitudes of the Southern Hemisphere. Poleward of the
Antarctic Circumpolar Trough the surface pressure rises, resulting in a
narrow ring of easterly winds around the Antarctic coast. Over the
continent itself, there is a persistent surface anticyclone which helps to
maintain predominantly clear skies.

At higher levels in the atmosphere the air flow is smoother than
near the surface, so the daily upper-air charts often resemble the pattern
of the time-averaged surface flow (Figure 8.4a). The stratosphere is an
atmospheric layer extending from about 10 to 50 kilometres above the
earth, in which the maximum ozone concentrations are found. Ozone is
a strong absorber of solar radiation, so that as the sun's angle varies with
the seasons there are dramatic changes in the temperature and
circulation of the polar stratosphere. In winter the stratospheric
circulation consists of westerly winds encircling a low-pressure area
centred over the pole. This "polar vortex" forms rapidly in March when
the sun moves northward leaving Antarctica in darkness. Stratospheric
temperatures decrease by about 50°C between March and August, with
the level of greatest cooling coinciding approximately with that of the
maximum ozone concentration at 15–20 kilometres height. As

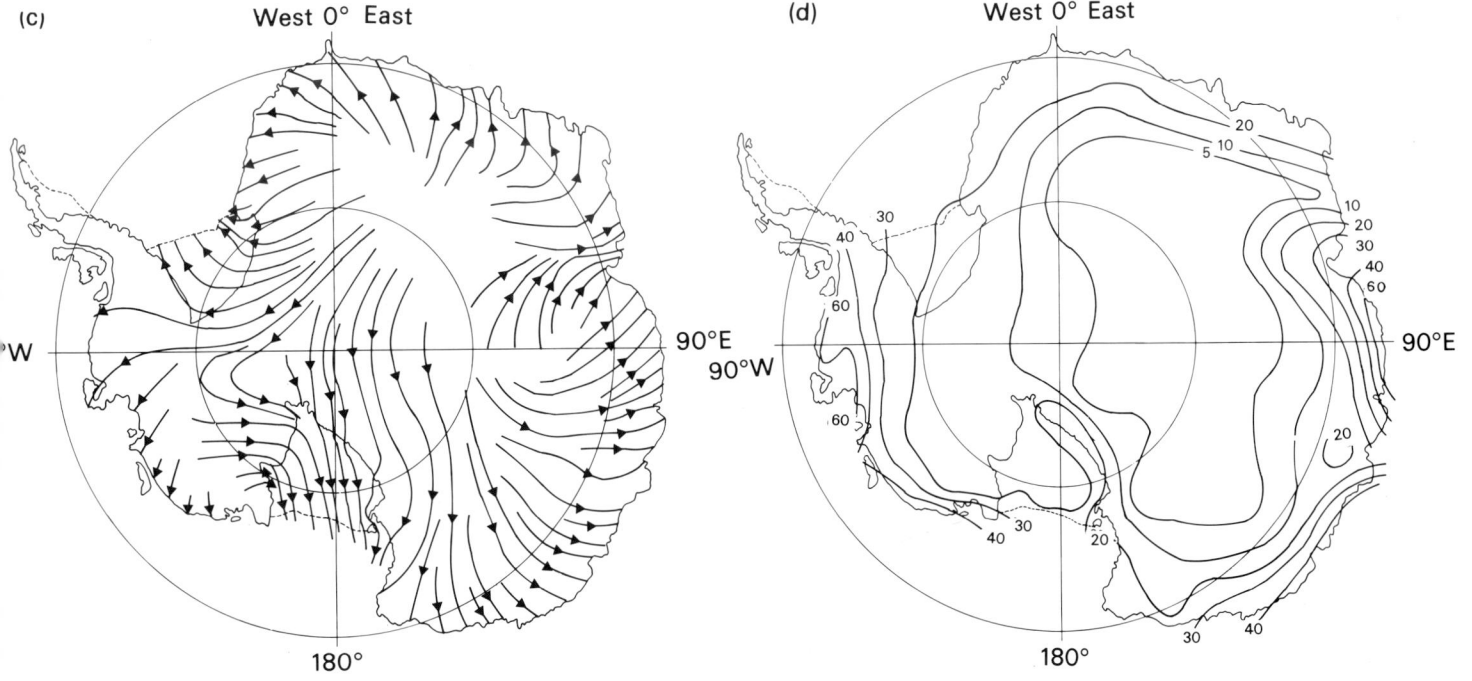

temperatures fall during winter the polar vortex intensifies, and westerlies of over 100 metres per second (the "polar-night jet") can be found circulating around the vortex. There is also a vertical air circulation over Antarctica, implied by high ozone concentrations measured near the surface around the coast in winter. Slow subsidence in the cooling air carries ozone down from the stratosphere to the Polar Plateau, where it is finally swept off the continent in a shallow surface layer.

With the return of the sun at the end of winter, the polar-night jet weakens and allows a greater influx of warm air from lower latitudes. This influx of heat, together with local heating in the ozone layer, often produces explosive warming of the lower stratosphere; temperatures can rise by up to 50°C in a few weeks during October–November. In summer the Antarctic stratosphere is dominated by a stable anticyclonic circulation (i.e., a high pressure centre encircled by easterlies) of weak winds and uniform temperatures.

Although the mean circulation over the Southern Oceans in the lower atmosphere appears to be a simple westerly flow (Figure 8.4a), daily weather charts show that there are sequences of high and low pressure systems embedded within the westerlies (Figure 8.4b). The low pressure centres (or "cyclones") have winds rotating clockwise around them, and are associated with broad bands of cloudiness ("fronts") and generally bad weather. As the cyclones track eastward around the globe, they also have a gradual poleward movement, so that many cyclones eventually dissipate in the "graveyard" of the Antarctic Circumpolar Trough (Plate 8.1). Within the Trough, there is a tendency for distinct pressure minima to persist in three or four longitude sectors, one of which is the Ross Sea region. The wind direction experienced will depend on the observer's location with respect to the cyclone centre. The passage of a cyclone to the north of the observer will be heralded by increasing north-easterly winds and followed by a change to south-easterlies. The high altitude of the Polar Plateau presents a formidable barrier to the

Figure 8.4. *Mean sea-level pressure (in hectoPascals) over the Southern Hemisphere. (a) annual average for 1986; and (b) daily chart for 1 May 1987.*

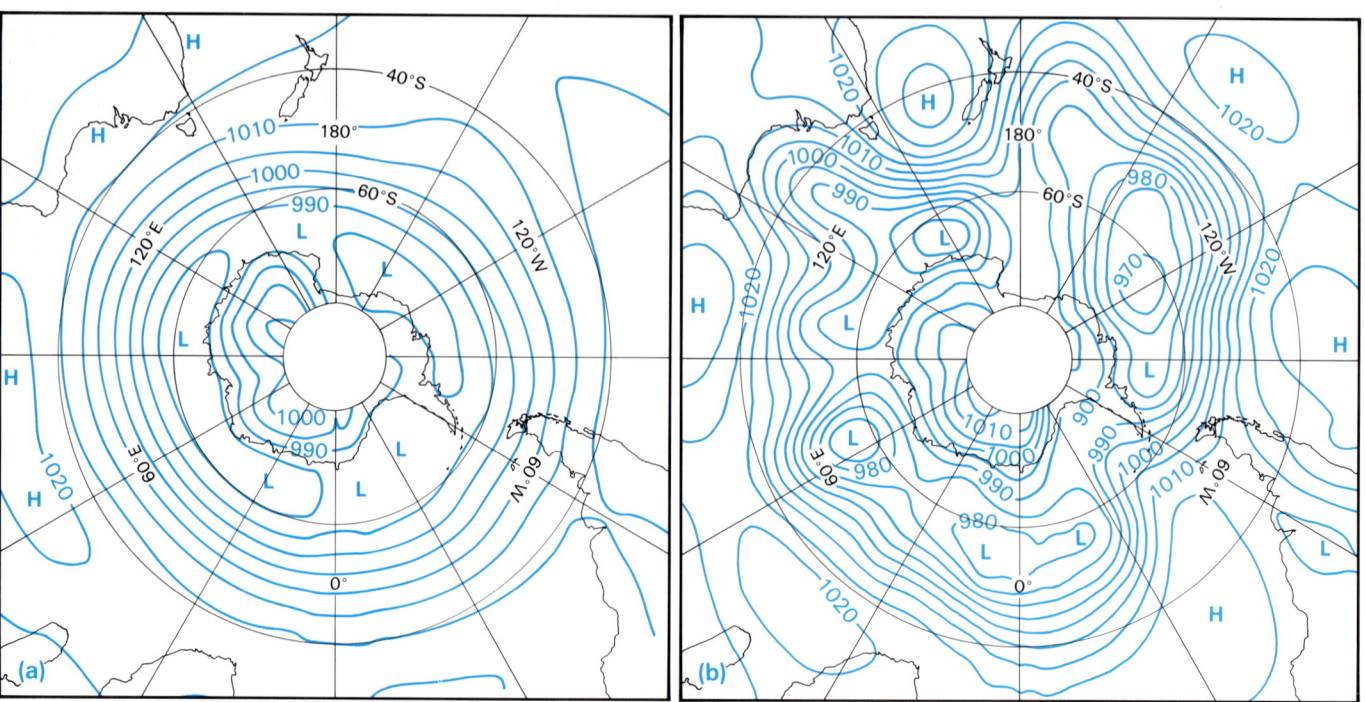

mobile pressure systems, and only rarely do the cyclones travel any distance inland. It is these cyclones that provide the main mechanism for exchange of air (and heat) between lower latitudes and Antarctica, accomplishing this exchange more efficiently than would a direct meridional flow.

The cyclones also transport moisture as well as heat into polar latitudes. Over the seas to the north of Antarctica the annual precipitation is about 100 centimetres. (Compare this value with that of approximately 125 centimetres for Wellington, New Zealand, and 65 centimetres for Christchurch). However, because the cyclones do not penetrate far inland over the continent, there is a strong precipitation gradient around the coast. Most of the precipitation, which virtually all falls as snow, occurs in a coastal belt 200–300 kilometres wide, where the mean annual accumulation of 60–150 centimetres of snow is equivalent to 20–50 centimetres of water (Figure 8.3d). The accumulated precipitation decreases markedly in the interior, partly because the cyclonic storms do not reach the high plateau regions, and partly because the amount of water vapour that can be held in the atmosphere decreases rapidly at colder temperatures. Mean annual accumulation over much of East Antarctica is below 5 centimetres water-equivalent, making this area almost as dry as the Sahara Desert!

Marked seasonal changes occur in the high latitude circulation. Because of the larger equator-to-pole temperature gradients in winter, the high latitude westerlies are stronger and extend over a greater latitude range at this time of year. The winter cyclones are generally larger and more intense, and they travel further south-east into the Ross Sea. Hence, there are more westerlies near the Antarctic coast in winter than in summer. In the Ross Sea region, surface pressures are highest in summer, with more settled anticyclonic conditions, and lowest in winter.

The Antarctic Circumpolar Trough (A.C.T.) also oscillates in intensity and position with the seasons, but this oscillation is mainly a 6-

Plate 8.2. *When the wind blows constantly from one direction for days at a time, a pattern of ripples called "sastrugi" forms in the snow surface. The sastrugi are aligned at right angles to the predominant wind direction. Photo: W. Fowlie.*

Plate 8.1. *This satellite photograph shows a series of cyclones circling clockwise around the Antarctic Continent. It is an infra-red image, i.e., the camera senses temperature. Light-coloured areas indicate colder temperatures than do dark areas. The cold dome of Antarctica is seen to have a temperature similar to that of clouds high in the atmosphere. Photo: National Climate Centre, U.S. Department of Commerce.*

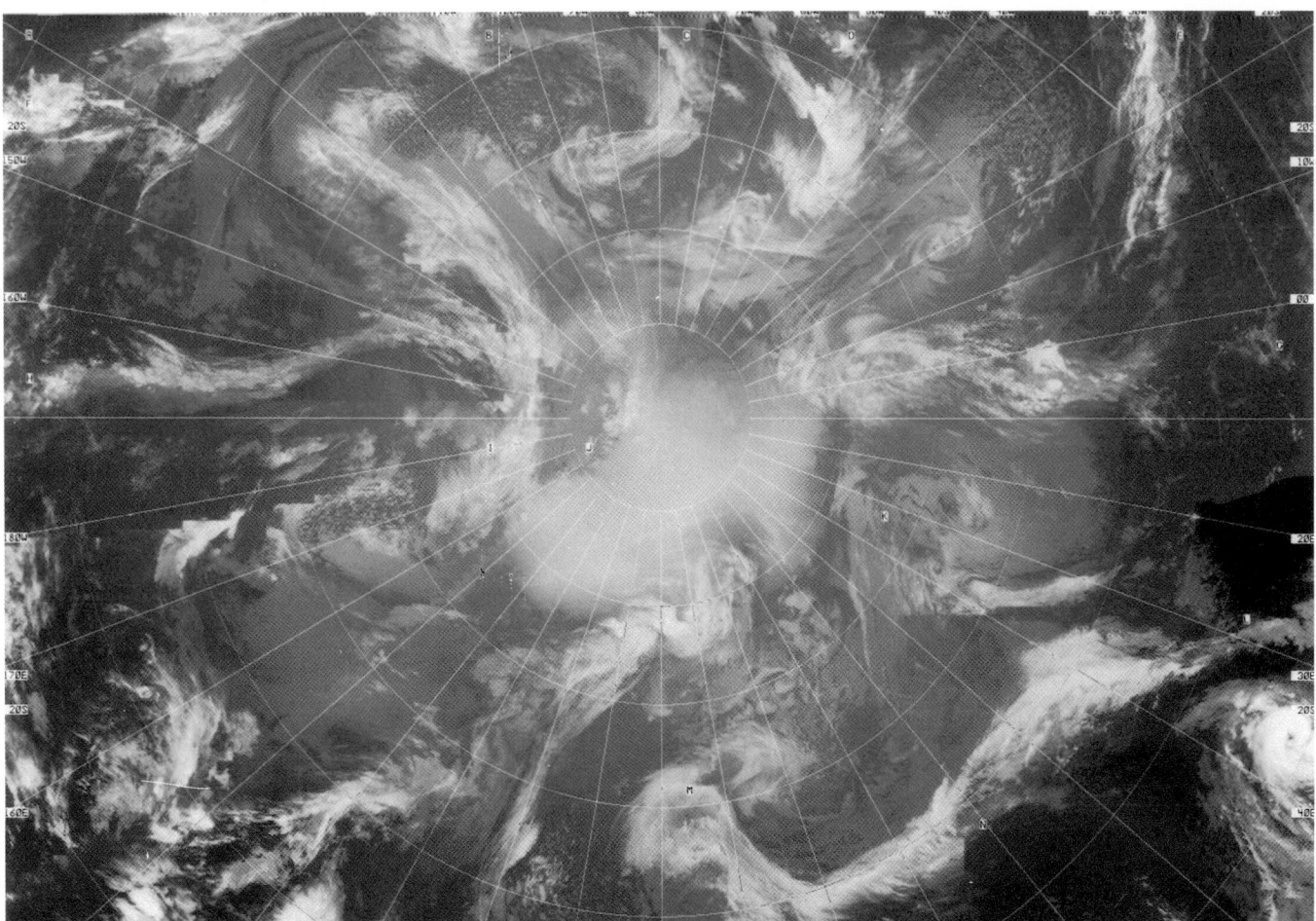

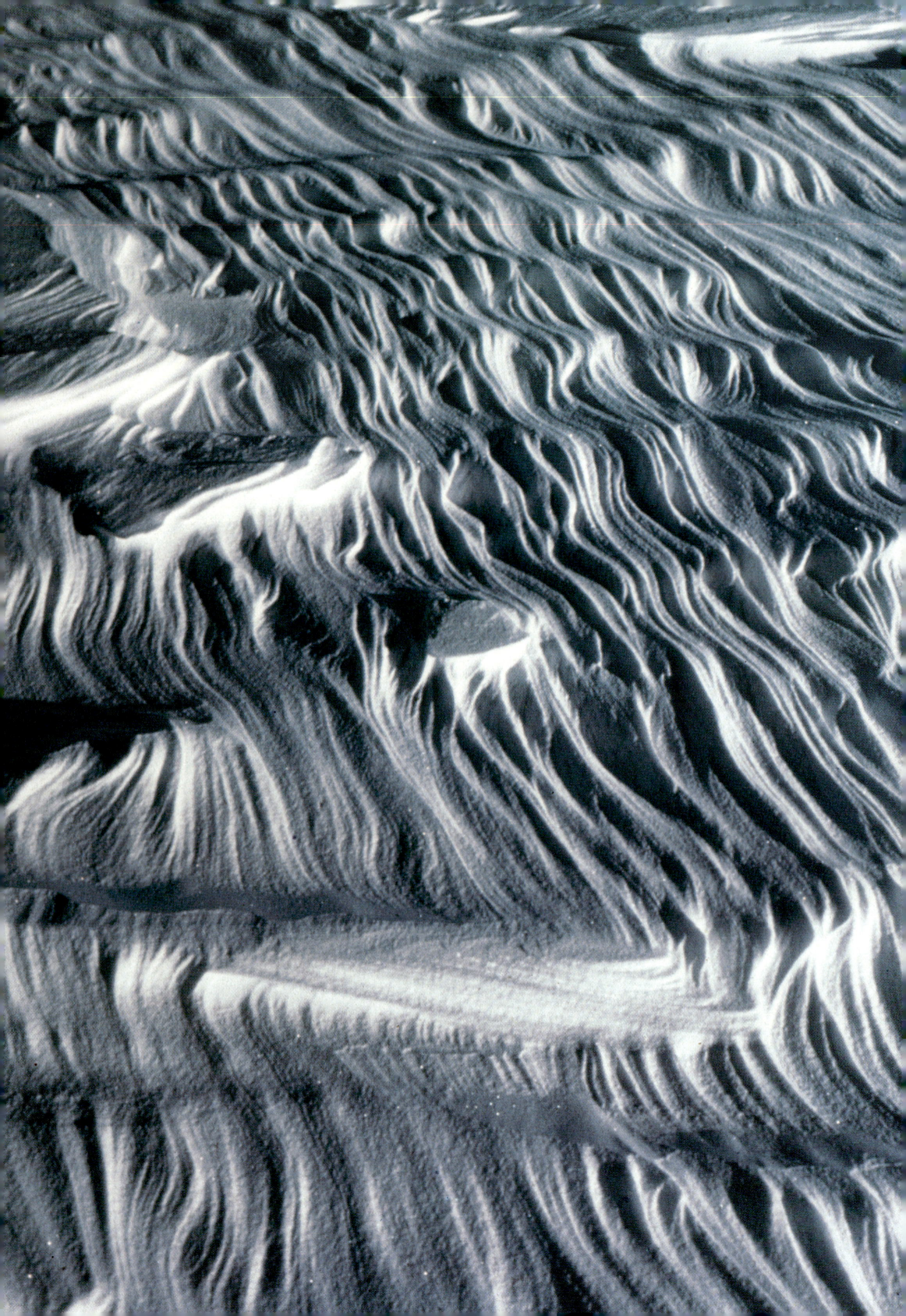

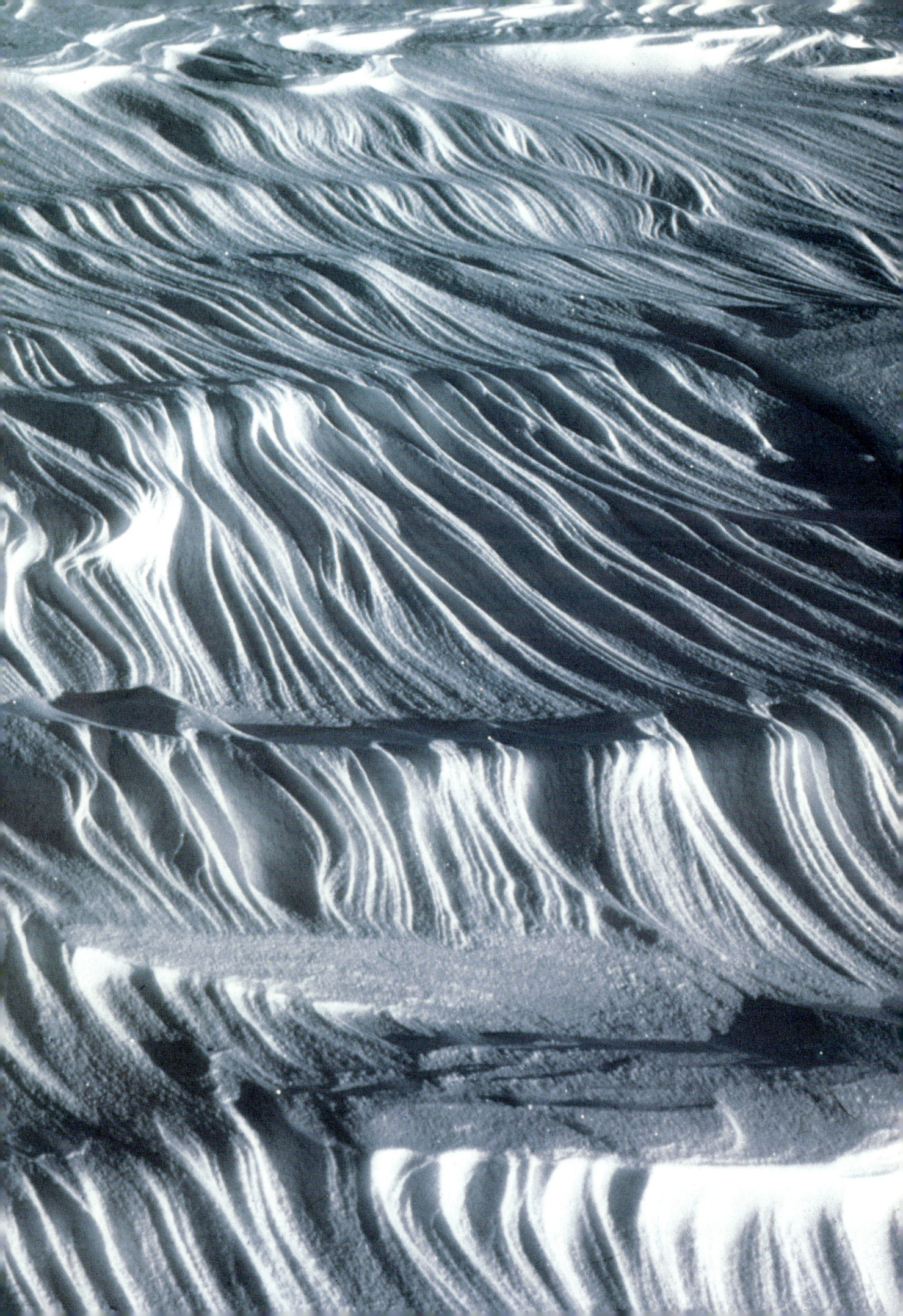

month cycle rather than a 12-month one. The A.C.T. deepens and simultaneously shifts southward in January–April, then weakens and retreats northward slightly from April to July, before returning south and attaining its minimum pressure in October. Thus, the lowest pressures are found in the equinoctial months of April and October. The 6-month cycle in the A.C.T. is also reflected in the strength of the westerlies to the north of the Trough (i.e., the westerlies in the 40–60°S belt are strongest in autumn and spring).

Over the Antarctic Continent, the migratory weather systems that play such a big part in the weather of the middle latitudes have only a secondary role in determining the low-level wind circulation. Instead, the patterns of prevailing surface-wind flow (Figure 8.3c) are closely related to the topography. Over the gently sloping plateau of inland Antarctica, the factors that determine the air flow are the slope of the terrain and the strength of the temperature inversion. Consequently, these winds are called "inversion winds" and blow in a remarkably constant direction (Plate 8.2). The inversion wind direction is fixed relative to the line of maximum slope of the terrain, in the sense that air moves downslope at an angle of about 45° to the left of the fall line, whereas its speed is proportional to the steepness of the slope. Continual radiational cooling replenishes the supply of cold air and maintains a steady drainage off the Polar Plateau. Within the inversion, mixing of different layers of air is strongly inhibited because the coldest (most dense) air is adjacent to the surface. Thus, the inversion wind is detached from the air flow above the inversion layer, which varies according to the changing patterns of atmospheric pressure.

A different situation occurs at the escarpment at the edge of the polar "dome". There, because of the steeper slopes, the cold air drains more rapidly and may temporarily deplete the supply of cold air upstream. Because of this law of supply and demand, the resulting "katabatic" winds (Greek, katabasis, going down) are characteristically spasmodic and more violent than their plateau counterparts. The violent onset of a katabatic event has been likened to the "sudden rush of water from a lock gate in a stream". The wind speed often jumps instantaneously from calm to 30–40 knots (15.5–20.5 metres per second) (Figure 8.5), and equally sudden lulls may occur. Usually, temperature rises at the onset of a katabatic event due to the sweeping away of the cold layer that forms under calm conditions. The katabatic wind is confined to the lowest kilometre of the atmosphere, and seldom extends more than a few kilometres from the foot of the ice slopes. Since cold air drainage depends on the surface layer cooling by radiation, katabatics tend to be stronger at night and are most persistent during winter.

Certain locations are particularly prone to katabatic winds. Cape Denison, home of Sir Douglas Mawson's Australasian Antarctic Expedition, is probably the windiest location on earth. In July 1913, the mean wind speed for the entire month was 24.9 metres per second (46 knots). For 1 hour during that month, the mean wind was 42.9 metres per second (80 knots). In "The Home of the Blizzard", Mawson wrote: "Picture drift so dense that daylight comes through dully . . ., the drift is hurled through space at a hundred miles an hour, and the temperature is below zero Fahrenheit".

Local winds and weather

Since the International Geophysical Year (1957–58), Ross Island has been continuously inhabited, and a meteorological programme has been maintained at both Scott Base and McMurdo Station. In addition, scientific parties have provided weather information from other sites within the Ross Sea region, as well as further afield. Between 1902

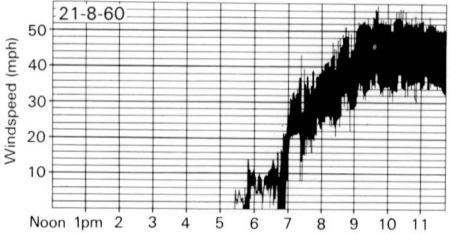

Figure 8.5. *It is thought that the spasmodic nature of katabatic winds is due to the laws of supply and demand of cold air. Katabatic onset is typically sudden, as in this example from Mawson.*

and 1917 the winter headquarters of four British Antarctic expeditions were also located on Ross Island. Many vivid descriptions of the furies of the Antarctic climate, written from the heart, emerged from this "heroic era" of polar exploration. It is from these diverse sources of information that the weather elements of the Ross Sea region, of such practical concern to Antarctic inhabitants, can be described.

To many people, Antarctica still brings to mind images of continual howling blizzards and freezing temperatures. Although blizzards are a fact of life in Antarctica, most areas of the Ross Sea region enjoy long periods of partly cloudy skies and light winds. During clear weather, the great beauty and stillness provide an experience that stands in striking contrast to the blizzards.

Of the various weather elements, the wind is probably the major factor that limits activity and comfort in Antarctica. One of the major hazards of strong wind is the chilling effect caused by increased heat loss from the body (Figure 8.6). To reduce this wind-chill effect, Antarctic clothing must not only conserve body heat but allow little wind to penetrate. Another problem is the blowing snow that accompanies strong wind. Gales are more frequent and severe in winter. Between March and October, wind gusting to over 18 metres per second (33 knots) occurs on average about 11 days per month at Scott Base, compared with only 3 days per month in summer (November to February). Virtually all these strong winds blow from the south. Gusts over 28 metres per second (51 knots) occur about 3 days per month in winter, but only 0.2 days per month in summer. Overall, the mean annual wind speed at Scott Base is about 6 metres per second (11 knots). At Cape Hallett, the mean speed is only 3.6 metres per second (7 knots), with a tendency for strong winds to blow from the south-east.

Wind is usually accompanied by drifting or blowing snow. Snow "drift" commences with winds of over 5 metres per second, although the exact speed of onset depends on the nature of the loose snow and the turbulence of the wind. The amount of snow picked up increases rapidly as the wind increases (varying as the fourth power of the wind speed). With winds less than about 10 metres per second, the drift is usually confined to a shallow (less than 1 metre) layer near the surface. However, as the wind increases, the layer of snow-drift becomes deeper, and is more aptly called "blowing" snow. This more unpleasant situation occurs with winds of over 20 metres per second. These stronger winds reduce visibility to near zero, making outdoor activity dangerous. Such a combination of gale-force wind and blowing snow is called a blizzard. Blizzards are often accompanied by precipitation, although the distinction between falling and blowing snow is usually academic! Sometimes, blizzards stop all outdoor activity for several days (Plate 8.3). Blowing snow is slightly less of a problem during the warmest part of summer, because some melting and consolidation of the loose surface layer of snow occurs. In the McMurdo area, blowing snow occurs on about 4 days per month during summer (November to February), but averages 13 days per month during the remainder of the year (Table 8.1). At Cape Hallett, blowing snow is less frequent, but with a winter maximum.

Practically all blizzards are caused by gale-force southerly winds that are usually associated with an intense cyclonic centre several hundred kilometres east or north-east of McMurdo. These southerly storms are most frequent during winter months, but can occur at any time of the year. Visual clues such as lens-shaped clouds (see Plate 8.4) over the mountains to the south, and the obscuring of Minna Bluff in low stratus cloud, are sometimes the first indications of deteriorating weather as viewed from McMurdo Sound. The appearance of clouds of blowing snow to the south often heralds the onset of a blizzard. The wind is

Figure 8.6. *Wind increases the rate of heat loss from the body. The graph shows how the effective temperature "felt" by the body decreases dramatically as the wind speed increases.*

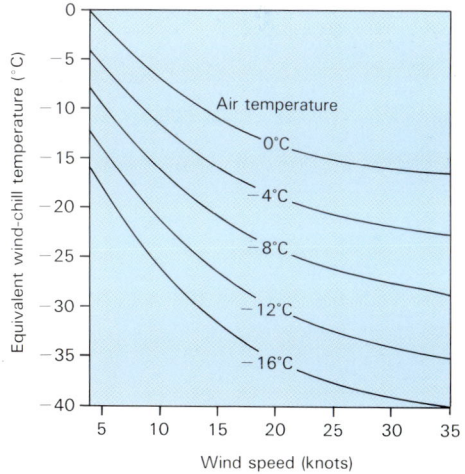

usually squally at its onset, rising rapidly to maximum speed, then decreasing gradually over an hour or so, or until the next squall. In winter, the first gust of wind is often accompanied by a sharp rise in temperature as the stagnant, surface cold layer is swept away. Strong local variations frequently occur; blizzards prevail at one site whereas a nearby location enjoys light winds.

Persistent blizzards played a tragic part in Scott's return from the Pole in 1912. In his "Message To The Public", penned during his last hours, Scott wrote: ". . . our wreck is certainly due to this sudden advent of severe weather, which does not seem to have any satisfactory cause. I do not think human beings ever came through such a month as we have come through, and we should have got through . . ., but for the storm which has fallen on us within 11 miles of the depot at which we hoped to secure our final supplies."

One fact that came to light after the race between Scott and Amundsen to the Pole in 1911–12 was that the blizzards that plagued Scott on the western side of the Ross Ice Shelf were not experienced by Amundsen, whose route to the Pole lay several hundred kilometres to the east of Scott's. Simultaneous observations taken at Cape Evans and at Amundsen's winter quarters at Framheim, 600 kilometres to the east, show that during 10 months, winds over 16 metres per second (30 knots) occurred 30 percent of the time at Cape Evans compared with only 2 percent at Framheim. It is now known that this general westward intensification of the winds is due to damming of air by the Transantarctic Mountains. Recent studies using automatic weather stations have shown that there is a current of air about 200–300 kilometres wide that sweeps north along the western coast of the Ross Ice Shelf, parallel to this mountain barrier. This current, called a "barrier wind", is a result of the deflection of cold stable air that is constrained to follow the configuration of the terrain. Because the air that encounters the mountain barrier is also forced to rise, there is a general increase in

Plate 8.3. *Blizzards are a combination of strong winds and blowing snow; after them it is often necessary to remove accumulations of snow "drift" from vehicles and buildings. Note the lens-shaped clouds that indicate that strong winds extend through a deep layer of the atmosphere. Photo: Antarctic Division, DSIR.*

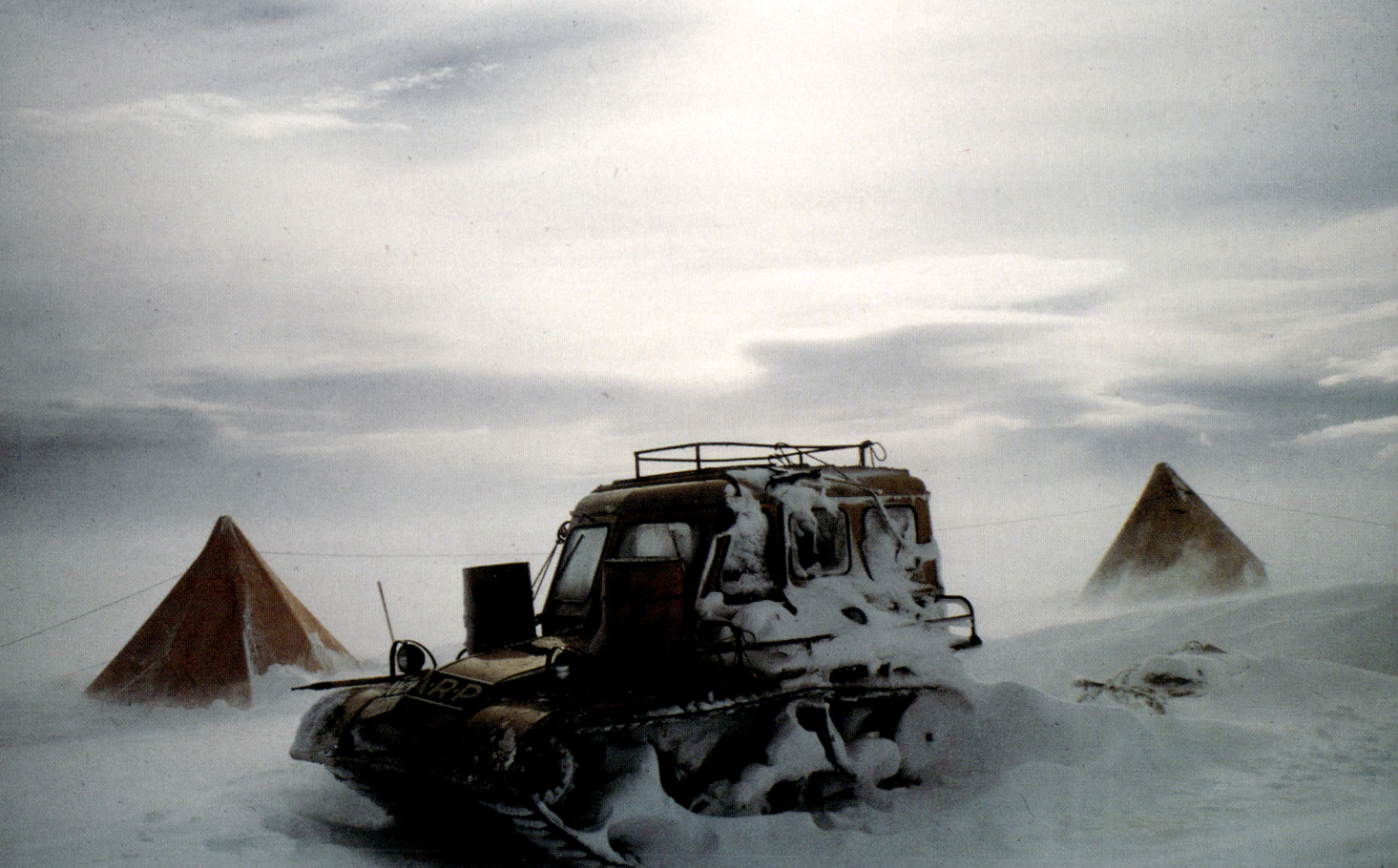

precipitation toward the west.

In most areas of Antarctica, the patterns of prevailing surface wind are closely related to terrain. We have already seen how the winds over the Polar Plateau drain toward the coasts. On a smaller scale, the layer of cold dense air near the surface requires less energy to go around rather than over islands, mountains, etc., in its path. Thus, the surface winds tend to closely follow the configuration of the terrain. This effect was noted as early as 1919 by G. C. Simpson, the meteorologist who analysed the 1910–13 British Antarctic Expedition's meteorological data. Figure 8.7 shows the streamlines of prevailing winds around Ross Island. At Scott Base the prevailing wind is from the north-east due to the deflection of the flow by the ridge of Hut Point Peninsula. At Cape Crozier, only 80 kilometres from Scott Base at the eastern edge of Ross Island, the prevailing wind is from the south-west. In between these two sites, where the southerly flow diverges, is an area known as the Windless Bight, which although exposed to the south does not experience strong winds.

Areas near the edge of the Polar Plateau are prone to localised valley winds. In the Dry Valleys there is a marked daily cycle of winds in summer. Easterlies blow inland during the day, reach a maximum in mid-afternoon, then decrease to a minimum early in the morning, or even turn to the west. Thus, the hours between midnight and 6 a.m. are often the most comfortable for outside activity. Winter in the Dry Valleys is characterised by lengthy calm spells followed by violent westerly gales. These westerlies are relatively warm, extremely dry, and are often accompanied by blowing grit and sand. Dust and snow whirls have occasionally caused considerable damage to outside stores and base buildings at Vanda Station. These vortices reach heights of 100–200 metres, and transport vast quantities of sand, grit, and small stones. Larger tornado-like vortices have been observed. During westerly gales, static charges build up on radio antennae, sometimes causing damage to

Figure 8.7. *Because of the presence of a cold layer at the surface, air tends to move around rather than over Ross Island. Consequently, winds at closely spaced locations near Ross Island can be quite different. For instance, winds at Cape Crozier and Scott Base can be blowing in opposite directions during the same weather.*

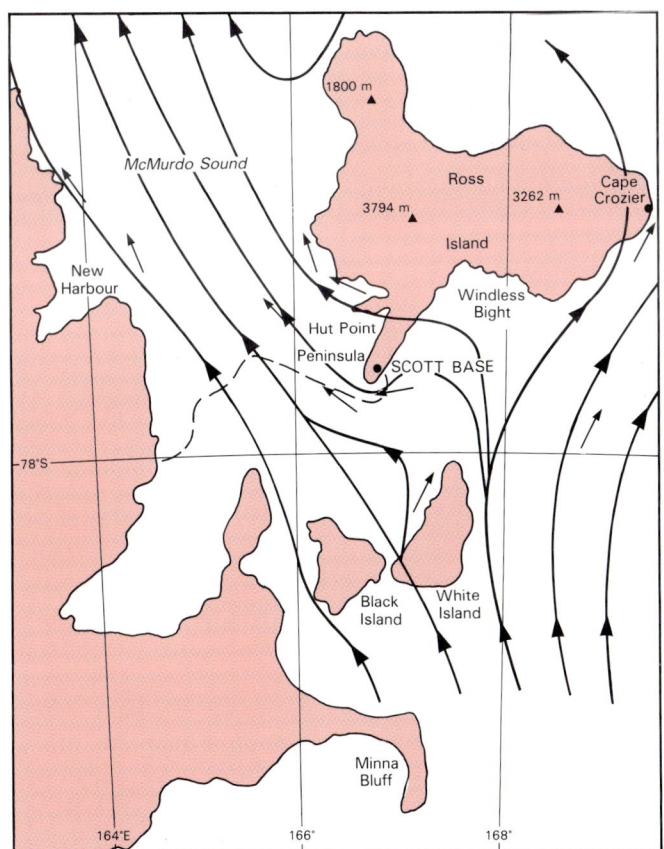

electronic equipment. Clouds of blowing snow on the Polar Plateau to
the west provide reliable advance warning of these westerly storms.

The various weather elements at Scott Base, McMurdo Station, and
Cape Hallett are summarised in Table 8.1. Precipitation (in the form of
snow) and cloudiness show a distinct autumn maximum at all three
locations. This coincides with the minimum extent of sea ice, which
ensures there is maximum moisture available for clouds (Plate 8.4 a, b, c,
d) and precipitation. In comparison with the coastal areas, precipitation
in the Dry Valleys is slight. When it does occur, the relative humidity is
usually so low that any snowfall ablates away within hours. Thus, apart
from the presence of several glaciers, the Dry Valleys remain essentially
snow-free all year. Fog occurs at Scott Base on average about 1–2 days
per month. When it occurs, especially in the summer months, it leaves
beautiful deposits of ice crystals on aerials and superstructures. Rain is
extremely rare, but it has been observed at Vanda on one or two
occasions, and at more-northern Antarctic stations.

One particularly insidious Antarctic hazard is "whiteout", which is
an optical phenomenon that occurs in uniformly overcast conditions
over a snow-covered surface. It is associated with diffuse (uniform),
shadowless illumination which causes a lack of surface definition and
reduced horizon definition. The effect has been likened to being inside a
(full) milk bottle. Because the ability to perceive snow-covered
topographic features depends on the shadows that they cast, such forms
become indistinguishable under whiteout conditions (as any skier, even
in more temperature latitudes, will confirm). Without any visual
stimulation it is common to incorrectly evaluate an incline. One may
even walk up and down hills without realising it. Furthermore, an
individual attempting to follow a straight path unaided under these
conditions will veer. Judgement of the distance and orientation of objects
in the field of view is severely handicapped. Such spatial disorientation is
enhanced inside a moving vehicle, particularly aircraft. It is important to
realise that whiteout conditions can occur while "visibility" (i.e., the
transparency of the air) remains good. Partially reduced horizon and
surface definition can occur under a broken cloud layer.

One important characteristic of the Antarctic climate that cannot be
overemphasised is the highly local and variable nature of the weather.
Calm conditions at a base do not mean calm conditions at a field site 20
kilometres away. Furthermore, the weather can change quickly. A
temperature of –20°C with no wind is vastly different from –20°C and a
moderate breeze. Blizzards can develop beneath a clear sky, and in open
snow-covered areas disorientation due to whiteout can occur if the sky
becomes overcast. Workers in the field need to be prepared for these
sudden changes.

Sea ice

Sea ice is a major component of the climate system of the Ross Sea
region but the growth pattern of sea ice can be quite different from one
year to the next. However, variations at one longitude are often
compensated by opposite changes at another longitude, so that the mean
latitude of the ice–ocean boundary (averaged around the Antarctic coast)
has a well-established seasonal variation. Growth and northward advance
of the sea ice over winter months occurs more slowly than its
subsequent decay in summer. The areal extent of sea ice thus varies by a
factor of almost 10 between February and September, so that at the time
of maximum ice extent the size of Antarctica is effectively doubled. This
naturally has a considerable influence on the heat budget over the
Southern Hemisphere—the sea ice not only reflects more solar radiation
than open water, but also considerably cuts down the amount of heat

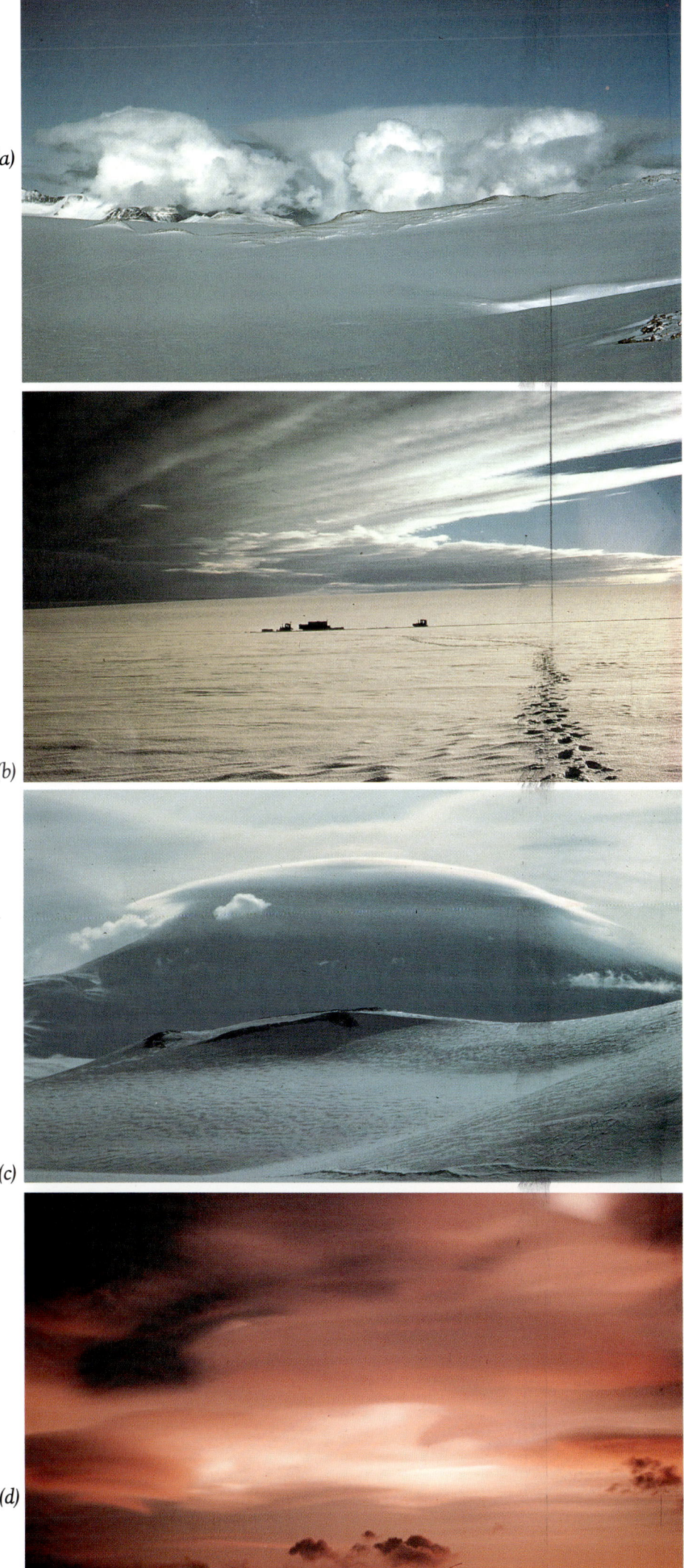

(a)

(b)

(c)

(d)

113

Plate 8.4. *Clouds in Antarctica.*
(a) Because cumulus clouds form when the atmosphere is moist and strongly heated from below they are rarely seen over the ice-covered surfaces. However, in mid-summer the heating that occurs from the ice-free Dry Valleys is sometimes sufficient to support convection. These small cumulo-nimbus clouds probably produced light snow showers. Photo: Antarctic Division, DSIR.
(b) Clouds in Antarctica usually form in layers of little vertical extent because of low moisture content and the absence of strong heat sources in the atmosphere. The low angle of the sun produces striking contrasts in this altostratus layer over the Polar Plateau. Photo: Antarctic Division, DSIR.
(c) "Hogsback", or lens-shaped clouds, shown over Mount Erebus, form near mountains when there is strong wind throughout a thick layer of the atmosphere. Such clouds often herald a deterioration in weather. Photo: Antarctic Division, DSIR.
(d) Nacreous or "mother of pearl" clouds form at altitudes of 20–30 kilometres and are rarely seen except in high latitudes during strong westerly flows. These clouds resemble pale cirrus by day, but a brilliant colour display can be produced at sunrise or sunset as here at Cape Hallett in early spring. Photo: Antarctic Division, DSIR.

exchanged between atmosphere and ocean.

There is also some interaction between sea ice and atmospheric circulation features. The position of the ice edge and the tracks of cyclones appear to be related, although it is not always clear which is cause and which is effect. Strong winds around a cyclone centre can produce an ice advance to the west of the centre, and likewise ice on the eastern side can be broken up or at least restrained from advancing. Other studies have linked year-to-year changes in tracks and frequencies of cyclones with interannual variations in ice growth.

Mankind and the Antarctic atmosphere

The inaccessibility of Antarctica has kept the continent relatively isolated, and therefore unexploited by man, until very recently. However, from the point of view of the meteorologist, no part of the planet is totally isolated from any other part. We have seen that Antarctica is a major heat sink for the globe, and the latitudinal temperature gradients that are set up generate cyclonic storms that carry heat and moisture southwards. The tracks of these cyclones are influenced by the presence of the ice cap and its associated cover of sea ice. Year-to-year variations in amounts of sea ice, as well as longer-term trends, can therefore have a significant effect on the climate of New Zealand and southern parts of Australia, South America, and Africa.

In turn, the Antarctic environment can also be affected by industrial activities far from the continent. Antarctica is a unique clean-air site for sampling the background concentration of trace gases because of the absence of local sources of these gases. It has been known for many years that the carbon dioxide concentration of the atmosphere is increasing, and this will ultimately result in higher surface temperatures worldwide. The South Pole station has one of the longest records of carbon dioxide measurements available, and the data show the same rate of increase of carbon dioxide (about 0.4 percent per year) found in other parts of the world. Computer models of the global climate predict that by the middle of next century, when the atmospheric content of carbon dioxide will have doubled from pre-industrial times, surface air temperatures may have increased by 2–3 °C as a global average, but perhaps up to 10 °C locally in polar latitudes. This greater sensitivity of high latitudes to a global warming could have an important impact on possible commercial activities in Antarctica. There is also speculation that the anticipated warming may cause changes to Antarctic ice sheets that in time could lead to rises in world sea levels, with catastrophic effects on many coastal cities.

Another trace gas that is sensitive to human activities is ozone. The ozone that is present in the atmosphere shields life forms on earth from the damaging effects of the ultraviolet radiation from the sun. The "ozone hole", which has received so much attention in the past few years, is an annually recurring decrease in ozone levels over the 12–20 kilometres height range during the Antarctic spring (Figure 8.8). The decreases apparently began in the late 1970s, and the hole rapidly became progressively wider and deeper, so that by 1987 ozone levels were more than 50 percent down over the entire continent. The sudden appearance of the hole, and its fine structure (such aspects as sharp edges, and virtually 100 percent depletion in layers of 100 metres or so), strongly suggested a chemical cause, although the biennial variation in the amplitude of the hole indicated that atmospheric circulation also played a role. Recent aircraft and ground-based measurements have confirmed that chlorine chemistry, arising from the breakdown of chlorofluorocarbons present in the atmosphere, is the primary cause of the hole. These chemicals are used commercially in large quantities as

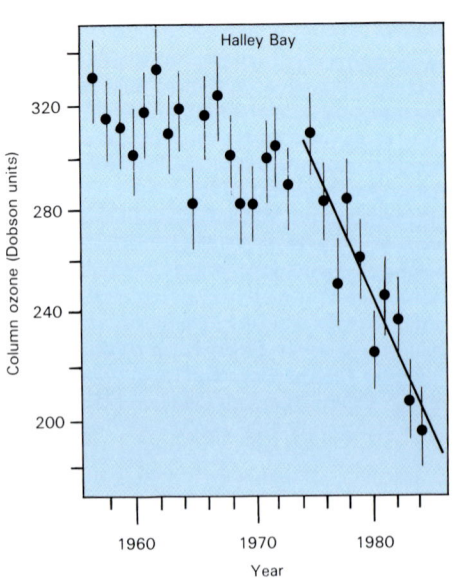

Figure 8.8. *Ozone concentrations in the Antarctic stratosphere remain relatively constant throughout winter, and then typically increase in late spring when the breakdown of the polar vortex allows the influx of fresh ozone-rich air from lower latitudes. Since the 1970s, however, a dramatic drop in ozone levels has been observed before the breakdown of the vortex.*

aerosol propellants, refrigerants, foam-plastic blowing agents, and solvents. They are relatively inert gases chemically, and so remain in the atmosphere for a long time, eventually reaching the stratosphere where, activated by sunlight, they react catalytically with ozone. It has been estimated that a 1 percent decline in global ozone levels could lead to a 4–6 percent rise in the incidence of some skin cancers through an increase in the amount of damaging ultraviolet radiation reaching the earth's surface.

Because Antarctica is sensitive to global changes in the atmosphere, there is a continual need to monitor this fragile environment. If we understand Antarctica's role as the "refrigerator" of the earth we may gain insight into future climatic changes. Records of past climates are contained in the ice layers of the Polar Plateau. Improved knowledge of daily weather variations is necessary to determine the viability of future Antarctic exploitation. We therefore need an ongoing commitment to meteorological research in Antarctica to improve our understanding of past climates and possible future climatic changes, as well as making it safer for those who work in the Ross Sea region.

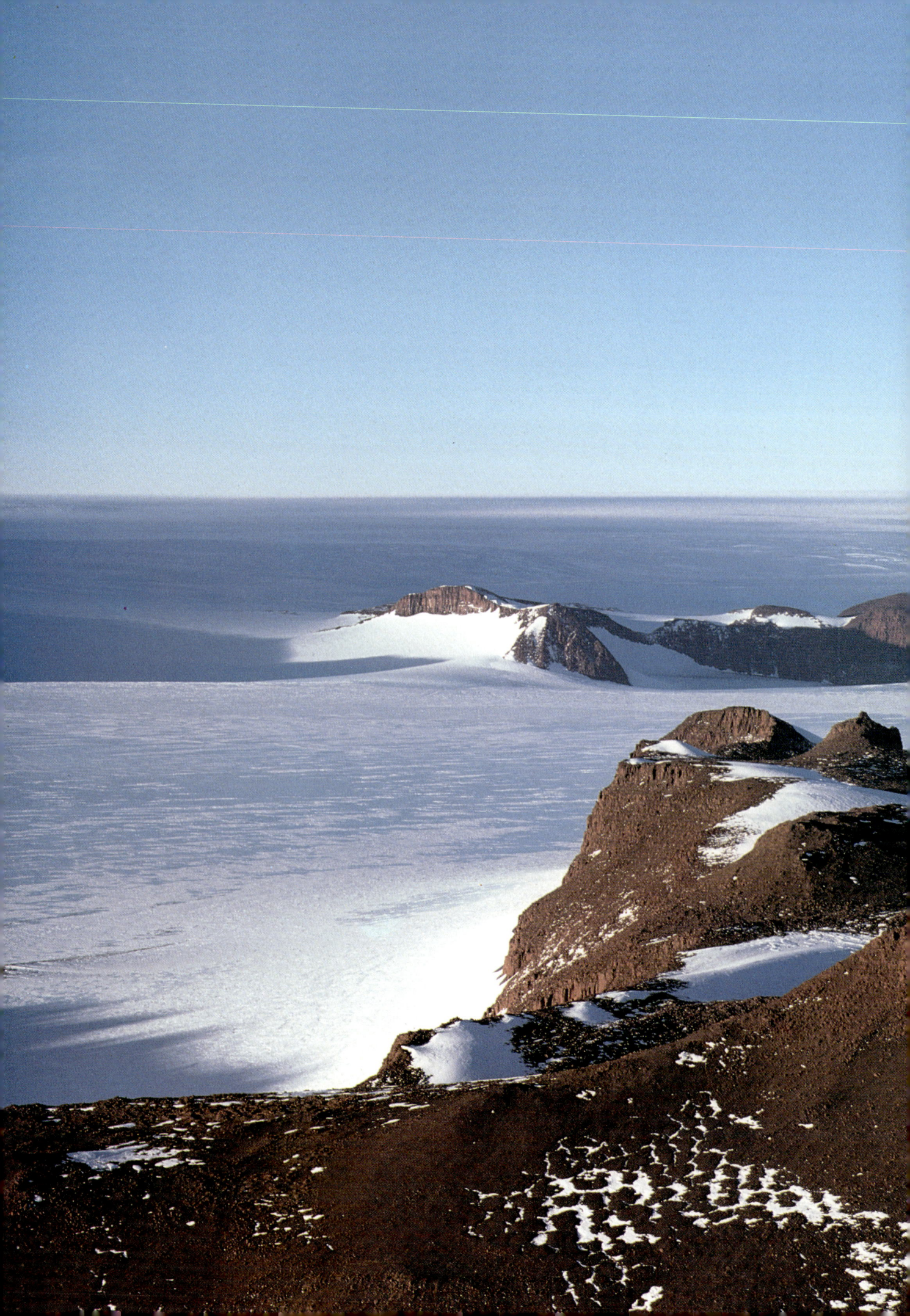

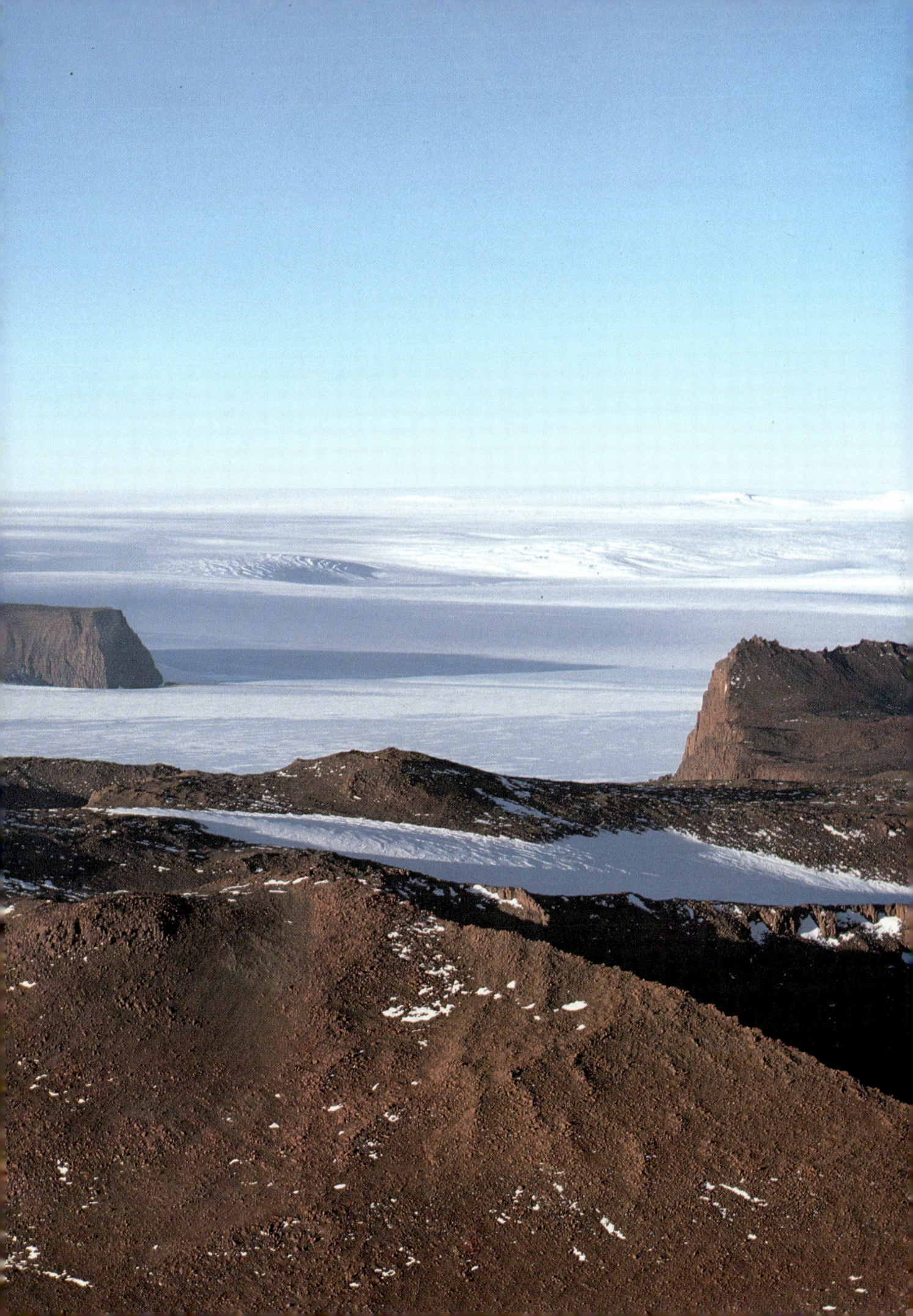

9. The ice forms

Plate 9.1. *Beyond the flat-topped Depot Nunatak the East Antarctic Ice Sheet stretches for thousands of kilometres, uninterrupted by rock. Photo: H. Keys.*

Ice in many distinctive forms gives the Antarctic its main character. There are two major types of this ice. Glacial ice is formed from accumulated snow, itself composed of crystals and grains of ice. Sea ice is formed from the freezing of sea water. Other forms of ice are found only locally and include water ice formed from liquid water (e.g., Lake Vanda, see Plate 10.3) and fumarole ice formed from geothermal steam (e.g., on Mounts Erebus and Melbourne, see Plate 6.3).

Glacial ice obviously dominates the land. An immense ice sheet up to 4500 kilometres wide, 12 million square kilometres in area, and nearly 30 million cubic kilometres in volume covers almost 98 percent of the continental land mass (Plate 9.1). Fallen and windblown snow accumulates on the surface and is compacted with increasing depth of burial. True glacial ice is formed at depths of about 40 metres near the coast and 160 metres inland. This ice moves under gravity generally outwards and downwards from the interior of Antarctica towards the coast. Movement is concentrated in ice streams, with margins of slower moving ice, and outlet valley glaciers (e.g., Byrd Glacier, see Plate 4.2) which flow through mountains. The fastest of these glaciers move at 1–10 metres per day, similar to the fastest moving glaciers in New Zealand. Highland ice and ice caps mantle high islands south of 60°S (e.g., Ross Island, Franklin Island, Balleny Islands) except for some summits and ridges where rock is exposed. Ice piedmonts (glaciers which cover low coastal land backed by mountains) are common in Victoria Land, whereas piedmont glaciers (lobe-shaped glaciers spreading out at the foot of mountains, Plate 9.2) are less common. Glaciers in cirques, alpine glaciers draining them, and small glaciers formed from snowdrifts are also widespread in the coastal ranges.

Glacial ice is also present offshore. The continental ice sheet is fringed by floating glacial ice over half of its perimeter in the form of ice shelves and other glaciers which add another 1.6 million square kilometres of ice. Icebergs calve (break) off the fronts of these glaciers to

Plate 9.2. *Piedmont glaciers formed from avalanche debris along the foot of the mountainous Adare Peninsula. Photo: R. Taylor.*

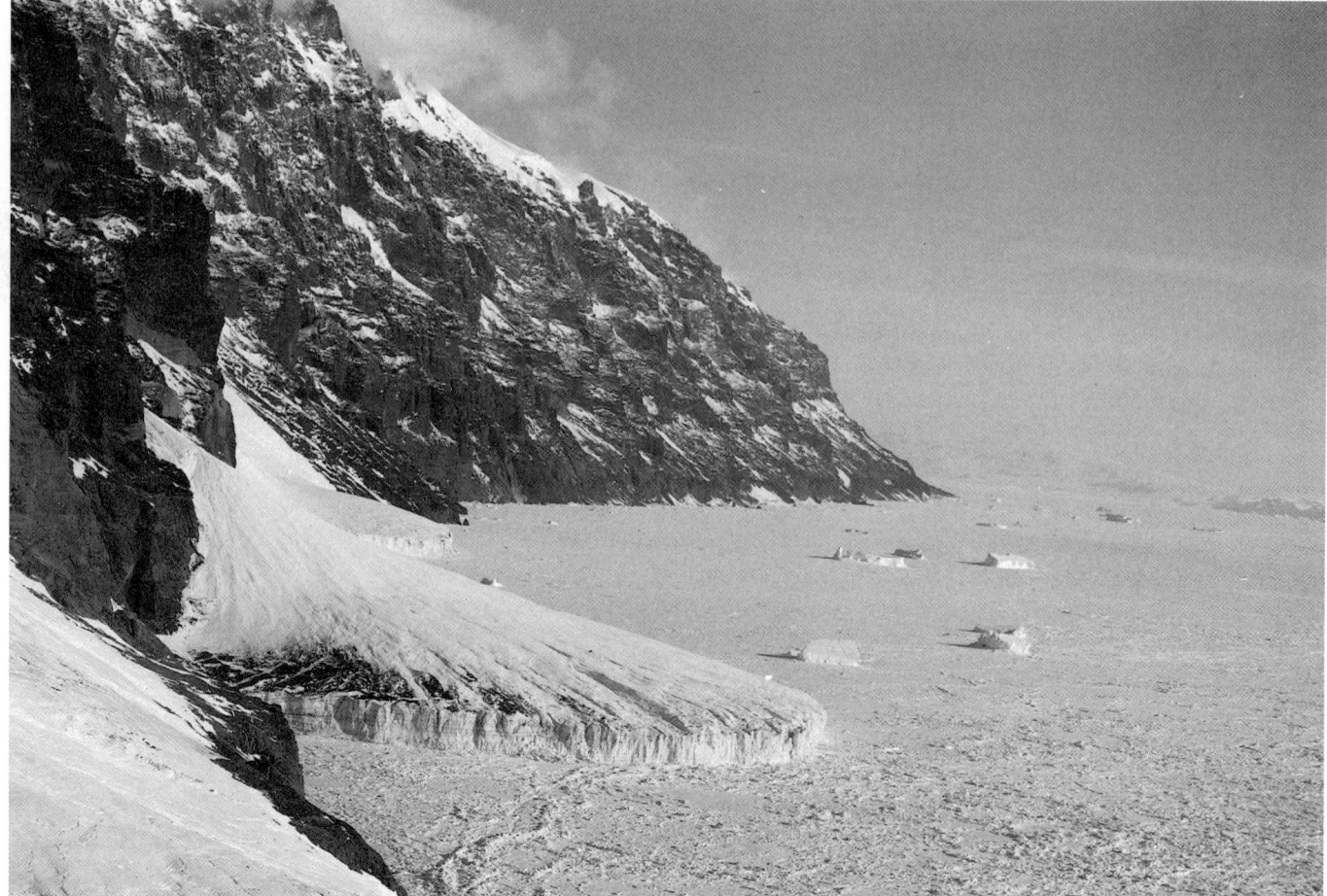

drift around the continent and north into the Southern Ocean as far as 30°S.

The Antarctic seas are dominated by sea ice which in winter more than doubles the area of Antarctic ice. At its maximum in late winter, the sea ice forms a belt 400–1900 kilometres wide from the coast of Antarctica to as far north as 53°S and extends over 20 million square kilometres of ocean. During spring, this ice recedes southwards to the coast in several places, to a minimum area averaging 3.6 million square kilometres in February. Most of this ice is pack ice, drifting slabs of ice usually less than 2 metres thick, though sometimes more than 30 kilometres across. Fast ice, which is sea ice attached to the coast, develops along most of the coastline during winter when it may extend over 100 kilometres offshore.

The Antarctic Ice Sheet

The Antarctic Ice Sheet, by far the largest body of ice on earth, is a dome-shaped mass of ice regarded as two coalesced ice sheets. The larger, thicker East Antarctic Ice Sheet, dammed to the west of the Ross Sea by the Transantarctic Mountains, is classified as terrestrial because its base rests on land which is mostly above sea level. The smaller, lower-elevation West Antarctic Ice Sheet, sited south-east of the Ross Sea, is a marine ice sheet because it is grounded mostly well below sea level. The ice reaches 4000 metres above sea level in East Antarctica, 2500 metres below sea level in West Antarctica, and is up to 4800 metres thick (Plate 9.3). The thickness and weight of the ice sheet are sufficient to cause zones to melt and sometimes even form lakes of water beneath this ice.

Because of its size, composition, and geographic position, the ice sheet is an important feature of the earth. It affects global energy and moisture budgets and, therefore, climate and sea level, as well as patterns of weather, sea ice, ocean currents, and life in the Southern Hemisphere.

At depth, the ice sheet contains a detailed record of past climate,

Plate 9.3. Radio echo-sounding profiles through the Antarctic Ice Sheet. Depth scales are shown on the left and 0 marks the ice sheet surface. (a) Rough mountainous bedrock indicates the base of the ice sheet. (b) A lake about 4 kilometres long can be seen at a depth of 3.5 kilometres below the ice surface. Photo: Scott Polar Research Institute.

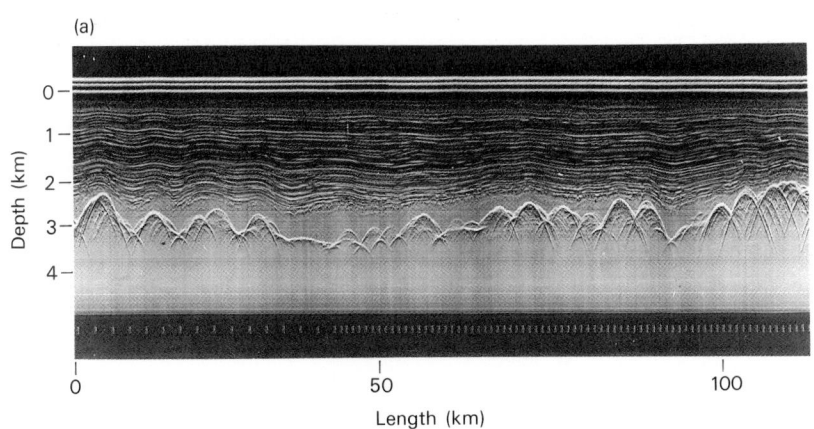

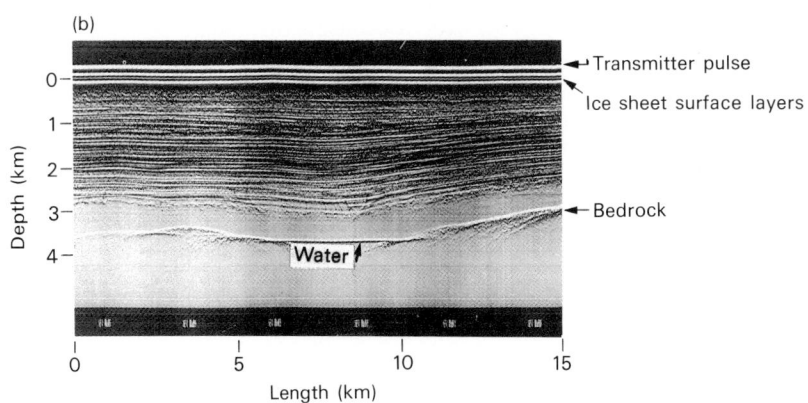

snowfall, air temperature, atmospheric gases and solids such as dust and volcanic products deposited over the last several hundred-thousand years. Measurement of impurities in the snow and ice (including methane, lead, and radioactive elements) provides a guide to the character, extent, and processes of global pollution. Ice plays an important role on our planet and Antarctica contains the only marine and terrestrial ice sheets of continental size, enabling us to study a wide range of processes that occurred before, during, and after world glaciations. Antarctica is also the largest terrestrial collector and preserver of extraterrestrial matter (e.g., meteorites and dust). Global sea levels are possibly the most spectacular linkage between Antarctica, the atmosphere, and the oceans. If the whole Antarctic Ice Sheet melted it would raise the sea level by about 60 metres, although such an event is probably most unlikely because of the terrestrial nature and slow response to change of most of the ice mass. The East Antarctic Ice Sheet is believed to have undergone extreme fluctuations over millions of years. The West Antarctic Ice Sheet, however, is regarded by some scientists as unstable, perhaps on time scales of a few hundred to tens of thousands of years because so much of its mass is below sea level.

We think that a future climatic warming, produced by an increased concentration of atmospheric gases causing a "greenhouse effect", will initiate a small rise in sea level mainly due to thermal expansion of the oceans. By 2100 A.D. this rise is expected to be almost 1 metre. A higher sea level would cause West Antarctic glacial ice to float further south than it does now. This may cause the ice sheet to discharge faster or retreat in those areas where it is resting on land far below sea level, depending on the ice flow and bottom topography of critical outlet glaciers. Once started, such a retreat may become irreversible — leading to the West Antarctic Ice Sheet disintegrating and sea level rising catastrophically by as much as 6 metres. Warmer sea and air temperatures could also accelerate the decay of ice shelves leading to a similar increase in discharge and disintegration.

However, the dynamics of such events or of counteracting processes, and the extent and time over which they might act are not clear. Bedrock scarps tens to hundreds of metres high at the heads of many outlet glaciers and ice streams may provide a stabilising buffer against the direct effects of any climatically-induced changes in sea level. Recent mathematical modelling, based on an increasingly precise understanding of glacial processes, suggests that an increased ice flow of West Antarctica would cause sea level to rise up to 1 metre 500 years after the fringing ice shelves disintegrate. However, a warmer Antarctic atmosphere would lead to a greater snowfall and storage of water as glacier ice on the Antarctic Continent, which would lower sea level. An increase of only 1 percent in the volume of the Antarctic Ice Sheet would cause a 0.7 metre drop in sea level. Therefore, the probability and time scale of such global sea level changes are still uncertain. There is currently debate as to whether or not increased discharge of ice has already started on the coast of the Amundsen Sea, to the east of the Ross Sea.

The vast size of Antarctica plus a sparsity of data on snow accumulation, glacier flow, and ice shelf wastage have made it difficult to accurately assess whether the Antarctic Ice Sheet is growing or shrinking. Calculations suggest that the total annual accumulation of snow and ice, about 2000 cubic kilometres (2 million million tonnes, or 1 percent of the amount of water present in all the world's lakes and rivers), is probably nearly balanced by the outflow and wastage with a possible discrepancy that would imply a growing ice sheet. However, recent work has suggested that the rate of iceberg formation may have

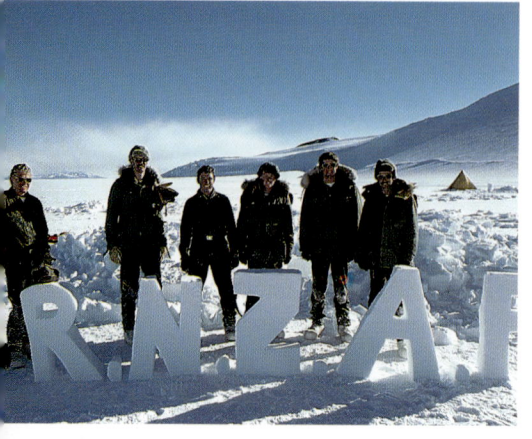

Plate 9.4. *Snow on ice shelves and the plateau is compacted by wind and can easily be cut and shaped. Photo: K. Williams.*

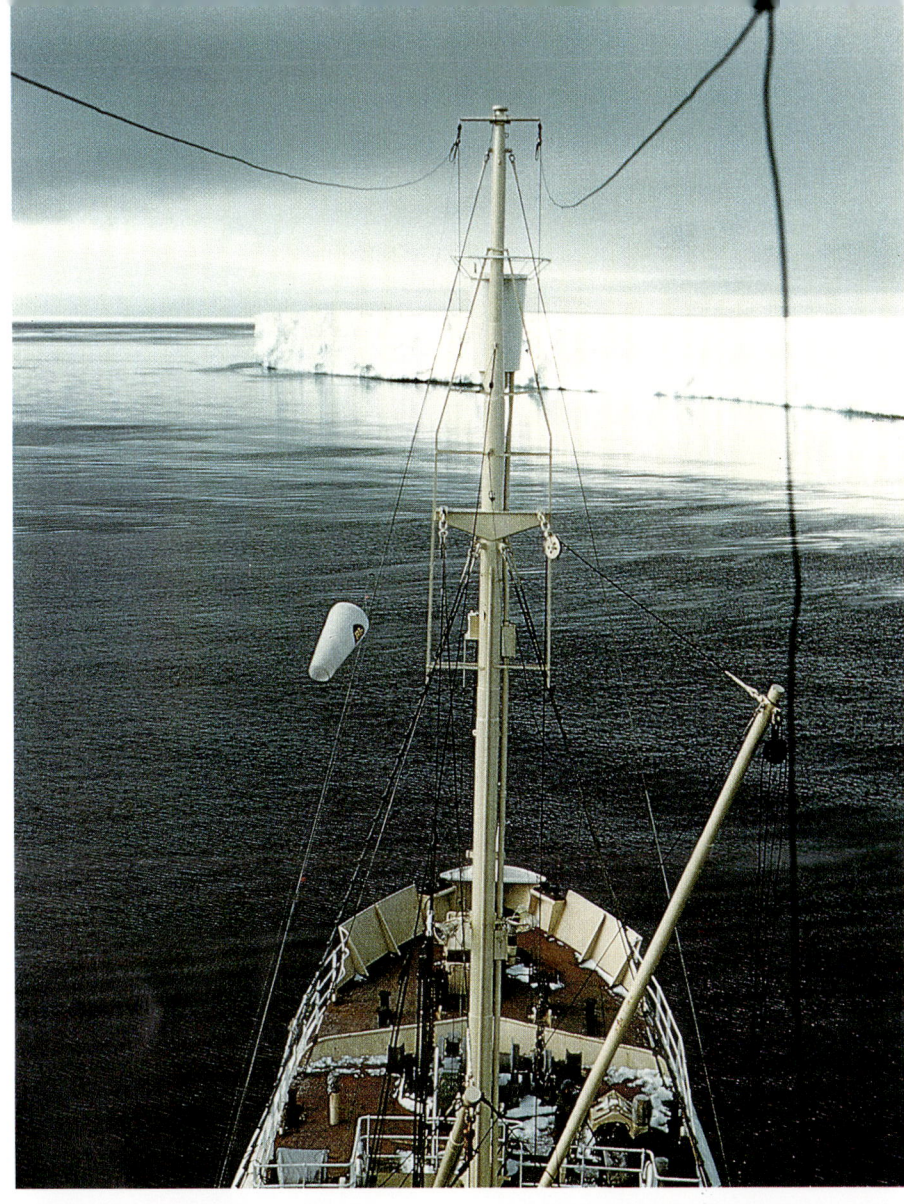

Plate 9.5. *The seaward edge of the "Great Ice Barrier" or Ross Ice Shelf. Photo: Antarctic Division, DSIR.*

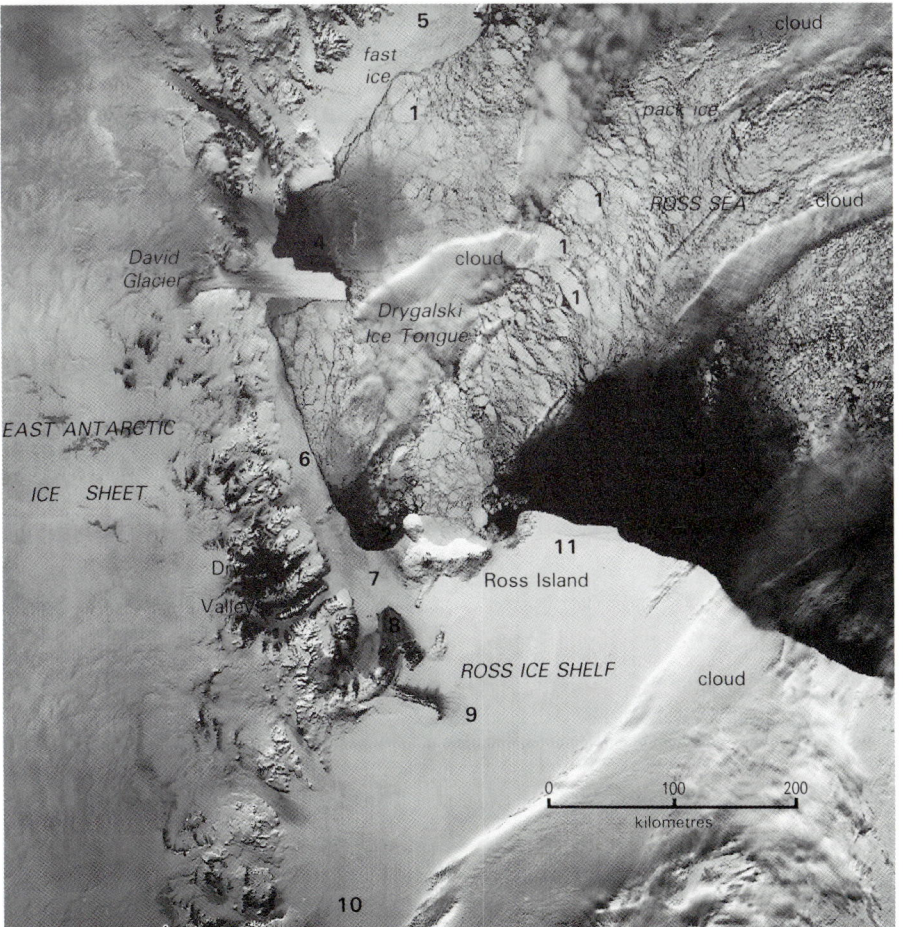

Plate 9.6. *Satellite photograph of the southern Ross Sea region in late November 1985, showing the main ice forms. Individual floes of sea ice, some more than 10 kilometres across (1), and dark leads between floes, constitute the ice pack which still covers most of the Ross Sea. The large dark areas within the pack are polynyas in McMurdo Sound (2), southern Ross Sea (3), and Terra Nova Bay (4). Fast ice extends across Lady Newnes Bay (5), off the coast of South Victoria Land (6), and covers most of McMurdo Sound (7) whose southern extremity is the McMurdo Ice Shelf (8). Large crevasses are visible in the ice shelf where it flows around Minna Bluff (9), where the Byrd Glacier thrusts into it (10), and near the ice front (11). Photo: U.S. National Science Foundation.*

been grossly underestimated. If so, Antarctica may be losing not gaining mass.

The Ross Ice Shelf

The Ross Ice Shelf is a flat-topped body of snow-covered glacier ice floating over most of its area but grounded along coastlines and over other shallow parts of the sea floor. It is the largest ice shelf on earth, covering 532 000 square kilometres (twice the area of New Zealand) and with a volume of 23 000 cubic kilometres. Because of its size it is believed to be important in restricting the discharge of the West Antarctic Ice Sheet. It also influences the marine ecosystem, sea currents, and sea ice distribution in the Ross Sea and is a major source of icebergs. Historically regarded as a barrier to human activity, the Ross Ice Shelf is now recognised as being logistically convenient for oversnow transport and landing aircraft.

The ice shelf is nourished by the inflow of glacier ice from the ice sheet, and by accumulation of snow and rime (a deposit of rough ice crystals formed from supercooled fog droplets below 0°C) measured at 0.1–0.3 metres per year water equivalent on its top surface (Plate 9.4). Seasonal stratifications in the compacting snow pack are normally conspicuous along the exposed front of the ice shelf and on crevasse walls. Driven by glacier flow of the feeder glaciers and its own weight, the ice spreads outwards over the sea at up to 1.1 kilometres per year. Wastage is mainly by calving from the ice front to produce an estimated 150 cubic kilometres of icebergs per year and by bottom melting at up to an estimated 2.5 metres per year, especially near the ice front. Loss by surface wastage, particularly wind action, is relatively small, but can look spectacular when blowing snow pours over the ice front.

The thickness of the Ross Ice Shelf is not uniform and its surface is not always flat. Broad intrusions of ice up to 1000 metres thick enter the shelf from West Antarctic ice streams and East Antarctic outlet glaciers,

Figure 9.1. *Ice thicknesses (in metres) and movements (arrows) in the McMurdo Ice Shelf area. The moraine patterns (dashed lines) show the extent of the "dirty ice" region. The dotted line (EL) is the boundary between the snow-covered and snow-free parts of the ice shelf. To the west of this line surface ablation exceeds the rate of snow accumulation whereas the opposite is the case to the east.*

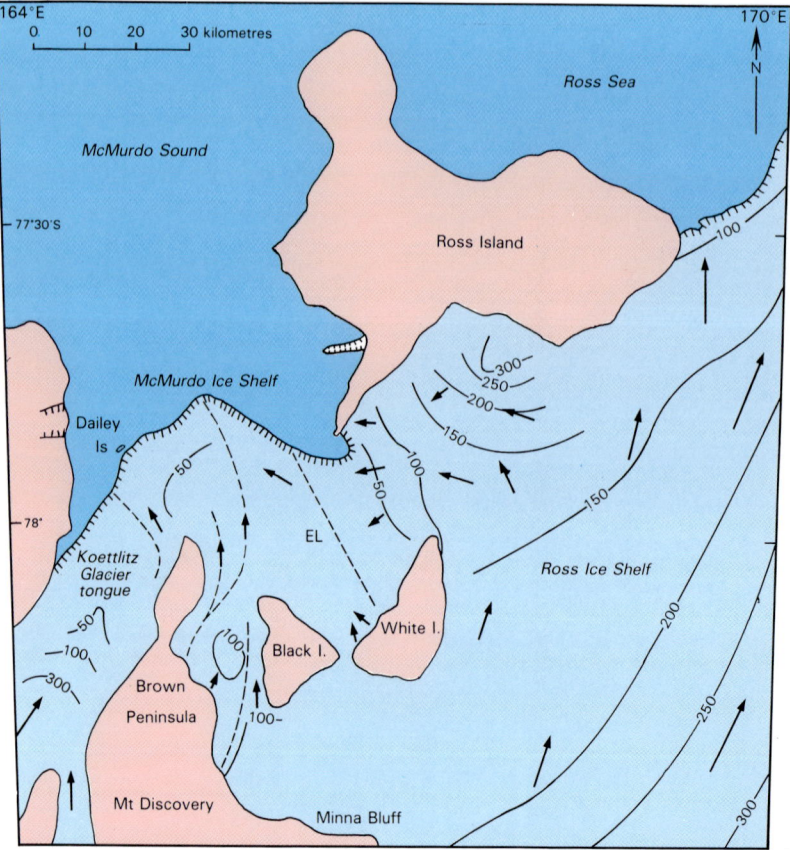

whereas the shelf thins to less than 100 metres at the ice front. Where the ice runs aground or ice streams merge, the ice tends to thicken. It may buckle as it deforms, to produce pressure rollers (ridges) with wavelengths of a few hundred metres or less and amplitudes of up to 15 metres and which are often crevassed. Other undulations or depressions with wavelengths as long as 10 kilometres and amplitudes less than 5 metres are widespread.

Roosevelt Island forces ice to flow around it, causing a thinner convergence zone downstream. In this zone, compression, rifting, and differential thinning cause probably the most complex topography on the shelf. However, large glaciers such as the Byrd also create complex crevassed topography where they enter the floating ice shelf.

The shelf is thinnest along the ice front (Plate 9.5) but here also thickness is variable. The ice front is presently less than 100 metres thick west of 173°E and similar thicknesses have been recorded north of Roosevelt Island at 165°W. The ice front is thickest in the central portion between 178°E and 167°W, but even here it may not be much thicker than 100 metres at times. The height of the ice front above sea level varies from a few metres to about 50 metres reflecting ice front thickness, ice shelf density, and hydrostatic forces.

Although the general trend of its coastline has been maintained, the position of the ice front has moved north over recent years. The ice front reached its most northern point (77°10'S) in recorded history (i.e., since 1841) in longitude 170°30'E. A bay-like feature (the Bay of Whales) persisted for most of this period at 165°W, north of Roosevelt Island. Major calving, when hundreds of cubic kilometres of ice shelf break off, is probably episodic and separated by intervals of several decades. For instance, in late September–early October 1987, a huge iceberg 154 kilometres long by 35 kilometres wide, termed B-9, broke off the eastern front of the ice shelf. The "coastline" of this part of the Antarctic changed significantly with this calving event. Clearly, the ice front does

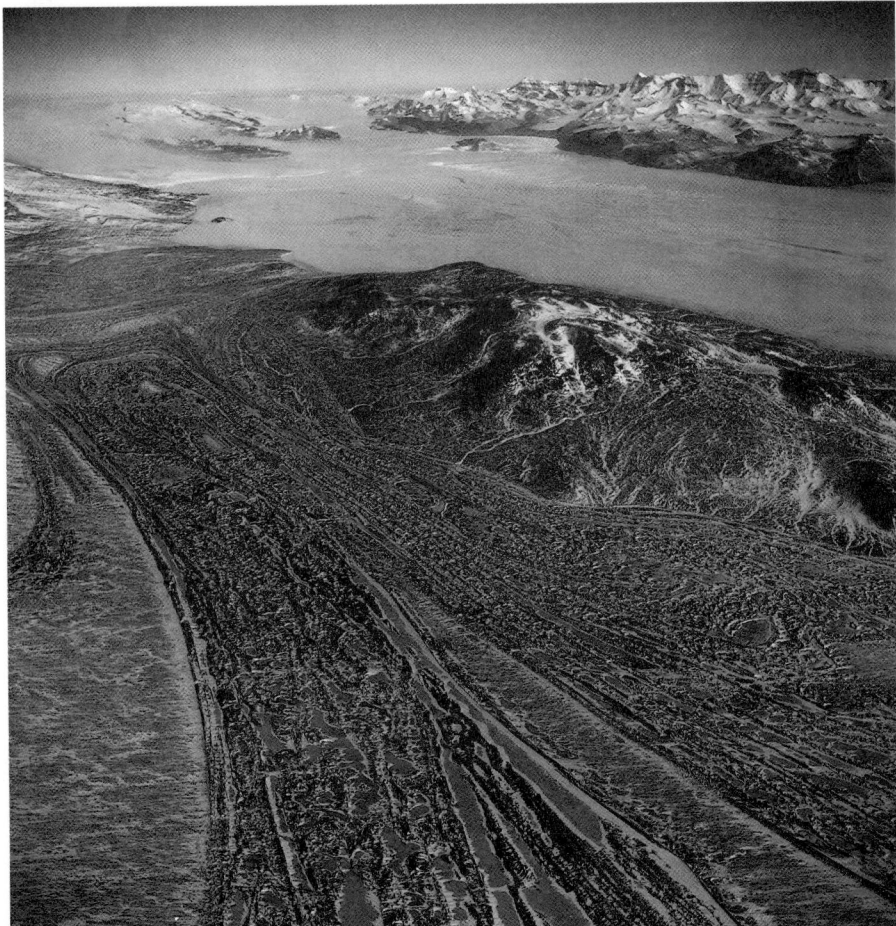

Plate 9.7. *Part of the McMurdo Ice Shelf, with the Koettlitz Glacier in the distance, showing the "dirty ice" region with its elongated melt pools on the eastern side of Brown Peninsula. Photo: U.S. Navy.*

not maintain a precise position.

However, minor calving occurs more frequently and relative advance rates suggest that this has been most prevalent west of 178°E. About 0.3 kilometres per year appears to calve off this portion of the ice front within 200 kilometres of Ross Island, which represents less than 10 cubic kilometres per year, but probably hundreds of icebergs.

Long linear features (Plate 9.6) run nearly parallel to the ice front in this area and show where future large icebergs will be formed. They could be an indication that major calving is relatively imminent in this, possibly over-thinned, portion of the ice shelf. Unusual activity recorded on seismographs has been attributed to "icequakes", caused by the calving of icebergs.

The state of equilibrium, mass balance, and seaward extent of ice shelves and other floating glaciers are influenced by a variety of glacial and non-glacial processes. These act over periods of up to a few hours (e.g., storm surges, waves, and tides) to many tens (e.g., calving, snowfall), or even tens of thousands of years (e.g., past temperatures and accumulation rates, sea level change, rising of the sea floor). Changes in the position and geometry of the grounding line, in ice thickness, and in the position of the seaward margin, all reflect a complex inter-related set of processes. Points of seaward protection or "anchoring", such as islands, headlands, or shoal areas, seem important for determining the seaward extent, particularly of some small ice shelves and fjord glaciers. Lateral stretching and thinning of the ice beyond these anchoring points leads to calving of icebergs.

McMurdo Ice Shelf–Koettlitz Glacier Complex

The floating glacier system that occupies the southern part of McMurdo Sound (Figure 9.1) is an unusual and complex feature composed of different ice masses. The eastern part of the McMurdo Ice Shelf is formed from snow accumulating on ice flowing westwards between Ross and White Islands from the Ross Ice Shelf. Much of this ice is removed by melting (up to 3 metres per year) at its base. There is little distributary flow from the Ross Ice Shelf between Mount Discovery and Black Island and negligible flow between Black and White Islands. In the west, the Koettlitz Glacier flows down from its upper part in the Royal Society Range out to McMurdo Sound. Surface ablation of its floating tongue (at about 0.4 metres per year) is compensated by growth of sea ice on the underside of the tongue. The continuation of this process causes the entire tongue north of Garwood Valley (and possibly even north of Miers Valley) to be composed of ice formed initially from sea water and intensely modified by ablation, including melting.

The central portion of the glacier complex, between the Dailey Islands and Black Island, also seems to be formed from frozen sea water. This central zone protrudes farthest into the sound because it is thickest, and is clearly marked by moraine patterns on the surface (Figure 9.1). Apparently, ice in the central zone is formed in the tide cracks along the shores of Brown Peninsula, Mount Discovery, Bratina and Black Islands, in the water column or on the sea floor ("anchor ice") beneath the ice shelf, or directly on to the bottom of the shelf. Fresh or brackish water draining off the ice shelf through holes in it may also contribute after it freezes.

Widespread deposits of fossiliferous sediments, marine fauna, and non-marine unicellular algae (diatoms) occur on the surface of the central ice shelf. Sediments and marine fauna and flora (such as sponges, shells, fish, and diatoms) are thought to be incorporated mainly by anchor ice or other freezing on to the base of the ice shelf in shallow areas where it is nearly aground or just touching the sea floor. As anchor ice rises from the sea floor in such areas, it carries rock debris and organisms into the layer of platelet ice beneath the shelf and thence on to the base of the shelf. Surface ablation eventually brings the rock and biotic material to the surface of the ice shelf where it accumulates. In addition, non-marine diatoms are abundant in surface melt ponds and sediments, some of which may have been blown from nearby ice-free land.

The result is a mass of ice isotopically similar to sea water but containing some dirt and biological material concentrated in bands or patches on the surface. This ice flows north with a curve towards the west as it is pushed from the south and east.

Extensive calving of the ice front occurs followed by freezing of sea water and, hence, growth *in situ* of new ice shelf. In 1947 the ice front between Dailey and Ross Islands curved south to the latitude almost of Bratina Island and less than 10 kilometres from it. Within a few years the ice front was again well to the north as new ice shelf had replaced what was sea ice and open water in 1947.

Lakes and ponds of meltwater up to 45 kilometres inland from the ice front, and aligned meltwater pools (Plate 9.7), add to the complexity of ice formation. In early summer, serried ranks of ice pinnacles rising up to 12 metres over smooth lakes of ice are the culmination of intense differential ablation south towards Mount Discovery. Prevailing southerly winds, low in moisture and warmed slightly while descending over Mount Discovery, probably accelerate ablation in this area and elongate meltwater features downwind.

Ice velocities and thicknesses are much less than on the Ross Ice Shelf. Along the 9–20-metre-thick ice front, speeds range from 100 metres per year in the east to only 5–10 metres per year in the west. Recent calving of the eastern part has been between 100 and 1700 metres per year, on average faster than ice velocities there. The western ice front has been calving back over recent years also and now is only just reaching the Dailey Islands. A rapid retreat might normally be expected

Plate 9.8. *The grounded ice wall at the snout of the Barne Glacier on Ross Island. Photo: H. Keys.*

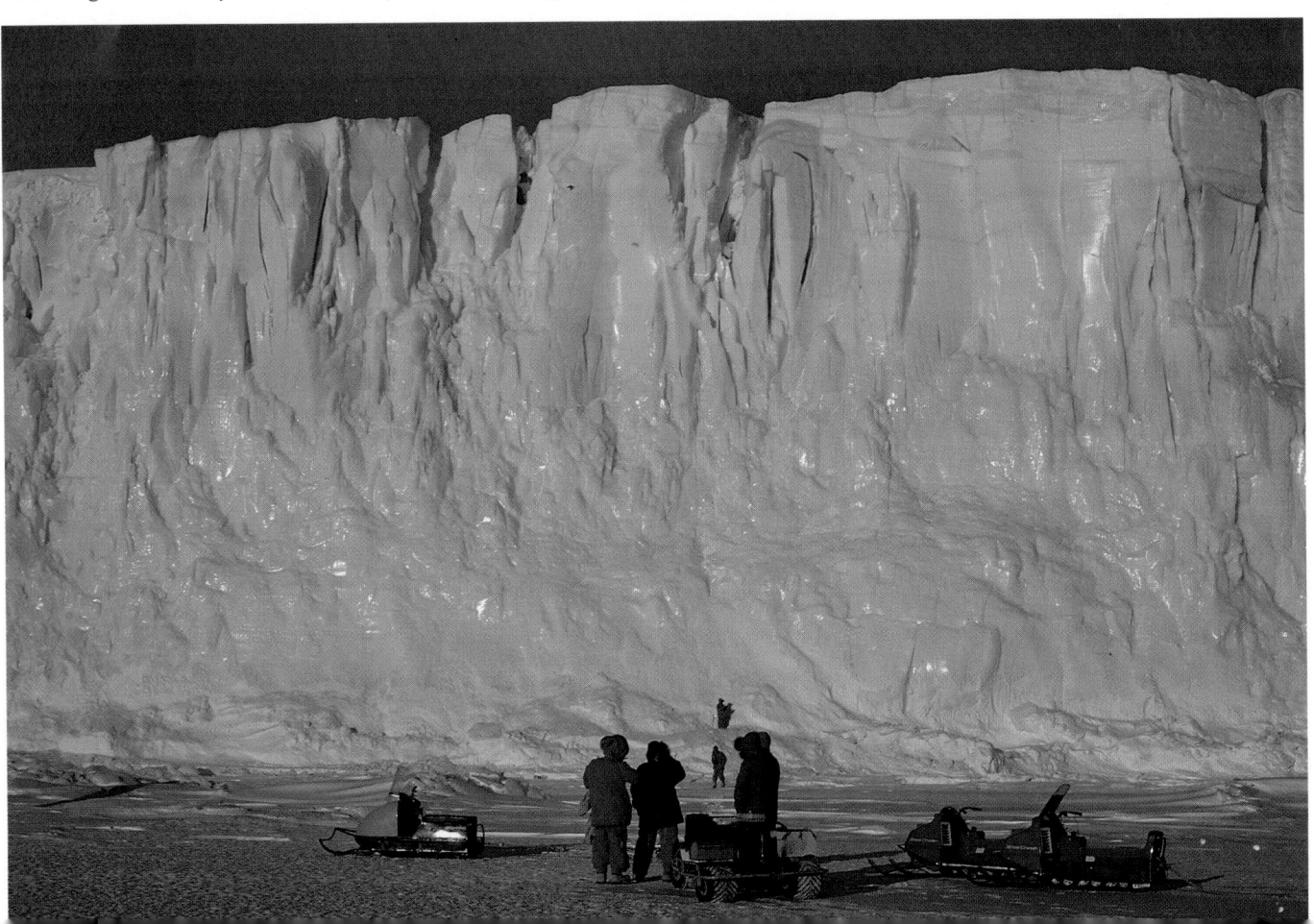

to follow the final severing of the anchoring linkage to Dailey Islands. However, the ice may be sufficiently thick up-flow of the islands to prevent extensive calving there. Nevertheless, this thin, slow moving ice shelf is probably in a delicate balance with its environment and may be useful in monitoring any effects of climatic warming in the region.

The Nansen Ice Shelf in Terra Nova Bay is also a complex but little studied floating glacier. This 1000 square kilometres of ice shelf is erroneously referred to on maps as an ice sheet but is fed by the Reeves and Priestley Glaciers which drain part of the East Antarctic Ice Sheet. Much of the surface of the ice shelf is ablating due partly to strong winds that persist in the area, but only between Inexpressible Island and the Northern Foothills are there significant amounts of surficial moraine. The ice front south of Inexpressible Island is moving at 100–400 metres per year and has partly recovered the 180 square kilometres of ice shelf lost in a large calving which occurred between 1957 and 1973.

Glacier tongues

Glacier (or ice) tongues are protrusions of floating glacier ice formed usually by ice streams or glaciers converging at the coast and discharging into the sea. These tongues extend seaward because discharge of ice across their landward grounding line (the junction between the grounded feeder glacier and the floating ice) is faster than the rate at which icebergs break off. The tongues thin progressively seaward of the grounding line and towards the sides of the tongue. Wastage is mainly by calving and bottom melting. Some glacier tongues are actually partly grounded on the sea floor and may be crevassed. Other glaciers that reach the sea do not extend past their valley sides (Plate 9.8) or rarely even past the high tide level, because their rate of decay (due to the combined rates of wave erosion, calving of ice, melting below the waterline, and surface ablation) equals the forward motion of the ice.

Plate 9.9. *Drygalski Ice Tongue extending into the western Ross Sea. Photo: C. Rudge, Antarctic Division, DSIR.*

The Drygalski Ice Tongue (Plate 9.9) is the longest tongue in the Ross Sea region. This 70 kilometre-long, 20 kilometre-wide tongue has continued to advance into the western Ross Sea in recent years at a rate similar to that of its ice, i.e., several hundred metres per year. The tongue blocks northward-moving sea ice from replacing that blown by katabatic winds out of Terra Nova Bay immediately to the north. This creates a stable, recurring area of thinner pack ice or open water (polynya) in Terra Nova Bay (Plate 9.6). In addition, the continuing westwards advance of the tongue could be increasing the amount of sea ice and its persistence south towards McMurdo Sound.

The Erebus Glacier Tongue (Plate 9.10) in McMurdo Sound is also growing longer. Its most recent calving was in the early 1940s since when it has lengthened at a rate similar to the velocity of ice at its snout (150 metres per year). Presently it is 14 kilometres long and tapers from 300 metres to 50 metres thick. This tongue has been intensively studied for its response to ocean waves; it is thought that wave-induced vibration may be a cause of calving and that a major calving is overdue.

In contrast, the 5 kilometres long by 3 kilometres wide, snow-free Mackay Glacier Tongue in Granite Harbour was shortened in February 1983 when about 3 square kilometres of its seaward portion broke off. One large 3-kilometre-long iceberg and several smaller ones were formed. The forward speed of the ice in the Mackay Tongue is about 300 metres per year, which would replace the lost portion in about 10 years. Aerial photos show that a previous major calving of the tongue occurred in the 1950s.

Rarely, a protruding tongue composed of icebergs rather than a coherent mass of ice is formed where a glacier discharges into the sea. Icebergs in such tongues are held in place by grounding on the sea floor

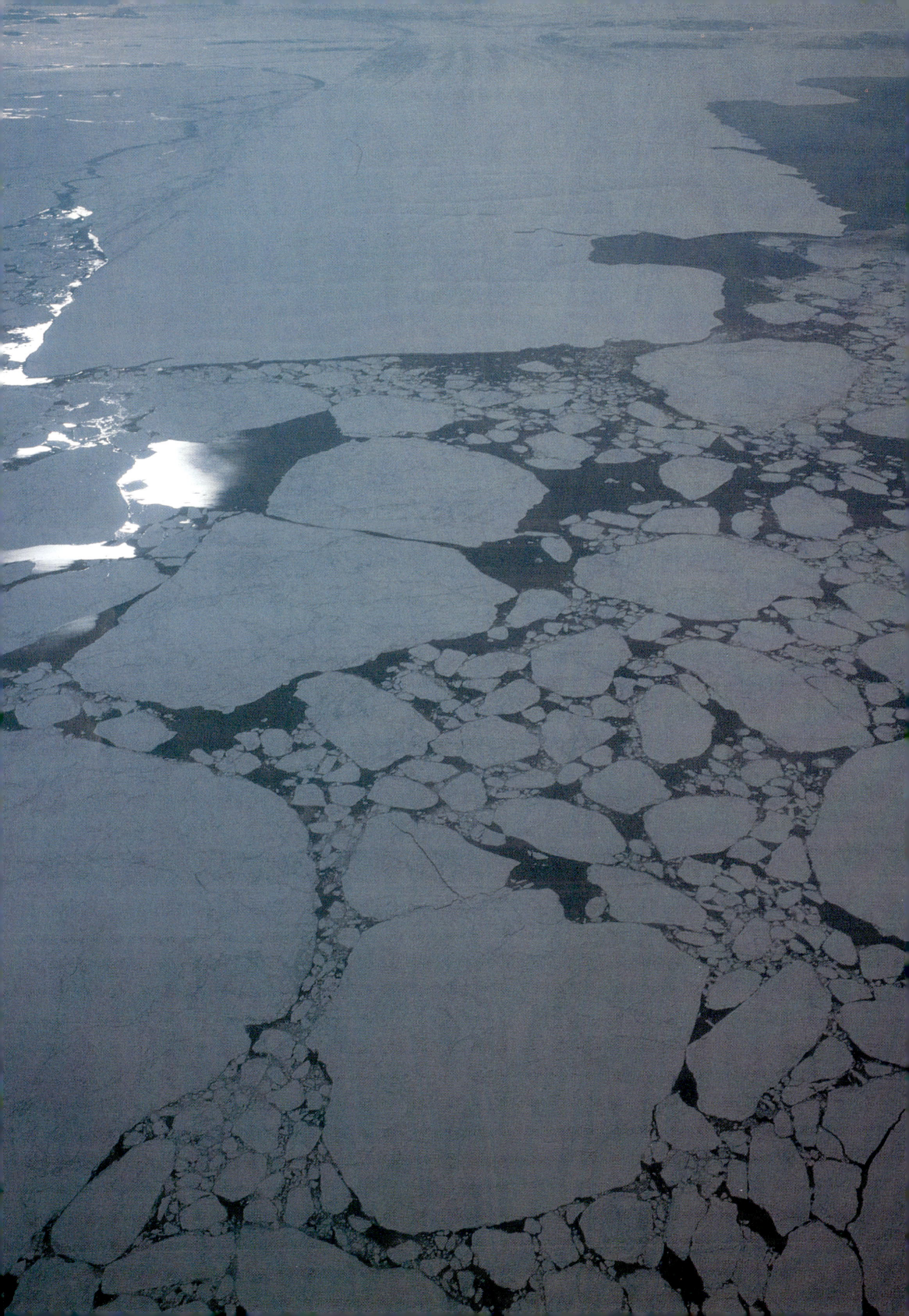

and by fast sea ice. Some of the icebergs in such tongues may have been derived elsewhere and drifted into the area before grounding at the tongue.

An iceberg tongue 2 kilometres wide and up to 1 kilometre long is located between the Erebus Glacier Tongue and Turks Head in eastern McMurdo Sound. Smaller pieces of glacial ice and old sea ice accumulate in the area between the iceberg tongue and the glacier tongue but are probably not grounded there. Aerial photos suggest that the iceberg tongue has been present for at least 30 years but has varied in size in that time. It presently contains about 50 icebergs longer than about 30 metres and some have been there for long periods. Four new icebergs were added to the tongue due to calving at the hinge zone between December 1987 and December 1988.

Icebergs

Icebergs are masses of ice that have broken away from ice shelves and other marine glaciers. They have a wide variety of shapes and sizes (Plates 9.11, 9.12), depending on their source and how they have decayed, and may be floating or grounded on the sea floor. The two largest icebergs recorded in or near the Ross Sea were respectively 154 kilometres long by 35 kilometres wide (when it calved from the Ross Ice Shelf in 1987) and 120 kilometres long and 70 kilometres wide. This second iceberg was seen about 200 kilometres west of Scott Island in 1956. Such large bergs have a mass of about 1 million million tonnes and contain enough water to fill New Zealand's Lake Taupo about 40 times. The median width of all Antarctic icebergs is probably less than 100 metres, although tabular icebergs are larger. Some icebergs in the Ross Sea may be thicker than 300 metres but icebergs near the coast are much thinner. Contrary to popular belief, the above-water portion of icebergs can be more than one-half the thickness of the submerged portion when an iceberg is pinnacle-shaped or composed mostly of snow

Plate 9.10. *Erebus Glacier Tongue in 1964, when it was 11 kilometres long. The saw-tooth pattern at its sides is caused by the ice relaxing as it emerges from its confining valley. Photo: U.S. Navy.*

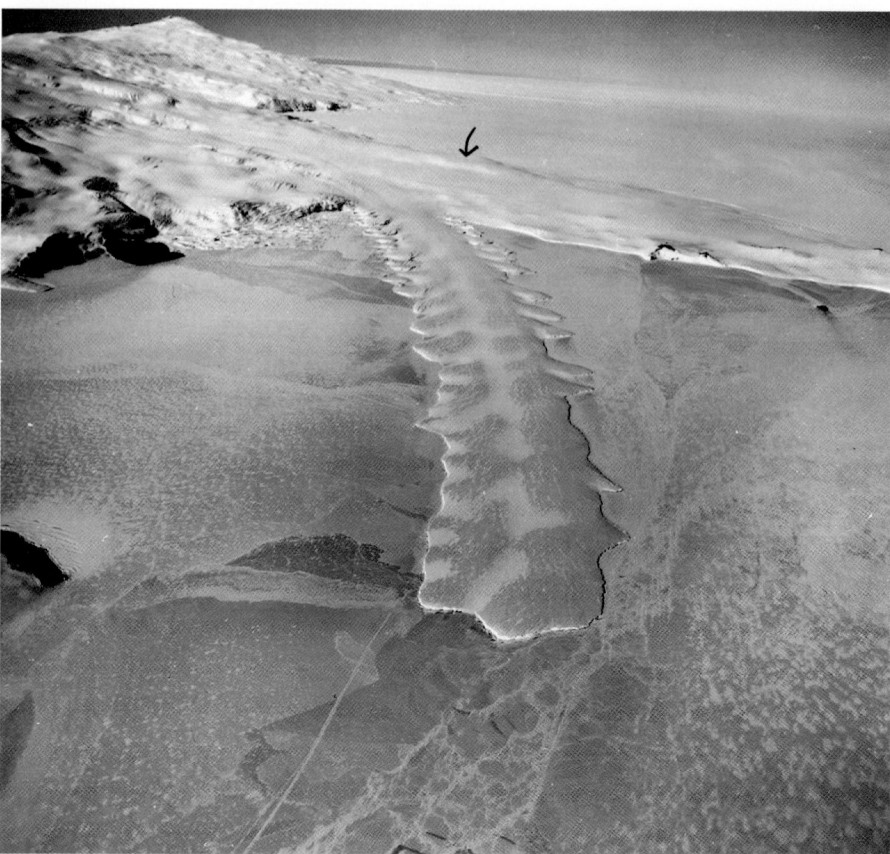

and firn (snow partly consolidated by alternate thawing and freezing), but little ice. More normally about one-third to one-sixth of Antarctic icebergs protrudes above water, depending on their density and shape.

Icebergs, the largest floating objects, are a serious hazard to shipping. Several ships have been sunk, damaged, or threatened by Antarctic icebergs in southern waters. Speculation persists about their possible use as sources of fresh water. Icebergs would be expected to increase in number after a climatic warming or some other event grossly affecting the Ross Ice Shelf.

Icebergs are calved ("berged") by different mechanisms. These are not completely understood but are believed to be affected by a variety of processes: weaknesses in ice shelves or glacier tongues including surface and bottom crevasses and hinge lines; lateral and vertical stresses developed during glacial thinning, flotation, vertical vibration, horizontal bending, and the impact of colliding icebergs; ocean swells, storm surges, tsunamis, and tides; and seafloor or shore topography. The giant iceberg B-9 broke off along a giant chasm which in 1971 was 100 kilometres long and up to 5 kilometres wide. Most icebergs (perhaps 60–80 percent by volume) separate from the flat-topped ice shelves, and the Ross Ice Shelf may produce about 10 percent of all Antarctic icebergs. Tabular icebergs, therefore, are common.

About one-third of the icebergs that run aground or otherwise get trapped in fast ice near the coast of south-west Ross Sea are tabular. Of the remaining two-thirds, one-half are irregular-shaped and one-half are rounded and water-worn. One-half of these tabular bergs are featureless—they have no distinguishing upper surface that could be used to trace their source. However, many of these featureless tabular bergs are less than 100 metres thick, similar to the thickness of the front of the Ross Ice Shelf west of 173°E. The composition and density of icebergs and ice shelf are also similar and a wind-blown dirt layer in one iceberg is consistent with a volcanic-dominated dirt mixture from the eastern

Plate 9.11. *A tabular berg about 40 metres high and 900 metres long floating amongst open pack ice. Photo: K. Williams.*

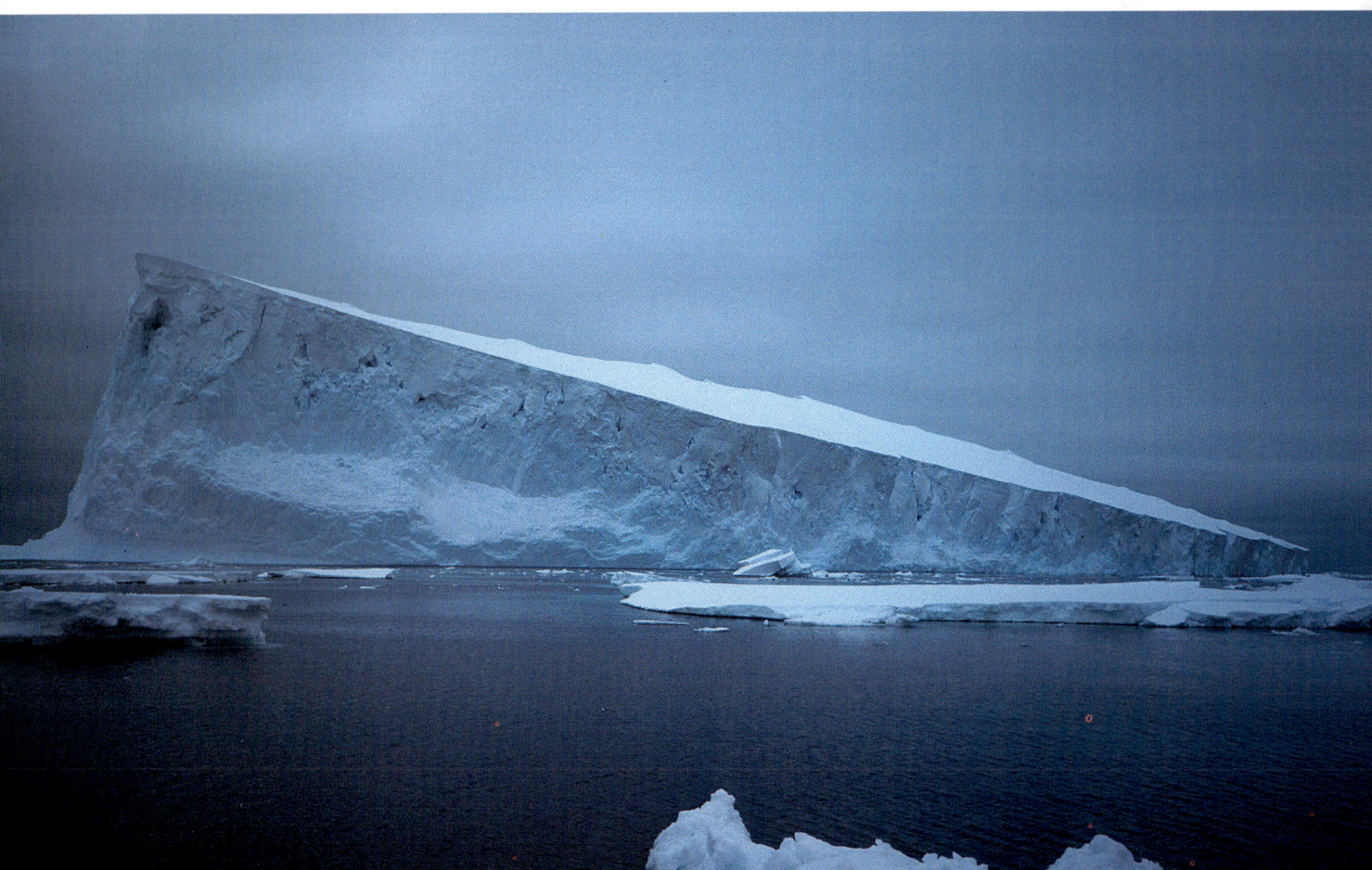

part of Ross Island. In recent years calving of the ice shelf has been most prevalent west of 178°E. Thus, evidence suggests that most of the featureless tabular icebergs probably come from the Ross Ice Shelf west of 173°E. The combined volume of all the featureless tabular icebergs and irregular-shaped icebergs also thought to come from western Ross Ice Shelf is less than 0.1 cubic kilometres. This is less than 1 percent of the 10 cubic kilometres thought to be calved annually from this part of the Ross Ice Shelf. As there are up to 500 icebergs trapped annually in the 4000 square kilometres of fast ice north of the McMurdo Ice Shelf, there must be several tens of thousands which do not get trapped. Thus, we can assume that most icebergs drift with the pack ice rather than getting trapped in fast ice and staying aground for months or years.

Other tabular icebergs trapped near the coast have uneven upper surfaces and many of these are composed of stratified firn and ice without snow. Most of these icebergs calved from glacier tongues, piedmont, and other marine glaciers along the Victoria Land coast.

Icebergs are frequently encountered in Antarctic seas especially near the coast. Iceberg distribution is complex, varying greatly in space and time due to different sources, varying rates of production, variable sea currents, sea floor topography, and seasonal changes in sea ice and climate. In the Ross Sea, in general, the average concentration of icebergs increases south of Coulman Island and westwards towards the coast of Victoria Land. However, concentrations are variable in the western Ross Sea and can range within a few days from none to clusters with over 20 icebergs per 1000 square kilometres.

A model to explain this distribution would have icebergs produced from the Ross Ice Shelf and Victoria Land marine glaciers and transported westwards and northwards respectively in the clockwise Ross Sea current gyre (a great closed circular motion of water) (Figure 9.2). Mixing and possibly some accumulation of icebergs would occur in the west from Ross Island to the Drygalski Ice Tongue, depending on local

Plate 9.12. *Tilting of this iceberg reveals fine ice forms previously under water. Below a once vertical icicle can be seen a broad groove which was the original waterline of the berg. Below this are once vertical flutings or narrow grooves caused by bubbles escaping from the glacier ice as it slowly melted in the sea. Photo: H. Keys.*

Figure 9.2. *The main paths of icebergs drifting in the Ross Sea, indicated by the colour-shaded areas.*

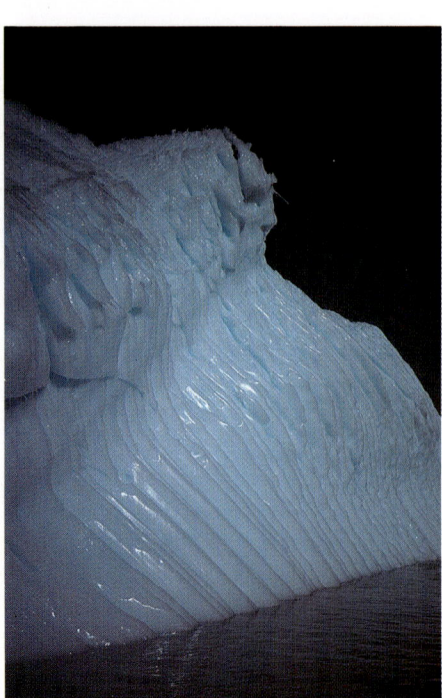

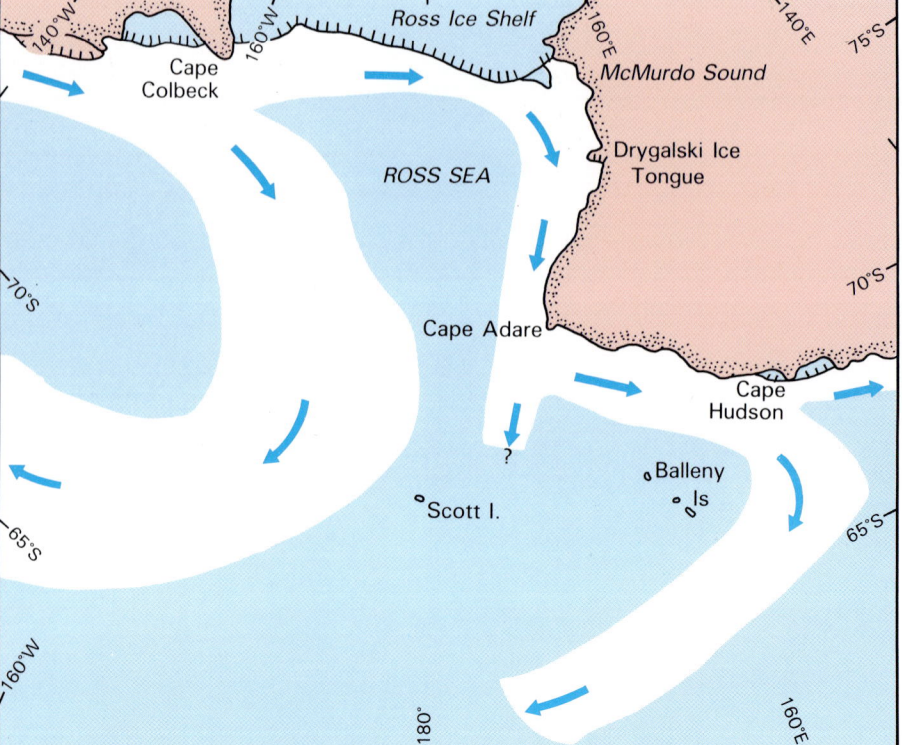

current patterns. Sporadic bursts of iceberg production, or either breakage or accumulation in eddies could explain the "clusters" of icebergs which appear in the pack ice in the western Ross Sea. Onshore currents and current eddies together with suitably oriented topography of the shallow sea floor cause icebergs to become trapped in specific and predictable places in the fast ice strip.

There are several places around the continent where drifting icebergs turn north from the coastal zone. In the general Ross Sea area the best defined of these appears to be north of Cape Colbeck on the east side of the Ross Sea (Figure 9.2), where the continental shelf break starts trending north-west and so subdivides the coastal, East Wind Drift current. Consequently, relatively few icebergs seem to enter the Ross Sea from the east, most instead are carried northwards at speeds about half that of the current or up to 0.5 kilometres per hour. Thereafter, they are carried into the great Antarctic Circumpolar Current (West Wind Drift) with sinuous but generally eastward drifts. A similar north-turning zone is located north of Cape Hudson, west of Cape Adare on the Oates Coast (Figure 9.2). Capes Colbeck and Hudson may be places where iceberg populations are subdivided by regional current patterns and those icebergs driven by local and tidal currents are mainly the only ones to mix across them.

The extreme northern limit reached by icebergs in historical times (about 200 years) is approximately 40°S in the Pacific Ocean near the Subtropical Convergence. Furrows gouged on the Chatham Rise east of New Zealand by grounding icebergs imply that icebergs were common in New Zealand waters during the Pleistocene Glaciation. There were several sightings of icebergs, including largish ones, in New Zealand waters in the 1890s. And in 1931 an iceberg reached the South Otago coast. However, icebergs cannot survive long in the temperate ocean north of the present Antarctic Convergence or even north of the sea ice belt. Therefore, relatively few icebergs have been sighted in recent times

Plate 9.13. Gotland II, *a Federal Republic of Germany expedition vessel, was trapped and crushed between coastal fast-ice (left) and loose pack-ice (right) when an 8 kilometre-wide polynya of open water closed in rapidly. The ship sank after 2 days, the crew being ferried to land by the expedition helicopters. Photo: F. Tessensohn.*

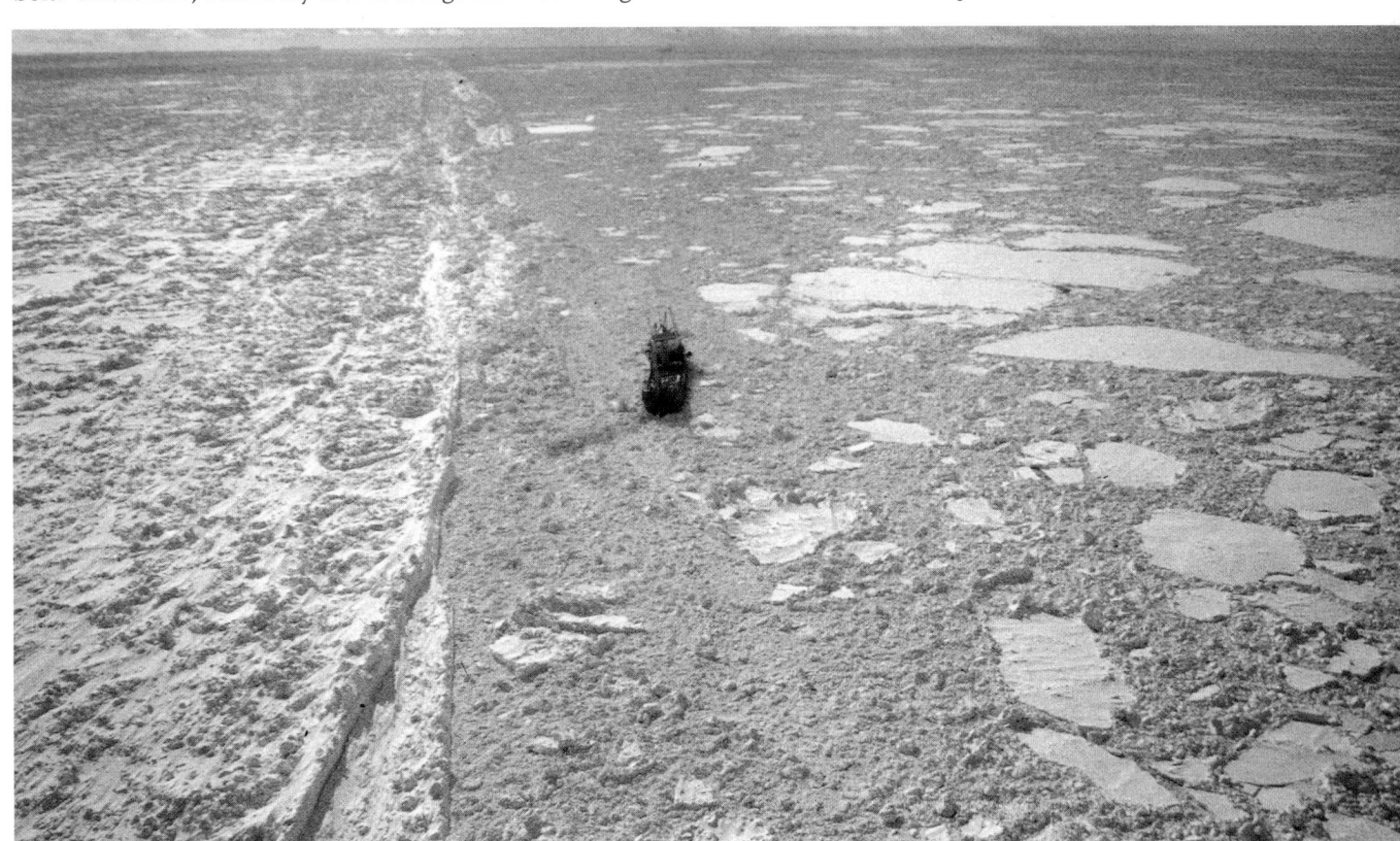

north of about 55°S in the Pacific Ocean.

Icebergs decay mainly by breaking apart, by wave action at the waterline, by calving off the sides above the waterline, and by melting below the waterline. Melting above the waterline is negligible. Observations near the coast of south-west Ross Sea suggest that breakage is caused mainly by stresses developed during grounding on the sea floor, collisions with other icebergs, and upward buoyancy on underwater portions. Offshore, ocean waves and buoyancy forces are probably more important. Splitting is probably the principal mechanism of decay, followed by melting below the waterline although ablation at the waterline is often spectacular. Melting is enhanced by roll-over (Plate 9.12) and wallowing of icebergs, strong currents, and higher sea temperatures. Iceberg melt rates, which have been estimated to be about 20 metres per year in sea water of –1°C, are therefore much faster than basal melt rates of ice shelves in water of the same temperature.

The life expectancy of icebergs depends on their initial size, and on the mechanisms and rates of decay. Tabular icebergs have been tracked using satellites for up to 2 years before they either rolled over, broke up, or radio transmitters placed on them stopped working. Icebergs of a particular size near the coast are thought to decay to half their original numbers every 2–5 years, whereas a 1000-metre-long iceberg in the Antarctic Circumpolar Current would take probably less than 2 years to decay completely. However, icebergs grounded near the coast have been known to survive for many years, perhaps decades, in the cold water there, especially when they are locked in fast sea ice for most of the year and protected from the destructive influence of waves.

Sea ice

Antarctic sea ice has a significant effect on the climate, weather, seas, and ecosystems of the region and human activities there. Pack ice (typically snow covered) reflects most of the incoming solar radiation, thereby amplifying the cooling effect of the region as well as delaying and limiting the summer warming of the continent and its surrounding seas. The ice pack extent, latitudinal temperature gradients, and the generation and paths of storms in the Southern Ocean are all related to each other. Sea ice formation increases the salinity of the sea water beneath causing vertical mixing in the water column. Polynyas and leads are thought to contribute to the high salinity Antarctic Bottom Water found throughout the world's oceans. Melting in summer dilutes and can increase the stability of the upper part of the water column, increasing phytoplankton activity there. The ice algae that grow in sea ice (see Figure 11.7) are believed to be an important source of food for krill and other zooplankton as indicated by increased productivity at the pack ice edge. Polynyas in the pack are thought to be important biologically as they may contribute to ecosystem productivity and provide winter refuges.

Pack ice effectively limits Antarctic shipping to less than 4 months of the year from mid-November and has recently caused two ships to sink in the Ross Sea (Plate 9.13). Fast ice, conversely, provides a very convenient logistic platform from as early as May to as late as January (Plate 9.14). Large wheeled aircraft fly south from New Zealand to land on fast ice in McMurdo Sound during spring and early summer (see Plate 2.6).

Sea ice in the Ross Sea, like elsewhere in the Antarctic, is mostly mobile pack ice up to 1 year old. When the sea surface begins to freeze, small needles or platelets of ice appear, growing and coalescing to form a slush called grease ice. As it thickens, this slush coagulates and tends to break up under the influence of waves and wind into intermediate soft and hard plates which collide with each other to form pancake ice. Still thickening, these alternatively freeze together and break up forming even larger, further-thickening slabs of ice called floes. Colliding floes may raft over one another when still thin, or form pressure ridges of ice rubble along the colliding margin (Plate 9.15).

Wind and currents are continually forcing the floes of ice to drift apart. This creates dark leads of open (Plate 9.6) water between or through floes that are rapidly covered by new ice in the colder months. The amount of open water within the pack is minimal (3–5%) in late winter.

The whole accumulation of broken sea ice is termed pack ice and collectively is referred to as the ice pack (or simply pack). Because most of this ice melts before the following winter, it is first-year pack ice and grows to about 2 metres thick.

In the Ross Sea sector (160°E to 130°W) the pack ice reaches a maximum extent averaging 4 million square kilometres (one-fifth of the total extent of the Antarctic ice pack) usually by late September (Figure 9.3). Extensive freezing occurs from the south from late February to March due to rapidly cooling air and sea temperatures. This freezing, together with ice moved by wind and currents, leads to rapid increase in sea ice cover from early April to about late July. Thereafter, the pack continues to expand more slowly until September to October. From October to February, the pack decays by breaking (Plate 9.16), dispersal, and melting to a minimum of about 700 000 square kilometres in the Ross Sea sector (Figure 9.3). However, because of divergent motion, sea ice covers an average of only 500 000 square kilometres of sea at the minimum, usually around mid to late February. The rapid decrease in ice cover from mid November to January is thought to be due to upwelling of relatively warm deep water, particularly in the southern Ross Sea and along the Antarctic Divergence (a poorly defined wind-induced feature at about 65°S) and to exchange of heat between sea and air.

Plate 9.14. *Sea ice fast to the shore makes a good surface for travel and other work even when the wind has blown the snow off the ice, making it slippery. Photo: H. Keys.*

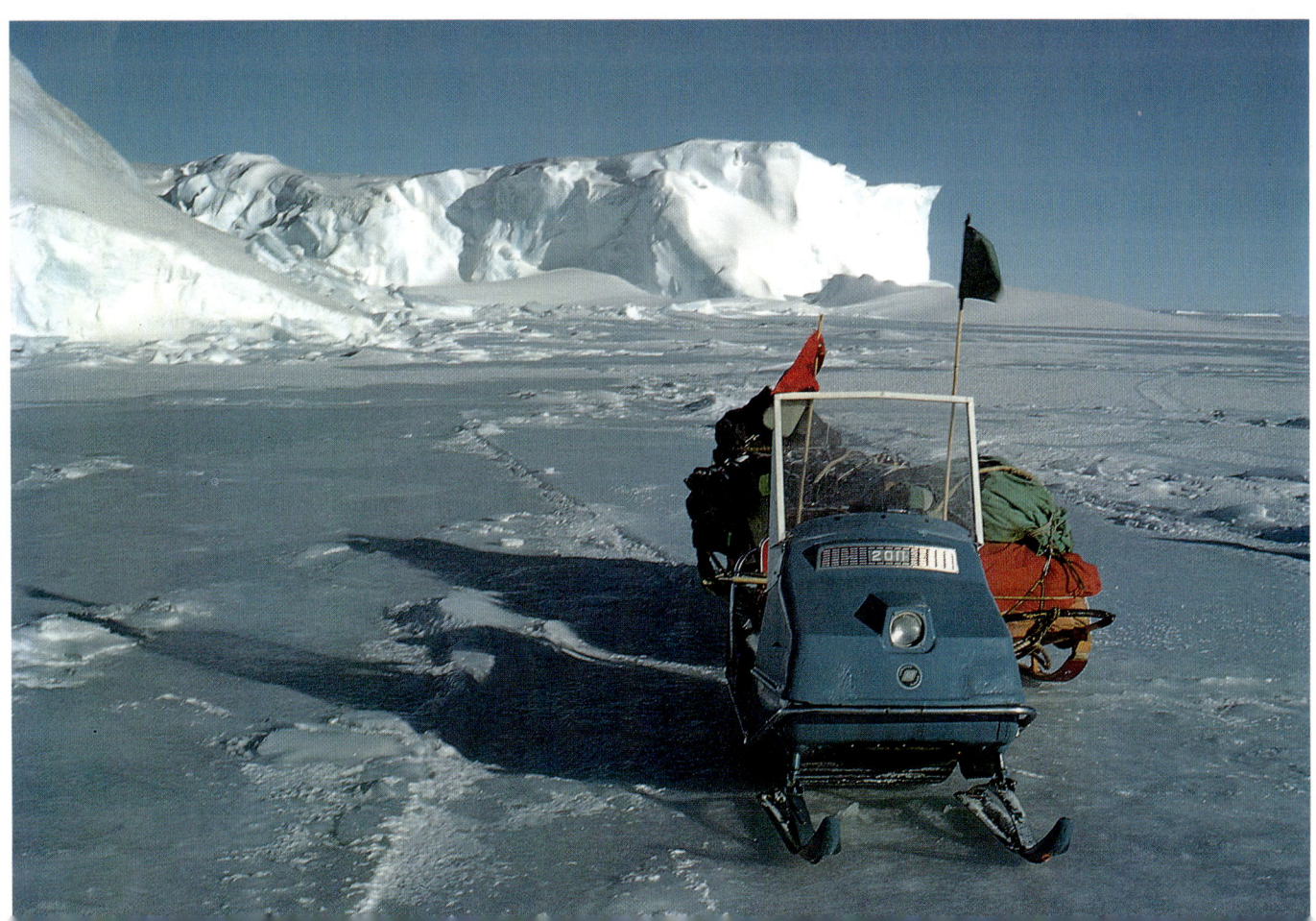

Second-year (or multi-year) pack ice is pack which survives the summer melting. It therefore tends to be thicker than first-year ice and takes longer to melt. Both these characteristics make areas of multi-year pack ice significant because they are a hindrance to shipping (especially early or late in the navigation season) yet provide breeding areas for the world's most populous seal, the crabeater. Multi-year sea ice is normally encountered in north-east Ross Sea, and off the Oates Coast west of Cape Adare. In addition, some pack ice often persists into a second winter in western Ross Sea south of the Drygalski Ice Tongue and in outer Lady Newnes Bay northwards.

Of the other forms of sea ice that also occur in the Ross Sea area fast ice is the most important. This is sea ice that forms a sheet attached to the shore. It develops along most of the coastline during winter, especially in coastal indentations and where the sea is less than about 200 metres deep. Although there may be large expanses of smooth ice, broken ice is created during the freezing period when pack ice floes freeze together or when ice is piled up by ice pressure or waves. The fast ice sheet may move up to about 1 metre per day due to wind, moving glaciers, and probably tides and new ice forming in cracks. Stresses develop, causing cracks to form where the ice is pinned to coastal promontories and around grounded icebergs. Most of this fast ice has broken up and dispersed by February, except occasionally in some coastal indentations.

An ice foot consisting of frozen sea water with blocks of sea ice and glacial ice, or ice-push ridges or heaped-up sea ice, can develop along many rocky shores throughout the year but most melt in summer. Submarine ice cliffs found along the coast in western McMurdo Sound are exposed faces of ice-cored moraines or permafrost with sea water frozen directly on to them.

Two important and related features of pack ice distribution are the differences in area the pack ice covers from year to year, and polynyas (see below). Variability is most pronounced in the summer when the area of the pack ice at any time may vary by up to 50 percent between years. The Ross Sea pack is especially variable. Some summers are heavy ice seasons when the pack is very slow to dissipate, and others have extensive and long (e.g., 3 months) open-water seasons. The large variability suggests that any effect of man on sea ice distribution or colour in the Antarctic would be negligible in relation to natural variations.

The development of polynyas strongly influences pack ice distribution and dissipation, including the length of the open-water season in coastal areas. There are at least three recurring polynyas in the Ross Sea (Plate 9.6). These areas of reduced ice concentration and thickness or open water are thought to be caused by winds acting together with coastal ice formations and suitable topography of land and seafloor. For example, sea ice continually forming in the 1000 square kilometres of polynya in Terra Nova Bay, for probably 9–10 months of the year, is moved eastwards by strong westerly winds, contributing to the persistent, pack-ice zone in north-western Ross Sea. The Drygalski Ice Tongue to the south blocks out pack ice moving up from the south. However, the same polynya also helps to disperse the pack in the western Ross Sea in late summer (Figure 9.3), due probably to the increased ice divergence and upwelling of deeper warmer water.

The large Ross Sea polynya appears to be driven by some upwelling of warmer or more saline water probably assisted by southerly through easterly winds. Advection of sea ice into the southern Ross Sea is limited because of the current directions and the existence of the Ross Ice Shelf, whereas the winds and upwelling act to drive and melt ice out of this area. The polynya therefore tends to open the Ross Sea from the south.

The polynya in McMurdo Sound appears to be a typical flaw polynya which develops between fast ice and pack ice. This recurring polynya is caused mainly by prevailing south-east winds blowing off the fast ice and moving the pack ice away from the fast ice edge. The curving "bight" in the fast ice is a typical feature.

The patterns of pack ice movement are known in general. In most of the Ross Sea, drift appears to be towards the north-west, especially in the north. However, in the southern Ross Sea, westward drift also occurs turning to the north near Ross Island then north-west around Cape Adare. North of the Ross Sea, northward movement appears to be most common until the latitudes of the West Wind Drift are reached, where eastward drift predominates.

This pattern of ice movement is similar to the pattern and direction of predominant winds, surface currents, the drifts of ships beset in the pack and also those of icebergs (Figure 9.2). This implies that movement of sea ice can be attributed qualitatively to both wind and current. However, the direction of ice drift can be highly variable over short periods. Drift rates are normally less than 17 kilometres per day (0.2 metres per second) but maximum speeds probably exceed 1.2 metres per second when drift is driven by strong winds or tides. Sea ice probably drifts faster than icebergs on average due to the greater influence of wind on the former.

The pattern of ice movement also affects ice distribution and persistence. The higher concentrations in the eastern Ross Sea and north of the Oates Coast are probably due to slower ice movement, convergence, and accumulation of ice in these areas throughout the year. Advection of ice and cold water in regional currents is also probably a strong contributing cause. Persistent pack ice in the western Ross Sea is thought to be due to convergence caused by both offshore (katabatic) winds off Victoria Land and southerly (and south-easterly) winds further

Plate 9.15. *A pressure ridge of broken sea ice within fast ice. The ridges are formed by ice floes moving together, breaking and piling up along the join. Photo: H. Keys.*

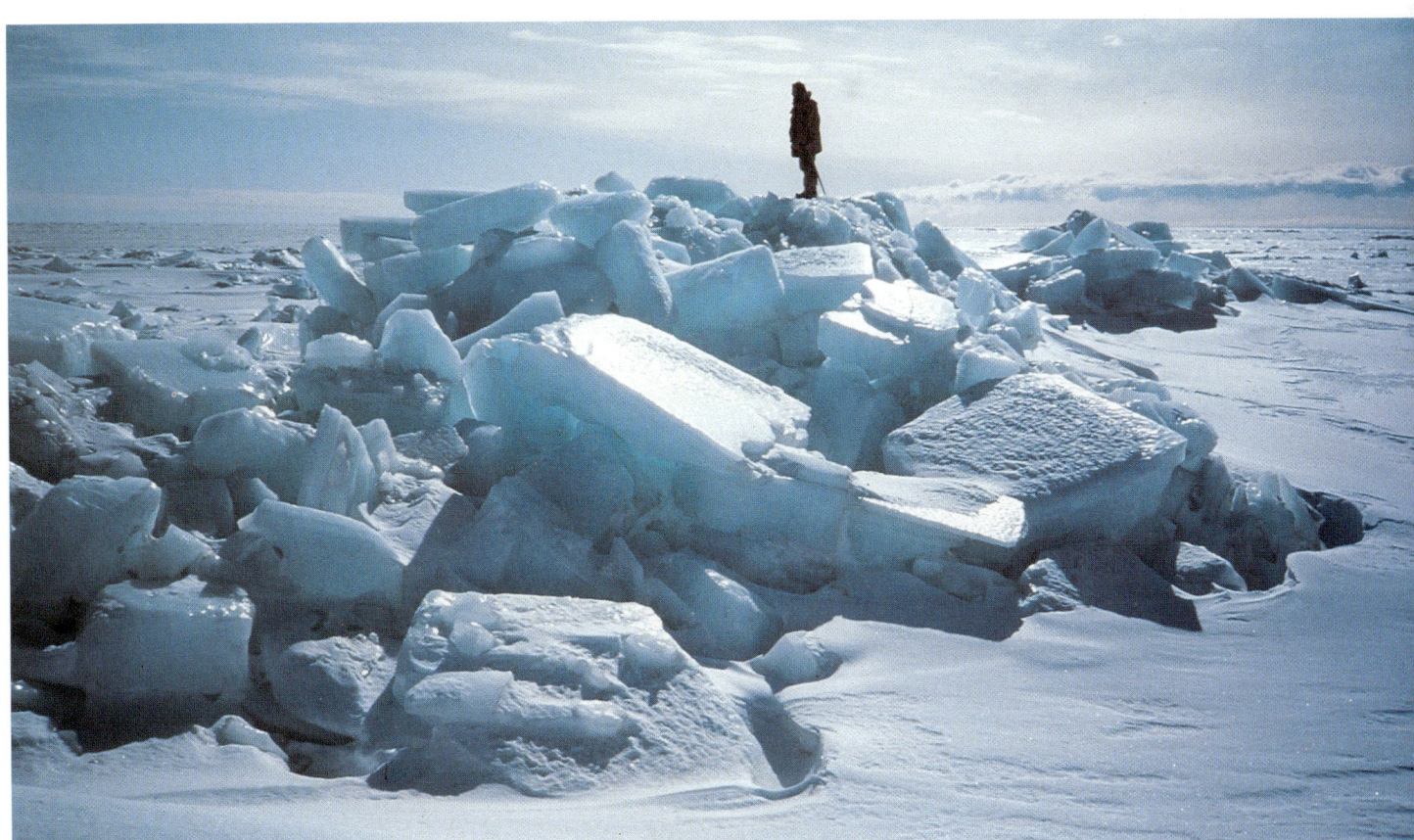

out to sea. Coastal topography contributes to this convergence, for example, in Lady Newnes Bay where the coastline trends north-east, approximately normal to south-easterly winds and the regional current in the western Ross Sea. An increased frequency of low ridges of ice debris in much of the fast ice and pack ice along the western shore of the Ross Sea demonstrates the local importance of convergence in a region otherwise dominated by divergent motion of first-year pack ice.

Figure 9.3. *Distribution of pack ice in the Ross Sea region in mid-February when the pack is at its minimum (light blue) and in early September when the pack is at its maximum (dark blue). In February open water covers most of the Ross Sea, an advantage exploited by the early explorers seeking the most southerly access to the continent.*

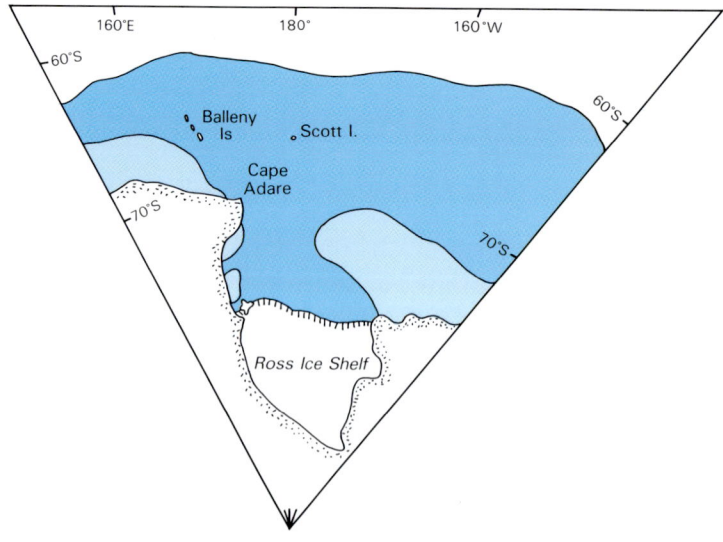

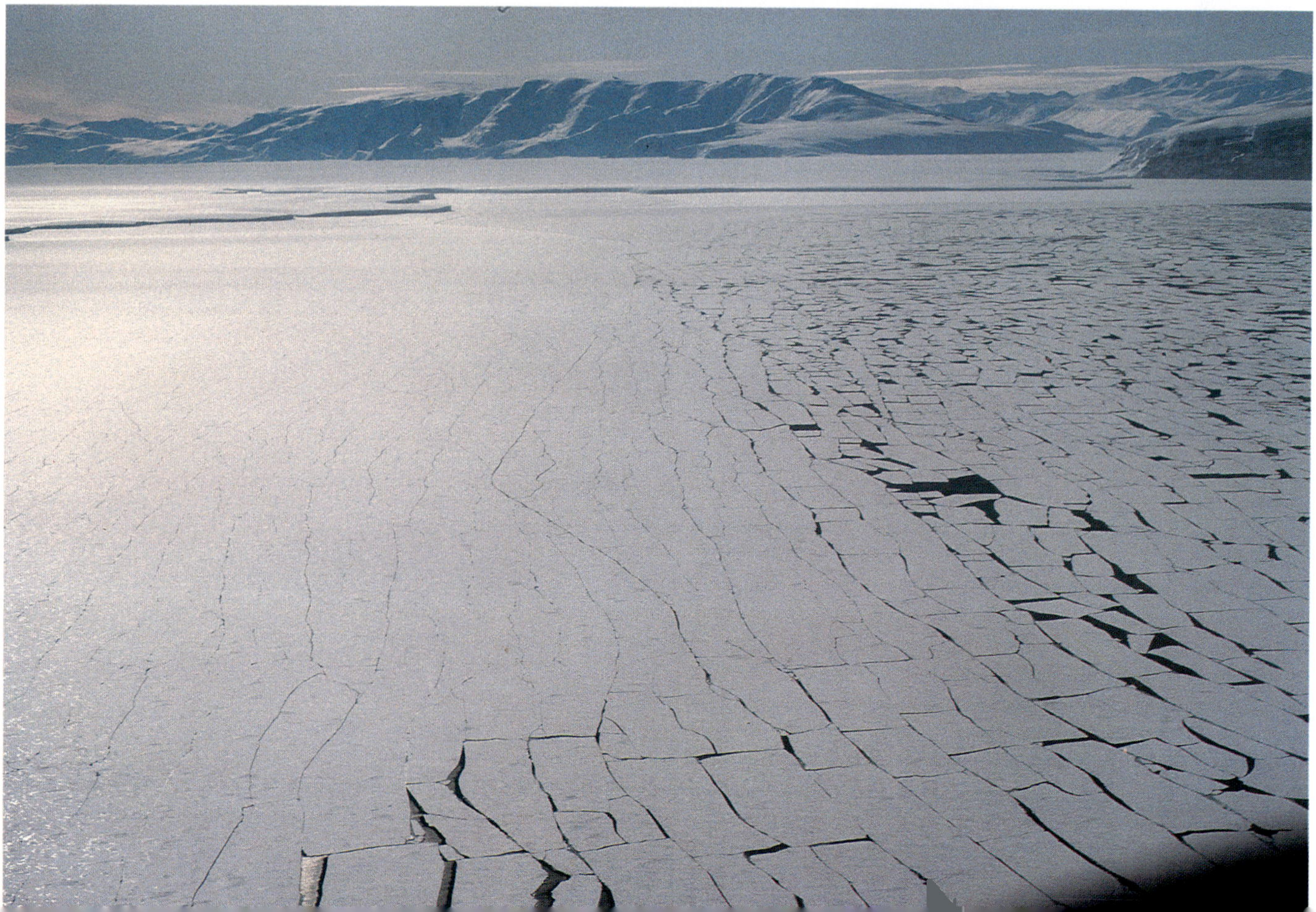

Plate 9.16. *Fast ice breaking up first into strips and then into rectangular floes under the influence of ocean swell, in Glacier Strait. Photo: H. Keys.*

There is a hole in the edge of the East Antarctic Ice Sheet containing a desert as arid as the Sahara, where the bare rocky ground was shaped before mankind had evolved from his simian ancestors, and which has since remained almost unaltered. Here the sun never sets in summer, never rises in winter, and the ground is frozen to a depth of nearly half a kilometre. The most abundant living organisms dwell inside rocks. Sand dunes lie alongside glaciers, and the glaciers (which evaporate rather than melt) are dry to the touch. Of the rare meltwater streams that flow briefly in mid summer, the largest begins near the coast and flows inland for 30 kilometres before entering a permanently ice-covered lake which has no outlet. The surface water of the lake is fresh, whereas the bottom waters are three times saltier than the sea. The lake contains layers of water of increasing saltiness and density which are warmest at the bottom, where the year-round temperature is 25°C, in a climate where the surface temperatures drop to –50°C in winter. There also is a pond with salts so concentrated that it cannot freeze in winter, and around its shores a mineral is precipitated which is not found elsewhere on earth. Mankind first set foot in this unique domain barely more than 80 years ago.

The McMurdo "oasis"

This dry rocky area of about 2500 square kilometres has been called an "oasis" in a desert of ice (Plate 10.1). It contains three large valleys: Taylor, Wright, and Victoria Valleys, separated by mountain ranges of heights up to 2500 metres. All of the valley floors are bare rocky ground. The ground is not snow-covered because the air is so dry it can evaporate more snow than falls. Exceptions occur at higher elevations on the ranges where heavier snowfalls are redistributed by strong winds, to accumulate enough snow in sheltered positions to form alpine-type glaciers. The Dry Valleys region, although the largest, is not the only ice-free area in Antarctica—at least another 20 "oases" exist.

The low annual precipitation of most of the Antarctic continent makes it technically a desert, with less than 50 millimetres water equivalent of snow accumulating in high central areas. Precipitation is greater towards the coast and Scott Base receives an estimated 200 millimetres water equivalent of snow, whereas at Vanda Station in the Dry Valleys the total annual snowfall is only about 10 millimetres water equivalent. The mean loss of water to the air from the surface of Lake Vanda is 300 millimetres per year, which gives the area a precipitation deficit of at least 290 millimetres per year. Annual precipitation at Vanda Station is about 35 times less than the annual loss to sublimation (transition of snow directly from the solid state to water vapour without passing through the intermediate, liquid water stage).

The aridity of the region is accentuated by a precipitation shadow. Snow-bearing cyclonic systems tend to track southward over the Ross Sea, and their clockwise movement brings onshore winds from the south and east. Snowfalls from this direction feed the piedmont glaciers, but inland snow penetration is largely blocked by mountains. Ross Island is the first obstruction, then the high Royal Society Range, followed by each of the east-west ranges aligned across the area. The northernmost Victoria Valley may be predicted to be the most arid of the valleys.

Inland of the Dry Valleys region, high bedrock necks prevent the ice sheet from overflowing into the valleys and flooding them with ice as happens in the north and south. If we consider the full area of precipitation deficit, then climatically the McMurdo "oasis" stretches 200 kilometres from Black Island in the south to Fry Glacier in the north.

The dark, exposed ground modifies the local climate. It absorbs more heat during summer, and loses more heat in winter than do the

Plate 10.1. *Satellite image of the McMurdo Sound region showing the extent of snow-free ground. Note the Taylor, Wright, and Victoria Valleys, the three principal dry valleys. Photo: Physics and Engineering Laboratory, DSIR.*

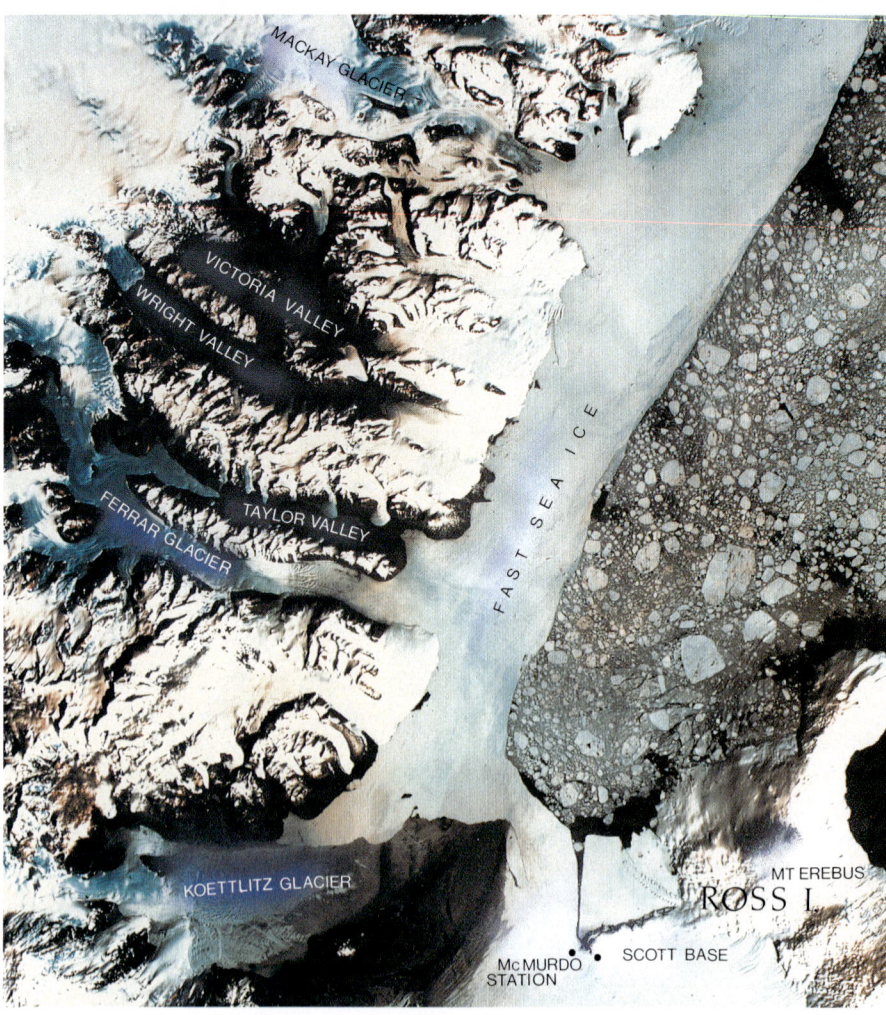

Figure 10.1. *Stages in the development of the Dry Valleys. (a) The Dry Valleys submerged by one of the temperate "robust" ice sheets which alternatively grew and collapsed during Miocene times, 35 to 15 million years ago. (b) Collapsed ice sheet at about 4 million years ago. (c) The Dry Valleys and the present polar ice sheet.*

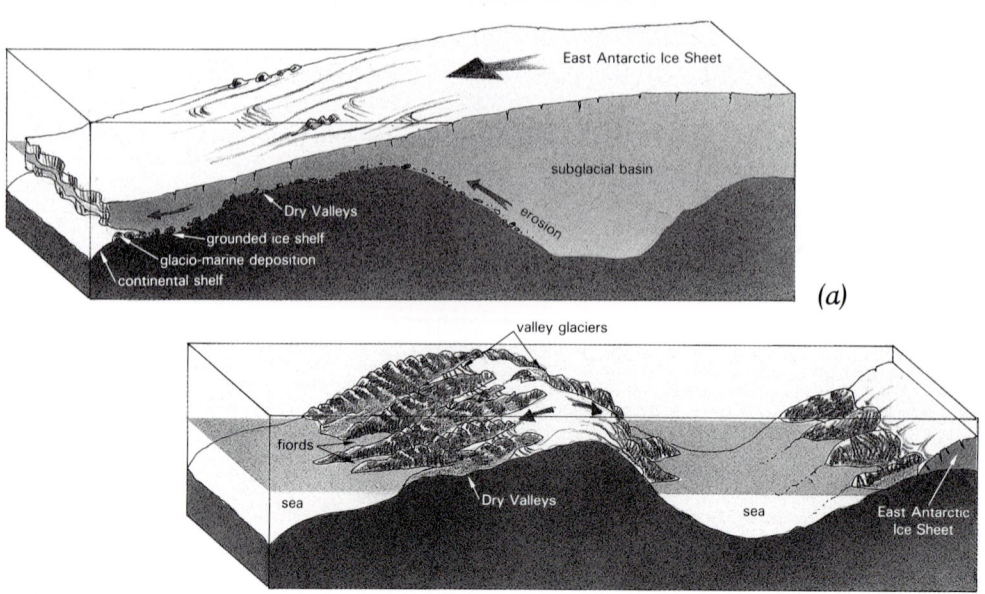

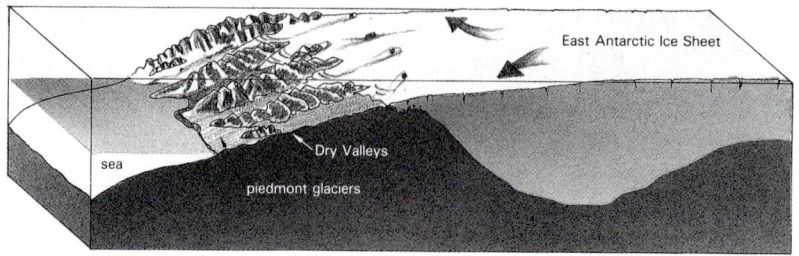

surrounding snow-covered areas. Summers average 5°C warmer, and
winters 5°C colder than at Scott Base and other snow-covered areas at
similar low altitudes. Summer heating also warms the air at ground level,
causing it to rise to form an easterly "sea breeze". This thermally forced
easterly wind, which commences at about midday and persists until
midnight, reaches an unpleasant 20 knots (10 metres per second) and is a
daily feature of the valleys. It is a shallow wind which may be escaped by
moving some hundreds of metres up the valley sides. The other
dominant wind of the valleys is a "warm" westerly föhn wind, associated
with a surface layer of cold air draining off the ice sheet. These
"katabatic" winds are intermittent and last a few days; they are the
strongest winds during winter.

Discovery of the Dry Valleys

The first of the Dry Valleys was discovered by accident during
Scott's Discovery Expedition of 1901–04. A party led by Albert Armitage
had proved a route up the Ferrar Glacier to the Polar Plateau
in 1902–03, which was followed in 1903–04 by Scott, Feather, and Edgar
Evans on a reconnaissance journey to the East Antarctic Ice Sheet.
During their return in thick cloud, they found that they had descended
into the wrong valley. Hoping that the glacier they were following would
lead to McMurdo Sound, they continued until they were stopped when
this, the Taylor Glacier, terminated at Lake Bonney. Because they were
sledging and not equipped for travel over rocky ground, they retraced
their steps to the Ferrar Glacier.

On Shackleton's first expedition of 1907–09, Priestley, Bertram
Armytage, and Brocklehurst visited the coastal areas of Taylor Valley
after a geological survey of the Ferrar and Upper Taylor Glaciers. The
connection between this valley and the one Scott's party had seen
remained to be verified.

Central Taylor Valley was not visited until February 1911, when
Taylor, Debenham, Gran, and Forde from Scott's last expedition sledged
up the Ferrar Glacier and walked down the Taylor Valley from a camp
on the snout of Taylor Glacier. They passed through the defile of Suess
Glacier and camped at the outlet from Lake Chad. Before their return,
views from a hill opposite Commonwealth Glacier assured them that this
was the same valley that had been seen from the coast.

The ice-free areas of the Wright and Victoria Valleys remained
unseen for a further 43 years. It was not until United States and New
Zealand aircraft flew over the area between 1955 and 1958 that the true
nature of the valleys became known. Scientific parties were soon in the
area, which has become the most intensively investigated part of the
continent. In the 30 years since the beginning of routine scientific visits,
the area has evolved from one of pristine desolation to one of possible
scientific despoilation.

The shaping of the Dry Valleys

The present topography and surfaces of the valleys are surprisingly
ancient and the main features are of mid-Tertiary origin. At some time
before Antarctic glaciation (about 50 million years ago) the outlines of
the present valleys would have been defined by river erosion. With the
onset of glaciation, local alpine-type glaciers would have formed on both
western and eastern flanks of the Transantarctic Mountains, which were
then only low hills. We can only speculate on when these glaciers
expanded and coalesced to form an ice sheet, although it is likely that
they formed two separate sheets over East and West Antarctica which
have decayed and reformed over the millenia.

When sea level temperatures were about 0°C, an intermediate-sized,

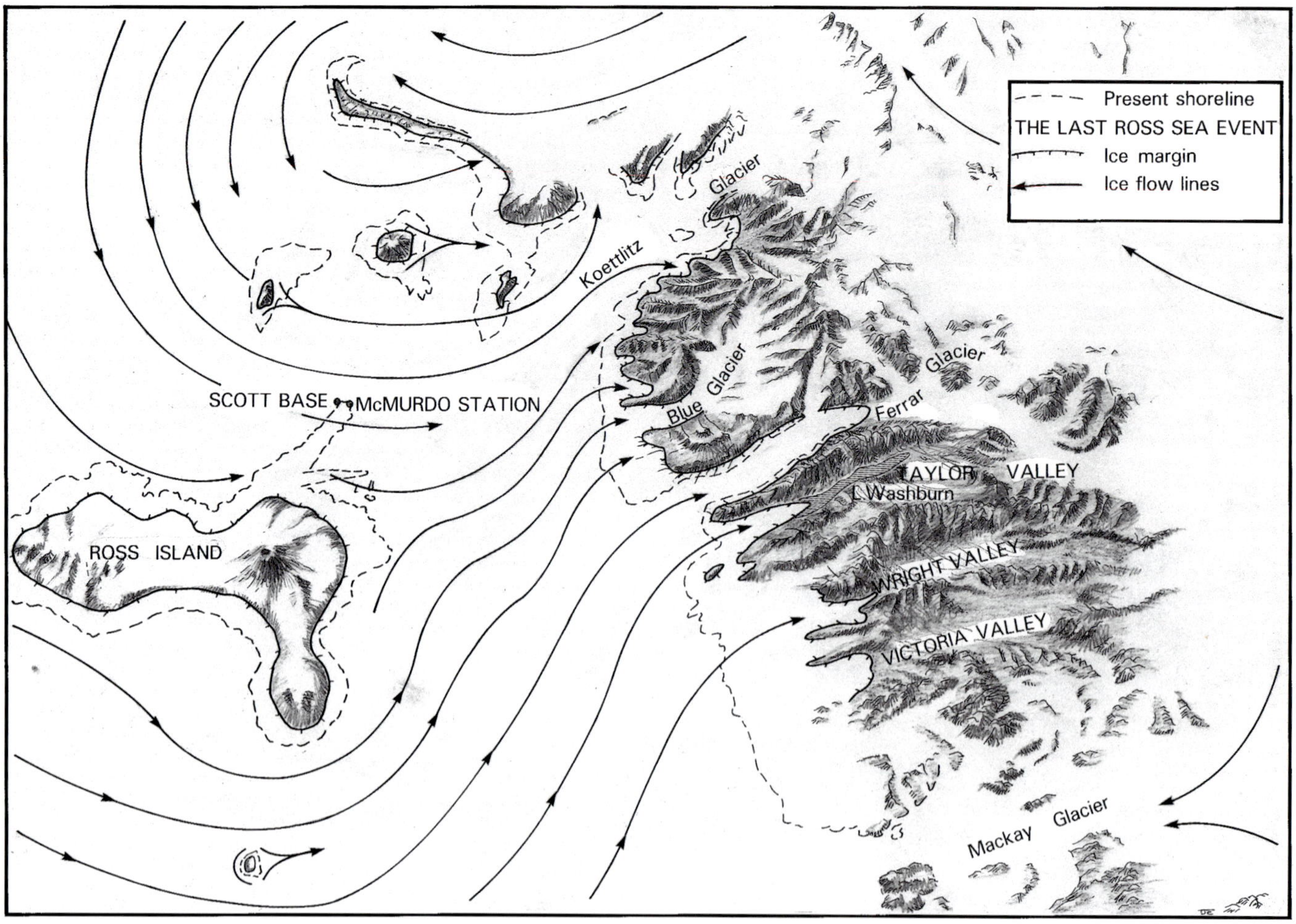

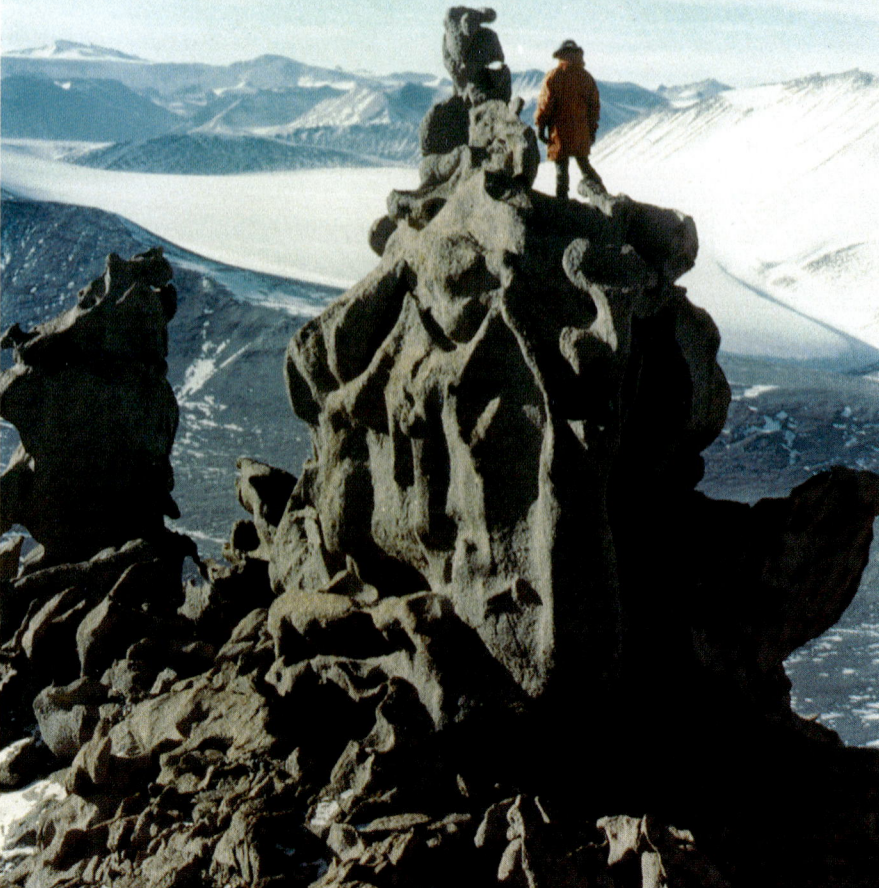

Figure 10.2. *The Dry Valleys and McMurdo Sound at the maximum of the most recent Ross Sea Glaciation, about 20 000 years ago.*

Plate 10.2. *Cavernous weathering at a ridge crest of coarse dolerite, Asgaard Range. Photo: T. Chinn.*

dynamic, temperate ice sheet must have formed. Such ice sheets change quickly in response to the climate, move fast, and have abundant meltwater streams. During warmer intervals, when mean sea level temperatures rose to about 5°C the East Antarctic Ice Sheet would have collapsed, with West Antarctica possibly reduced to a lightly glaciated island archipelago. Cooling to –5°C at sea level provided optimum conditions for the formation of a large, robust ice sheet which covered the entire continent and extended to the continental shelf. Further cooling to today's average –14°C at sea level drastically changed the ice sheet from a dynamic, fast moving, temperate type to one of sluggish polar character. The Miocene Period (24–5 million years ago) was dominated by wild fluctuations of a robust, temperate ice sheet which reached the continental shelf and shaped the present topography of the Dry Valleys (Figure 10.1a). It inundated the Transantarctic Mountains to above 4000 metres, for the Royal Society Range has glacially striated surfaces under protecting deposits of Sirius Formation till. But marine beds in the Wright Valley, dated at 4.7 million years ago, indicate that the Dry Valleys were then fjords. Marine diatoms found in tills high in the Transantarctic Mountains suggest that the collapse of the ice sheet was of such magnitude that open water occurred to the west of the mountains (Figure 10.1b). The diatoms deposited in this water were later elevated in tills and deposited near the tops of the ranges by the subsequent expansion of the ice sheet.

The change from temperate to sub-polar character of the ice sheet and its outlet glaciers seems to have occurred about 2.5 million years ago, coinciding with the start of a period of global ice ages. Moraines left by advances of alpine-type glaciers which have been dated at between 2 and 3.3 million years ago are polar in character. The presence of these undisturbed moraines in the Wright Valley, only a few hundred metres in front of the present glaciers, indicates that there have since been no very large advances of the outlet glaciers.

As the outlet glaciers withdrew from the Dry Valleys the sea followed the retreating ice up the valleys, forming a fjord landscape. The Dry Valleys as we know them today (Figure 10.1c) date from the time, about 4.7 million years ago, when uplift of the Transantarctic Mountains raised the valley outlets above sea level, transforming the fjords into salt lakes. Victoria Valley, which has the highest threshold barrier between the valley and the sea, contains no evidence of a fjord period, and may have escaped many of the invasions of an expanded Ross Ice Shelf. However, deposits in the Lower Taylor Valley record advances of Taylor Glacier to the coastline and beyond on at least six occasions between 4.2 million and 850 000 years ago, though on one occasion a grounded Ross Ice Shelf entered the valley. From 850 000 years ago, the glacial history changed from a dominance of valley, or outlet glaciers, to a dominance of invasions by an expanded and thickened Ross Ice Shelf.

Over time complex interfingering of the local valley and alpine glaciers with the Ross Sea ice, in response to climate changes, resulted from the opposing behaviour of temperate and fully polar glaciers, together with falling sea levels which triggered the Ross Sea glaciations. During the Pleistocene, at times when global temperatures fell and glaciers expanded world-wide, the local glaciers (both the Alpine and the Taylor, Upper Wright, and Upper Victoria Glaciers) of the Dry Valleys were starved of snow and receded. At the same time, the growth of the Northern Hemisphere ice sheets withdrew large volumes of water from the oceans, and lowered the world's sea level by many tens of metres. The low sea levels grounded the floating Ross Ice Shelf, which thickened because of increased basal friction and made extensive seaward advances. These events, known as the Ross Sea glaciations, were not necessarily associated with a thickening of the East Antarctic Ice Sheet; evidence suggests that the ice sheet became thinner. The Ross Ice Shelf expanded seaward and along the coast of North Victoria Land and thickened by hundreds of metres above its present level to overtop the thresholds at the mouths of the Dry Valleys and fed tongues of ice inland on several occasions. During these times, the Dry Valleys were completely surrounded by ice and became even more arid than today. Ice tongues and summer meltwater streams from the encircling ice entered this "ablation hollow" from all directions. Over the past 850 000 years, Ross Sea ice has apparently entered the Taylor Valley during at least seven cool periods. Between these events, warmer interglacial advances of the Taylor Glacier interfingered from the opposite direction.

The most recent of these incursions was the Ross Sea I Glaciation (Figure 10.2), which reached a maximum about 20 000 years ago, leaving an extensive sheet of glacial drift along coastal areas, and extensive glacial deposits in the Taylor Valley. Large volumes of meltwater were channelled into the Taylor Valley across the surface of the expanded ice shelf and along its margin, in a manner similar to that at the Koettlitz Glacier today. This meltwater formed Lake Washburn (Figure 10.2) which rose to 300 metres above present sea level, and had a maximum area of about 80 square kilometres. At today's evaporation rates, a lake of this size would require 13 times the present mean annual flow of the Onyx River to maintain itself. Surprisingly, the ice tongues into the Wright and possibly the Victoria Valleys do not appear to have had sufficient height and gradients to provide enough melt to significantly raise the levels of Lakes Vanda and Vida. Neither of these lakes show evidence of high levels at this time.

As the Ross Sea ice withdrew over a long period between 20 000 and 8000 years ago, the level of Lake Washburn slowly lowered and separated into today's individual lakes (Lakes Bonney, Henderson, Popplewell, Chad, Hoare, and Fryxell). As the lake lowered, all of the dissolved salts in the waters were concentrated, and remain in the valley today in both the sediments and the lakes.

During the past 10 000 years, the local glaciers have moved out of their withdrawn Pleistocene positions in the high valley cirques in a series of small fluctuations. An advance occurred about 3000 years ago, which at the low activity of these glaciers (5000 years is an average time for ice to flow through an alpine glacier) is geologically only yesterday. Today some of the glaciers are advancing, some are receding, and all of the enclosed lakes are rising.

The rocks of the Dry Valleys have been exposed for a very long time, but the common weathering agent—water—has been almost absent. This has permitted other weathering processes, normally obliterated by the various forms of water action, to become the dominant landscape-forming agents. Mechanical disintegration of rock by temperature change and frost action (where water occurs) assist in breaking down the landscape, but the most spectacular forms and features have been sculptured by salt weathering and wind action (Plate 10.2). These processes have been discussed in Chapter 4.

Glaciers of the Dry Valleys

The glaciers of the Dry Valleys have been classified into three types according to size and location: piedmont, outlet, and the smaller alpine glaciers. Coastal snowfalls maintain the low, broad piedmont glaciers covering most of the seaward margins, whereas large outlet glaciers descend into the valleys from the west. Only the Ferrar and Mackay Glaciers traverse the valleys to enter the sea; between these, the Taylor and Upper Wright Glaciers terminate on land. The latter two glaciers

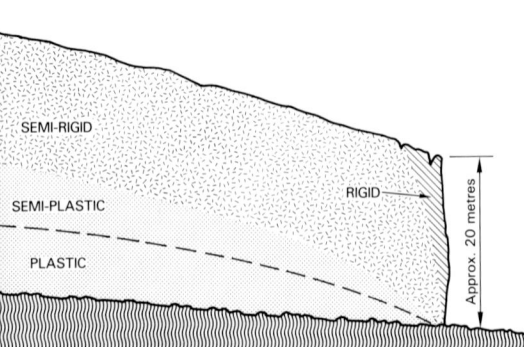

SEMI-RIGID

SEMI-PLASTIC

PLASTIC

RIGID

Approx. 20 metres

Figure 10.3. *Section of an ice-cliffed glacier tongue, showing zones of ice of differing behaviour.*

are not strictly outlet glaciers as they flow from a local ice dome and fluctuate in unison with the local glaciers, rather than with the level of the East Antarctic Ice Sheet. On the ranges between these inland and coastal glaciers, numerous alpine glaciers are scattered amongst cirques of mid-Tertiary age. Towards the coast, the cirques overflow with ice but further inland and at higher elevations they are either underfilled or empty.

The alpine glaciers are striking features which strongly contrast with the surrounding, bare rocky ground (Plate 10.3). They are generally smooth surfaced, free of surface moraine, and terminate in a variety of forms from gentle ice-ramps to abrupt cliffs about 20 metres in height. They are cold, or dry-based glaciers, in contrast to the better known temperate, or wet-based glaciers. The low temperatures of the area (mean annual temperatures are below –20°C) keep the total thickness of these glaciers below freezing at all seasons so that no basal meltwater or water from within the glacier can form. At the sole, the ice of a dry-based glacier is frozen to the bed and movement by slipping is negligible to non-existent. The bulk of the ice deformation due to flow occurs in a zone of dirty, salt-rich ice within the first metre above the bed. Thus, these glaciers are almost incapable of basal erosion and move very slowly because there is no basal sliding.

Because of their low erosion rates, the dry-based glaciers protect the bare rocky slopes against wind erosion. Thus, the ground under the glacier may be higher than the adjacent ground which has been eroded by wind.

In contrast, the larger and thicker outlet glaciers can generate sufficient heat by internal friction to reach pressure melting point at the base. Liquid water at the base detaches the ice from the rock, permitting temperate-glacier behaviour. Flow velocities are much faster, because in addition to plastic flow, they also move by basal slipping; and extensive erosion of the bed pulls much debris into the glacier.

Plate 10.3. *Lake Vanda with Vanda Station at the far (eastern) end. Alpine-type glaciers descend into the Wright Valley from the Asgaard Range. From left to right: Goodspeed, Hart, Meserve, and Hartley Glaciers. Photo: Antarctic Division, DSIR.*

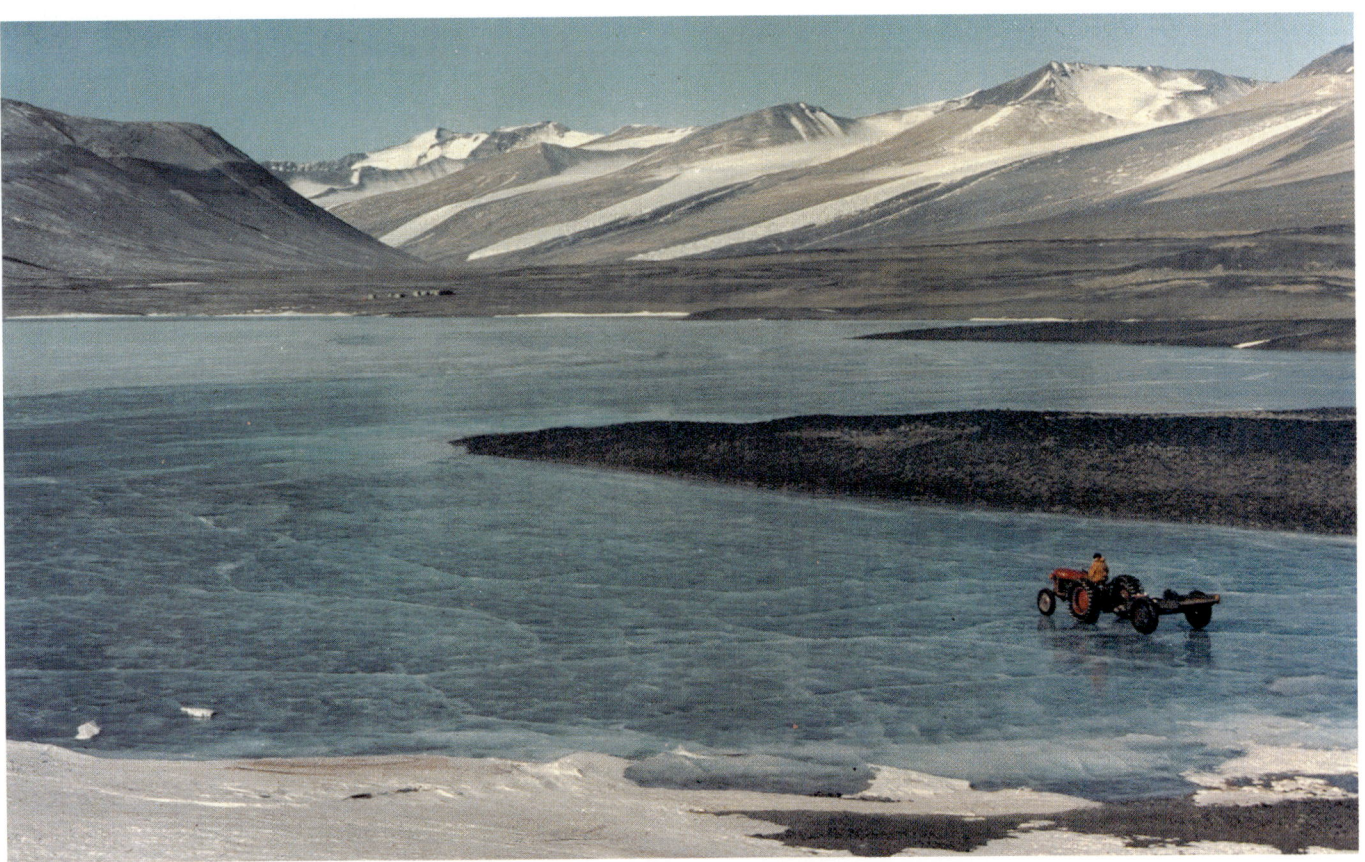

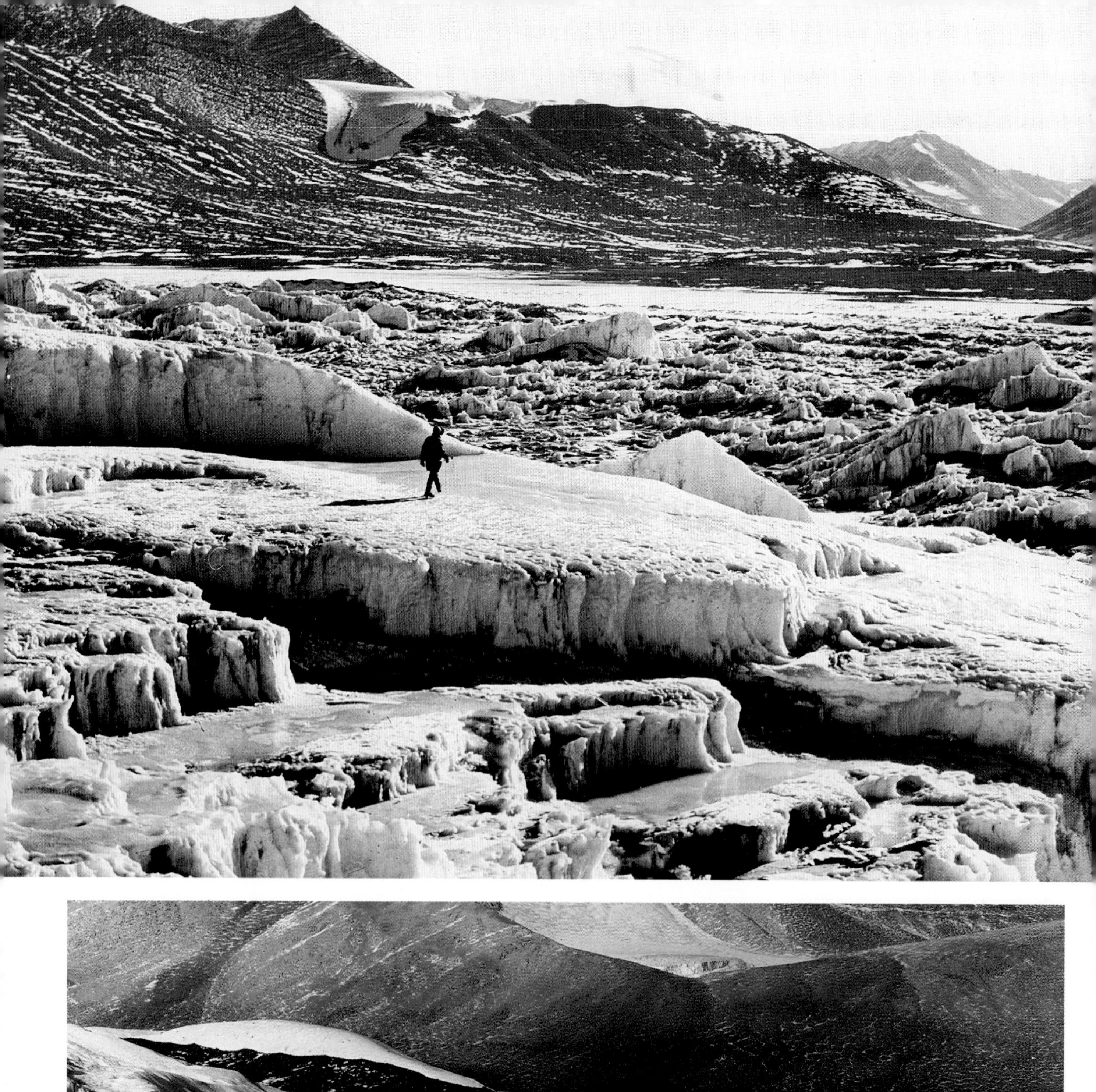

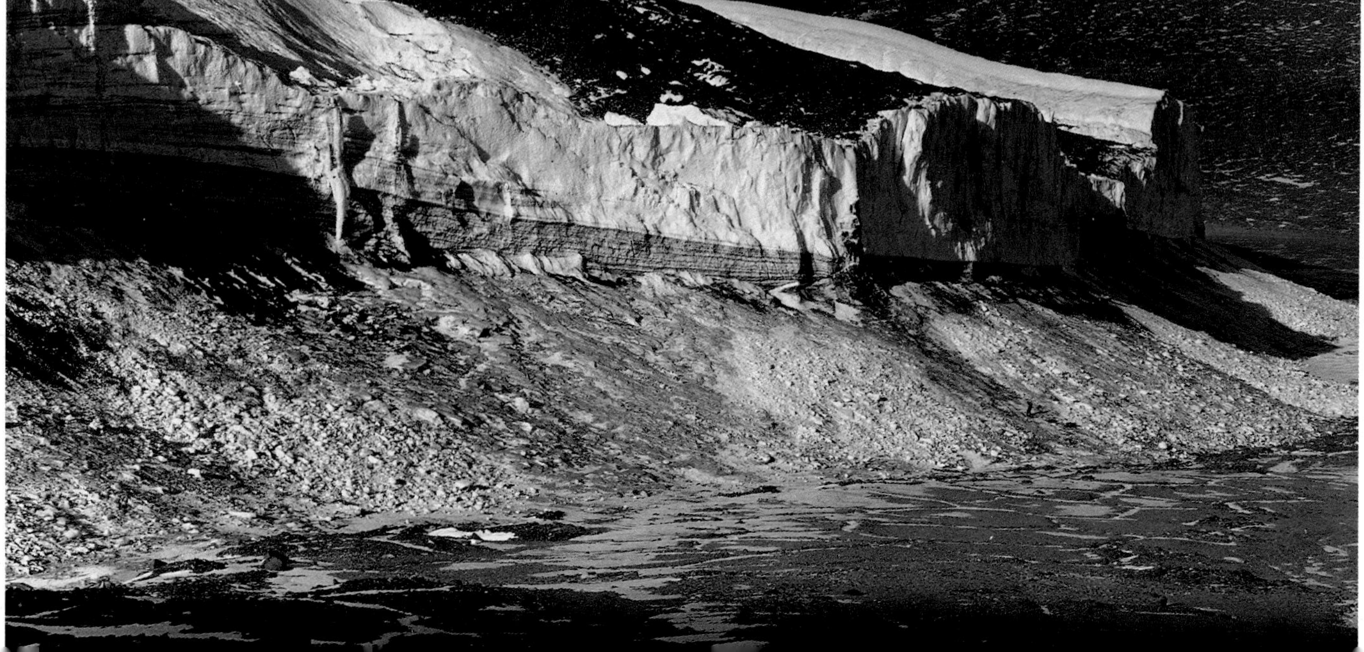

The shape of the glaciers

"Why are all the ice-cliffs about 20 metres high?" Ice-cliffs form because there is a change in the behaviour of ice with depth, or pressure of overlying ice. The deeper it is within a glacier, the more easily ice flows. In a section through a dry-based glacier tongue several zones of differing ease of flow can be recognised (Figure 10.3). Below a depth of 20–30 metres, the ice is semi-plastic and flows and deforms relatively easily. A semi-rigid layer, approximately 20 metres thick, which extends across the glacier surface behaves more like a solid than a plastic. The glacier structure is crudely analogous to a layer of icing on treacle, and where the glacier surface is bent, distorted, or stretched, the "icing" may fracture into crevasses. Thus, crevasses are rarely deeper than 20–30 metres.

Near the glacier margin where the ice thins to about 20 metres thick, the semi-rigid zone or "icing" becomes grounded. In a wet-based glacier this is of little consequence because the ice, lubricated by water, slides along the bed as melt thins the glacier front to a wedge. But under a dry-based glacier, the semi-rigid ice is frozen to the ground and obstructs the flow of the thicker, upstream ice. This obstruction may cause the ice to buckle and fold, but more commonly the ice from the glacier sole is upwarped towards the surface to over-ride the thin obstructing apron of margin ice. This shear zone brings basal debris to the surface to form the "inner moraines" common at the margins of the smaller glaciers. A wedge of drifted snow and ice usually forms in the inflection between the glacier and the ground to form an "apron" around the glacier margin. The heights of ice-cliffs are, therefore, a reflection of the thickness of the semi-rigid zone.

Snow accumulation

Dry Valleys glaciers do not have the same seasons of snowfall and thaw as occurs in temperate climates. Snowfalls occur at any time of the

Plate 10.4. *Terraces developed by differential ablation on the western margin of the Lower Wright Glacier. Photo: T. Chinn.*

Plate 10.5. *An advancing ice-cliff, terminus of Upper Victoria Glacier. Photo: T. Chinn.*

Plate 10.6. *A receding ice-cliffed glacier, Schlatter Glacier, Taylor Valley. Photo: T. Chinn.*

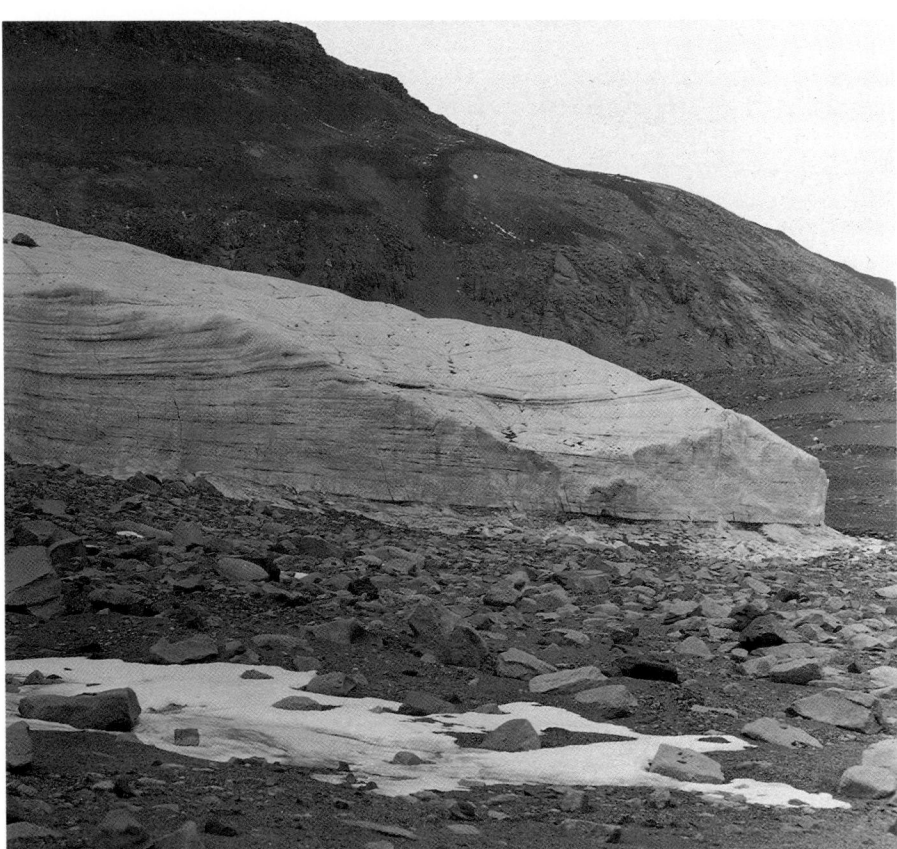

year. Because the warmer air of summer can carry more moisture, snowfalls are heaviest in this season. Furthermore, extremely light snowfalls and low temperatures permit strong winds to determine where snow accumulates. Without wind-drifted accumulation of snow into névé basins, it is unlikely that any of the glaciers could exist, as overall wastage losses exceed precipitation at all altitudes.

Snowfalls are light, and annual snow depths rarely exceed 0.5 metre (compared with 10 metres on the Upper Tasman Glacier in New Zealand). Measurements of mass balance (the annual sum of gains by snowfalls compared with losses by evaporation and melting over the glacier surface) made on several alpine glaciers have shown that the balances of these glaciers are at present approximately zero.

Ablation

Loss from the Dry Valleys glaciers is predominantly by evaporation of ice and snow, and meltwater is rarely seen above 1500 metres on even the warmest summer days. Below 1500 metres meltwater streams flow intermittently beside the tongues of the lower glaciers for about 2 months each year, but the proportion of the total losses from the glacier by melt is very small. On the Meserve Glacier, melt amounts to only 2–3 percent of the loss from the tongue, the remainder is blown away as water vapour. The rarity of meltwater above 1500 metres may be seen from the effects of one extreme event. During the first week of January 1974, the Dry Valleys experienced abnormally high temperatures (a maximum temperature of 15°C was recorded at Lake Vanda). Meltwater from Jeremy Sykes Glacier formed a pond about 20 metres in length at the glacier margin where previously no pond had existed. The surface of this ice-pond at 2500 metres is lowering by about 80 millimetres per year. At present, 13 years after the event, the pond still exists—demonstrating the lasting effects of infrequent extreme events.

Ice and snow losses increase directly with both temperature and exposure to wind. Because the maximum elevation reached by the sun at midday is less than 45°, vertical north-facing ice-cliffs receive more radiation than the horizontal upper glacier surface, and ablation (combined wastage processes) losses from cliff faces can be 2–8 times greater than those from upper glacier surfaces. This difference in ablation losses accentuates irregularities in the ice slopes at glacier margins by selectively ablating the steeper slopes to create a series of terraces. Meltwater from the risers flows to the crest of the next lower terrace, where wind exposure is greater, and the meltwater freezes at the margin of the trend. A spectacular set of terraces has developed on the western margin of Lower Wright Glacier (Plate 10.4). Because of these differences in ablation rates, cliff faces rather than glacier surfaces are the dominant sources of meltwater for the summer streams.

Glacier flow

Dry Valleys glaciers move sluggishly because they are frozen to the glacier bed, and cannot move by sliding. In addition, cold ice is stiffer than temperate ice which is at freezing point. Generally, on similar gradients, the smaller the glacier and the shallower the ice, the more slowly the glacier moves. The small alpine glaciers move as slowly as 10–50 centimetres per year, and would be called "stagnant ice" at a temperate glacier.

Ice of the Heimdall Glacier, at its present mean velocity of about 1 metre per year, would take about 5000 years to travel the distance from headwall to glacier snout, more than 10 times slower than in a temperate glacier. Thus, a response to a climate change would take about this length of time to affect the glacier terminus. The fastest recorded velocities are those of the larger glaciers, yet the steep trunk of Meserve Glacier reaches only 3 metres per year. The large Taylor Glacier, which flows across salt-rich sediments of a former lake bed, is apparently wet-based and has surface movements of up to 14 metres per year at a narrow section of the trunk, and movements of 2–4.8 metres per year near the snout.

Debris

The alpine and piedmont glaciers carry very little debris—mainly some rocks from rockfalls from the headwalls, together with wind-blown sand. Wind-blown sand is a major fraction of surface moraines, particularly on the lower altitude glaciers. Beside these glaciers, interbedded sand and snow layers are common.

Sand blown by the prevailing westerly winds collects on and against the glaciers toward the eastern ends of the valleys and significantly increases their summer ablation rates. Sandy Glacier, above Bull Pass, comprises up to 50 percent by volume of wind-blown sand. A little debris is carried at the base of cold or dry-based glaciers, in the layer where the ice shears over the bedrock. Towards the terminus of the glacier, an outer encircling rim of thin ice obstructs the glacier flow and causes the basal ice to rise to the surface. Here, basal moraine emerges to form a line of debris known as an "inner moraine". Where an ice-cliff is present, this inner moraine occurs near the foot of the cliff at the top of a surrounding "apron" of snow and ice, although the inner moraine is often buried under the apron.

Are the Dry Valleys glaciers advancing or retreating?

Measurements of the positions of the frontal ice-cliffs of the Dry Valleys glaciers, repeated after intervals of a few years, have shown that the glaciers are not in equilibrium. Some are advancing while others are receding. The very slow movement of these glaciers means that these

Plate 10.7. *The Onyx River flowing to Lake Vanda in the Wright Valley. Photo: T. Chinn.*

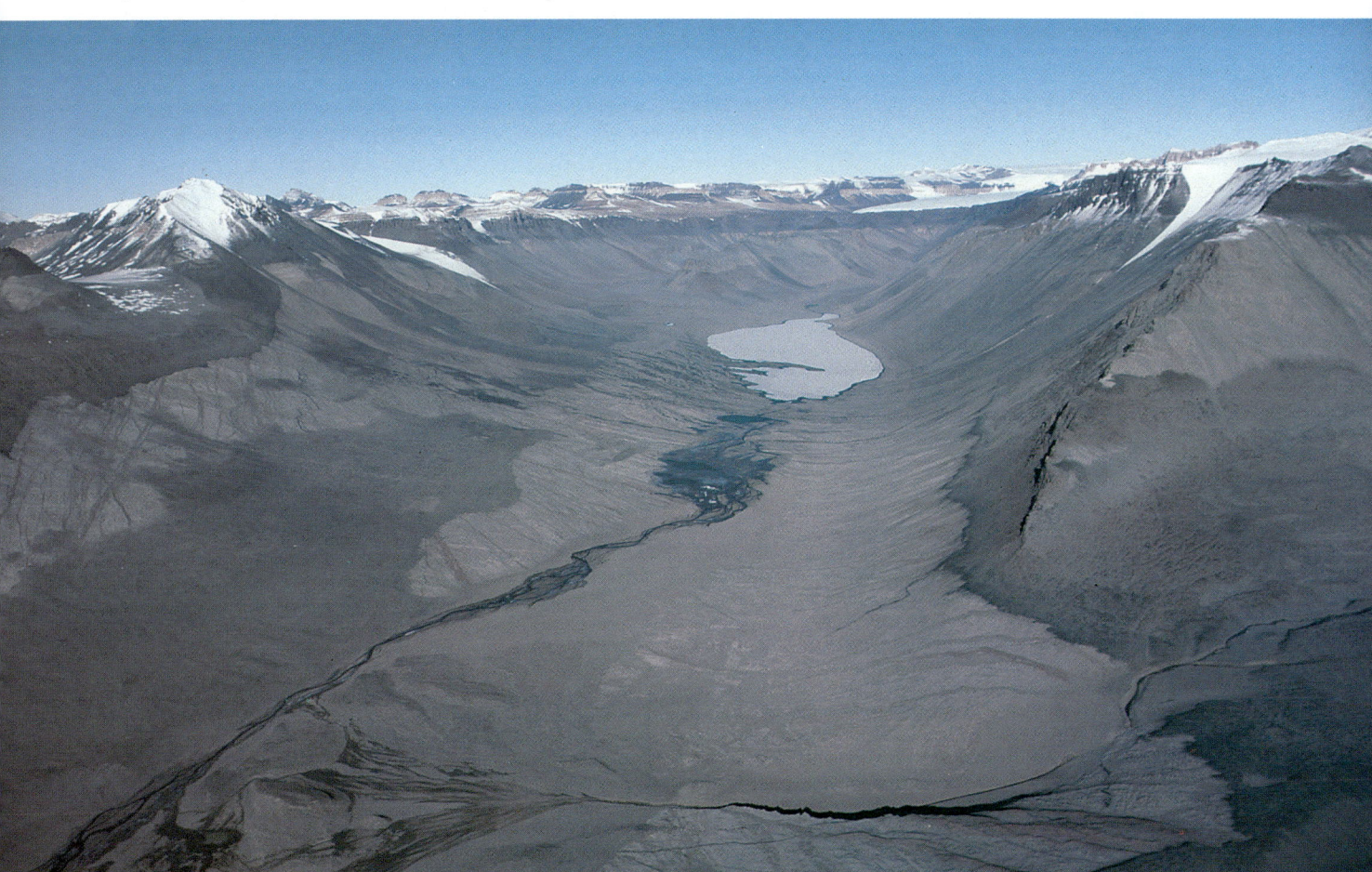

fluctuations are not necessarily a response to climate changes in the recent past, but rather are a reflection of past climates up to a few thousand years ago. Consequently, the Dry Valleys glaciers should not be expected to follow the general glacier recession which has been a worldwide trend over the past century.

Fortunately, the geometry of the ice-cliffs and their relation to internal ice flow makes it possible to distinguish between advancing and receding glaciers, based on the dimensions of the ice-cliff alone.

Ice-cliffs, whether calving or stable, have a similar basic structure where a vertical cliff of clean ice rests on a band of discoloured basal ice containing sediment. From approximately this level, a surrounding apron of ice talus fragments, water, and ice and snow extends at the angle of repose, out to the ground surface. Heights of both ice-cliffs and aprons vary considerably, together with the proportion of the total cliff height occupied by the apron. There is a simple relationship between glacier equilibrium and ice-cliff dimensions: where the total cliff height is 20 metres or more, and the ratio of apron height to total cliff height is greater than 1:5, then the glacier margin is advancing. Conversely, where the total cliff height is 20 metres or less and the apron smaller (with apron height to cliff height ratio equal to or less than 1:5), the glacier is receding.

The result is a very useful and easy method of estimating the state of the glacier with ice-cliffs (Plates 10.5, 10.6). It also removes the temptation to equate glacier equilibrium with the degree of activity of the cliff face, although cliffs of retreating glaciers generally show less calving.

Water in the Dry Valleys

Throughout the Dry Valleys losses from both snow and ice areas are almost entirely by sublimation. The little melt that occurs is significant only below 1500 metres for about 2 months in summer, and is mainly from glacier cliffs and margins as temporary snow patches sublimate rather than melt.

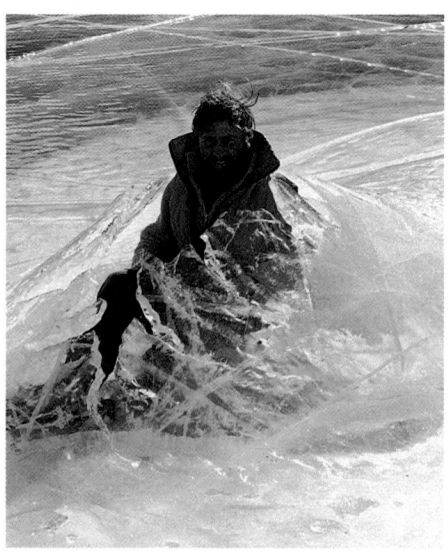

Streams and rivers

Numerous perennial meltwater streams and seeps occur during the warmest summer weeks. Thaws occur first in damp salt-laden sediments of the valley floors and within pockets and enclaves in ice, where the "greenhouse effect" is a very efficient melting process. Surface melt by direct radiation amongst terraced glacier faces can occur at −15°C. All small meltwater streams flow intermittently with a pronounced daily pattern. Highest flows occur with highest temperatures, but are strongly affected by cloudiness and, to a lesser extent, windiness. Almost all of the streams flow in internal drainage systems and ultimately feed enclosed lakes.

Snowfalls do not increase the flow in meltwater streams, but have the opposite effect. A summer snowfall which blankets the ground of the Dry Valleys increases the reflectivity of the areas, so that solar energy is reflected and lost and temperatures fall. Such a snowfall occurred in late October 1977, and that was the only year on record that flow in the Onyx River did not reach Lake Vanda.

The Onyx River

The largest stream in the valleys is the Onyx River (Plate 10.7), where flows have been recorded since 1969. It has been referred to as the largest river in Antarctica. Although it may be the longest at 30 kilometres, other streams, such as the Alph River flowing on the moraines of the Koettlitz Glacier, may rival it for discharge.

The Onyx River arises in the Wilson Piedmont–Lower Wright

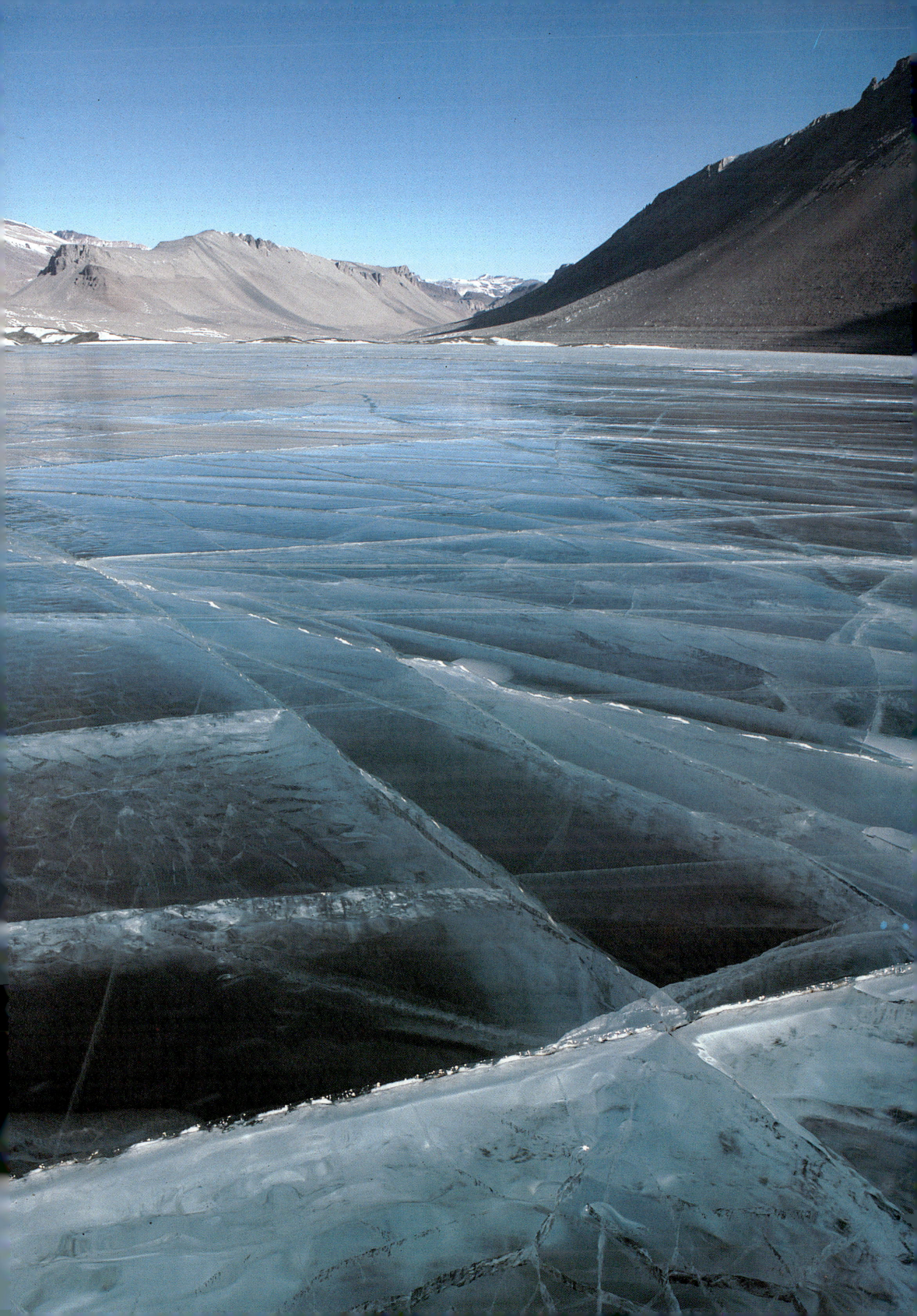

Glacier area, passing through a proglacial lake (one in a basin in front of a glacier) and continues inland for about 28 kilometres to a flow-measuring weir near the point of discharge into Lake Vanda. Minor tributary streams from alpine glaciers along the south side of the Wright Valley contribute about 10 percent of the Onyx River flows.

Water from the Onyx River normally reaches Lake Vanda in the first week of December, but this depends on the weather. Early summer snow cover, cloud cover, and temperature dictate the time when flow commences and similarly determine its variation during the season.

Summer discharges into Lake Vanda vary from 0 to 15 million cubic metres, with an average volume of 3.1 million cubic metres. Over the 1970-71 summer nearly 15 million cubic metres of water poured into Lake Vanda. This is about 4.8 times the average annual discharge and is well above other recorded discharges.

Annual discharges have been increasing and Lake Vanda is rising at a corresponding rate, suggesting that Vanda Station on its shore may be inundated within 3-10 years. The increased flows are caused by warming of the summer climate, but, since meltwater production is extremely sensitive to small climatic changes, this warming has been subtle.

The lakes

The glaciated topography of the Dry Valleys has left numerous depressions which contain a diverse group of remarkable lakes and ponds. Most of the lakes are capped by a permanent ice cover 2.5-5 metres thick which melts free of the shore to leave a moat of water during summer. Chemically, the lakes range from almost distilled water of crystal clarity, to salt-rich brines which do not freeze over the frigid winters. Those lakes with outlets have a through-flow and are usually fresh, whereas the numerous enclosed lakes contain salts concentrated by evaporation.

Plate 10.10. *Don Juan Pond in the Wright Valley. Calcium chloride precipitates show as light areas in the foreground and on the far shore-line. Photo: T. Chinn.*

The enclosed lakes are fed by streams of summer glacial meltwater, and lose water by year-round sublimation and evaporation from the permanent ice cover, and surrounding moat of meltwater in summer. The delicate balance between rise from inflow and lowering by ablation makes the lake levels very sensitive indicators of climate variations. Changes in level have been dominated by changes in inflow. The levels of all of the enclosed lakes have been rising over recent decades and this trend is accelerating, indicating that summers are becoming warmer. Lake Bonney has been rising at least since February 1911, when it was visited and measured by a party from Scott's Expedition.

The depths of the lakes vary from a few centimetres to over 60 metres. Most of the smaller lakes and a few of the larger ones, e.g., Lake Vida, are frozen through to their beds. Like the glaciers, the lakes may be grouped into "wet-based" and "dry-based" categories. Dry-based lakes rise by surface flooding which freezes during winter, and consequently they have a smooth ice cover. Wet-based lakes rise by water flooding in a sheet under the permanent ice cover. Winter freezing adds this water as ice to the underside of the ice cover, while ablation removes the ice from the surface. The perennial ice cover, therefore, undergoes a slow vertical turnover which takes about 10 years for one complete cycle. Ponds, shallower than the winter freezing front, freeze completely through each winter. During this process, the freezing front commonly spreads through the ground beneath the ponds before it is completely frozen, to enclose a pool of water. This water expands into ice during final freezing, and erupts towards the surface to construct an ice dome. Such domes are common, and indicate a shallow pond at a depth within the perennial ice-cover thickness. The surface of the ice covering the lakes is generally rough, being scalloped by ablation and patterned by cracks and

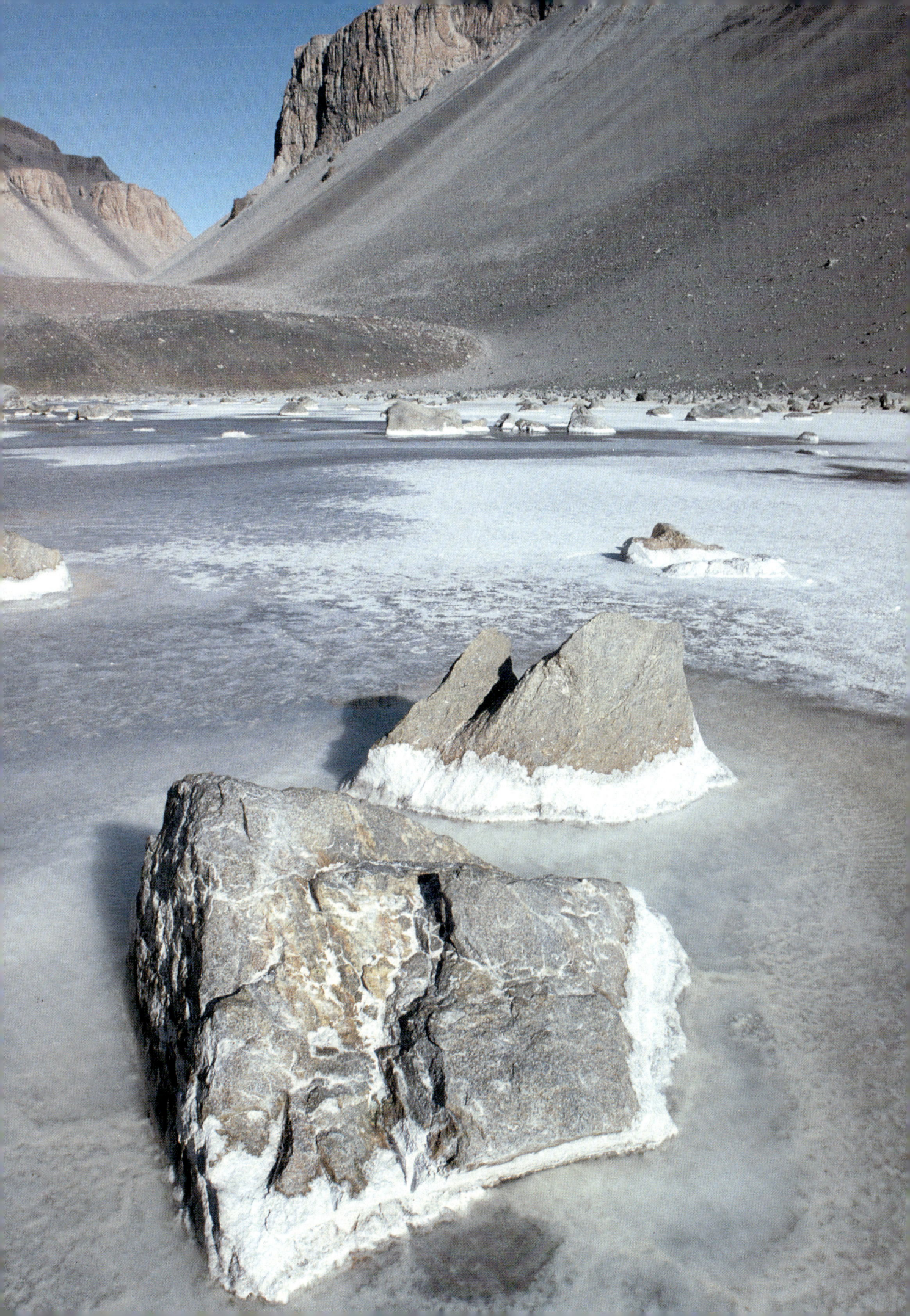

fractures and may even form "blisters" during an early summer sudden warming (Plate 10.8). A most unusual feature of many of the lakes is their curious internal structure of stepped saline and temperature gradients. Lake Vanda (Plate 10.9, Figure 10.4) has at least 12 density stratified layers, each more saline than the overlying layer, and each at a higher temperature. At the bottom of the deepest part of the lake, the water is three times saltier than the sea and reaches a remarkably tepid 25°C.

The saline stratification derives from past low levels of the lake. During periods of lake lowering, pure water is lost to the atmosphere while the salts remain to be concentrated into a brine. When the lake rises during periods of high inflow, the fresh water floods across the lake in a sheet between the brine and the ice cover. The presence of the ice cover is critical as it prevents wind from mixing the two layers. Subsequent rises and falls construct further layers, with the top of each layer marking a past lower level. Although each layer is cooler than the underlying layer, there is no mixing by convection as the downward increase in salt content ensures that the salinity density change is greater than the thermal density change. Within each layer, slow convection has mixed the water into a uniform composition. Between layers, mixing is only by very slow chemical diffusion.

What is the origin of the heat for the high temperatures at the lake bottoms? Measurements made in the sediments at the bottom of Lake Vanda have shown that temperatures begin cooling below the lake bottom. It is now agreed that solar heat warms these lakes. During summer, 24 hours of daylight allows continuous entry of solar energy through the ice cover. The permanent ice cover has vertically orientated ice crystals which pass about 15 percent of the sun's incident radiant energy through the ice into the clear water. These crystals are evident at the moat edge towards the end of summer when they easily separate into shards known as "candle ice". Some radiant energy can apparently penetrate 60 metres to the bottom of Lake Vanda, warming the water on the way. This energy can only be lost by the very slow process of conduction (in the absence of convection), so within these stratified lakes there is a net gain of energy over summer which cannot escape over winter. This system has been duplicated outside Antarctica to create artificial thermo-haline lakes for storing heat.

Don Juan Pond

One of the more interesting of the "lakes" is Don Juan Pond in the Upper Wright Valley (Plate 10.10). The shallow clear water of this pond is a concentrated solution of calcium chloride, so dense that the surface does not ripple in a light breeze, and it does not freeze over winter when temperatures reach −50°C. The air is so dry that hydrated crystals of calcium chloride are frequently precipitated around the pond edge in a mineral named antarcticite. This mineral is unstable in warmer temperatures and higher humidities and is unique to this pond. Water-level records made by a recorder left operating over winter showed a rise in level over this period. Since surface inflow is impossible in winter, the rise in level indicates that underground water seeps into the lake. The high salt concentrations have caused a hole to be melted through the permafrost to give the only known, deep ground-water outlet in Antarctica.

Life in the Dry Valleys

Apart from the prolific and minute ecosystems described in Chapter 12, which include insects and isopods amongst the mosses and algae of the wet flushes, and yeasts in the lakes, nothing appears to be living in the valleys. Occasionally, one comes across orange crustose lichens surviving on rocks high up in the ranges where temperatures never rise above freezing. These appear to be the only life forms capable of surviving in this environment, but micro-organisms living in rocks (see p. 182) are surprisingly widespread throughout the valleys and may be recognised by vague green staining of their rocks.

The Dry Valleys do have visitors other than humans, none of which live off the land. Crabeater seals, the most numerous of all seal species, live on broken sea-ice floes beyond the coastline, but occasionally one becomes disoriented and heads inland, where it finally dies of dehydration and starvation. These stragglers have travelled remarkable distances over the rocky ground and have been found 20–40 kilometres from the coast. When they die their carcasses become freeze-dried and would be preserved forever, except that they are eroded by windblasting—a process which takes 500–1000 years to completely destroy a carcase. The Dry Valleys are dotted with mummified seal carcasses, almost all crabeaters, in all stages of erosion (Plate 11.10). Superficially, the valleys look like the abode of suicidal animals. But, from the age distribution and the number of carcasses it requires less than one wayward seal per year to head inland, from the million or so out on the pack ice, to account for the number of dead seals in the valleys.

Adélie penguins, a breed of determined walkers, also periodically enter the Dry Valleys. Occasional mummified carcasses have been found, even in the Labyrinth in the Wright Valley, 50 kilometres from the sea.

Skua gulls routinely visit the valleys on scavenging excursions, but they have the mobility to return to the coast, or elsewhere, as they please.

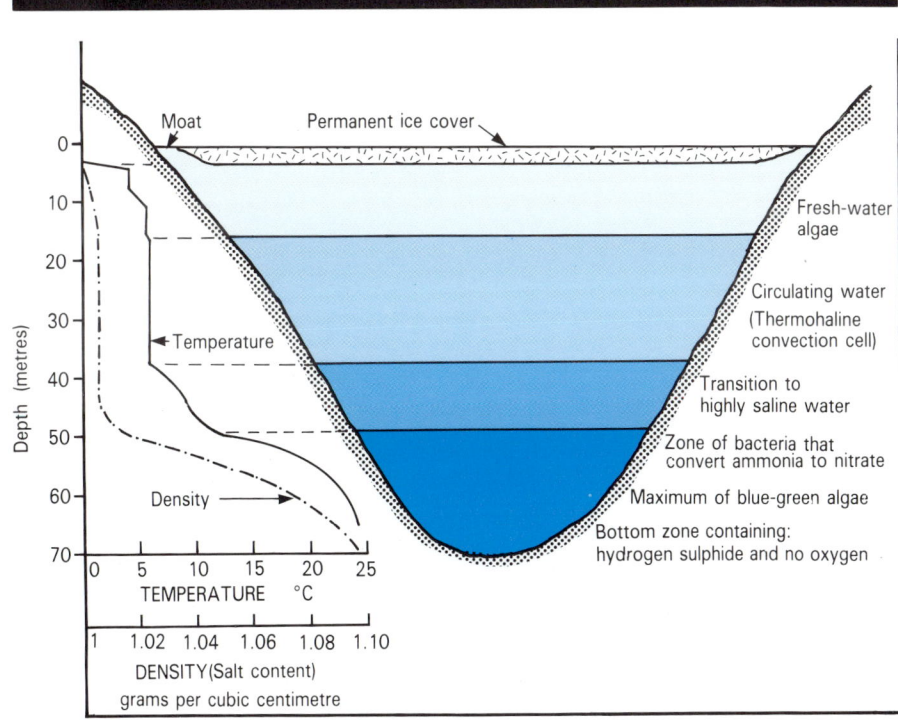

Figure 10.4. *Internal temperature and density structure of Lake Vanda.*

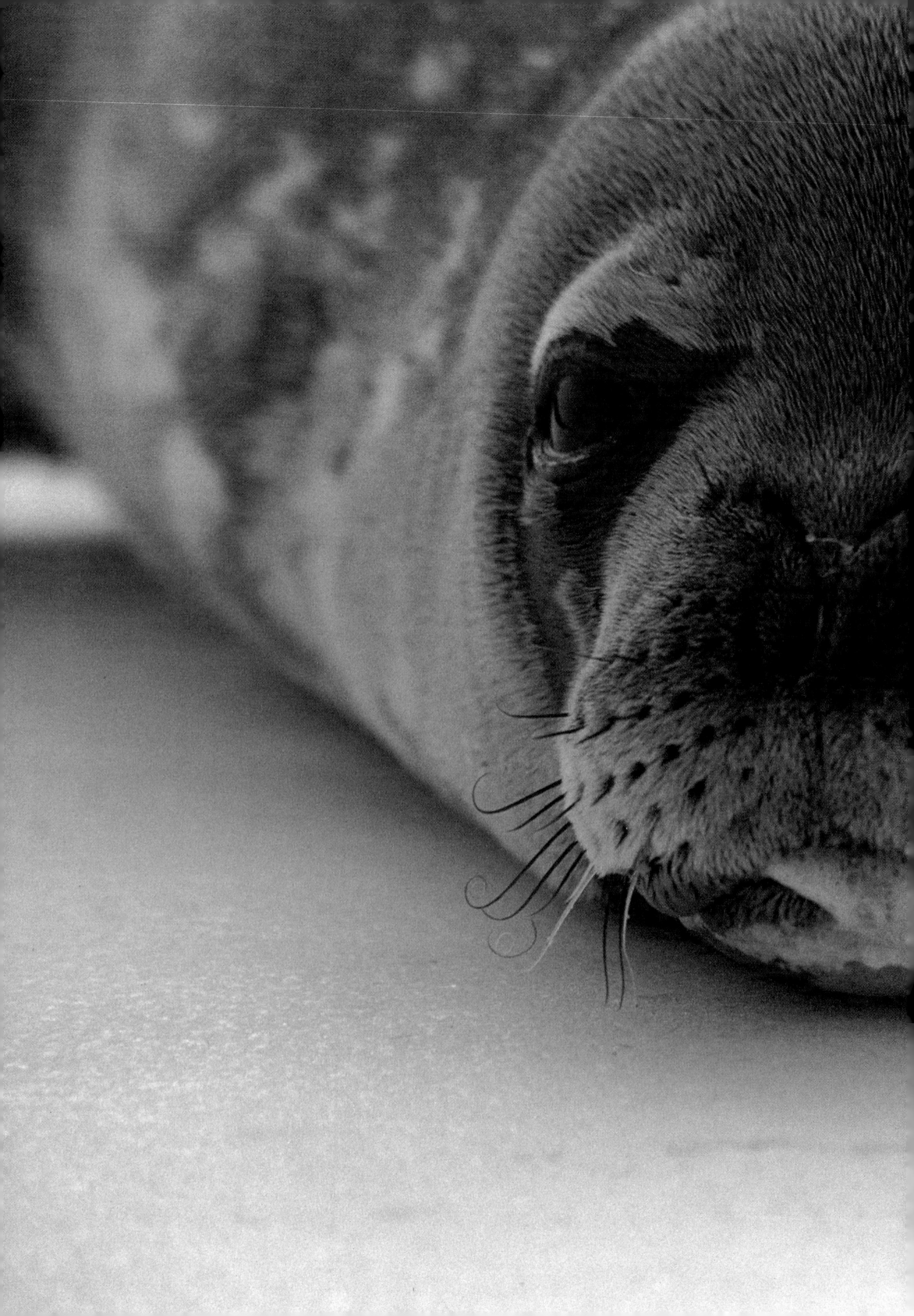

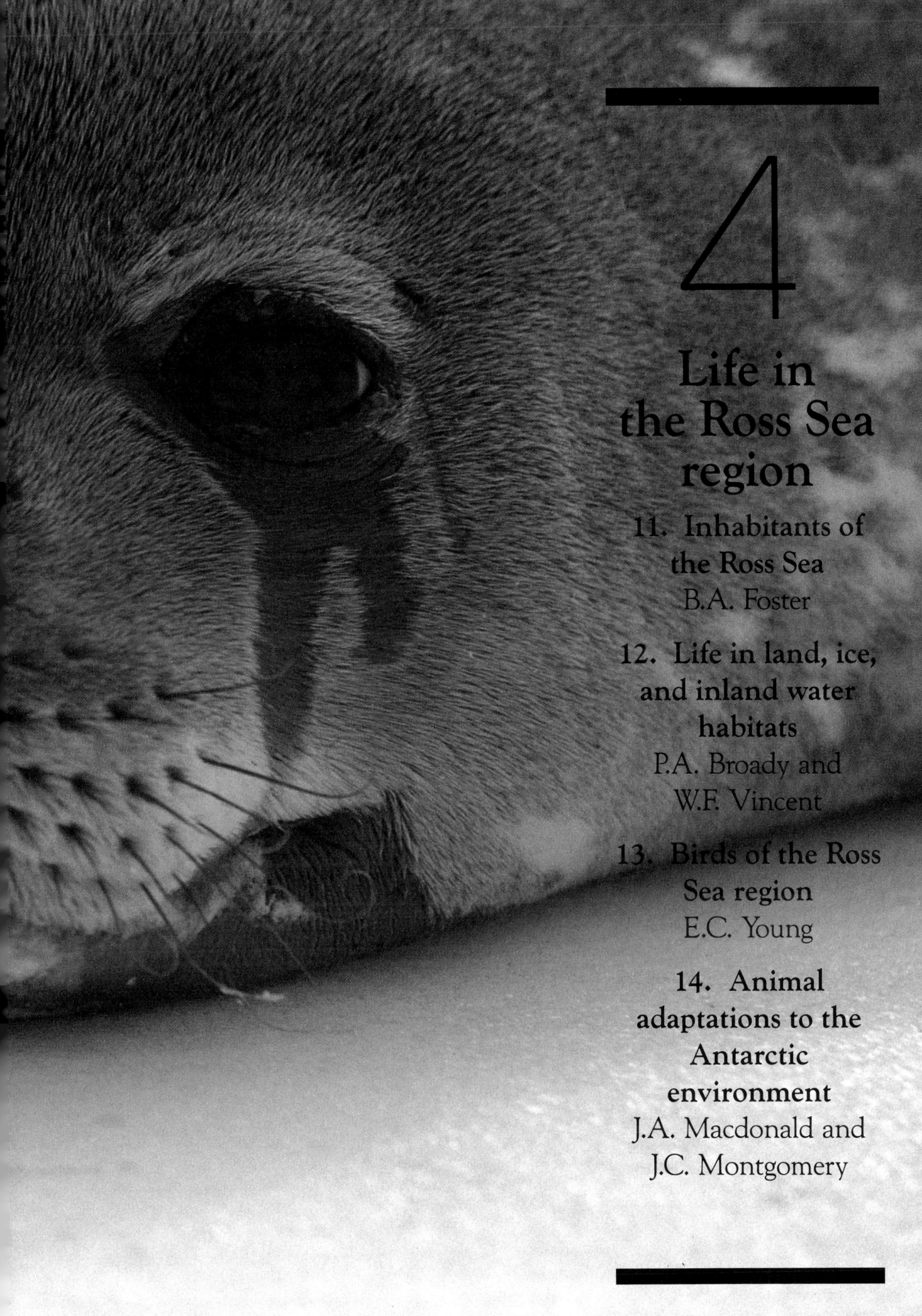

4

Life in the Ross Sea region

11. Inhabitants of the Ross Sea

Penguins and seals on the ice are the familiar signposts that marine life exists under the ice in McMurdo Sound. Their survival depends on the productivity of the seas around Antarctica. It is important to understand the marine biology, if only in the interests of conservation of these warm-blooded vertebrates that have to live there. We have a good knowledge of the kinds of marine life, but we also need to know how they are all interrelated and how the ecosystem (a community of organisms, interacting with one another, plus the environment in which they live and with which they also interact) works. Much of our understanding of the Antarctic marine ecosystem is in the form of conceptual models, depicting energy flow from phytoplankton through zooplankton and fish to the birds and the mammals we more readily identify. Some of our information is sparse, particularly about which species are linked to which in the food web, and little is known about their stocks and productivity.

The Southern Ocean is popularly conceived to be everywhere teeming with life. This is based on knowledge of the flow of water masses and the upwelling of nutrients which enhance diatom growth in the summer when light levels are high enough; on observations of enormous swarms of shrimp-like euphausiids (krill, in part) which feed on the diatoms produced in summer; and on knowledge about conspicuous resident populations of penguins and seals, and summer aggregations of whales, which feed directly or indirectly on krill.

We know these phenomena exist. What may be in doubt is the general applicability of isolated patches of high stock or production of diatoms, krill, penguins, or seals to the whole Southern Ocean, or to any part of it that has not been adequately sampled. The Ross Sea is part of the Southern Ocean, but even now, 80 years since it was first investigated, there is very little information on many important points concerning stocks and productivity of the marine life in its waters.

Life in the Ross Sea is constrained because:

- The water temperature is always low, from about 2.0°C in summer to −2.0°C at its coolest as the ice freezes in winter.
- Floating ice physically interrupts light penetrating into the water, preventing photosynthesis of phytoplankton, and along the edges of the coast causes mechanical abrasion and prevents development of intertidal life.
- Sunlight is absent during mid-winter months, but continuously present over mid-summer months, imposing a strong seasonality on primary production, and necessitating over-wintering strategies in both primary producers and consumers.
- The Ross Sea, particularly in the south and south-west parts, is partially isolated from the oceanic influences that circulate around the continent, and the vast Ross Ice Shelf imposes its permanent bulk over about one-half of the sea bed of the Ross Sea embayment.

Because of these unusual conditions, study of the marine life of the Ross Sea is of interest in itself, as well as being significant for the survival and well-being of penguins and seals, or for providing information about harvestable fish, krill, and whale resources that some claim to be there. An overview follows of the diversity of life-forms and how they are distributed in the sea and structured within the food webs of the marine ecosystem.

Physical aspects

Bathymetry

The Ross Sea is a wide and deep embayment of the continental shelf between Cape Adare and Cape Colbeck (Figure 11.1). Its average depth is 550 metres, carved out by glacial action in times of extended ice cover and lowered sea levels. As with the Weddell Sea, it contrasts with the rest of the Antarctic Continent where the shelf is narrow and steeply shelving. The Ross Sea is marked by a relatively shallow rim which runs near the northern edge from Cape Colbeck in the east to the Pennell Bank, possibly a vast terminal moraine formed by a grounded ice shelf of earlier ice ages. Troughs deeper than 900 metres occur immediately north of Ross Island; others extend southward into McMurdo Sound and beneath the Ross Ice Shelf. The break in the continental shelf where it steepens towards the deep ocean occurs at about 800 metres, more than twice the depth found elsewhere in the world.

Currents and water properties

The major water flow around and close to Antarctica is the East Wind Drift which flows from east to west at average speeds of less than 1 knot (0.5 metres per second). Part of this current flows south-west at Cape Colbeck into the Ross Sea and is the major supplier of the clockwise Ross Sea gyre. West-setting currents of 1–3 knots (0.5–1.5 metres per second) along the edge of the Ross Ice Shelf carry drifting ice and subsurface water towards Victoria Land on the western boundary. Some of this water flows under the Ross Ice Shelf to circulate in largely unknown ways under the permanent ice. Despite these general current sets, there are many localised and short-term variations such as eddies and meanders caused by winds, islands, headlands, glacier tongues, and tides.

For the Ross Sea as a whole, tides are diurnal and the maximum spring tidal range is 2.2 metres. Tidal currents are strongly superimposed on the mean flow, with tidal flow reaching 5 knots (2.6 metres per second) along parts of the Victoria Land coast. Within McMurdo Sound there are complex surface and subsurface flows, often stronger than 2 knots (1 metre per second), due mainly to tidal movements. Water

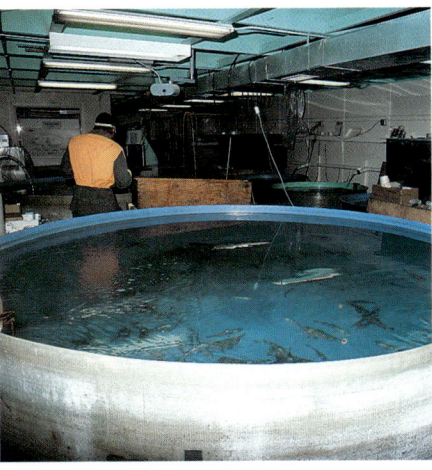

Plate 11.1. *The aquarium at McMurdo Station where live fish are kept for study in controlled conditions. Photo: C. Rudge, Antarctic Division, DSIR.*

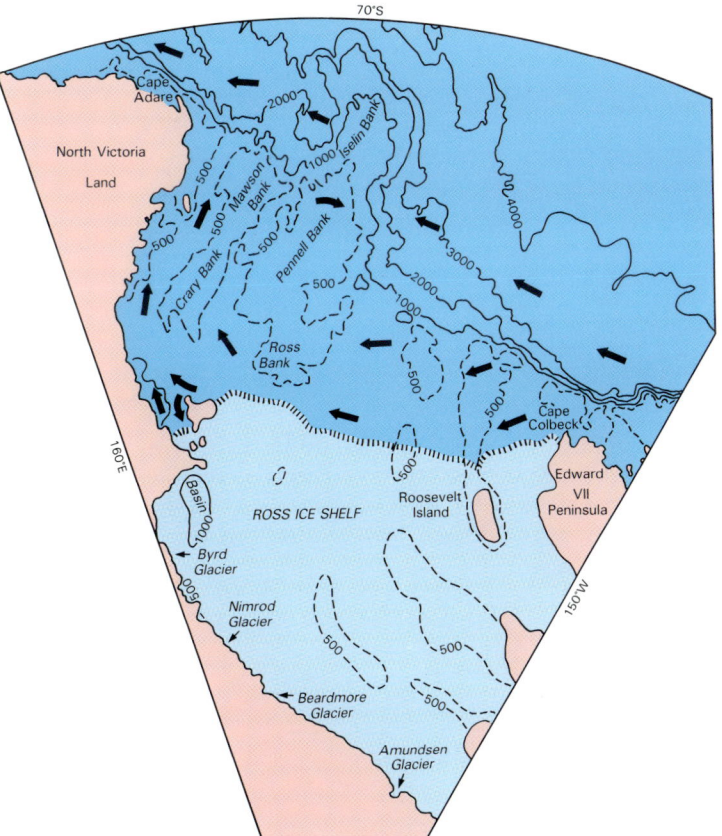

Figure 11.1. *The Ross Sea showing principal features of bathymetry and circulation (solid arrows). Depths are in metres.*

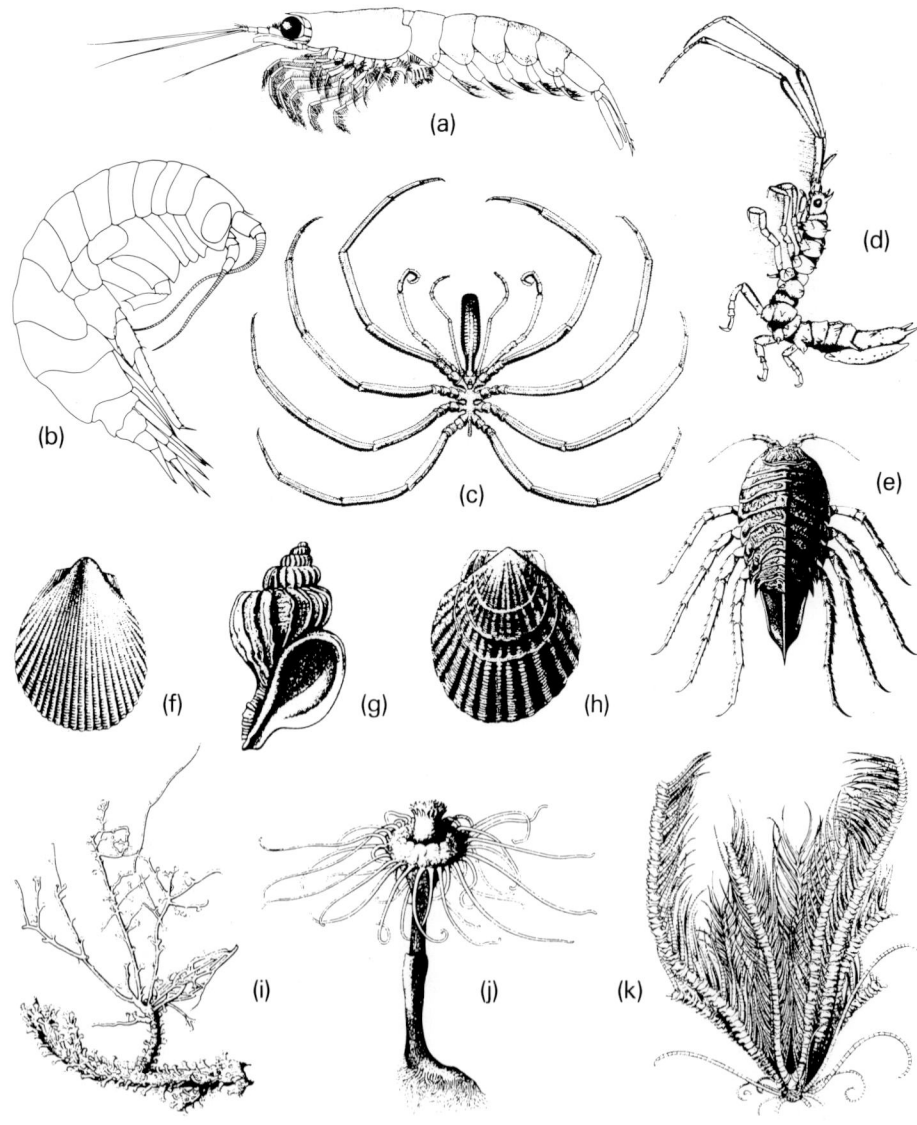

Figure 11.3. *Invertebrates from McMurdo Sound from reports of the early expeditions. (a) Euphausia crystallorophia (crustacean, euphausiid, krill); (b) Epimeriella macronyx (crustacean, amphipod); (c) Collosendeis australis (pycnogonid, sea spider); (d) Antarcturus polaris (crustacean, isopod); (e) Glyptonotus antarcticus (crustacean, isopod); (f) Limatula hodgsoni (mollusc, bivalve, file shell); (g) Trophon longstaffi (mollusc, gastropod, whelk); (h) Adamussium colbecki (mollusc, bivalve, scallop); (i) Hydractinia dendritica (coelenterate, hydroid); (j) Lampra microrhiza (coelenterate, hydroid); (k) Anthometra adriana (echinoderm, crinoid, brittle star).*

Figure 11.2. *Invertebrate communities in McMurdo Sound: a section through the Hut Point Peninsula area, showing some of the life-forms.*

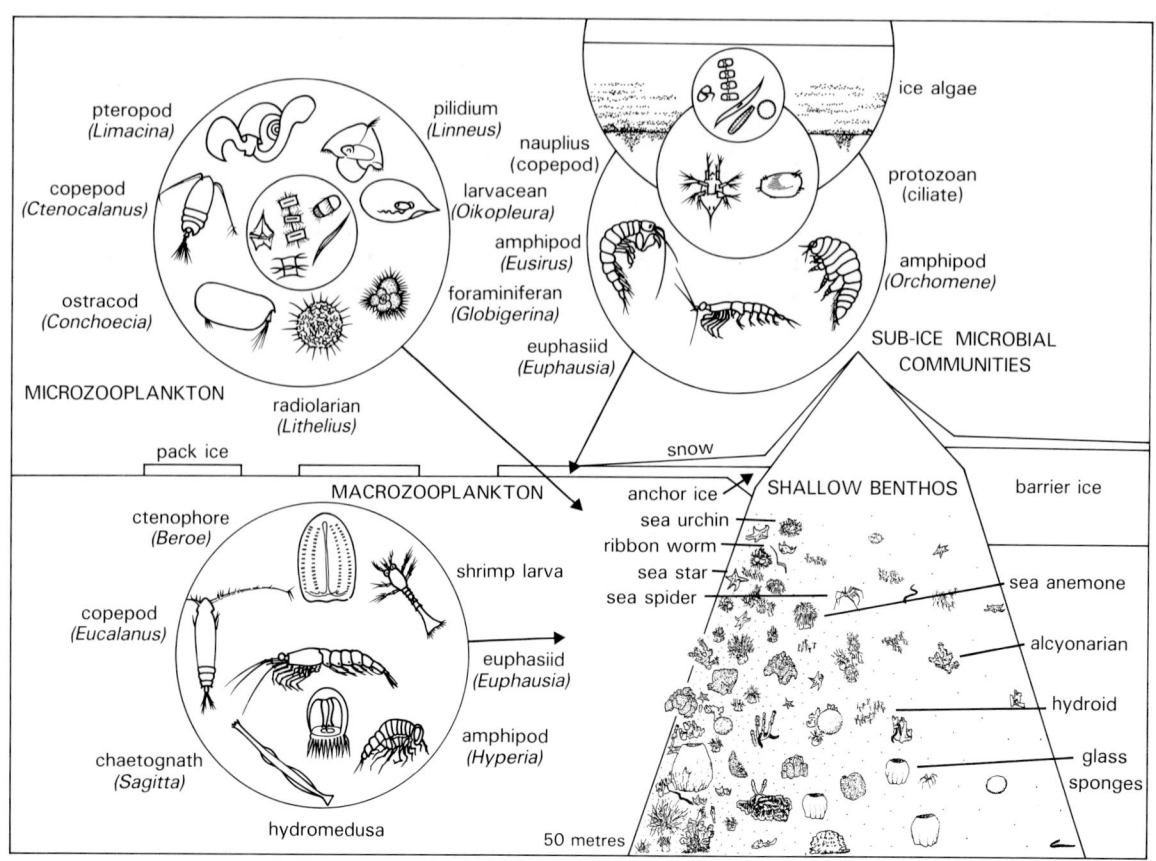

temperatures and salinities vary from over 2.0°C to about –2.05°C, and 33.5–35.0 parts per thousand, respectively. The precise temperature and salinity combinations reflect the origin of several recognisable types of water: surface water is diluted by ice melt; deep water is of low latitude origins; and shelf water is the coldest, most saline, and densest water of the Ross Sea and is found mostly in the south-western part during the freeze-out of fresh water into surface ice. Plant nutrient salts of phosphorous, nitrogen, and silica are generally present in relatively high concentrations in the Ross Sea, possibly mixing in from upwelling deep water along the shelf edge. High nutrient concentrations may be maintained because they are not depleted by the phytoplankton, which can only grow during summer. Water clarity is also generally high because very little suspended sediment enters from the land, plankton growth is seasonally limited, and sea ice dampens wave action which would stir up bottom sediments.

Ice

Sea ice (see Chapter 9) affects life-forms in several ways. For example, it provides resting platforms for penguins and seals, and acts on its undersurface as a substratum for the growth of microalgae which possibly equal, and locally exceed, the production of the free-floating phytoplankton in the water column below. Physically, the ice dampens turbulence due to wind action, and it must also provide a barrier to gaseous exchange between sea water and the atmosphere. Further, as the ice forms it extracts fresh water from the sea surface, increasing the ocean water salinity to over 34.5 parts per thousand. Most importantly for the ecosystem, the ice decreases the amount of light penetrating into the sea, almost completely if there is any snow on the ice.

Seasonally, light is the most variable factor. As seasons progress towards continual darkness of winter, light levels fall below those necessary to support photosynthesis of microalgae. Without photosynthesis there can be no growth. In winter the ecosystem has to run on the reserves accumulated in the previous summer.

Biology

In the 200 000 square kilometres of cold, clear sea water of the Ross Sea, averaging 500 metres deep and mostly covered with ice, there is a diversity of life-forms, from bacteria to whales. Our knowledge about these varies according to ease of access to them. The low temperatures create special logistic problems for research. Ice hinders access to the water below, and threatens ship operations on the surface. Ice crystals form on lines, cables, and nets suspended for more than an hour in near-surface waters in spring and winter, impairing the efficiency of instruments and fine nets. Samples brought to the surface are in danger of immediate freezing and disruption, but if taken into heated vehicles or a laboratory they also need to be prevented from warming up and disruption.

SCUBA divers in some shallow parts of McMurdo Sound have provided a good understanding of how the organisms brought up in dredges and grabs are arrayed on the sea bed. Nowadays at McMurdo Station, an aquarium with sea water pumped from beneath the ice keeps marine fish and invertebrates alive and visible without observers being exposed to subzero conditions on the ice (Plate 11.1). In contrast to the rigours facing surface inhabitants, the marine life in the sea water lies in a reasonably uniform environment, the most stable on earth. The freezing sea waters of McMurdo Sound produce the world's most saline ocean-connected sea.

Although tropical or temperate species would immediately perish in

Plate 11.2. *Taking in the fish trap off Cape Evans, 28 May 1911. Photo: Canterbury Museum, Christchurch, N.Z.*

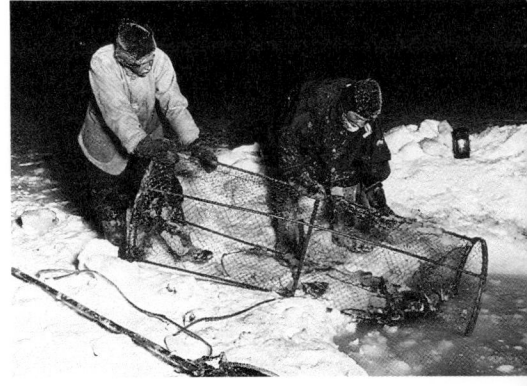

the low temperatures of the Ross Sea, Antarctic marine species are clearly adapted (see Chapter 14). Penguins and seals are equipped with effective insulation of feathers or blubber to maintain warm body temperatures, and cold-blooded animals are physiologically adapted to enable them to function at similar rates to non-polar forms. McMurdo copepods swim just as fast at $-1.5°C$ as temperate ones do at $20°C$. Sea spiders at McMurdo crawl no more slowly than do temperate ones. Antarctic scallops grow at similar rates to temperate water scallops. During spring, diatoms divide once each day as diatoms do in temperate waters. What is different though is that the range of tolerance to temperature change is much narrower in the polar forms. Some of the invertebrates and fish can tolerate an amount of hibernation if frozen in the ice at temperatures below $-2°C$, as long as the temperature does not go too low or last too long, but they cannot tolerate being warmed up to much more than $5°C$. This small temperature range is only a fraction of the temperature range that temperate species experience annually.

Different life-forms are associated with different habitats, and scientists tend to use different techniques for studying the latter. So our knowledge about marine life can be conveniently considered in categories of *benthos* (organisms living on or near the sea bottom), *plankton* (organisms freely suspended in the water and unable to swim against currents), and *nekton* (larger free-swimming organisms such as fish that can overcome the influence of currents). This approach is used below. But remember that the continuum of the water enables much interchange of different parts of life cycles and in the feeding of individuals, and there is a continuum of energy flow that connects all parts of the ecosystem.

Generally, the biota of the Ross Sea is now fairly well documented, although new species will continue to be recorded. The general form of the organisms are like those anywhere else in the world but most species are unique to Antarctica. The prevalence of species names such as

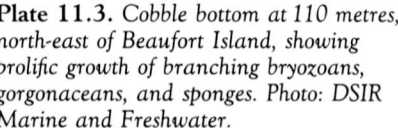

Plate 11.3. *Cobble bottom at 110 metres, north-east of Beaufort Island, showing prolific growth of branching bryozoans, gorgonaceans, and sponges. Photo: DSIR Marine and Freshwater.*

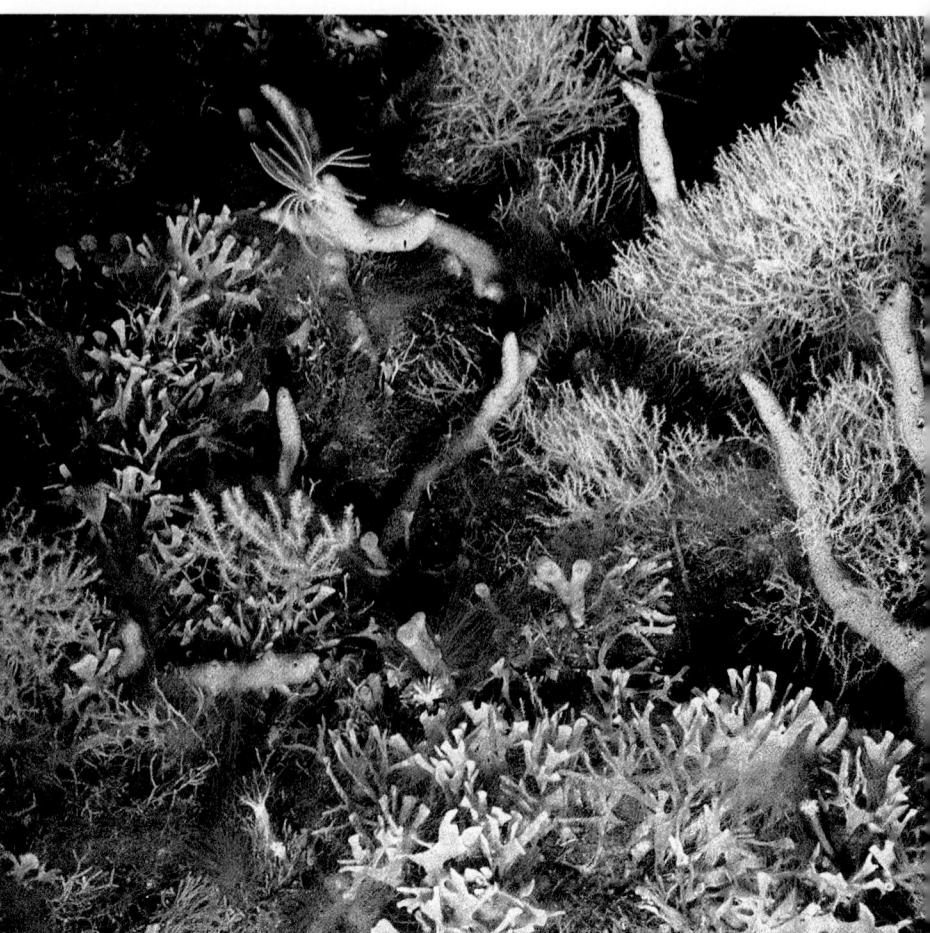

antarcticus, *polaris*, *frigida*, and *meridionalis* acknowledges the geographic uniqueness of the biota. The names *scotti*, *wilsoni*, *hodgsoni*, *mawsoni*, *shackletoni*, and *lilliei* similarly acknowledge the efforts of early explorers.

Benthos

Of the 1000 or so different invertebrate species and 35 species of fish that have been documented from the Ross Sea, most are benthic forms. This is about one-third of the number of benthic forms known from the whole of Antarctica. Diversity increases further north, particularly along the Scotia Ridge islands towards South America in slightly warmer waters, closer to immigration routes from that continent, and where there is decreased ice abrasion on the coastline. In the Ross Sea, freezing of the shoreline during winter, and movement of ice along the shore with tides and winds during summer, prevents the development of intertidal life, in marked contrast to the situation in temperate latitudes. Where ice forms fast against the shore there can be a 15-metre-thick wedge of "anchor ice" against the cold subtidal rocks, and the formation and movement of this causes some impoverishment of immediately subtidal biotas. Below the abrasive effects of ice, however, a rich and diverse benthic fauna exists.

McMurdo Sound benthos

The sea bottom in parts of McMurdo Sound has been sampled most intensively because of the focus of scientific expeditions and programmes there. The relatively recent use of SCUBA in diving studies has revealed a degree of subtidal zonation (Figure 11.2). To depths of 15 metres, on substrates of volcanic cobble, gravel and sand, affected by anchor ice, there is a foraging fauna of sea anemones, sea urchins (*Sterechinus newmayeri*), starfish of numerous species (but dominantly *Odontaster validus* and *Diplasterias brucei*), sea spiders, and multitudes of smaller crustaceans and worms. With shelter from currents the

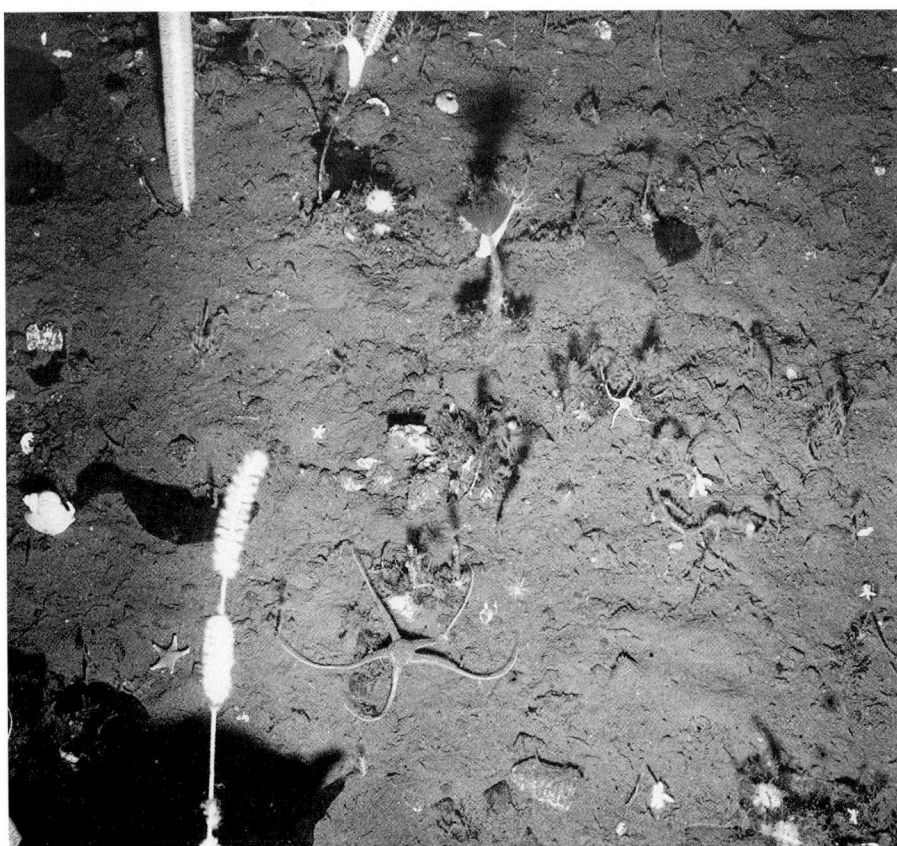

Plate 11.4. *Mud bottom at 75 metres, just west of Beaufort Island, showing a diverse but not prolific community of brittle stars, bryozoans, tube worms, gorgonaceans, and a whelk. Photo: DSIR Marine and Freshwater.*

substratum is more muddy, and a large red nemertean worm *(Lineus corrugatus)*, yellow sea anemones, and several shellfish species are characteristic. Among the shellfish is the Antarctic scallop *Adamussium colbecki* (Figure 11.3h), abundant in some places, and reported to have a delicate, slightly sweet flavour when eaten raw or cooked.

Below 15 metres, but where silt and mud do not accumulate, there are numerous sessile and bushy coelenterates, and below 30 metres there is a thick mat of glass sponges and bivalve molluscs and their debris. Shells are mostly of *Limatula hodgsoni* (Figure 11.3f), a species which lies in burrows among the sponge debris. On rock and boulders down to 200 metres there are rich communities of siliceous sponges, encrusting and bushy colonies of bryozoans, hydroids, and alcyonarians, as well as tube worms. Starfish, brittle stars, mobile worms, crustaceans, and many small tree-like foraminifera occur among these sessile animals.

The benthos of McMurdo Sound has long been admired for its continuous carpet of life-forms, conspicuously sponges and coral-like organisms, which testifies to there being sufficient water-borne plankton, detritus, and bacteria for the sessile benthic animals to feed on. These benthic communities, in turn, provide food for fish that are able to move about in and over the bottom, and all are important nutrient recyclers.

Fish of the characteristic Antarctic suborder Notothenioidei are common throughout the Ross Sea, where 35 species have been recorded. Most of our knowledge of Ross Sea fishes comes from capturing them with baited traps (Plate 11.2), but this has been augmented with observations made by SCUBA divers. Early expeditionists used fish traps, often not so much to add to scientific knowledge as to replenish the larder, but found that most of the fish that could be caught in shallow waters were small and ugly, with big heads, and were not appreciated when cooked. Further, an area under a fishing hole could be quickly fished out, reflecting the generally sluggish behaviour of the species and site tenacity of the individuals. Most of the species stay near the bottom, foraging on a variety of benthic invertebrates or on each other, and occupy different depth ranges. The largest species is the "giant Antarctic cod", *Dissostichus mawsoni*, up to 170 centimetres long and 70 kilograms in weight, that occurs at depths over 300 metres near the ice barrier.

As well as the fish that can be caught in wire mesh traps baited with seal meat, other life-forms can be recovered. Among the various invertebrate scavengers that are adept at locating carrion are amphipods, mostly *Orchomene plebs* and related lysianassids, which can be collected in tens of thousands (a bucketful) within one trap if it is left down too long. The large nemertean worm *Lineus corrugatus* comes up hanging from the traps like long contracted ribbons. Whelks, starfish, and sea spiders also readily crawl into baited traps, and in shallow waters starfish congregate on falling organic detritus under cracks in the ice where algal growth occurs. In very deep waters, organic input from above (where all productivity originates) may be infrequent, widely spaced, and substantial. Fine organic remains sink slowly and are eaten or dissolved in the upper layers of the sea. But seal and whale carcasses sink rapidly and many deep sea animals can quickly locate such occasional feasts amongst wider famine.

Plate 11.5. Brittle star and sea spider off Cape Bird. Photo: Antarctic Division, DSIR.

Life in Ross Sea ocean basins

Offshore, out in the Ross Sea, from the evidence of grab and dredge samples supplemented with underwater photography, there is a mosaic of assemblages of benthic animals correlated with the types of substrate. The sea floor is poorly sorted glacial deposits, interspersed with boulders and gravel transported by icebergs. On coarse sediments with scattered erratic cobbles and boulders, there is a mixture of tube worms, bushy bryozoans and coelenterates, individual corals, brachiopods, and numerous echinoderms, brittle stars, sea spiders, and some molluscs (Plate 11.3). This community is well developed on the shallow Pennell Bank. Isolated current-swept rocks carry an acorn barnacle, the shells of which on death form extensive sediments. On finer sediments (sandy mud) (Plate 11.4) the benthos is not such a continuous carpet of life as that in the glass sponge community of McMurdo Sound, but nonetheless is rich in comparison with the continental shelf benthos in temperate latitudes. Tubeworms, sipunculid worms, arenaceous foraminiferans, and scattered mobile brittlestars, sea cucumbers, and other echinoderms are characteristic of the muddier sea bottom. Below 1000 metres at the far northern edge of the Ross Sea, the bottom is sandy diatomaceous ooze with sparse echinoderms and polychaete worms like anywhere else in the world's deep seas.

Life under the Ross Ice Shelf

With increasing distance under the floating Ross Ice Shelf, the diversity of life decreases in a similar manner to that with increasing depth to the abyssal ocean floor. But there have been limited studies of the life under the barrier ice of the Ross Ice Shelf because of obvious logistic difficulties of penetrating the 200–1300-metre thick ice. At the most southern hole (82°S) some of the fauna has been seen by remote-controlled camera and collected in baited traps. It is comparable with that of the deep-water ocean, with sparse populations of bacteria and no living fauna in the sediments. Larger, scavenging amphipods, isopods, and fish occur there, 450 kilometres from the sources of primary production in the Ross Sea. Under the barrier ice near White Island and the Koettlitz Glacier, closer to the open sea, there are more species than at 82°S, but fewer than are known from the McMurdo Sound benthos. There is no light under the snow-covered ice, and food of whatever form is probably conveyed by south-flowing currents under the ice.

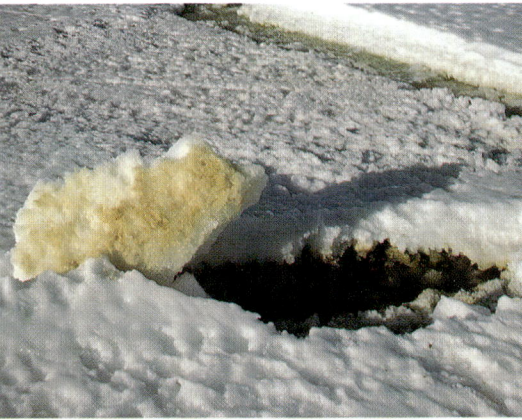

Plate 11.6. *Piece of sea ice removed and upturned to show algae (yellow-green stain) within the ice. Photo: P. Broady.*

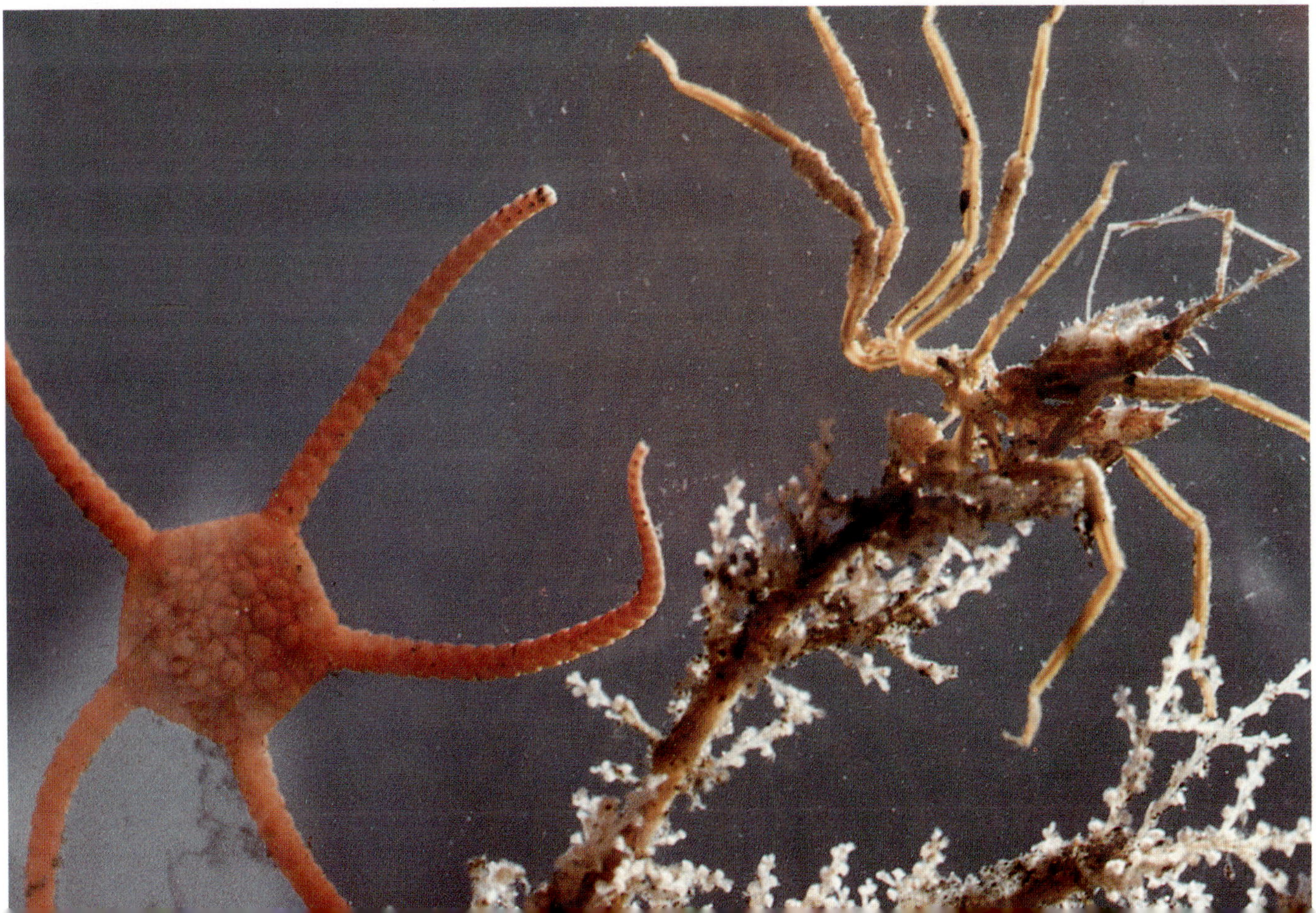

Ice transport of marine life

Shallow-water animals can get caught in winter-forming ice as it anchors on to the land. With subsequent ice break-up, the frozen-in animals (sea spiders, sea urchins, starfish, worms, etc.) are transported for considerable distances before being dropped when the ice melts from below. Consequently, starfish and sea spiders (Plate 11.5) common in very shallow depths are also found living at the greatest depths of the Ross Sea.

These ice-trapped animals can also become stranded on top of the ice where the moving ice sheets buckle and pile up with pressures from tides and the advancing barrier ice. Sun ablation of the "ice rows" strands the organisms. In Pleistocene ice ages, the ice level was higher in McMurdo Sound, and ice movements then stranded shells and invertebrate debris on Ross Island, White Island, and along the Victoria Land coast at elevations up to 100 metres above present sea-level.

These ice-uplift phenomena are different from fossilisation processes where organisms and their debris are incorporated into the sediments on the sea floor. Later glacial action can redistribute the consolidated sediments, leaving fossiliferous conglomerates among the ice-uplifted debris of later times and different processes. Some of the marine animals that were present in the Pliocene no longer occur in Antarctica; the ecosystem as seen today is a changing one in geological time.

Unique features of the benthos

As well as the unique species, the benthic fauna as a whole is characterised by the large diversity and biomass of sessile animals associated with the coarse sediments and ice-rafted cobbles and boulders that occur to relatively deep waters (500 metres). Certain special features are the extensive developments of siliceous sponges and bryozoans, the preponderance of sea spiders or pycnogonids (50 species), the sparseness of large algae (11 species), and the absence of crabs. Apart from the

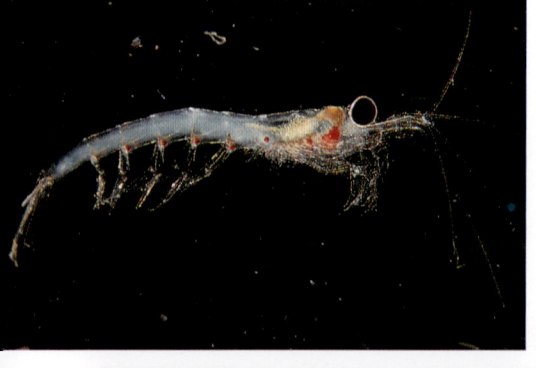

Plate 11.7. Euphausia crystallorophias, *the most common species of krill in the southern Ross Sea. Photo: B. Foster.*

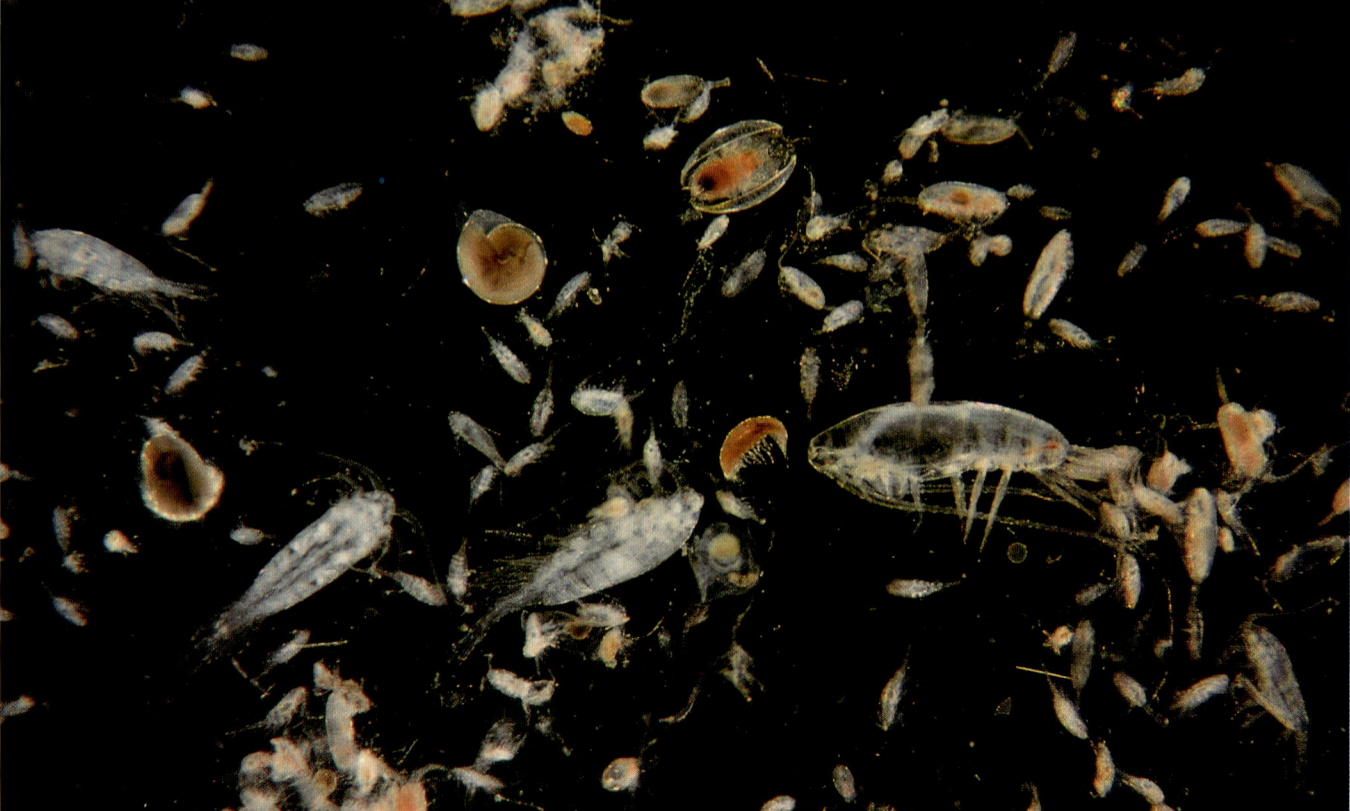

shallow-water occurrences of the scallop *Adamussium colbecki* and the file shell *Limatula hodgsoni*, in high densities in some places, molluscs are inconspicuous. Some invertebrate species attain relatively large sizes for their groups; the isopod *Glyptonotus antarcticus* (see Figure 11.3e), 8 centimetres long, is the largest known from anywhere, and pycnogonids such as *Colossendeis robusta* (20 centimetres leg span) are impressive when compared with the few millimetres-wide sea spiders of temperate shallow-sea habitats.

However, relative richness and high biomass of benthos, or large sizes of animals, do not necessarily indicate high productivity. Some forms are long-lived, and the community as a whole may take a long time to grow and be delicately structured in the relatively stable environment on the sea floor below the abrasive effects of ice. Disruption of these habitats and communities may have long-term effects.

Biogeography

The Antarctic Continent has had its shelf areas isolated from the rest of the world since the Miocene era, 25 million years ago, when the last links with Australia and South America were broken. Because of their closeness to the sea bed, benthic organisms are more geographically restricted than the free-swimming animals. The absence of crabs, and the presence of only one species of acorn barnacle, to cite just two instances, may be a result of the isolation from the mainstream of Tertiary radiations elsewhere around the globe. On the other hand, radiation in the isolation of Antarctica has occurred in some groups, such as the notothenioid fishes (80 species), the lysianassid amphipods (93 species), and the colossendeid pycnogonids (21 species). The extreme cold in itself is clearly not limiting for marine life, given the powers of adaptation and sufficient time for range spread and evolution to occur. After all, deep water faunas everywhere experience very low, and constant, temperatures, and many deep water species are cosmopolitan. In polar seas we can witness "high latitude emergence" of life-forms normally found in very deep and cold waters in temperate and tropic latitudes.

Plankton

Planktonic life-forms are generally microscopic and not recognisable by non-biologists in the way that many benthic forms are. Further, plankton in Ross Sea waters is generally sparse, so that the very clear waters seem mostly empty of life. Blooms of phytoplankton and aggregations of krill have been occasionally seen from the surface, but there is also an ever-present diversity of sparser planktonic life-forms that are more crucial in food chains leading to penguins and seals than are the rich benthic communities.

Phytoplankton

Blooms of phytoplankton have been reported in association with the summer break-up and melt of the ice in the south-western Ross Sea, reducing underwater visibility from about 200 metres to 5 metres. In the western Ross Sea the sea floor sediments contain skeletons of the diatom *Nitzschia curta*, a species which blooms in the water as the ice melts in the summer. From the thickness of the sediments, it can be deduced that this blooming has been a regular feature for 18 000 years. As well as diatoms, the chrysophycean (golden brown alga) *Phaeocystis* sometimes blooms in the southern Ross Sea. The blooms tend to occur only in December–February, in places staining the sea water brown for a few days.

The species of net-caught phytoplankton that have been described from the Ross Sea occur all round Antarctica and some also occur

Plate 11.8. *Zooplankton from under the sea ice of McMurdo Sound. Most of the organisms are copepods; the larger ones in the centre are the phosphorescent* Metridia gerlachei. *Photo: B. Foster.*

further north. Over 100 species of diatoms and 60 species of dinoflagellates (mobile, unicellular flagellate algae) have been documented, with the diatom *Corethron criophilum* particularly dominant in samples taken with plankton nets. Even though nutrient levels are never low enough to limit phytoplankton growth, light penetration into the sea is sufficient for algal growth only in summer, and then only in open water or under snow-free ice. With only a thin layer of snow, less than 0.1 percent of available light reaches the sea water below 1–3 metres of ice.

Sub-ice microbial communities

The bottom layers of sea ice are usually yellow to dark brown, due to diatoms enclosed in the spaces amongst the ice. These ice algae are trapped phytoplankton cells (Plate 11.6), adapted to low levels of light and temperature. They are able to sustain themselves over winter, and in spring grow in the ice densely enough to considerably further reduce light penetration through the ice. As the ice melts from below in summer, the algae may hang as festoons beneath the ice or break free into the water. Associated with these algae is a community of micro-organisms, including bacteria and single-celled animals (protozoans), which utilise the algal food source and its exudates.

Much attention is being given to these sub-ice microbial communities because of their suspected importance in food chains that lead ultimately to diving birds and mammals which keep close to the ice edges. The link is suspected to be through larger, multicellular planktonic animals (metazoans) such as rotifers, copepods, amphipods, euphausiids, and larval fish which aggregate and feed on the algae or the microbial particles associated with them. Reports of wriggling pink "shrimps" on the undersurfaces of pack ice turned up by ice-breaking ships also indicate that euphausiids aggregate at the ice–water interface. There are extensive areas of pack ice in the Ross Sea, and the sub-ice microbial communities may thus represent a high proportion of the

Figure 11.4. *Feeding relationships among the marine life of the Ross Sea.*

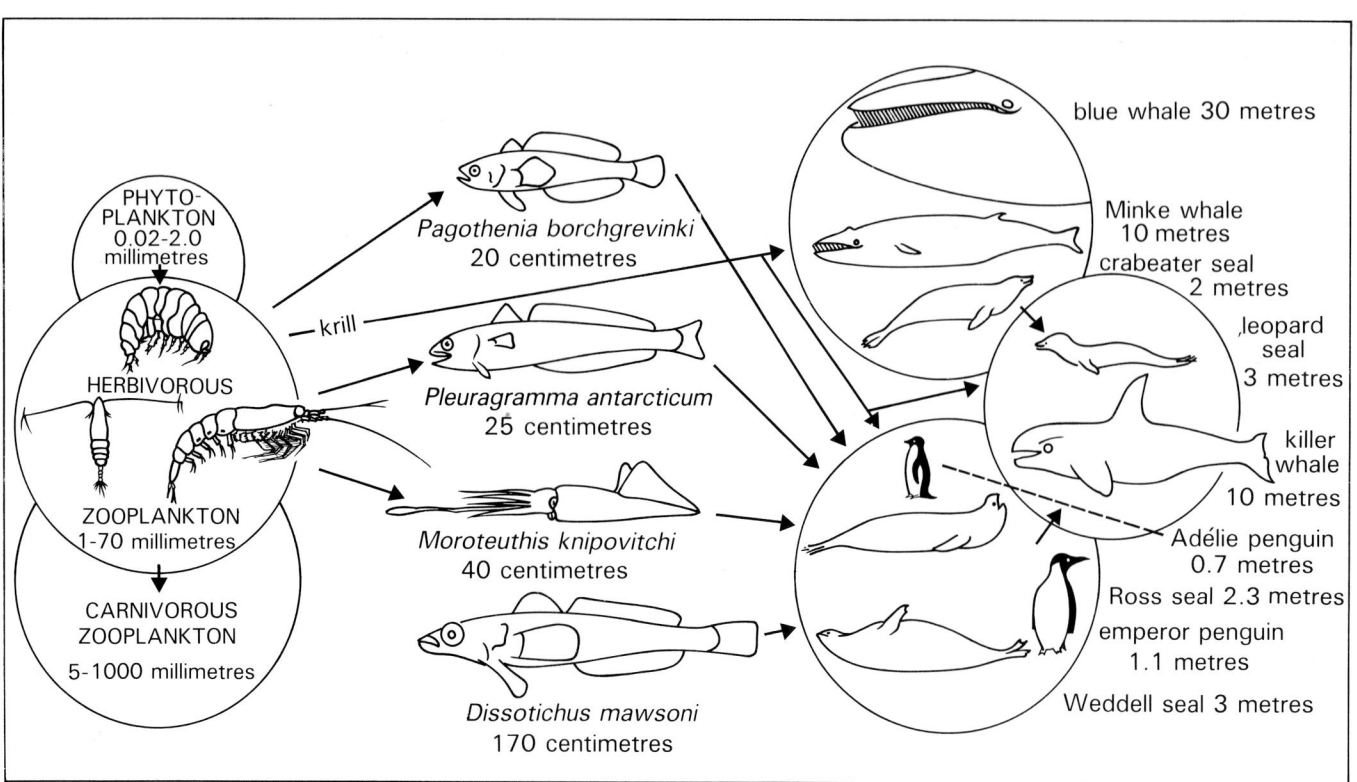

overall productivity. Penguins and seals may be targeting the marine life associated with the sub-ice interface. In the northern Ross Sea, where pack ice is more prevalent, there are large populations of crabeater seals whose principal diet is euphausiids. In the austral winter, Adélie penguins migrate north to the pack ice, and remain there for 8 months of the year. They also feed on euphausiids, one of which is *Euphausia superba*, or Antarctic krill.

Krill

Some biologists define krill as all species of *Euphausia*, or in Antarctica just *E. superba*. Whereas others regard krill as being all zooplanktonic crustaceans that are caught by the filtering teeth and baleen of seals and whales. From the predator's point of view, it may make no difference what the species of plankton are so long as they occur abundantly enough and are large enough to be seen or otherwise detected.

No single planktonic animal has received more attention in Antarctic marine biology than *Euphausia superba*. It occurs in swarms kilometres long in some places in the Southern Oceans, and it is a major component of the diet not only of crabeater seals, leopard seals, and Adélie penguins, but also of baleen whales, squid, and some fish. *E. superba* plays a major role in the marine ecosystem of much of the Southern Ocean, between the primary production of diatoms in surface layers of the sea on the one hand, and vertebrate predators on the other. Its useful features, as far as its predators are concerned, are its individually large size (7 centimetres adult length, the largest of all euphausiids) and the way it aggregates into large swarms. The swarms are frequently made up of individuals of the same size, representing groups derived from the simultaneous summer spawning of eggs from the adults.

The life history of *Euphausia superba* is being intensively studied because some nations regard krill as a resource worth harvesting, particularly since the great reduction of the populations of baleen whales in the first half of this century has, they argue, made more krill available. It is rich in protein (15 percent by weight) and vitamin A. Estimates of stocks indicate that the harvestable resource (up to 150 million tonnes per annum) exceeds that currently taken from all the rest of the world's fisheries.

Yet, *Euphausia superba* does not occur everywhere in the Southern Ocean, and swarms of it have not always been found by ship searches. There is no certainty about the magnitude of the krill resource. In the Ross Sea, *E. superba* occurs only in the very north. Further south its place is taken by a smaller species, *E. crystallorophias* (Plate 11.7). This is a coastal species, the euphausiid of the Ross Sea pack ice and the species which also occurs under the fast ice of McMurdo Sound. Near Ross Island it is taken by the Adélie penguins when they are there in summer. We know even less about the biology of *E. crystallorophias* than we do of *E. superba*.

Other zooplankton

Euphausiids are only a part of the zooplankton of the Ross Sea. Like the phytoplankton, the species of zooplankton occur all round Antarctica, although not all Antarctic species are found in the Ross Sea's shallow shelf waters. Most species of the zooplankton are copepods of which about 90 species are recorded from the Ross Sea, ranging in size from herbivores 1 millimetre in length to predators of 15 millimetres length (Plate 11.8). Large copepods, and amphipods, are also consumed by whales that eat krill.

The commoner euphausiids and copepods are herbivores, feeding on the diatoms and dinoflagellates of the phytoplankton, and perhaps thereby contributing to the generally low stocks of the phytoplankton. Other herbivores are crustaceans such as ostracods and mysids and the larvae of the three shrimp species known from the Ross Sea. A pteropod mollusc, *Limacina helicina*, and the few larval forms of benthic starfish and worms that have been collected in McMurdo Sound plankton are microherbivores that feed on smaller phytoplankton than the diatoms that can be caught in nets. Very little is known about this extremely small or "nano" plankton, but there may be proportionally less of it than in tropic and temperate waters.

Zooplankton that eat zooplankton include the larger copepods and hyperiid amphipods, and another pteropod mollusc called *Clione antarcticus*. Various gelatinous jellyfish from a few millimetres to 1 metre in diameter are all predators on zooplankton. There are also exquisite, elongate siphonophores and chaetognaths, and large (up to 15 centimetres long) delicate brownish ctenophores *(Beroe)*. Larval fish are also zooplankton, subject to being eaten by predators, and themselves feeding on various animals of the plankton, bigger ones as they themselves get bigger.

Within the zooplankton there are complex, certainly poorly understood, feeding relationships. Only some of the species utilise the phytoplankton, and only some of them are utilised directly by larger vertebrate predators. The zooplankton of the Ross Sea is probably the least known part of the marine ecosystem, with surprisingly little data about the biomass and productivity of its component species. Information from limited sampling suggests that the density of zooplankton in McMurdo Sound is not great when compared with biomasses in other parts of the world. Even the higher densities of the northern Ross Sea are low when *Euphausia superba* is excluded.

At all sampling stations greater volumes of zooplankton are obtained by netting from the deep dark levels of the water column, where the

Plate 11.9. *Weddell seal and pup on the ice basking in the sun. Photo: C. Rudge, Antarctic Division, DSIR.*

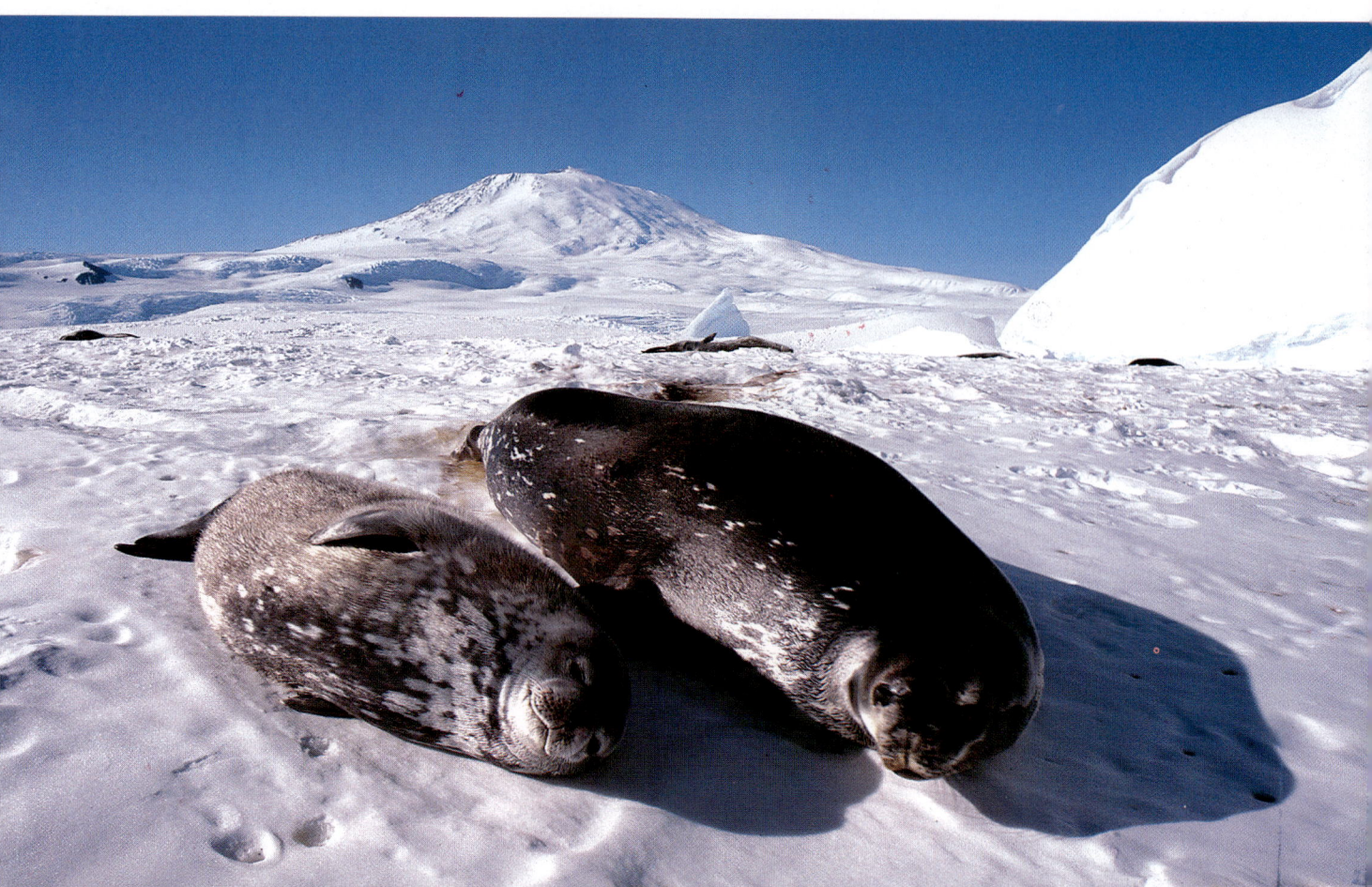

larger copepods occur more abundantly. In non-polar seas, many zooplankton species daily move vertically in the water column, rising to the surface layers at nightime, sinking again in the morning. If these movements are related to changes in light levels, as they are believed to be, then the zooplankton of McMurdo Sound may make their vertical migrations to other than diurnal cues, perhaps staying deep in mid-summer, but rising to shallow levels when swept under snow-covered ice.

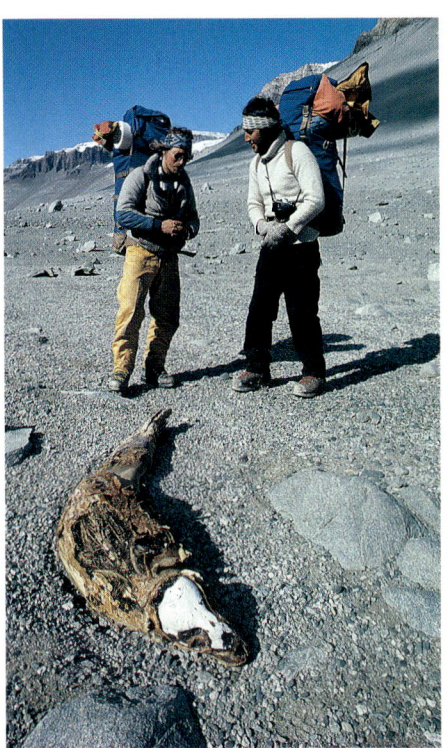

Plate 11.10. *Mummified crabeater seal, Wright Valley. These seals are seen rarely in McMurdo Sound and some become disorientated, perish, and mummify, as far as 60 kilometres from the sea. Photo: K. Westerskov, Antarctic Division, DSIR.*

Phosphorescence

Several marine organisms in McMurdo Sound are known to phosphoresce. One of the commonest of these is the planktonic copepod *Metridia gerlachei* (see Plate 11.8). When caught in nets it emits greenish-blue flashes of less than 1 second to glows of 20 seconds from light-organs in its tail, only visible if viewed in darkness. During the *Discovery* winter-over in 1903, the biologist Hodgson, while handling tow nets out on the ice, observed brilliant displays of emerald-green phosphoresence. These displays were probably as noteworthy during the prolonged winter night as the aurora lights in the sky.

The function of marine phosphorescence could include such advantages as confusing predators and molestors, or the organisms warning their own kind of danger. Among the larger animals that can phosphoresce are the euphausiids, each equipped along its body with 10 big phosphorescent organs which can be turned on spontaneously. Disturbance of krill swarms when whales feed may excite phosphorescence in the krill, but an alarm function may be of no use during the continuous daylight of summer when whales feed in Antarctic waters. It is known that the proclivity to glow in McMurdo Sound invertebrates decreases with advancing summer, so it is probable that any communicating functions are significant in winter, or in deep water, or under the perpetual darkness of the ice shelf.

Nekton

Fish and squid are nekton, and in Antarctica we can usefully include whales, seals, and even the diving penguins. In the Ross Sea there are two species of penguins (Adélie and emperor) and four species of seals (crabeater, Weddell, leopard, and Ross'). Eight species of summer-visiting whales, including both baleen and toothed species, have been observed in the Ross Sea, some of them rarely. The penguins have been extensively studied, and their biology is described in Chapter 13. Weddell seals and a few species of fish have also been well studied, and their adaptations are discussed in Chapter 14. All the penguins and mammals are linked to the plankton and benthos in the ecosystem through fish, squid, and krill (Figure 11.4). The intricacies of the food web are not fully known, certainly not quantified. The importance of certain key species of krill, pelagic fish, and squid to the lives of the warm-blooded vertebrates is recognised even though not much is known about their biology.

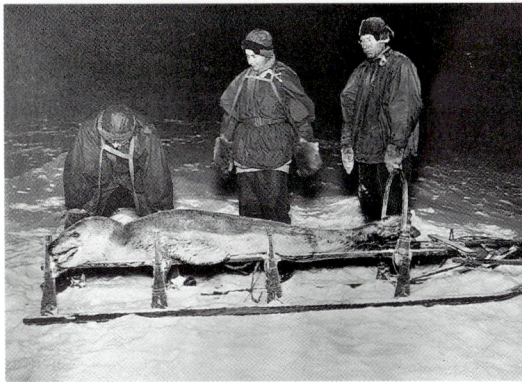

Plate 11.11. *The characteristic head, jaw, and neck of a leopard seal are seen here with Edward Wilson bent over the neck, Cherry-Garrard in the centre, and Forde on the right. The seal was killed near Inaccessible Island on 29 May 1911. Photo: Alexander Turnbull Library, Wellington.*

Fish

Of the 35 species of fish, most are bottom-dwelling forms. There are a few free-swimming forms which as adults feed directly on the larger zooplankton. One of these is *Pagothenia borchgrevinki* (see Figure 11.4) which keeps close up under the shore-fast ice, withdrawing into interstices in the ice when seals approach. This is not a fish of the open waters and pack ice. Occurring throughout the Ross Sea is the herring-like "Antarctic silverfish" *Pleuragramma antarcticum* (see Figure 11.4), also a notothenioid. It was once thought to occur only at mid-depths but we know very little about this fish because we have not been very adept at sampling them. Skuas on Ross Island exist on them over summer, and as skuas do not dive the fish must be near the surface. Weddell seals and Adélie penguins also extensively feed on them, but these predators can dive. The quantities of remains of these fish in seal and penguin faeces and stomachs indicate that *P. antarcticum* may be as important in the marine ecosystem off Ross Island as *Euphausia superba* is further north.

   McMurdo Sound fish have mostly been studied because of their physiological adaptations to life in near-freezing water. For experiments, scientists have caught fish through holes made in the sea ice, mostly not far from Scott Base and McMurdo Station. These holes have been variously cut, blasted, or drilled through the 1–3 metres of sea ice. When these holes are first made, the first fish to appear are the shallow-water *Pagothenia borchgrevinki*, hovering just below the ice and probably attracted to the increased light penetration (see Plate 14.1). These fish can be caught on hand-jigged lures simulating the jerky swimming movements of planktonic crustaceans. Before long, however, Weddell seals also locate the holes, largely it is thought by sight, and fearlessly use them for breathing, thereby extending their foraging ranges under the ice. They sometimes arrive apparently exhausted, judging from their very laboured breathing and disinclination to dive again.

   When the holes are positioned over deep (300 metres or more) water, the seals are able to dive deep to catch the large cod-like toothfish *Dissostichus mawsoni*, sometimes apparently running short of breath and returning to the hole with live fish held in their mouths. Scientists have in the past commandeered these fish for scientific, and later culinary, purposes. *D. mawsoni* can also be caught on long lines with hooks baited with whole *Pagothenia borchgrevinki*, so completing the evidence for a sub-ice food chain of: phytoplankton→herbivorous zooplankton→ predatory zooplankton→plankton-eating fish→fish-eating fish→seals.

   This 6-step food chain incorporating benthic fish contrasts with the short 3-step one of: diatoms→euphausiids→crabeater seals (or baleen whales) in open sea in the north of the Ross Sea, and serves to remind us that not all of the Antarctic ecosystem is made up of short food chains which have often been cited to account for enhanced biomass of marine vertebrates in Antarctic waters. Furthermore, there is considerable flexibility in the diet of many of the predators. Opportunism, that is taking what can be found, is clearly more advantageous than a rigid diet. Weddell seals are opportunists; as well as feeding on a variety of fish species they can also feed on small crustaceans on or near the sea bed, and these crustaceans (amphipods, isopods, mysids) are not directly linked with the plankton.

Weddell seals

   Weddell seals are very approachable when they lie out on the ice, where there are no natural enemies. They are generally unconcerned about humans and human activities unless actually touched or if newborn pups are approached too closely. This tolerance at close quarters has enabled scientists to study the seals of McMurdo Sound, so that we now know much about their physiology, reproduction, and social structure. In spring, Weddell seals form breeding assemblies along pressure cracks and tide cracks in the fast ice. The shortage of suitable cracks causes the seals to concentrate around them. Pupping occurs during September–November, and several females share a hauling-out hole (Plate 11.9). Breeding males spend most of the time in the water beneath the cracks defending aquatic territories with a variety of underwater sounds, gaining exclusive harem rights. Mating takes place under water, 1–2 months after the females have pupped. At birth the pups weigh 25 kilograms (cf. a fully grown female weighs 450 kilograms), and they are weaned at 6–7 weeks when they weigh 110 kilograms. The females abandon the pups shortly before the sea ice breaks out. Weddell seals are the world's best diving seals (see Chapter 14).

At White Island, only 20 kilometres from the barrier edge, there is a population of Weddell seals that gains access to the sea below the thick barrier ice through pressure cracks formed by the tides and the movement of the McMurdo Ice Shelf around the island. This is the most southern population of mammals in the world, with awesome climatic conditions above the ice. The seals are too far away from the seasonally open waters of McMurdo Sound to swim the distance under the ice. At the 5 knots (2.6 metres per second) they can attain, they can swim a radial distance of 2.5 nautical miles (4.6 kilometres) on one breath. There are no breathing holes on the McMurdo Ice Shelf, so the seals at White Island must be a self-contained population, feeding on fish including silverfish.

Other seals

Whereas Weddell seals are characteristic of the land-fast ice around the coast, the other species are only common in the pack ice and are thus not nearly so well known because they do not occur so handily or are less readily approached by humans and their machines. Crabeater seals (*Lobodon carcinophagus*) are the commonest, and mostly feed and breed out in the pack ice in the north of the Ross Sea. Young crabeater seals migrate south towards the shore of the continent, and some may continue, apparently in disorientation, some distance inland, only to perish and mummify there (Plate 11.10). Crabeater seals have teeth with complex series of cusps, such that when their jaws are closed a strainer is formed. They feed almost exclusively on krill, whereas Ross's seal (*Ommatophora rossi*), which also occurs in the pack ice, is predominantly a predator of fish and squid.

Plate 11.13. *A Minke whale in McMurdo Sound. Photo: C. Rudge, Antarctic Division, DSIR.*

Plate 11.12. *A killer whale looking for prey, a common sight where seals and penguins are resting on the sea ice. Photo: Antarctic Division, DSIR.*

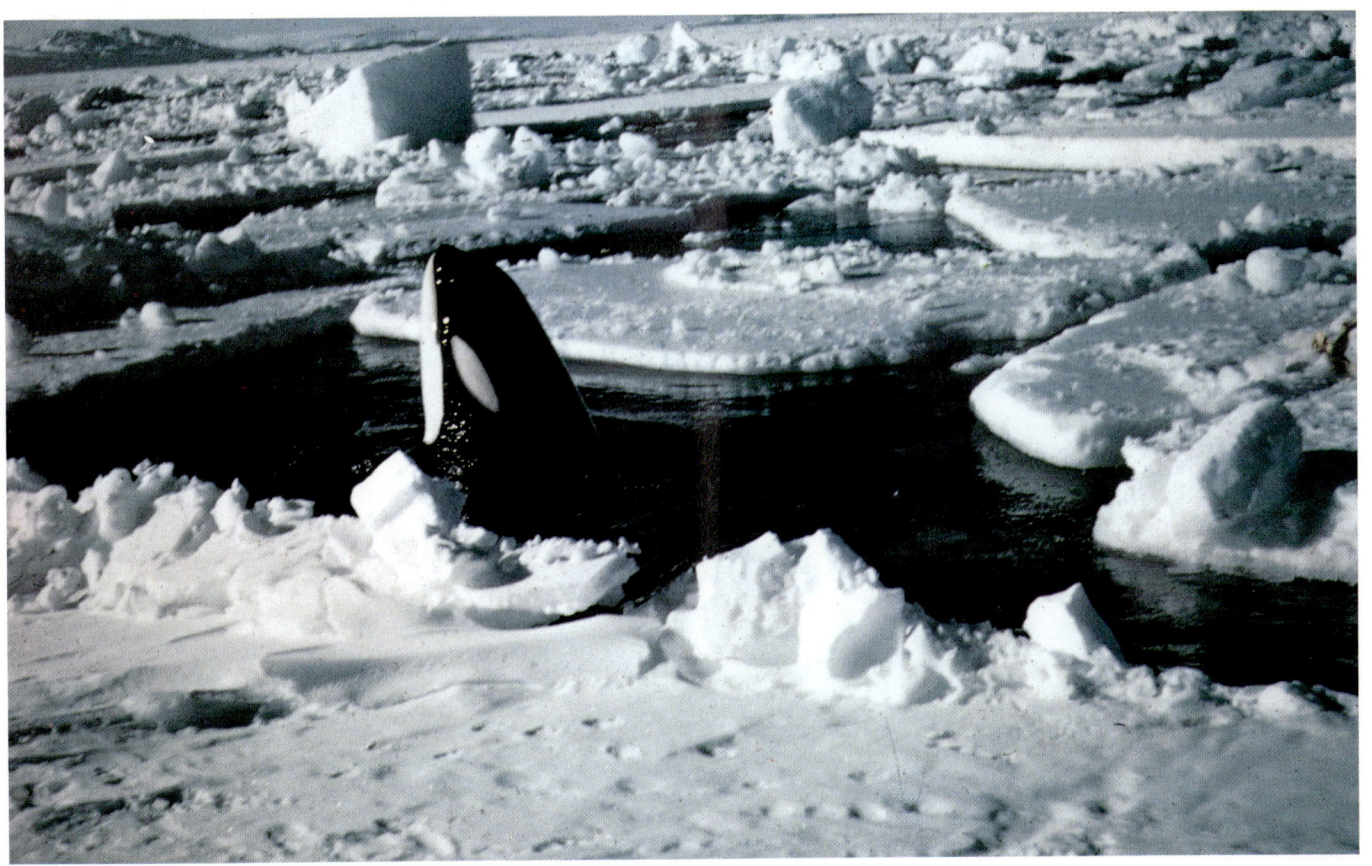

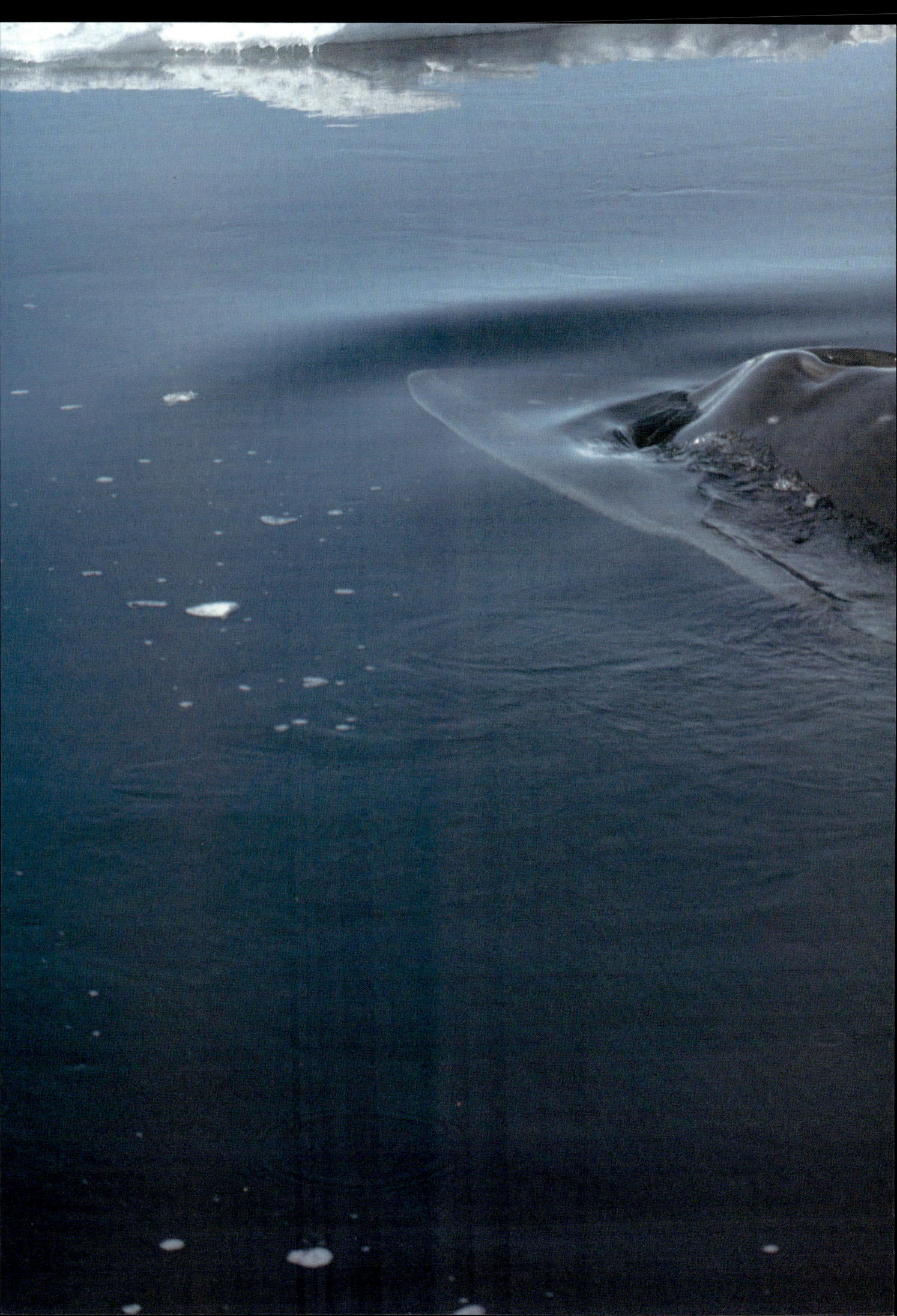

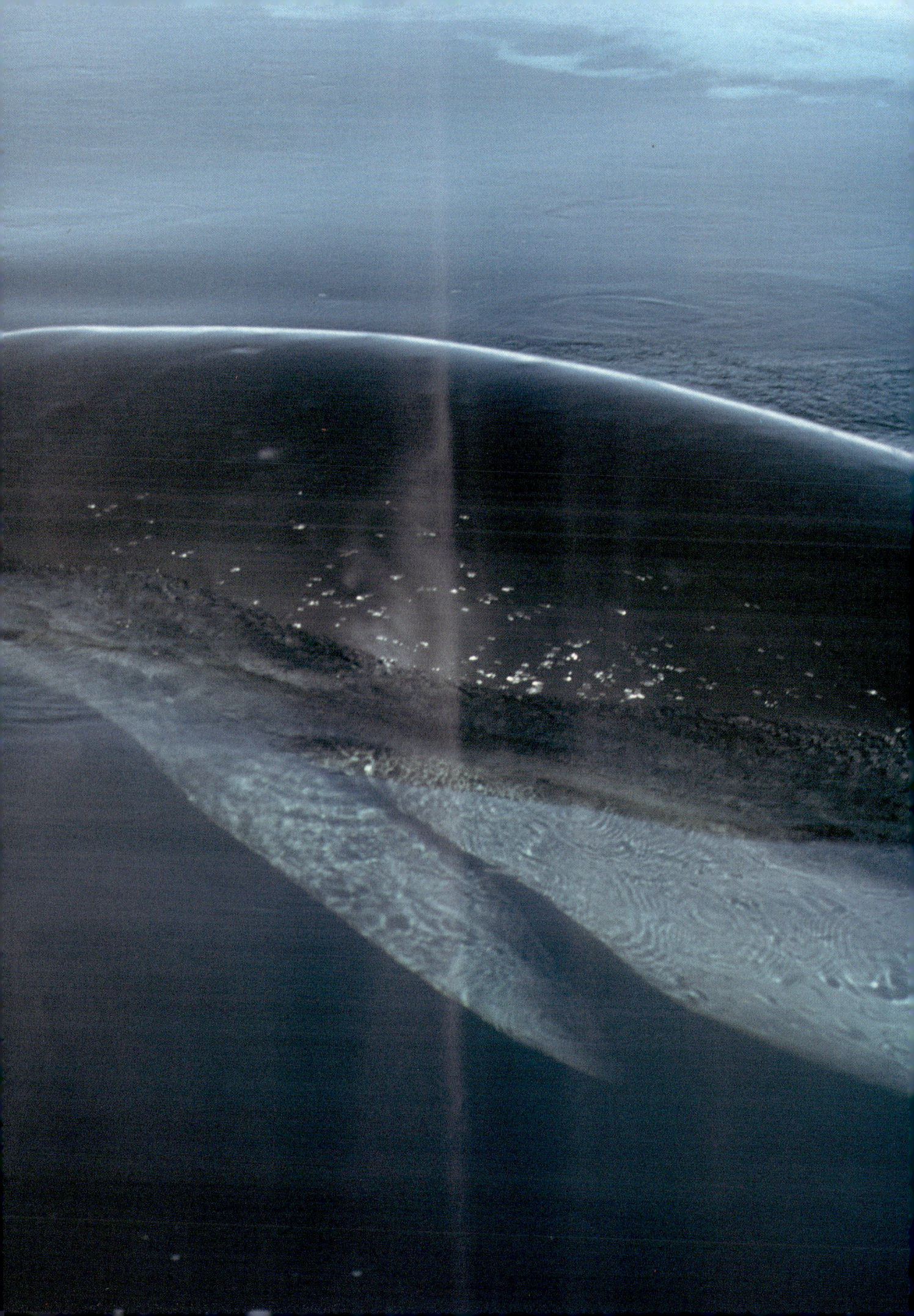

Leopard seals *(Hydrurga leptonyx)* (Plate 11.11) also have multi-cusped teeth and feed on krill. But they are also voracious predators which attack young crabeater seals, leaving wide parallel scars from their canine teeth on those lucky enough to escape. They also hunt fish, squid, and especially penguins. Leopard seals occur throughout the Ross Sea, mostly in the pack ice, but some venture south to Ross Island where they have been particularly noted for the attention they give to the comings and goings of penguins at rookeries (see Chapter 13).

Whales

The "top predator" of the Ross Sea must be the killer whale *(Orcinus orca)* (Plate 11.12). It roams the ocean in extended family pods of 5–20 individuals sharing a high degree of group co-operative activity, harassing and distracting their prey. They pursue squid, fish, seals, and penguins, and also take an interest in humans on ice flows. Feeding may involve scavenging, but there are reliable reports of groups attacking and killing Minke whales and seals. Probably any suitable prey, live or dead, is taken. Of the toothed whales other than killer whales, dolphins have been reported from the Ross Sea, but there is no information on their biology. Sperm whales are represented only by males in the Southern Ocean and rarely in the Ross Sea.

Most whales of the Ross Sea are baleen whales, with Minke whales *(Balaenoptera acutorostrata)* (Plate 11.13) and blue whales *(Balaenoptera musculus)* penetrating south in the summer to the continental ice edge in pursuit of aggregations of larger sized zooplankton, including krill. As heavier crops of zooplankton are to be found in the north of the Ross Sea, baleen whales are most commonly encountered there. The number of whales found together varies. As they are largely occupied with eating as much as possible, their groupings may reflect more the abundance and distribution of prey rather than social tendencies of the whales themselves. Baleen whales occur mostly in groups of 3–5 individuals, although single and larger groups have been observed. Aggregations of baleen whales often occur in the Bay of Whales in the eastern Ross Sea, and in the south-western sector, suggesting that these are significant feeding grounds. Minke whales are the commonest species, and were until recently harvested (by Japan and Russia) in limited numbers as far south as the Ross Ice Shelf. Southern right whales and humpback whales are infrequent in the Ross Sea, as are southern fin whales and southern sei whales which have on occasions been seen in McMurdo Sound.

Because of extensive whaling from the 1920s to 1978, several large populations of baleen whale have been drastically depleted, particularly blue and humpback whales. The report by Wilson on the *Discovery* voyage gives an impression of an abundance of whale life that once existed in the Ross Sea. The decline there is only a part of global consequences of whaling. Nevertheless, mammal biomass (whales and seals together) of the Ross Sea appears to be low in comparison with other Antarctic waters, perhaps reflecting an overall lower productivity of the marine ecosystem of the embayment.

Squid

Emperor penguins *(Aptenodytes forsteri)* can dive to 250 metres to pursue fish and squid, and squid have been reported from stomachs of emperor penguins at Ross Island. Squid are also known from stomachs of fish and seals in McMurdo Sound, so presumably they are part of the marine fauna of southern Ross Sea. But there is no report of a live squid being captured there, though an octopus has been caught in the Sound (Plate 11.14).

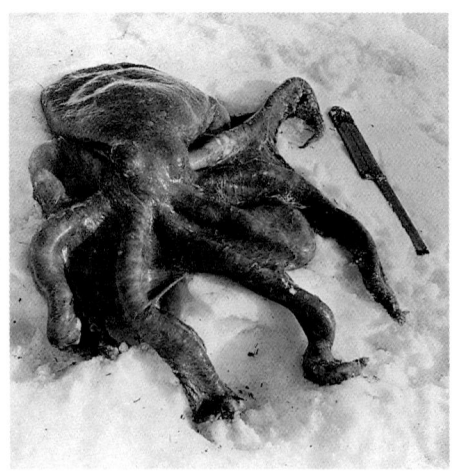

Plate 11.14. *An octopus, caught by Ponting and Lashly in Dead Horse Bay on 16 February 1911. It measured "about five feet from tip to tip" and seemed to be comatose as though half frozen. Photo: Alexander Turnbull Library, Wellington.*

Several species of squid occur in the Southern Ocean, mostly known from the occurrence of their beaks in the stomach contents of sperm whales, albatrosses, and seals in the northern parts. Undoubtedly squid are an important component of the open and deep sea ecosystem around Antarctica, perhaps occurring more frequently in shallower water during the long nights of winter. But we know hardly anything about their biology. From calculations involving the estimated biomasses of squid eaters in the Antarctic ecosystem as a whole, it has been estimated that there may be more biomass of squid in the Antarctic seas than there is of krill.

Prospects

Even though the water of the Ross Sea is very cold and bounded by much ice, there is a considerable diversity of life-forms in the sea below the ice to support the more obvious surface populations of penguins, seals, and whales. We are now beyond the stage of marvelling that there is life there, even that the life is relatively diverse, or that it is specifically distinctive. We are now interested in the adaptations to the low temperatures and to the seasonally light-limited and ice-constrained environment (see Chapter 14). We also seek to know more about how, and which, species interrelate through the food webs of the ecosystem—to support the vertebrate populations that have come to live among the ice, or the whales that migrate south every summer when conditions of light and broken ice permit sighted, air-breathing predators to partake of the resources that grow and overwinter there. It is a unique system populated by unique species and visited by only a few cosmopolitan species.

Human beings too are migrant visitors and represent an intrusion into the ecosystem. Sewage from shore bases may introduce local enrichment, and ships at sea carry their own threats and risks. But there is no way we can study the marine biology, particularly the pelagic organisms, without taking samples, in contrast to terrestrial biologists who can observe and measure without taking. To curtail marine biology studies might make us environmentally purer but would leave us scientifically poorer, considering the uniqueness of the environment and its life. We would certainly remain too ill-informed to make proper pronouncements in the interests of conservation and preservation of the penguins and seals that everyone likes.

12. Life in land, ice, and inland water habitats

To most observers the stark landscape of the Ross Sea region is a barren picture, seemingly a lifeless desert of ice or rock. On other continents even the harshest deserts may have prominent vegetation, for instance, cacti in North America and shrublands in central Australia. However, in Antarctica it is often only the accustomed eye and the aware mind that discovers the minute organisms of the unique flora and fauna that exists on and within snow, soils, and rocks. If you have eyes that search the ground rather than admiringly scan the grand scenery, and a patience that allows cracks and crevices to be investigated and stones to be overturned, you will soon discover a surprising variety of life in unusual places.

As with deserts elsewhere, the cold arid catchments of the Ross Sea region also contain isolated "oases" where aquatic life-forms can briefly flourish each year. For a few weeks in summer, meltwater streams flow from the glaciers and discharge into lakes or to the ocean. Coastal ponds thaw and are recharged with melted snow. In the large ice-capped lakes further inland, solar radiation rises during spring and early summer to levels that can support the photosynthesis of microscopic plants beneath the ice.

Life on land

The plants

Algae are the predominant plants. These mostly require the highest magnifications of a light microscope for details of their structure to be seen, as their cells may be as small as one thousandth of a millimetre in diameter. However, where they occur in abundance they form crusts, leathery sheets, or slimey masses ranging from green, through blue-green to orange (Plate 12.1), brown, and even purple. The dominant species are cyanobacteria (also known as "blue-green algae") (Plate 12.2), the most primitive of oxygen-producing photosynthetic organisms; several of them are similar to fossils 2000–3000 million years old. Next in abundance are the green algae which range in structure from single spherical cells (Plate 12.3) to unbranched and branched microscopic filaments to sheet-like aggregations of cells. Also present are single-celled diatoms (Plate 12.4) which occur in much larger quantities and diversity in the Southern Ocean, and the related yellow-green algae which superficially can closely resemble the green algae.

Algae and the microscopic, filamentous decomposer organisms called fungi combine in a mutualistic partnership to form lichens. The term mutualistic is used because each member of the partnership benefits from the other. The plants generally look completely unlike either partner growing alone. Where they occur on rock surfaces they form thin crusts which may be a dull dark grey or a vivid yellow or orange (Plate 12.5). Less frequently observed are small, bushy, richly branched lichens, up to a few centimetres in height, and others which produce leaf-like growths.

The plants with the most complex structure are still very small, attaining a height of a centimetre at the most. These are the mosses and liverworts, which collectively are termed bryophytes. They have a central stem from which tiny leaf-like lobes of tissue are produced, and they are attached to soil and rock by "rhizoids", microscopic threads which emerge from the bottom of the stem. Where hundreds of these individual plants (Figure 12.1) are packed together they form aggregates called cushions, and where growth conditions are particularly favourable the cushions merge to form carpets (Plate 12.6).

All these photosynthetic plants comprise the "primary producers", that is the organisms capable of forming organic material from carbon dioxide. Living amongst them and in the underlying soil are the decomposer micro-organisms which feed on the organic material made by the primary producers. These decomposers include microscopic filamentous fungi, although in the Ross Sea region and continental Antarctica as a whole the filaments do not aggregate to form the reproductive structures commonly known as mushrooms or toadstools. Related organisms are the unicellular yeasts. Functioning in a similar manner, but of a much simpler structure, are bacteria and the related filamentous actinomycetes.

The animals

Grazing on either the primary producers, the decomposer micro-organisms, or the dead remains of these is the animal life (Figure 12.2). The simplest forms are the unicellular protozoa, living in thin water films which coat the vegetation and soil particles. Rotifers (wheel animalcules), nematode (eel) worms, and tardigrades (bear animalcules) are multicellular but still microscopic and are encountered within the richer and wetter growths of algae and bryophytes. The "kings of the beasts" of Antarctic terrestrial life are the mites and collembolans. The former are eight-legged spider-like animals. The latter are six-legged primitive insects commonly called springtails (see Plate 14.6), due to their possession below their abdomen of an organ which when activated can flick them through the air. The approximately 1 millimetre or less stature of these is a huge contrast with the dimensions of the largest animals of the Southern Ocean, the whales.

Life on rock, soils, and snow

Only in certain particularly favourable localities do these organisms occur in relative abundance and even there the Antarctic environment provides a harsh existence. All terrestrial life experiences temperatures well below freezing for about 11 months of the year and during summer may be frozen and thawed several times in 1 day. Only during the brief summer, when dark rock, soil, and vegetation surfaces are warmed by a

Plate 12.1. *A large area of orange lichens (a mutualistic algal – fungal partnership) on a hillside at Cape Crozier. Photo: P. Broady.*

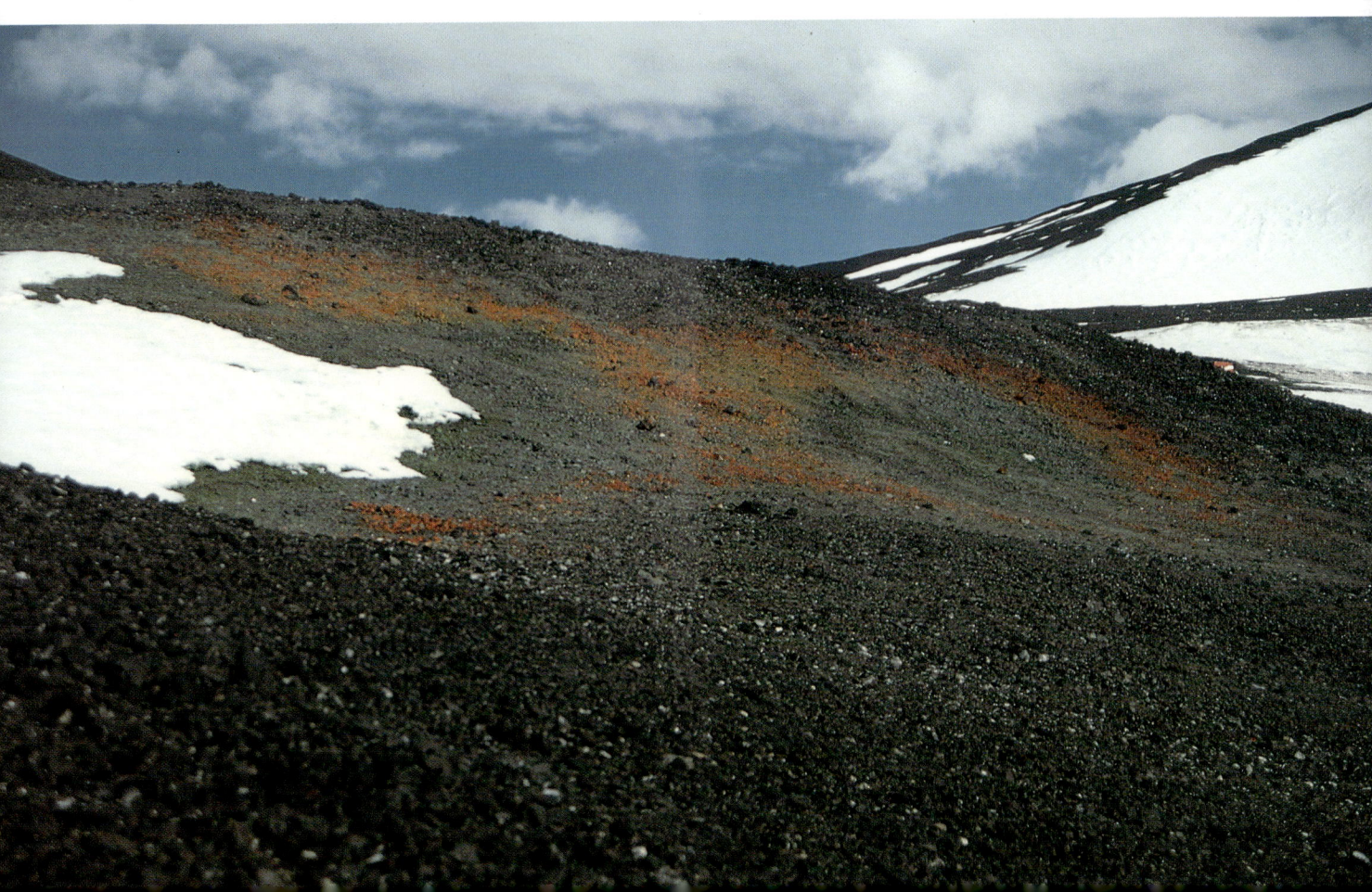

Plate 12.2. *Photomicrograph of a cyanobacterium. Photo: P. Broady.*

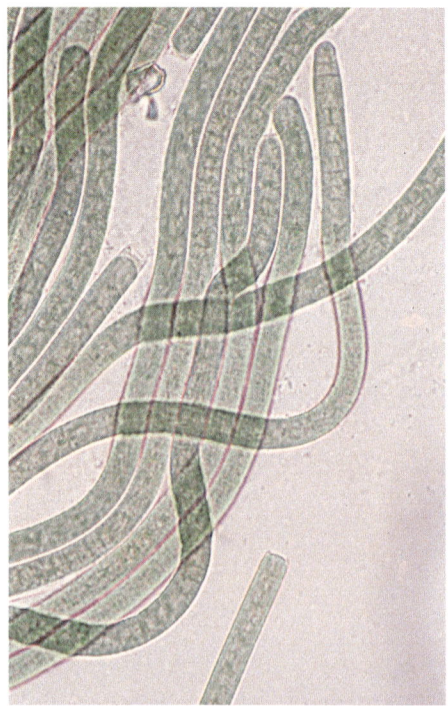

Plate 12.3. *Photomicrograph of a green alga from soil at the top of Mount Erebus. Photo: P. Broady.*

bright sun, will their temperatures rise to perhaps 15°C. This effect is most noticeable on north-facing slopes which collect the sun's rays more directly. Water supply can be highly erratic. The soils may dry out completely or freeze, and so organisms must also be able to survive desiccation. Salt accumulations in and on the surfaces of soils can also have a profound effect and where present in high concentrations can prevent plant growth. Frequent strong winds have an important influence on patterns of plant distribution. Hill slopes and rock surfaces exposed to the prevailing wind often lack vegetation due to the damaging action of ice crystals and sand grains carried by the winds. However, protection is provided on downwind sides of boulders, hummocks, and hills, in ground hollows and between fractured rocks. Also, it is here that snowdrifts accumulate and these subsequently melt and supply vital water. Therefore, it is in such sheltered places that the richest growths are likely to be encountered.

Most of the vast expanses of snow and ice of the continental ice cap, glaciers, and ice shelves largely lack living organisms (but see the discussion of "cryoconite holes" below). Over much of their area, there is no more than a very low concentration of resistant spores and cells of micro-organisms which survive in the snow after being deposited from the atmosphere. Only at rare locations are microscopic algae present in sufficient abundance to colour the snow fields green. Such sites have been observed at the Possession Islands, off the northern tip of Victoria Land, and at the southernmost known limit of their range, at Cape Crozier (Plate 12.7) and Cape Royds on Ross Island. The snow fields are coastal and adjacent to penguin colonies and provide algal growth with favourable conditions, these being percolating meltwater during summer and nutrients derived from the penguins.

The maximum development of vegetation is found in the more moist ice-free areas where, during summer, water flows from snowdrifts, snow fields, and glaciers, and where snowfalls are more frequent.

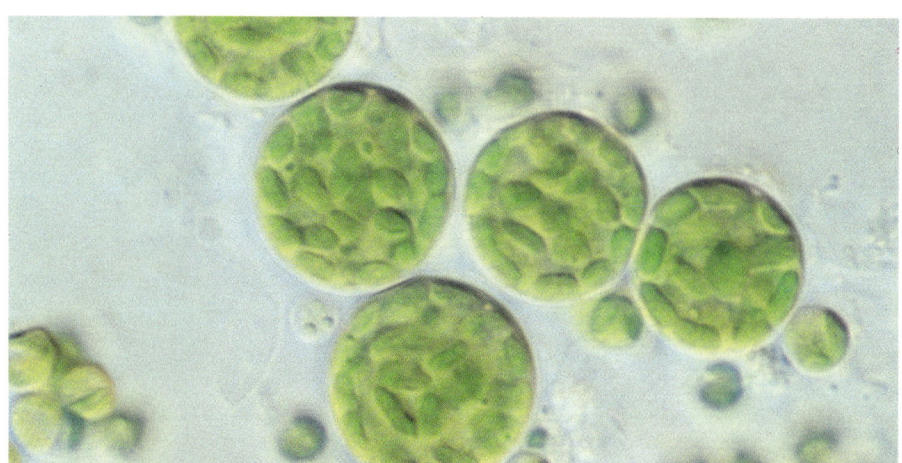

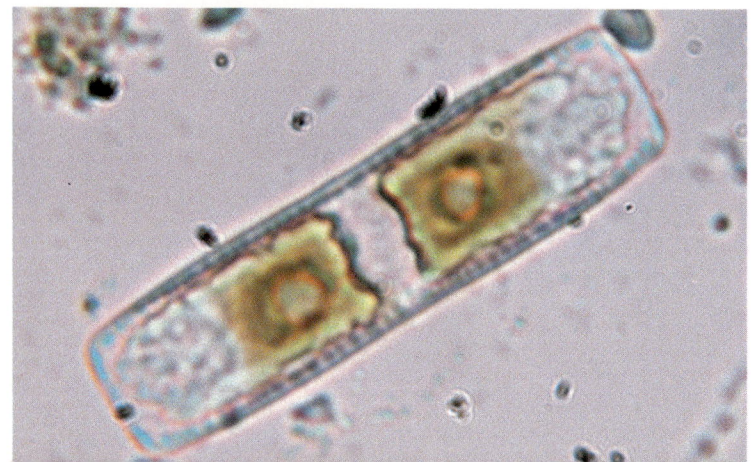

Plate 12.4. *Photomicrograph of a diatom. Photo: P. Broady.*

However, this vegetation is still usually very sparse. Examples of the lushest growths of moss occur close to the terminus of the Canada Glacier in the Taylor Valley and at Edmondson Point, a coastal ice-free area in northern Victoria Land. At each location the hundred or so square metres of continuous moss are oases surrounded by ground which supports a much sparser vegetation. Encrusting growths of algae are often found covering the wettest mosses. Some of these, for instance, the cyanobacteria *Nostoc* and *Calothrix*, are able to use nitrogen from the atmosphere as a nutrient (Figure 12.3), and some of this nourishment may then be transferred to the moss and stimulate its growth. Concentric brown rings scattered over the moss surface reveal the presence of fungal infections. Moisture trapped between the moss plants provides a suitable environment for protozoa, tardigrades, and nematodes. The drier patches can be encrusted with lichens.

On Ross Island, a factor which distinctly prevents the growth of mosses and lichens is windswept sea-spray. Where this blows over land during gales, such plants are absent although other conditions suitable for their growth might occur. This pattern is particularly obvious at Cape Royds.

There is considerable variation between different ice-free areas, and within any one area, in whether algae, lichens, or mosses predominate. For instance, at Cape Bird in an ice-free area close to the northern tip of Ross Island, lichens are extremely sparse although there are good examples of moss growths, particularly along the edges of melt runnels. In addition, there are many expansive growths of filamentous cyanobacteria forming leathery sheets on the surfaces of water-soaked soils downslope from melting snowdrifts. Amongst these can be found brownish, irregular mucilaginous clumps of the "nitrogen-fixing" alga *Nostoc*. In contrast, at Cape Crozier, at the eastern end of Ross Island, moss cushions are sparse but lichens are widely distributed and occasionally abundant. On the slopes above the penguin colony the

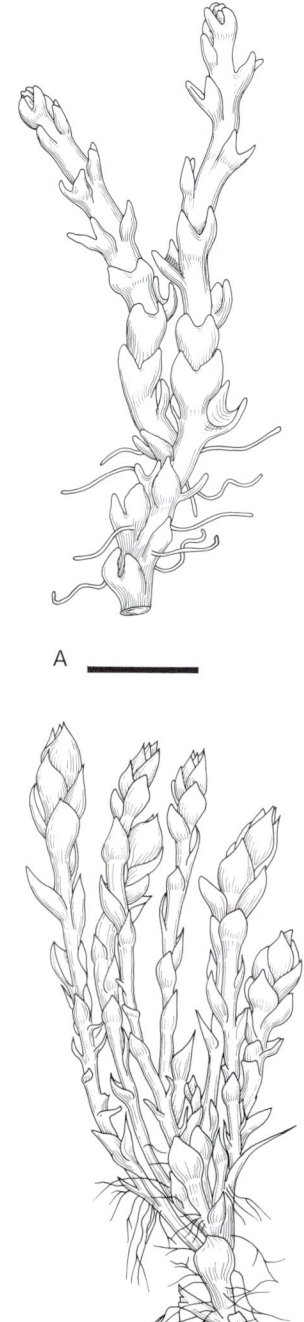

Figure 12.1. *Individual plants removed from cushions of a leafy liverwort (a) and a moss (b). The moss, Bryum argenteum, is widespread in the Ross Sea region but the liverwort, Cephaloziella exiliflora, has been found only at the summit of the volcano Mount Melbourne in northern Victoria Land. The scales are equivalent to 1 millimetre.*

A ▬▬▬▬

B ▬▬▬▬

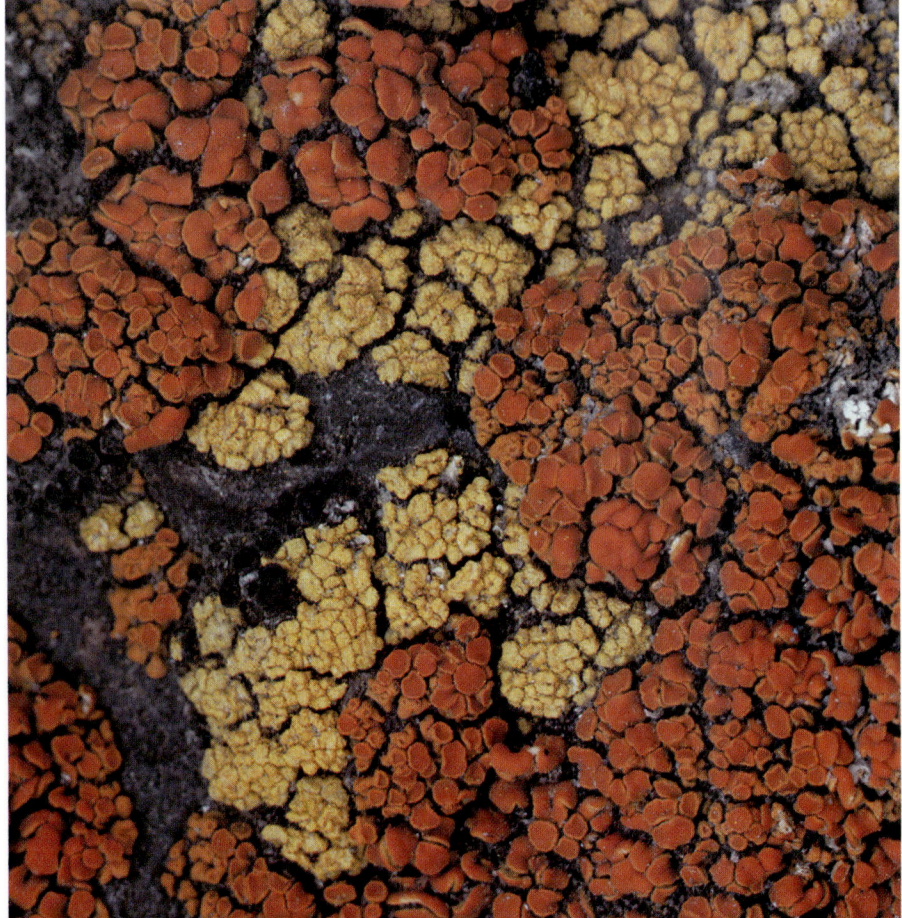

Plate 12.5. *Close-up view of a crustose lichen, red and yellow. Photo: P. Broady.*

Figure 12.2. *Microscopic animals found amongst vegetation and in moist soils. (a) a worm-like nematode; (b) an eight-legged tartigrade; (c) a rotifer, a filter-feeding animal inhabiting water films; (d, e, and f) three single-celled protozoa, all of which possess whip-like appendages which are used for motility and feeding; (g) a spider-like, eight-legged mite; (h and i) six-legged collembola or springtails. All the scales, except that for d, e, and f, are equivalent to 0.1 millimetres; scale for d, e, and f is equivalent to 0.05 millimetres.*

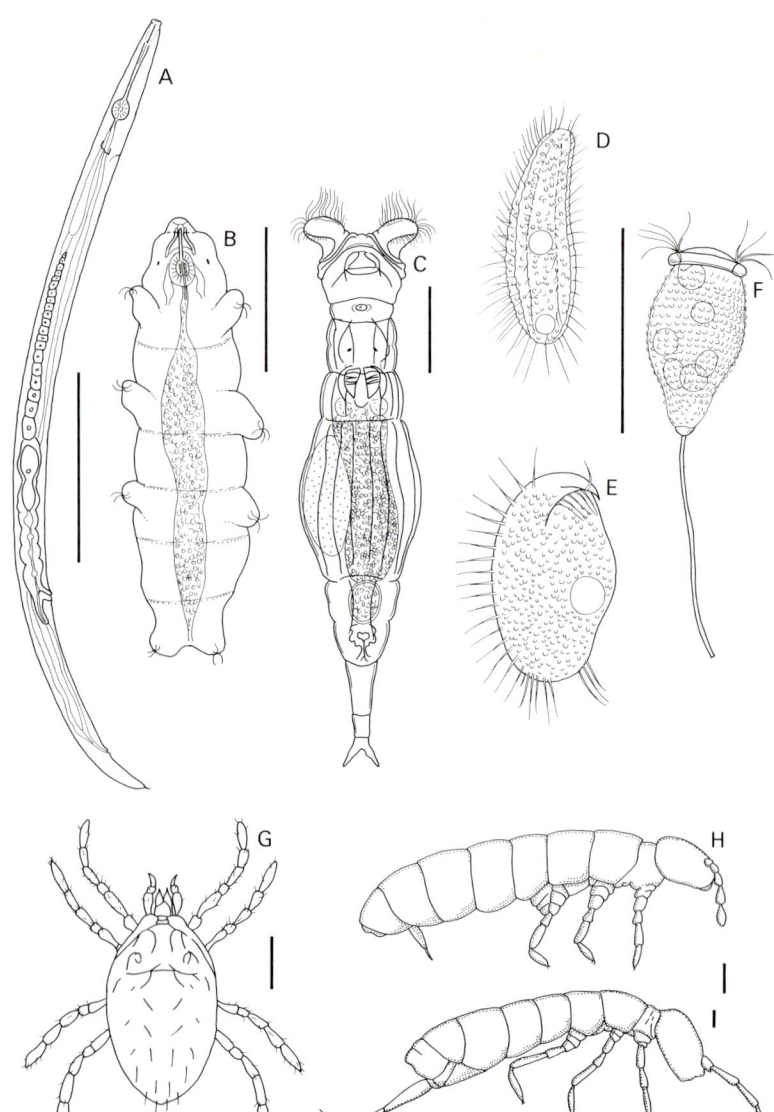

Plate 12.6. *Close-up view of the surface of a moss carpet in northern Victoria Land. Photo: P. Broady.*

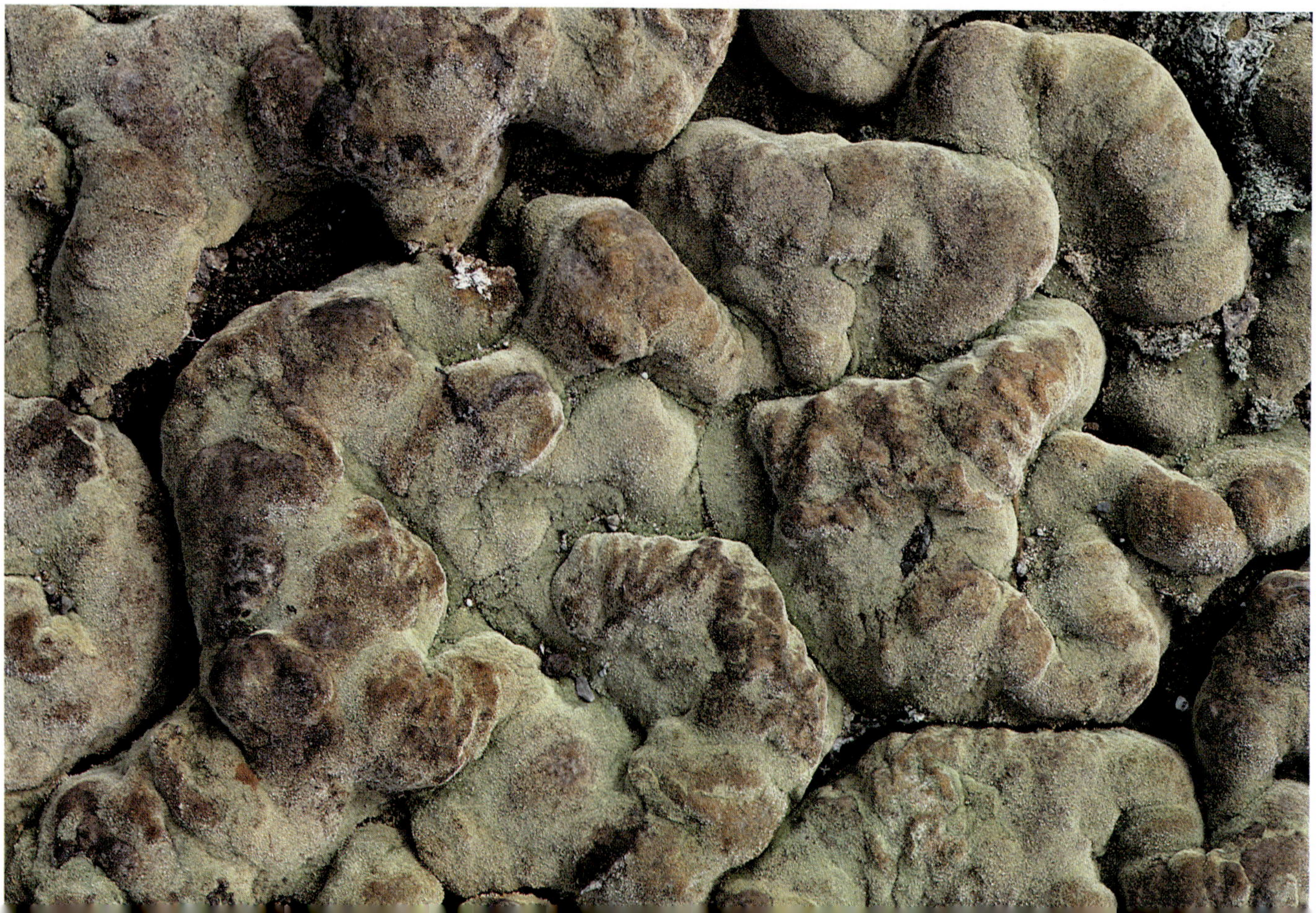

rocks and stones bear bright orange splashes of colour provided by tens of square metres of encrusting lichens. In the southerly part of the ice-free area different lichens occur. There are excellent examples of the more dully coloured, but more structurally complex, bushy (fruticose) lichen *Usnea* and the leaf-like (foliose) lichen *Umbilicaria*. These broad patterns in the distribution of the different species of algae, mosses, and lichens have yet to be fully explained.

Soil conditions amongst the penguin colonies are very different from those found elsewhere in the ice-free areas. The result of thousands of birds feeding at sea and then returning to their breeding areas is the accumulation of large quantities of nutrient-rich guano, broken egg shells, and dead carcasses. When this material is combined with meltwater percolations from a snowdrift, abundant growths of algae can carpet the ground (Plate 12.8). These growths comprise species which are well adapted to levels of nitrogen and phosphorus nutrients which would be toxic to other flora. Especially prominent are the lush green, sheet-like carpets of a green alga *(Prasiola crispa)* which grows where the ground is not too disturbed by the tramplings of multitudes of tiny feet. During summer the air can be pervaded by a strong smell of ammonia, as bacteria break down nitrogen-rich organic compounds in the guano. A few centimetres below the surface the guano can be black where bacteria are active in the absence of oxygen. Fungi grow amongst accumulations of moulted feathers and are able to specifically use the feathers' major component, the protein keratin. Grazing on the abundant decomposer micro-organisms are considerable populations of protozoa.

A fascinating habitat of extremely limited area is provided by ground that is heated by volcanic activity. Close to the summits of the volcanoes Mount Erebus on Ross Island and Mount Melbourne in northern Victoria Land, at altitudes over 2500 metres, are a few hundreds of square metres of steaming, ice-free ground. Over the surface of this ground thin crusts of algae and bryophytes grow at temperatures between 10 and 50°C. This is in a situation where the summer air temperatures vary between about –10 and –35°C! Without the geothermal supply of heat these plant communities would not exist as their water supply would freeze.

Each volcano has plant life with unique features. On Mount Erebus is the only Antarctic alga known to be "thermophilic", i.e., requiring high temperatures for growth. This alga will not grow at temperatures below 25°C but thrives at 45°C. Also, there is a highly unusual moss which occurs only as microscopic branching threads, a condition which is an immature stage in most other mosses. In some aspects the vegetation on Mount Melbourne is very different. A typical "leafy" mature moss is found, but the species occurs nowhere else in Antarctica. Also, there is a liverwort, a very rare occurrence in the Ross Sea region; the only other record is a specimen found near Cape Hallett. An additional contrast with Mount Erebus is the presence on Mount Melbourne of a single species of protozoan, a "testate" amoeba, i.e., an amoeba that lives within a shell.

Many interesting questions are posed by these communities. How are the organisms transported to these extremely isolated localities? How long have they been there? Were previous communities completely obliterated by the most recent eruptions, which have probably occurred during the last few centuries? How do the plants survive the long periods of darkness over winter, especially as they probably do not freeze? Why are there different organisms on each volcano, growing under conditions which appear to be superficially very similar?

In Antarctica, vegetation can grow over the exposed surfaces of rock and soil only where water is sufficient and frequent enough, and salt concentrations are suitable. Wide areas of ice-free land do not receive such a water supply. This is particularly true of the southern Victoria Land Dry Valley region, and especially Wright and Victoria Valleys and their associated branch valleys and surrounding mountain slopes. In this frigid desert the rock surfaces usually lack even the hardy encrusting lichens.

However, there is a hidden habitat where conditions are not so harsh as in the external environment and where microscopic algae, lichens, and bacteria are widespread. This habitat is found in a narrow zone, up to 7 mm thick, and commencing 1–2 mm below the surface of suitable rock types. The micro-organisms that occur there are termed "endolithic", i.e., they live inside rocks. Suitable rocks are ones which are translucent and allow the transmission of light. The major example in the Dry Valleys is sandstone. The humidity between the rock crystals is greater than that outside the rock. So, with light and moisture, together with a supply of nutrient salts from the rock, and the warming of the rock by the sun's rays, conditions are provided which are favourable for the growth of a limited range of algae and associated fungi and bacteria. The only way to reveal these communities is to hit the rock surface with a hammer to flake off the outer few millimetres, so exposing bright blue-green and green zones of algae (Plate 12.9). In some rocks the sole inhabitants are cyanobacteria whereas in others there may be a relatively complex layering of endolithic lichens and a green unicellular alga living separately and below the lichen (Figure 12.4). On the rare occasions when the lichens are able to form small growths at the rock surface they can be identified as ones which commonly form surface-encrusting growths elsewhere in Antarctica where conditions are more humid. However, in the Dry Valleys they are a dominating, though hidden, vegetation solely because they are flexible enough to be able to substantially change their structure and grow in the minute capillaries between rock crystals. The presence of these micro-organisms contributes to the weathering and fragmentation of rock outcrops and boulders, when, due to the biological activity, flakes of rock become separated and dislodged (Figure 12.5). These flakes drop to the ground where they continue to disintegrate and the micro-organisms present a small quantity of organic material to the sandy soils.

Early investigators of these soils in the driest parts of the Dry Valleys suggested that although micro-organisms could often be detected they had probably been brought in by the winds and were only surviving in frozen soil without actively growing. However, it is now known that there are some extremely unusual decomposer organisms, including yeast species, which so far have been found only in the Dry Valleys. In great contrast to the thermophilic micro-organisms found in the heated soils of Mount Erebus, the yeasts are "psychrophilic", i.e., they grow best at low temperatures around 4–10°C and are capable of growth down to at least –3°C. This, combined with an ability to survive deep-freezing and extreme desiccation and to be able to make use of the very low levels of organic material in the Dry Valley soils, allows these organisms to grow in one of the most extreme environments on earth.

Life in ice and inland water habitats

The land habitats described so far, Antarctic rock, soils, and snow, pose a severe test for all life-forms, and only the hardiest species are able to survive these conditions of extreme cold and desiccation. However, the catchments of the Ross Sea region also contain a range of aquatic environments that offer more favourable opportunities for survival and growth. These non-marine waters include small ponds distributed along the coast that melt for a few weeks each year; warm, but permanently ice-capped lakes further inland; and rivers and streams that are fed by summer meltwaters from the alpine glaciers and ice sheets.

Ponds

Even in these more hospitable environments, however, the diversity of species is very low. In part this reflects the extremely brief growing season in which organisms must complete their life-cycle, as well as the unusual physical and chemical conditions that they may often experience. However, the paucity of species in these catchments may also be related to the enormous distances that a plant or animal from temperate latitudes would need to travel to colonise these southernmost aquatic habitats.

Ponds are the most common inland water environment in the Ross Sea region. These waters are fed mostly by melting snow and occupy natural depressions in the rock or sand. Many are influenced by the nearby ocean by receiving seaspray or nutrient fertilisation by birds and seals that feed at sea. They are typically shallow, rarely deeper than 50 centimetres, and are completely frozen through for all but 1 or 2 months of the year.

The first biologist to become fascinated by the pond life of the McMurdo Sound region was James Murray who was based at Cape Royds during Shackleton's British Antarctic Expedition of 1907–09. Murray and his colleagues excavated through the ice of Blue Lake during winter and found thick pieces of algal vegetation that on careful thawing released "multitudes of living things for study" (Plate 12.10).

This high level of tolerance to changing chemical and physical conditions appears to be an essential attribute for all species in the ponds. The habitat is highly unstable, with abrupt changes in area, depth, nutrient levels, temperature, and salt content over short periods. For example, one pond at Cape Bird, melted in late November, doubled its salinity over December–January, and then completely dried up. A

Plate 12.7. *Green algae in a snowfield at Cape Crozier. Photo: P. Broady.*

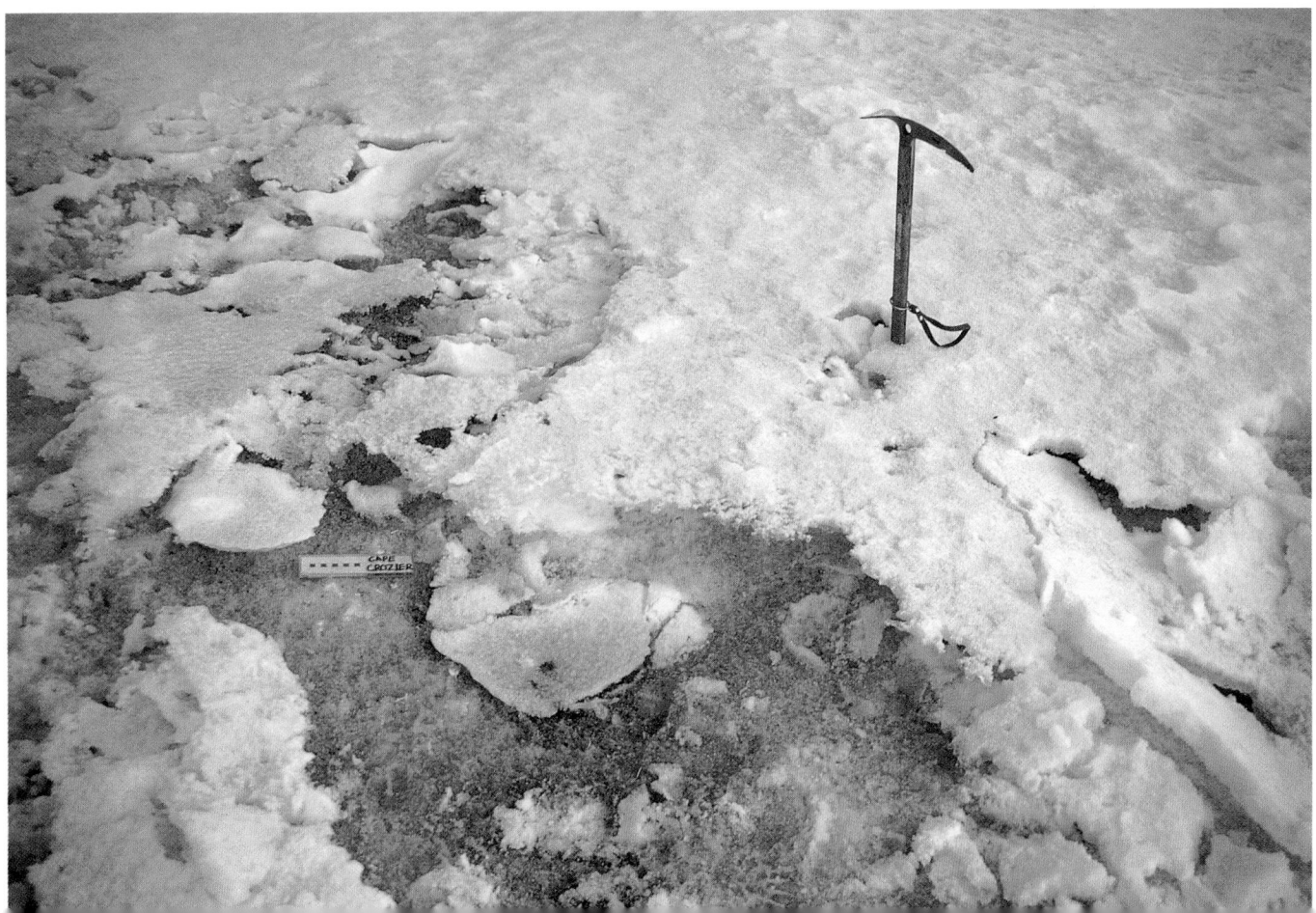

Plate 12.8. *Rich growths of green algae at the Adélie penguin rookery, Cape Crozier. Photo: P. Broady.*

nearby deeper pond, still frozen in early December, melted by mid-December, rose to temperatures as high as 15°C, and then refroze in late January.

Two types of community live in the ponds: benthic populations, i.e., those on the pond bottom, which cover the sediments with a thick film of cyanobacteria (Plate 12.11), and planktonic populations that live within the overlying water. The benthic cyanobacteria contain blue and green pigments which capture sunlight for use in photosynthesis. However, they often appear bright pink or orange as they also contain pigments called carotenoids that help protect the light-gathering pigments from damaging levels of continuously bright sunshine in summer. The benthic films probably experience a longer growing season than the plankton because the bottom region of the ponds is the first to melt each summer and the last to refreeze. They contain large quantities of jelly-like mucilage, presumably produced by the cyanobacteria, which may help the community survive the freezing process each year by slowing the rate at which water leaves the cells during the formation of ice crystals.

The planktonic community in the overlying water is dominated by single-celled algae that on calm days can adjust their depth by swimming up and down. In the ponds that are nutrient-enriched by seals, penguins, or skuas these mobile algae grow in large numbers and the water resembles pea soup. For example, Pony Lake at Cape Royds is enriched by runoff from the adjacent Adélie penguin rookery and each year it becomes highly turbid and green (Plate 12.12).

Lakes

Many decades after James Murray's pioneering work at Cape Royds, scientists in the Dry Valleys of the Ross Sea region described a group of lakes with some most unusual properties. The lakes have permanent caps of ice. Lake Vanda in the Wright Valley was found to be capped by a

Plate 12.9. *Green and blue "endolithic" algae revealed under the rock surface after it was split by a hammer blow. Photo: P. Broady.*

Figure 12.3. *Cyanobacteria found amongst mosses and which are capable of removing nitrogen gas from the atmosphere for use as a nutrient for their growth. (a) Nostoc; (b) Nodularia; (c) Calothrix; (d) Tolypothrix; (e) Stigonema; (f) Mastigocladus. The latter two have a very restricted distribution being found only on steam-warmed ground on the volcanoes Mounts Erebus and Melbourne. The scale is equivalent to 0.01 millimetres; "h" indicates special cells called heterocysts which are active in the removal of atmospheric nitrogen.*

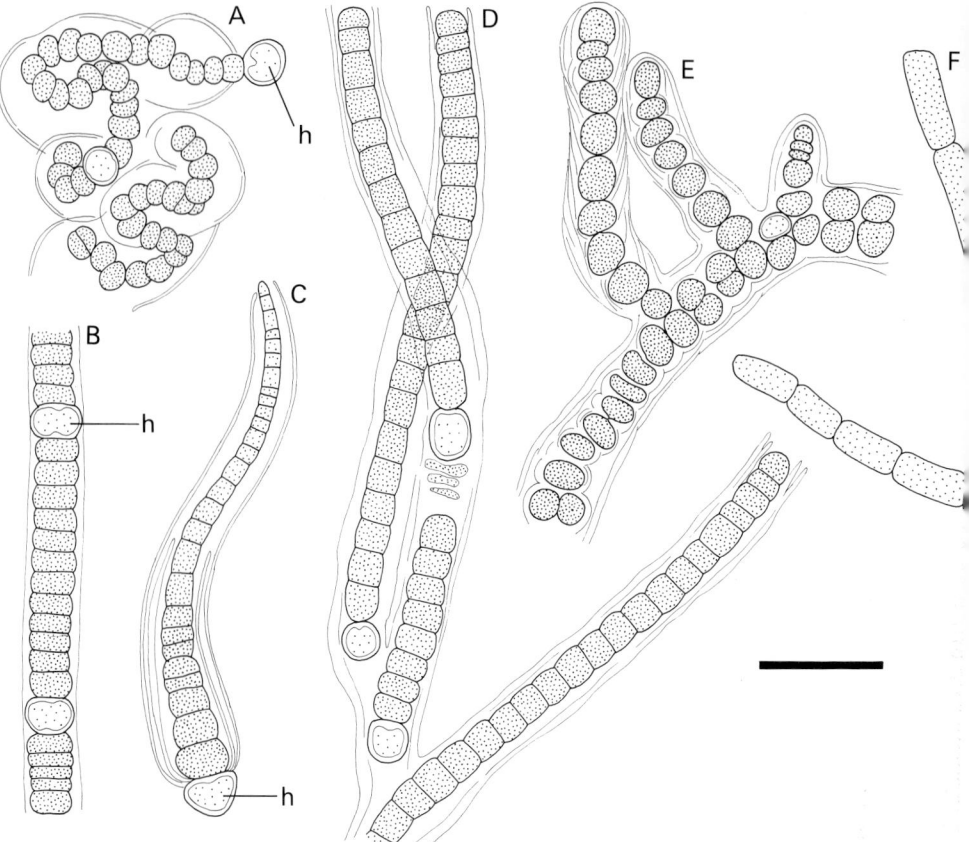

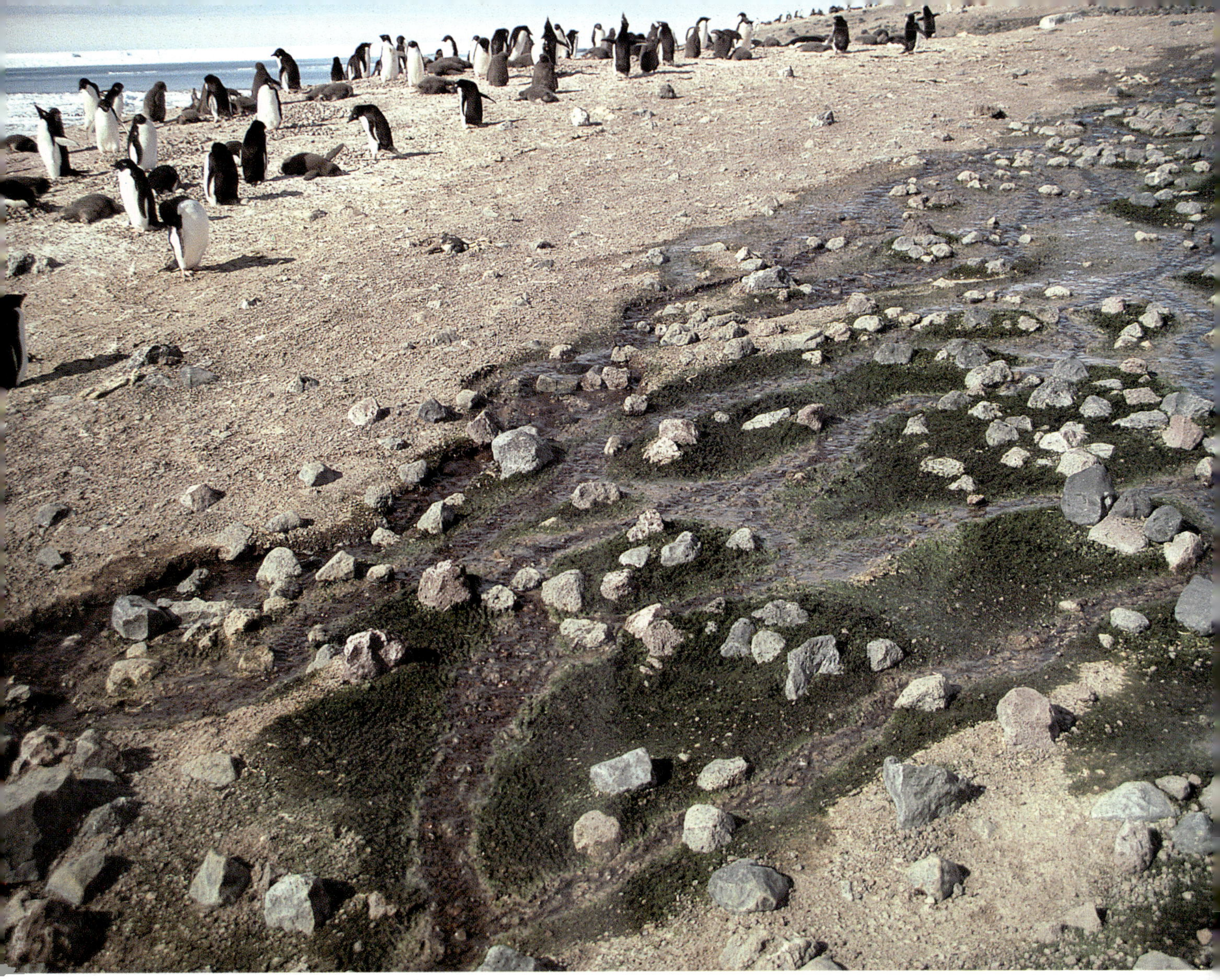

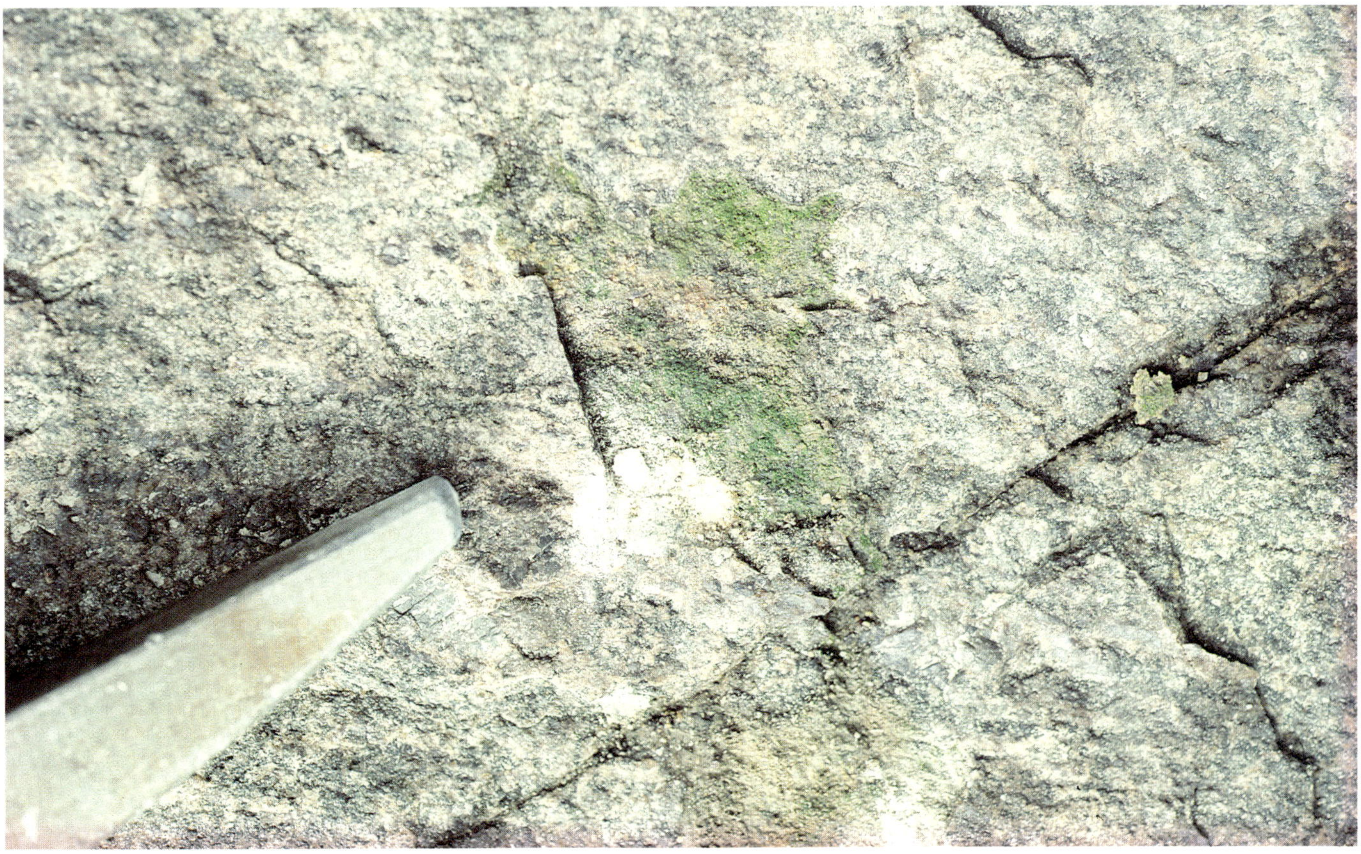

permanent, floating ice sheet, 3–4 metres thick. The water immediately
beneath the ice was cold and fresh, but with depth it became increasingly
warm and saline. At the bottom of the lake at about 68 metres depth,
the water was three times saltier than sea water and had a temperature of
25 °C, some 45 °C higher than the average annual air temperature in the
Wright Valley!

At least five other lakes in the Dry Valleys are similarly capped by
thick ice and have unusual temperature and salinity profiles, though less
so than Lake Vanda, and the reasons for their properties are discussed
on p. 152. Below the ice there is negligible mixing of the water. The
permanent ice-caps protect the waters from being stirred by the wind
and they are further stabilised by the water increasing in density with
depth due to the increasing salt content. The extreme physical stability
has allowed various micro-organisms in the lakes to form discrete layers
where light, temperature, and chemical conditions are exactly suited for
their growth. This highly layered structure was first revealed by studies
of the distribution and photosynthetic activity of planktonic algae.

When water samples were pumped out of Lake Miers, in one of the
southern Dry Valleys, and filtered through discs of plankton netting it
was immediately apparent from the green colour of the filters that more
algae were present in deep samples than from just under the 6-metre-
thick ice-cap. Most of the algae were collected from the layer between 15
and 18 metres which overlies a bottom layer of anoxic (oxygen-free)
water. Subsequent studies in other Dry Valley lakes have shown that the
greatest quantity of algae invariably occurs just above the nutrient-rich,
anoxic bottom waters of each lake. These algae carry out most of the
photosynthesis although they are in the poorly illuminated depths of the
lakes. They may also trap nutrients diffusing up from the anoxic zone
and thereby restrict the transfer of these nutrients to overlying water
where there is more light available for photosynthesis. This process of
nutrient removal may help maintain the low algal concentrations and
high transparency of the near-surface waters of these lakes.

Different species of algae occur in different strata of each lake. Each
species is able to swim, usually using whip-like appendages called flagella,
to the depth where it finds that light, temperature, salinity, and nutrient
supply are to its particular liking. In Lake Fryxell, for example, an algal
layer dominated by two species, Chlamydomonas sp. and Ochromonas
nannos, occurred immediately below the thick ice-cap. In a lower layer
there was an unidentified alga with two flagella, but the densest
population occurred lower still at 4.5 metres below the ice-cap and at the
base of the oxygenated zone. It comprised two algal species, a red
flagellate identified as Chroomonas lacustris, and a green alga with four
flagella, Pyramimonas sp. (Figure 12.6). In the much more transparent
waters of Lake Vanda the maximum population of algae lies at about 55
metres below the lower ice surface and is composed mainly of thin
filaments of cyanobacterium, Phormidium antarcticum.

This layered distribution is also a feature of other microbial
populations within the Dry Valley lakes. In Lake Fryxell a band of
photosynthetic bacteria forms a thin plate just below the depth where
most algae occur. These cells are capable of using extremely low levels of
light. During photosynthesis they produce sulphur, in contrast to the
algae, which like all other plants produce oxygen. At greater depths in
the lake, bacteria decompose the algal cells which die and sink into the
anoxic zone. In the process these bacteria produce hydrogen sulphide
(the gas with the stink of bad eggs).

In Lake Vanda different algae and bacteria that participate in the
natural cycling of nitrogen occupy different layers of the lake. Each of
these groups produces characteristic by-products, and these chemicals

have accumulated to extreme concentrations in thin layers within the lake. For example, the gas nitrous oxide ("laughing gas") is produced by an important group of bacteria that convert ammonia to nitrate. These bacteria are most active in Lake Vanda in the depths of 55–60 metres, and in this layer nitrous oxide can be found in concentrations that are 300 times the values typically observed in freshwater lakes and the sea.

The bottoms of the Dry Valley lakes are coated with thick dark-green or purple mats composed primarily of cyanobacteria. These mats vary in shape from smooth and flat to upright "pinnacle mats" which form columnar structures several centimetres high. Clumps of gas-filled mat can tear off and float up to the underside of the lake ice. Some of this material becomes frozen into newly forming ice each winter, and can gradually move through the ice-cap over several years until it reaches the upper ice surface where it blows away in the wind. It has been suggested that this "aerial escape" of algae and the chemicals they contain may be a major way in which nutrients and salts are lost from these lakes, but the quantitative importance of this mechanism is highly speculative.

Some of the mats contain thin layers of a type of limestone which is formed by the algae. These mats are thought to closely resemble the so-called "stromatolites" formed by micro-organisms living deep within the oceans of the young earth about 3000 million years ago. The algal mats also contain many species of unicellular protozoa as well as rotifers, tardigrades, and nematodes. However, there are no large grazing or burrowing animals which would disrupt the formation of these delicate structures.

Algal mats are also a feature of the pools which form on the ice shelves and glaciers. The most common type of this sort of environment in the Ross Sea region is the so-called "cryoconite hole". Cryoconite means cold rock dust and refers to the way in which these holes form on the glaciers. Wind-blown dust particles that settle on the ice absorb

Figure 12.4. *The vertical layering of zones of "endolithic" lichens and algae within a sandstone rock. (a) a cube of rock showing the vertical section in more detail to the right; (b) a thin crust of mineral material over the rock surface; (c) the upper layer of rock which is not colonised by micro-organisms; (d) the upper blackish layer of lichen growth in which clusters of algal cells are closely surrounded by filaments of fungi; (e) bacteria found associated with all lichens and algae; (f) a relatively wide zone which is white and contains filaments of fungi penetrating down into the rock; (g) a green zone of single-celled algae which are not associated with the lichen; (h) sometimes there is a final layer of small cyanobacteria forming a bluish-green zone.*

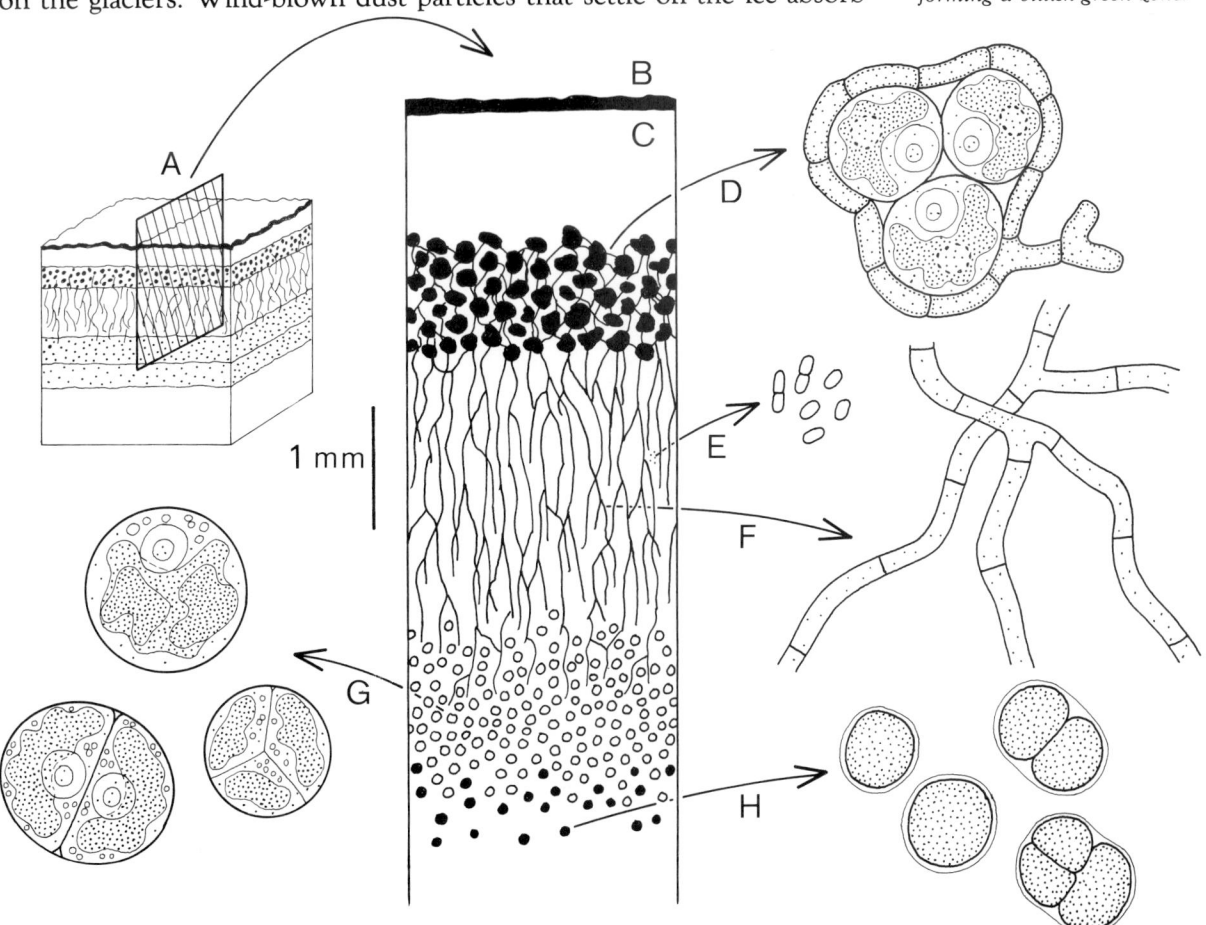

heat from the sun and gradually melt the underlying ice to form cylindrical holes. The holes gradually widen to a diameter of 10–50 centimetres and deepen to about 50 centimetres. During mid- to late-summer the holes partially fill with meltwater and provide a suitable habitat for microbial life. Cyanobacteria bind together the dust particles and form a mat at the base of these small pools; various other types of algae and bacteria grow amongst the mats and in the overlying water.

A more extensive example of this type of meltwater environment is the interconnected system of pools and streams on the McMurdo Ice Shelf (see Plate 9.7). This system covers over 2000 square kilometres with, in mid-summer, 10 – 60 percent of it being shallow open water flowing over ice and moraine. Sediment is distributed over this portion of the ice shelf, much of it derived from the seafloor. Like the smaller cryoconite environments, these pools within the ice are lined with mats of pink or orange cyanobacteria, and other microscopic algae and animals swim in the water above.

Perhaps the most extreme of all the aquatic environments in the Ross Sea region are the hypersaline lakes and pools. The best known example is Don Juan Pond which lies in the Lower Wright Valley. Evaporation in this very arid desert has resulted in dissolved salt gradually accumulating to a salinity 14 times that of sea water. At this concentration every kilogram of water from Don Juan Pond contains one-half kilogram of salts! The dominant salt is calcium chloride which precipitates out of solution as a mineral called antarctite. So far this mineral has been found only in the Don Juan Pond basin. This unusually high salinity is too high for aquatic life and even hardy microscopic forms are unable to survive and grow in this inhospitable place.

Rivers and streams

Rivers and streams are also a feature of many coastal and inland

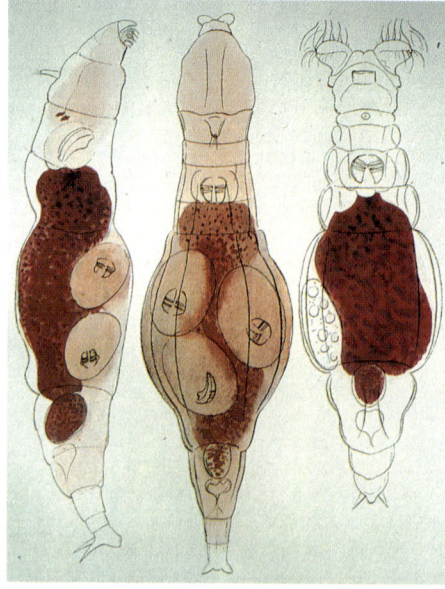

Plate 12.10. *Earliest drawings of the rotifers of Ross Island, by John Murray, biologist with Shackleton's 1907–09 expedition.*

Figure 12.5. *A diagrammatic vertical section through a rock to show weathering and fragmentation due to biological activity. (a) an intact crust with typical layering of lichens and algae (see Figure 12.4); (b) a flake of rock has fallen away having fractured through the zone of fungal filaments, leaving the fungi at the surface (c); (d) the micro-organisms move down into the rock and re-establish the typical, zonation pattern; (e and f) flaking occurs again as the rock continues to erode. The time scales involved are not known but are at least in the order of decades.*

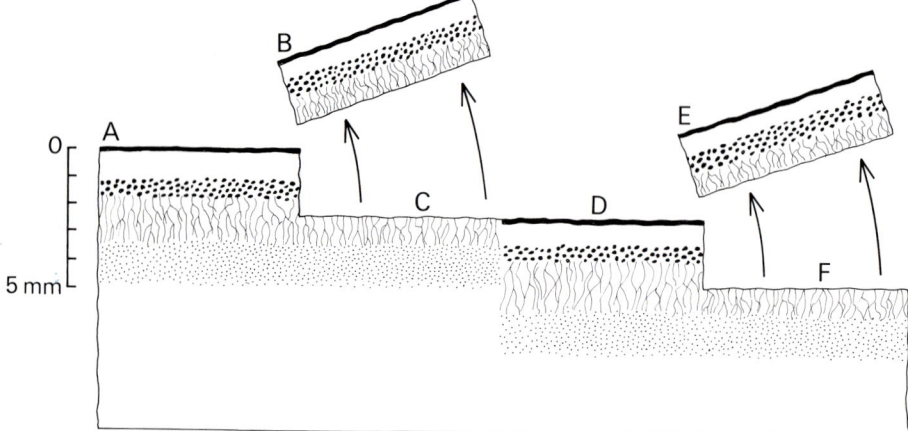

sites in the Ross Sea region. Near McMurdo Sound alone over 100 streams flow for a few weeks each year in perennial channels cut through ice-free ground. Most of these flowing waters are small, but some achieve substantial proportions. The Onyx River, for example, flows 30 kilometres down the Wright Valley, ultimately discharging into Lake Vanda. Under flood conditions it can be 15 metres wide, 1 metre deep, and flowing at 4 kilometres per hour. Further to the south, the Alph River occupies a channel 10 metres wide and 0.5 metres deep and flows several tens of kilometres along the edge of the Koettlitz Glacier towards the sea. This major stream system flows through a series of lakes and was discovered in 1911 by Scott's western sledging party led by Griffith Taylor. After manhauling their sledges across the appallingly broken ice terrain of the Koettlitz Glacier, and then southwards towards the Walcott Glacier in late February they were surprised to report that "we suddenly came across a steep gully about 100 feet deep at the bottom of which was a strongly flowing stream". More recent measurements on the Alph River indicate that its total annual discharge may exceed that of the Onyx River in some years.

Unlike temperate-latitude streams the discharge of Antarctic flowing waters is unrelated to rain or snowfall but instead is controlled by the melting of alpine glaciers and large ice sheets. During the summer period of discharge the air temperatures are around the melting point of ice and small changes in the weather can enormously alter the amount of flowing water.

Even over the 24-hour cycle the position of the sun relative to the glacier face drastically affects discharge. This is illustrated by a stream flowing through the penguin rookery at Cape Bird on Ross Island. The stream has a low discharge through most of the night and morning, but as the sun moves around to the west and shines directly on the ice face in the late afternoon the amount of flowing water increases by a factor of 100. Each day, streams can freeze completely or dry up during the

Plate 12.11. *Thick mats of cyanobacteria being removed from the base of a pool near the Koettlitz Glacier. Photo: C. Howard-Williams.*

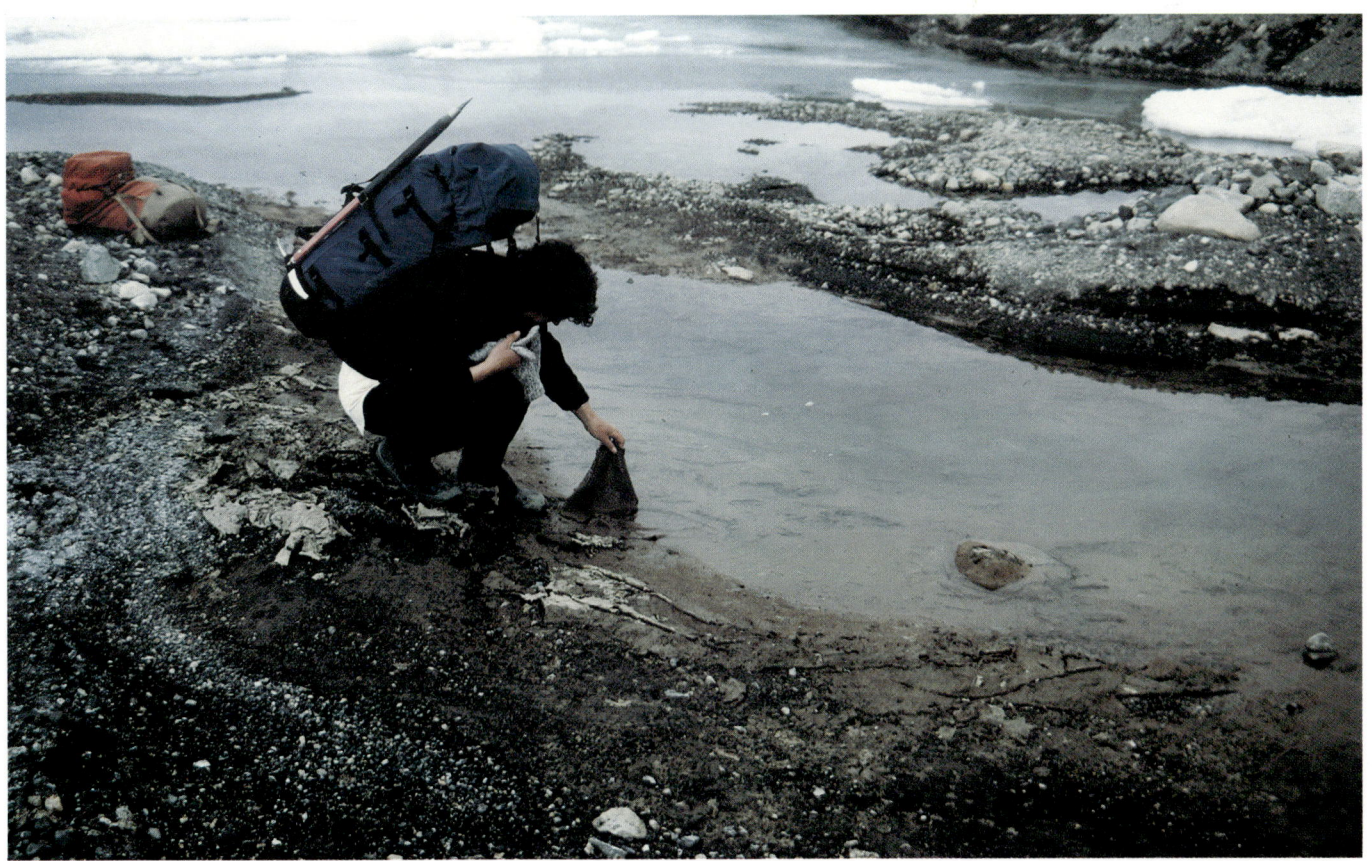

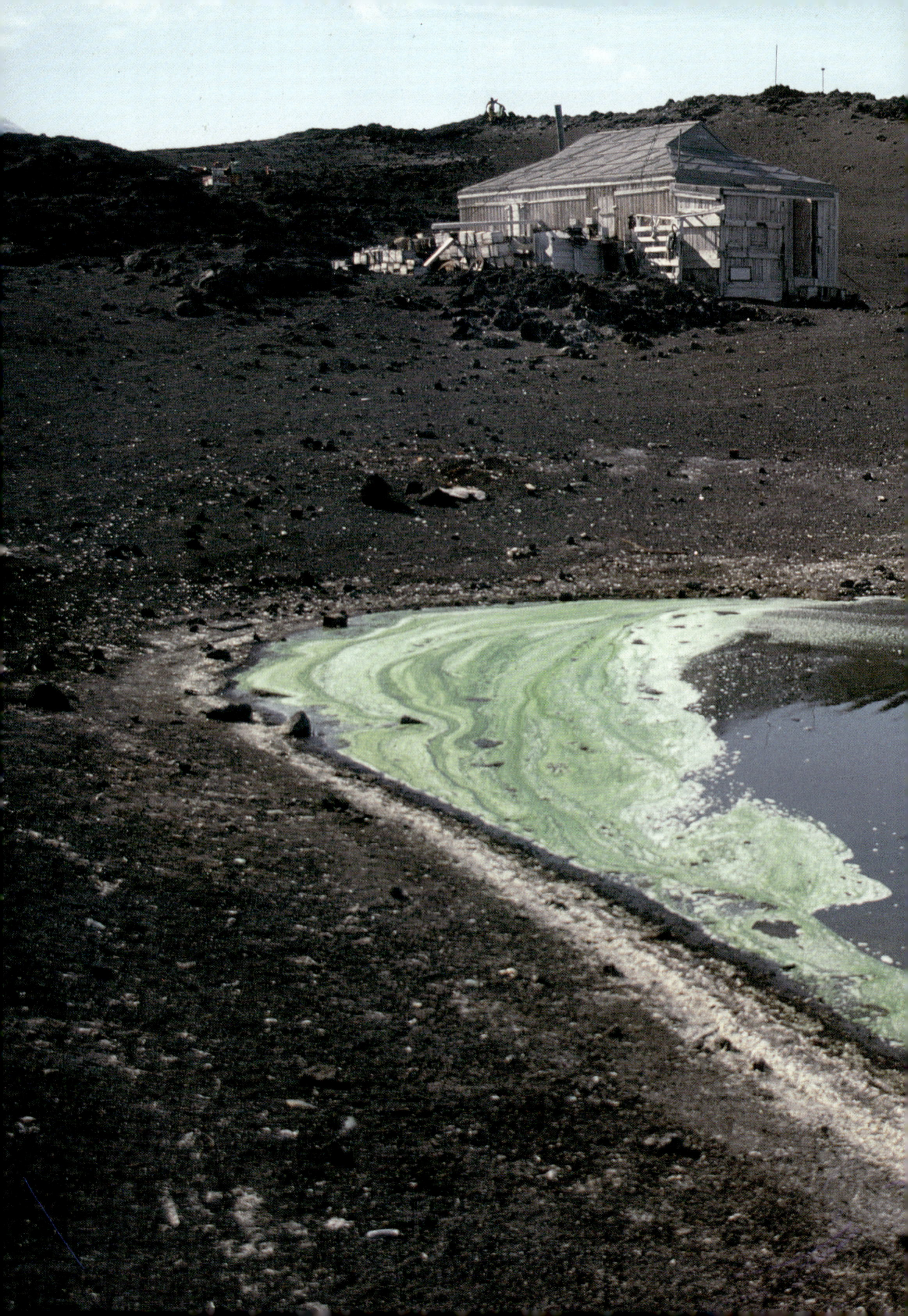

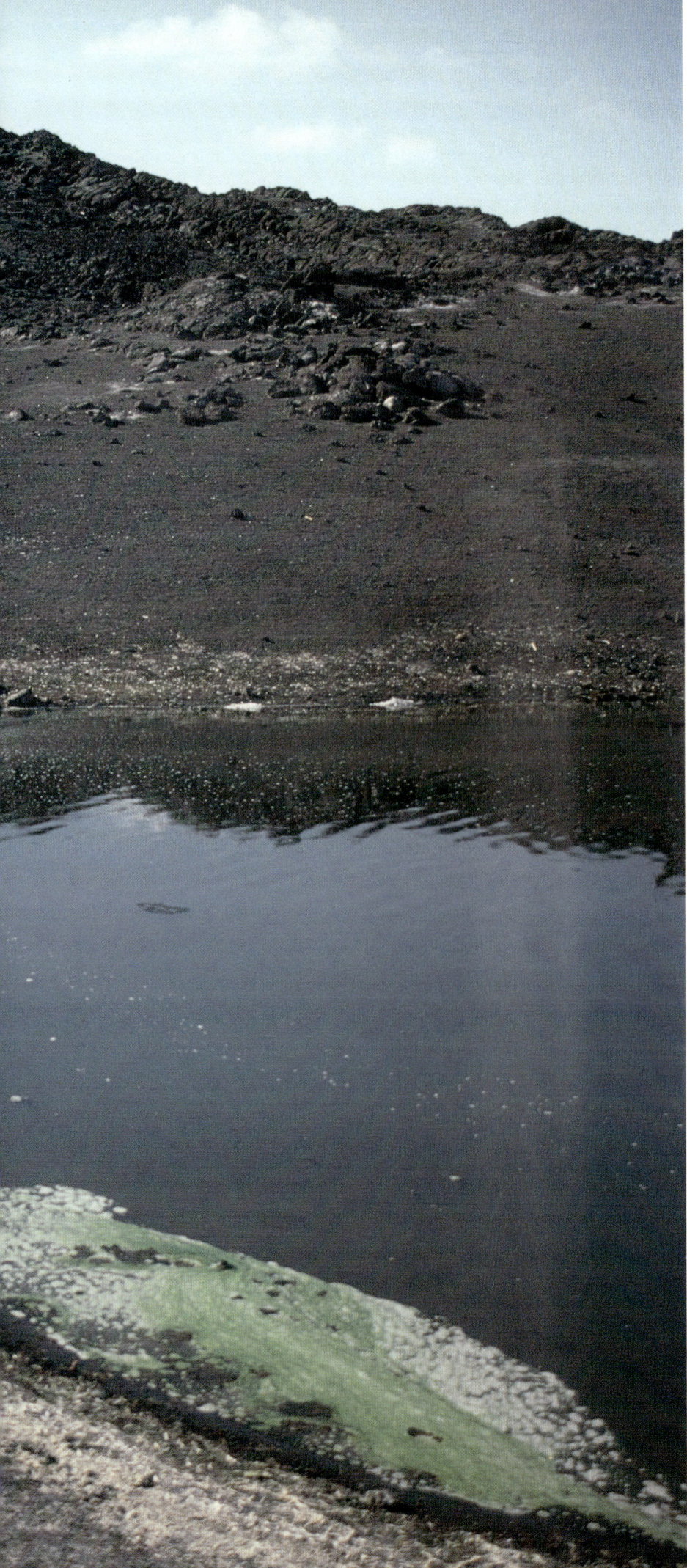

Plate 12.12. *A "bloom" of green algae in Pony Lake, Cape Royds. Shackleton's hut can be seen beyond the lake. Photo: P. Broady.*

hours when direct sunlight is not striking the glacier face, and field parties have been alarmed to find that their campsites on apparently dry river deltas can be rapidly inundated with water.

This highly variable environment is inhabited mostly by organisms that can survive periodic freezing and drying. The visually most abundant stream plants are cyanobacteria. Dark brown colonies of the nitrogen-fixing species *Nostoc commune* form particularly luxurious mats near the stream edge and other regions of low flow. These locations are less frequently inundated with water than the main body of the streams, but the dark pigmentation of *Nostoc* may allow the colonies to heat up above ambient air temperatures and thereby reduce the frequency and severity of freeze–thaw cycles. The central parts of the streambed are frequently coated with pink or orange mats dominated by the cyanobacterium *Phormidium* (Plate 12.13). Both the *Nostoc* and the *Phormidium* communities occur in quantities comparable with some of the richest algal growths in temperate latitude streams. These accumulations are probably the result of very slow rates of loss of algae from the streams, rather than rapid rates of growth. Unlike at temperate latitudes, there are no large grazing animals in the streams such as fish, insects, or crustaceans. Instead, there are low populations of nematodes, tardigrades, rotifers, and protozoans, but these minute herbivores are unlikely to graze a significant amount of the streambed algae.

The nutrients which feed these stream algal communities come from three sources: glacial ice which provides the meltwater, the highly weathered streambed sediments, and the plants and animals within the catchment. For most of the streams the catchment influence is small or negligible, except for certain coastal streams which receive nutrient wastes from seabirds. The latter type is shown by Cape Bird Stream which flows through a penguin rookery and as it does so its content of nitrogen and phosphorus salts increases by a factor of more than 20. In the poorly vegetated catchments of the Dry Valleys the glaciers provide

Figure 12.6. *Single-cell algae from Lake Fryxell. All possess whip-like flagella, which enable them to move through the water.* (a) Pyramimonas; (b) Chroomonas lacustris; (c) Ochromonas; (d) Chlamydomonas.

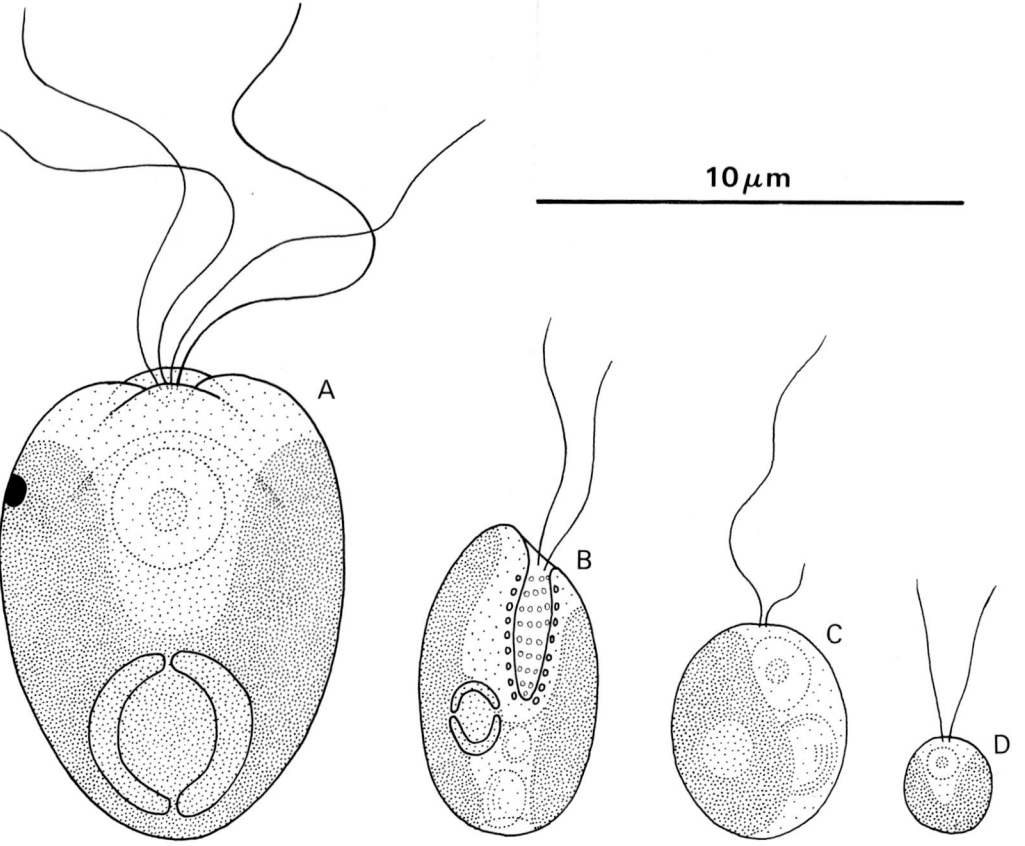

10 μm

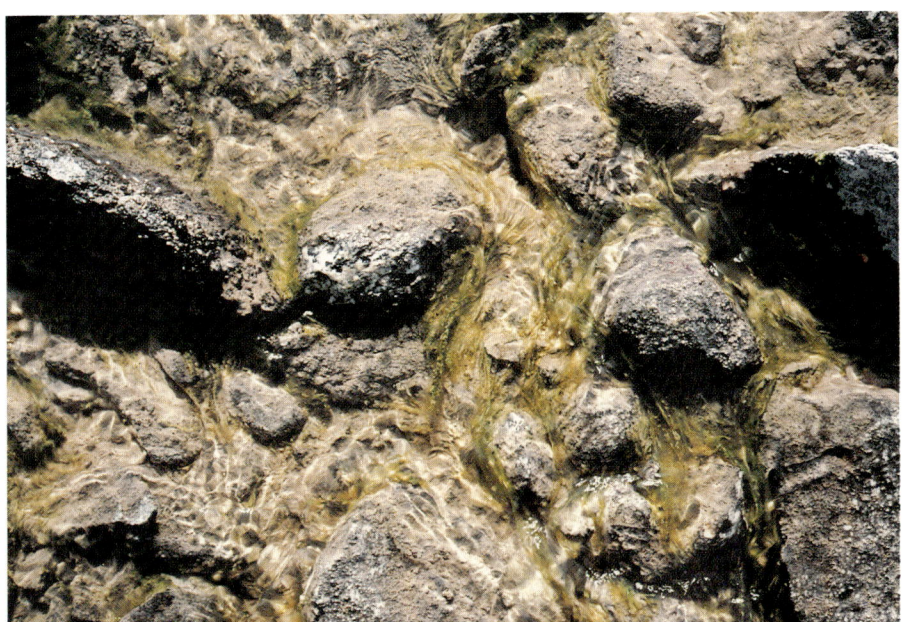

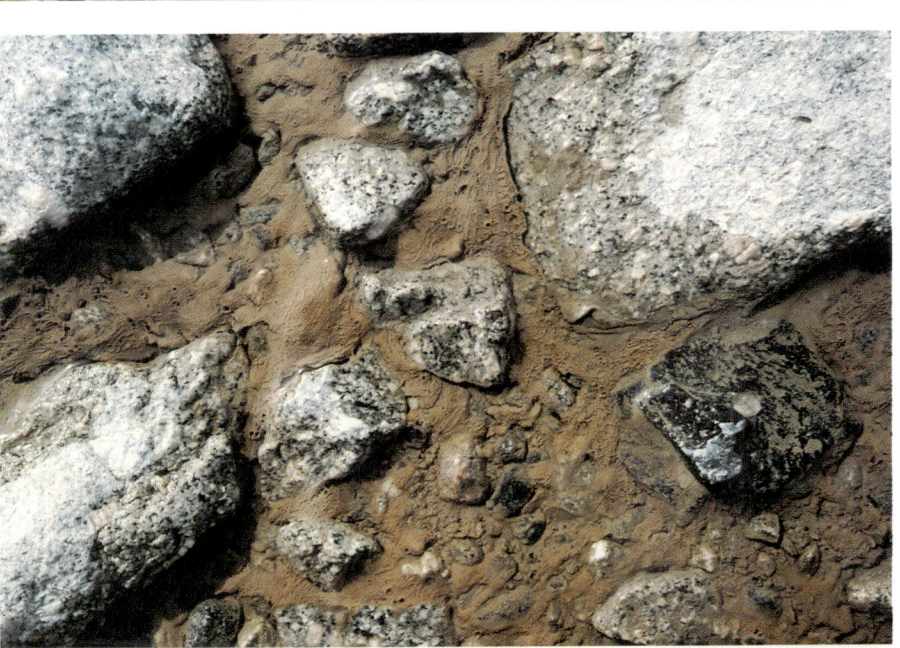

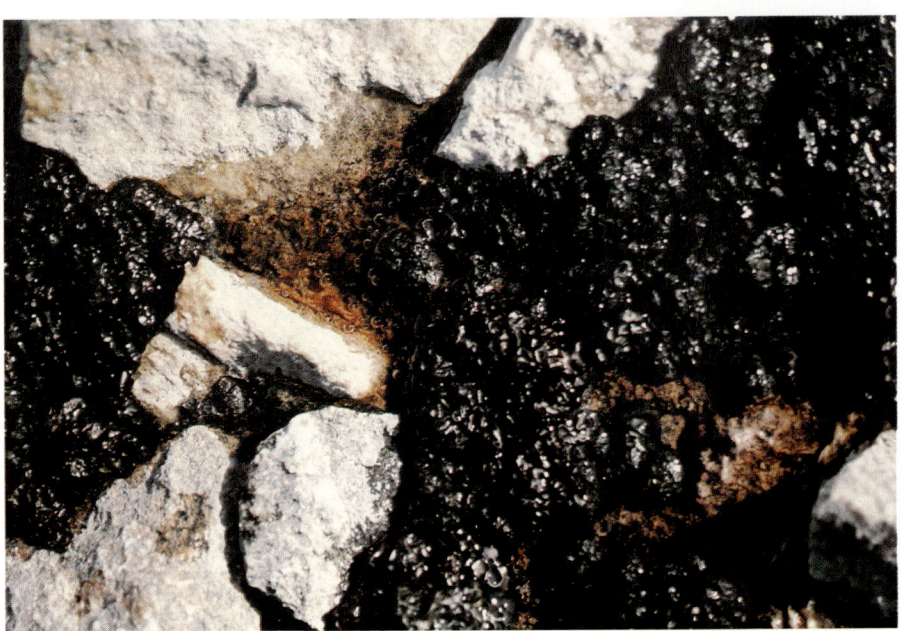

Plate 12.13. *Streambed algal communities: (a)* Binuclearia; *(b)* Phormidium; *(c)* Nostoc. *Photos: W. Vincent.*

(a)

(b)

(c)

the primary source of nutrients. The observation that algae grow within meltwater holes on these glaciers suggests there is a relatively high level of nutrients within the ice, and analysis of glacier samples has confirmed the presence of nitrogen and phosphorus. The first melts that flow from the glacier are particularly enriched in these nutrients, perhaps because they pick up the aerosol materials deposited during the preceding winter. This effect, in combination with streambed salts and newly weathered sediments, results in an initial flush of nutrient-rich water down the streambed with the first flows each season.

The Antarctic stream biota seem particularly well equipped to deal with this regime of episodic water flow and irregular nutrient supply. At the end of summer when the streams freeze solid, the ice gradually ablates leaving behind dry, frozen crusts of algae on the streambed. These algal mats remain freeze-dried throughout winter. On rehydration with the first flows of the next summer they can rapidly begin their cellular activities. The freeze-dried mats are capable of removing nutrients from solution almost immediately after being wetted, and respiration and photosynthesis are detectable within the first 20 minutes of rehydration. This large overwintering population of viable cells provides an important inoculum for the next season of streamflow. It may ensure the successful persistence of the meltstream cyanobacteria from year to year, and prevent the growth of potential competitors.

Human impact

With increasing human impact in areas of the Ross Sea region which until recently have been pristine, it is now important to protect a diversity of significantly large areas containing representative examples of all the terrestrial life-forms. Different plants, animals, and micro-organisms are found in different areas and even the most barren places can contain unusual and unique organisms. Damage and contamination can easily be caused by trampling, use of overland vehicles, wind-scattering of rubbish, construction of buildings and roads, inadequate disposal of human wastes, and even by the activities of scientific parties.

For example, the waters of the Dry Valleys include some of the most transparent, clean water lakes known in the world and these will be highly sensitive to any input of nutrients or other pollutants derived from human activity in their catchments. Similarly, the microscopic life-forms which live within Dry Valley soils can be grown only on very dilute culture media which contain very low concentrations of nutrients. Nutrient enrichment of the soils would completely destroy communities of these organisms.

Although, the indigenous life of these regions is extremely hardy when viewed in the context of its ability to grow and survive in a harsh natural environment, it is extremely delicate and prone to disturbance by the activities of human intruders. Growth rates are extremely slow and recovery from damage would be equally slow, if it happened at all. Even in the relatively mild and hospitable regions of the Antarctic Peninsula, where growth would be expected to proceed more rapidly, lichens a few centimetres in diameter may be up to 600 years old. We must ensure that our actions do not degrade the habitats of these organisms which, if we are careful, should remain a source of wonder and fruitful scientific enquiry.

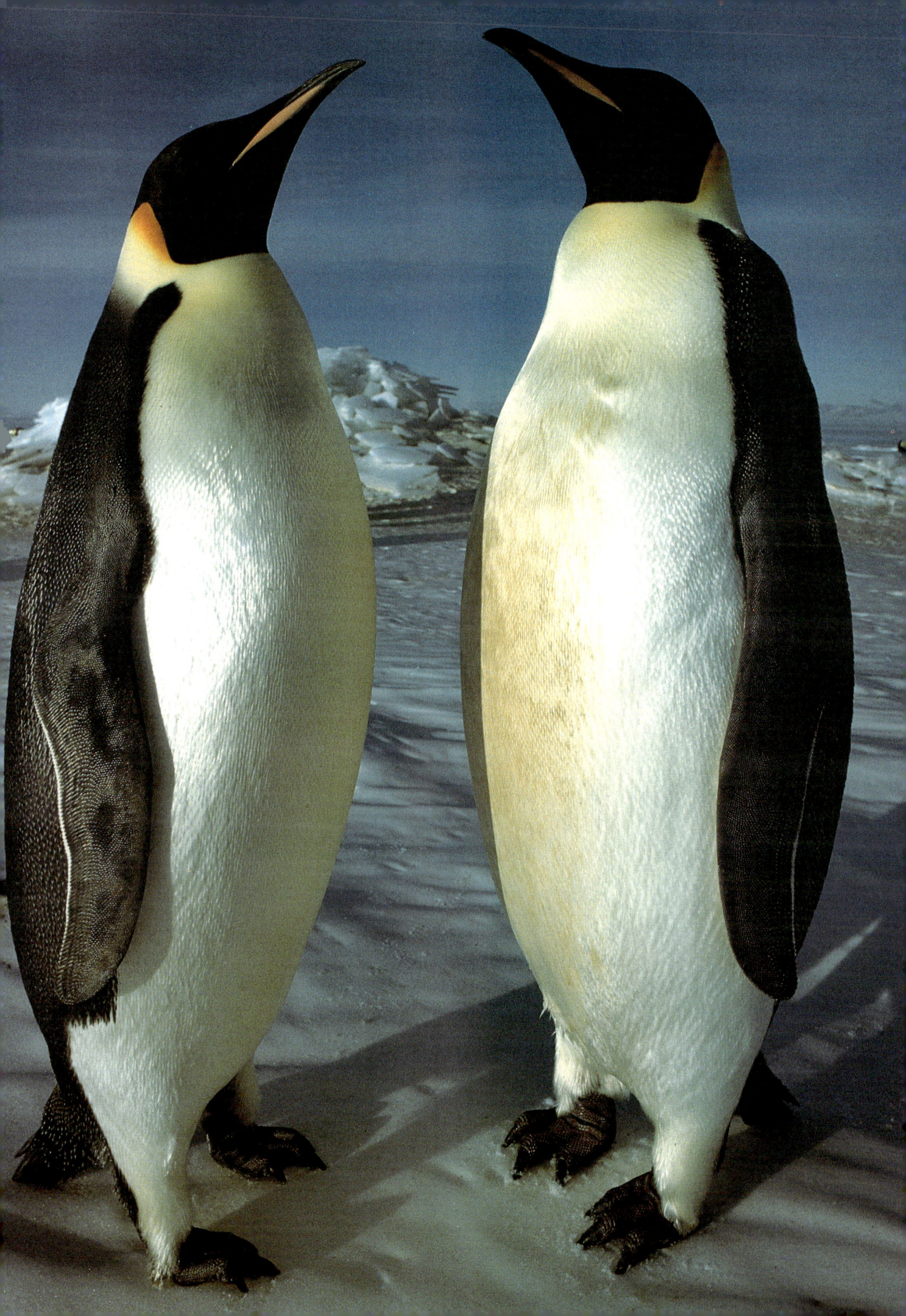

13. Birds of the Ross Sea region

Antarctica is one of the world's ornithological treasures, comparing well with any other natural area for spectacular scenery and abundant wildlife. But the only birds to be seen at Scott Base and McMurdo Station, the "townships" of the region, are the skuas. They are dark brown, drab gull-like birds very like juvenile black-backed gulls and with apparently the same habits—scavenging at rubbish, "loafing about", or, flying lazily along the horizon.

The continent is cold, far from other land, and lacks any food on its ice, snow, and scree. These factors broadly determine the structure of the Antarctic bird community. It has few species and they are all dependent ultimately on the sea for food; they are all sea birds. One other factor conditions their breeding habits. The absence of terrestrial predators, especially the range of mammals which characterise the Arctic, allows Antarctic birds to nest in large aggregations in the open both on the mainland and on the islands. It is not the same bird community as in the Arctic. Colonisation and evolution were not identical in the two places and the two groups have remained largely distinct. Penguins, albatrosses, and petrels inhabit the Southern Oceans; in the north are auks, puffins, and gulls. If one wanted to go from New Zealand to see enormous numbers of a wide range of sea birds one's destination would not be Antarctica but the wet and windy subantarctic islands—the Snares, Auckland, and Campbell Islands, or even the Chatham Islands.

If this is so, why has Antarctica attracted so much ornithological interest, not only from scientists but also from others not usually enthusiastic about birds? The main attraction is the penguins. In this part of Antarctica there are just two species; the Adélie penguin (Plate 13.1), a squat, noisy flightless bird and the much larger, winter-breeding emperor penguin (Plate 13.2), beautifully coloured on the head and neck but with the voice of a badly tuned organ.

Several factors contribute to the popularity of these two penguins. Undeniably they are attractively formed and coloured with a neat contrast of black and white feathering. In their few breeding places they are impressively abundant, concentrated into small areas of shelter, and are easily seen and photographed. They stand about on the ice and scree and nest at ground level whereas species breeding further north may be hidden in burrows or by vegetation. Finally, like other penguins they caricature the upright gait and exaggerated postures of humans. Of the two the Adélie penguins provoke the greater amusement and extravagant comment.

They are very rewarding subjects for scientific study, for three reasons. Firstly, they are part of a very straightforward food web; small plankton → fish and krill → birds. Secondly, they are so easily observable and their interactions with each other and the skua, their only predator–scavenger on land, are more readily recorded than for most other birds. Thirdly, their behaviour, morphology, and physiology is so well attuned to survival in this most severe of environments. Finally, for scientists working with the Adélie penguin, conditions are generally excellent with little rain, no smell, dry conditions underfoot, and 24 hours of sunshine every summer day. These factors, together with remarkably easy access, make Antarctica a much preferable working place to the closer, warmer, but isolated subantarctic islands. All this, however, applies only to the birds' life on shore. Only now, 90 years after ornithology began in Antarctica, are we beginning to comprehend the birds' life at sea—their foraging, the wintering months on the pack ice or on long migrations, and their pin-point navigation. These are the critical survival months of the annual cycle but they are the most difficult to study.

The Ross Sea penetrates so far south, to within 500 kilometres of

Plate 13.1. *An adult Adélie penguin. Photo: Antarctic Division, DSIR.*

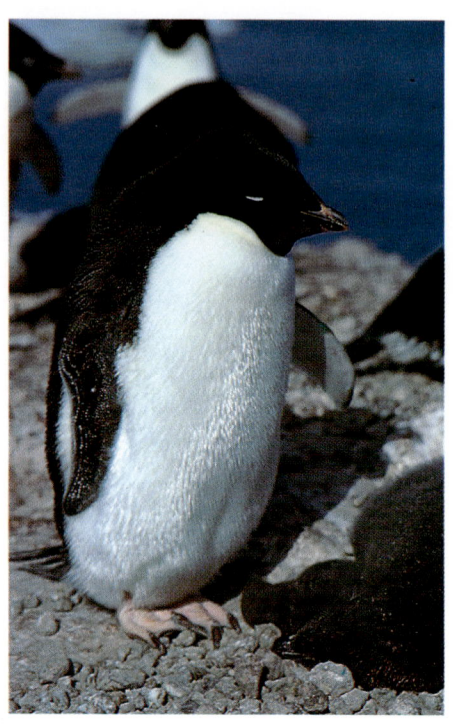

the South Pole, that it would be reasonable to assume that the birds on its shores are the southernmost breeding birds on earth. The extent of the sea is truncated, however, by the Ross Ice Shelf effectively placing a southern limit on Ross Sea birds at about 77°S, the latitude of Ross Island. Nevertheless, southern McMurdo Sound is the most southerly breeding locality on the continent for penguins. Skuas and petrels breed well inland in the Weddell Sea sector and no doubt could breed south of the Ross Sea if suitable terrain was present. The mountain ranges in Marie Byrd Land need to be explored in detail for breeding colonies.

In past warmer periods no doubt these birds, or ones similar to them, did breed much further south. Even further back in time an ice-free Antarctica would have had a very different fauna, a warm temperate one—like that of present New Zealand or Australia but with a surprisingly varied penguin fauna, including some very large species over twice the bulk of the today's emperor penguins. Recently an amazing penguin variety has been discovered as fossils at Seymour Island near the northern tip of the Antarctic Peninsula. There are 7 genera, with perhaps 14 species of late Eocene origin, about 40 million years before present—a time of very warm conditions in coastal Antarctica (and New Zealand) in which tropical palms grew. Penguins need not, therefore, be exclusively tied to colder areas.

The current group of birds, the penguins, petrels, and gulls all have an ancient evolutionary history—as recognisable forms for 100 million years at least. They existed, therefore, before New Zealand broke away from Gondwanaland 80 million years ago. Antarctic links since then have been with Australia until about 45 million years ago, and with South America up to about 25 million years ago. From this date on, and with the inception of the powerful westwind current of the Southern Ocean, bird links between Antarctica and the more northern lands have been exceedingly difficult and achieved only through long flights or by swimming.

The Pleistocene ice ages must have profoundly affected the Antarctic bird fauna. All analyses of the glacial episodes suggest that a much greater extent of sea ice ringed the continent and there were much bigger glaciers and snow fields. Consequently, there was less snow-free land. It is hard to imagine that bird species apparently at the limits of their survival in present conditions could persist in the Antarctic in the far worse conditions of the ice ages. From this climate comparison it is generally thought that the birds had to abandon Antarctica during the ice ages, returning there each time the climate improved. In this scenario, New Zealand could have had Adélie penguin colonies during ice ages, their remains now hidden by the rising sea level of the current interglacial period. Recent research by German scientists indicates that this account of the movements of birds to and from Antarctica may be too simplistic. Radiometric dating of the oil ejected by snow petrels shows that these birds were able to breed on the continent during the height of the last ice age 33 000 years ago. Breeding petrels require protected rock crevices for nesting and open sea within flight range for feeding. If these dates are substantiated they suggest that, even during these severe conditions, snow- and ice-free cliffs remained open to breeding birds and permanent stretches of water occurred in the enormously expanded sea ice cover. The occurrence of open water is not surprising. It is this feature that allows emperor penguins to breed in winter at present, when sea ice cover is about the same as in summer during the ice ages. Nevertheless, it is difficult to see how the penguins could have coped with Antarctic ice-age conditions. Broken sea ice is a formidable barrier to their movement and any increase in sea ice extent would surely prevent them occupying any ice-free coastal land. Certainly

McMurdo Sound was closed off. During the last ice age it was filled with a glacier extending to over 600 metres up the slopes of Ross Island and reaching past Franklin Island.

Estimates for the date when the emperor and Adélie penguins and the skuas were able to reoccupy the southern Ross Sea breeding grounds are becoming firmer. These estimates come from the dating of the raised beaches in the region, from the occurrence of marine fossils on the shorelines, and from the dated remains of penguin guano and skeletons at the bottom levels of penguin breeding sites. Open water in summer occurred before 4000 years ago and penguin skeletal material has been firmly dated for Franklin Island at about the same time.

Species of birds in the Ross Sea region

The Ross Sea sector from Victoria Land in the west to Marie Bryd Land in the east is a very significant part of the Antarctic marine ecosystem. The sector is a triangular segment from 71°S at Cape Adare on the northern edge of Victoria Land to 77°S at Ross Island, stretching for some 800 kilometres along this flank and for about the same distance from Cape Adare out to its eastern point. In all this area, however, the birds are confined for breeding to the few places along the narrow coastal margin of the continent where mountain slopes or ice- and snow-free moraine exist and on the islands dotting the western margin.

In the far north, the Balleny Islands have the most diverse variety of breeding bird species in the Ross Sea sector. Latitude may have less influence on diversity than access to open sea and the availability of nest sites, as a good diversity of species also occurs at the southern limit of the Ross Sea, 1000 kilometres to the south of the Ballenys. The U.S. Antarctic Service Expedition (1939–41) found breeding Antarctic petrels and snow petrels, with skuas, on the Rockefeller Mountains of the Edward VII Peninsula (78°S, 155°W)—about the same latitude as Ross Island.

The Ross Sea is a fertile food source for the breeding birds along its margin and for wide-ranging oceanic species penetrating from much further north. Turbulent currents sweeping across a broken sea floor and about the islands maintain ice-free places (polynyas) throughout the year—critical features for the survival of winter-breeding emperor penguins.

The birds of the Ross Sea are part of much larger populations occurring throughout Antarctica. None is unique to this locality. Table 13.1 illustrates the decline in diversity from the subantarctic islands to the Ross Sea. In McMurdo Sound there are just three species of breeding birds.

Table 13.1. Numbers of breeding birds in the different subantarctic and Antarctic regions.

Location	Number of species
Subantarctic islands	26
Southern islands of the Scotia Arc (South Shetlands, Orkneys, and Sandwich Islands)	18
Antarctic Peninsula	16
Antarctic Continent	12
Ross Sea region (excluding Balleny and Scott Islands)	6
McMurdo Sound	3

Table 13.2 lists the different breeding species in the Antarctic zone: the universality of the Ross Sea species is obvious. They all occur elsewhere on the continent. The number of species breeding in the Ross

Table 13.2. The breeding birds of the Antarctic zone.

	Continental Antarctica			Maritime subzone	
	McMurdo Sound/ Ross Island	Ross Sea sector	Antarctic Continent and islands	Antarctic Peninsula	Scotia Arc Islands
Spheniscidae (penguins)					
Emperor penguin *Aptenodytes forsteri*	●	●	●		
King penguin *Aptenodytes patagonicus*					●
Adélie penguin *Pygoscelis adeliae*	●	●	●	●	●
Chinstrap penguin *Pygoscelis antarctica*		●*	●	●	●
Gentoo penguin *Pygoscelis papua*				●	●
Macaroni penguin *Eudyptes chrysolophus*				●	●
Procellariidae (petrels)					
Southern giant petrel *Macronectes giganteus*			●	●	●
Cape pigeon *Daption capense*		●*	●	●	●
Southern fulmar *Fulmarus glacialoides*		●*	●	●	●
Snow petrel *Pagodroma nivea*		●	●	●	●
Antarctic petrel *Thalassoica antarctica*		●	●		
Antarctic prion *Pachyptila desolata*		●*	●	●	●
Hydrobatidae (storm petrels)					
Wilson's storm petrel *Oceanites oceanicus*		●	●	●	●
Black-bellied storm petrel *Fregetta tropica*					●
Phalacrocoracidae (shags)					
Blue-eyed shag *Phalacrocorax atriceps*				●	●
Stercorariidae (skuas and jaegers)					
Antarctic skua *Stercorarius maccormicki*	●	●	●		●
Southern skua *Stercorarius skua lonnbergi*		●*?		●	●
Laridae (gulls and terns)					
Black-backed gull *Larus dominicanus*			●	●	●
Antarctic tern *Sterna vittata*				●	●
Chionididae (sheathbills)					
American sheathbill *Chionis alba*				●	●

* Balleny and/or Scott Islands only. Breeding by the southern skua on Balleny Islands not yet confirmed.

Plate 13.3. *Emperor penguin colony at Cape Crozier, where the Ross Ice Shelf meets Ross Island. Photo: Antarctic Division, DSIR.*

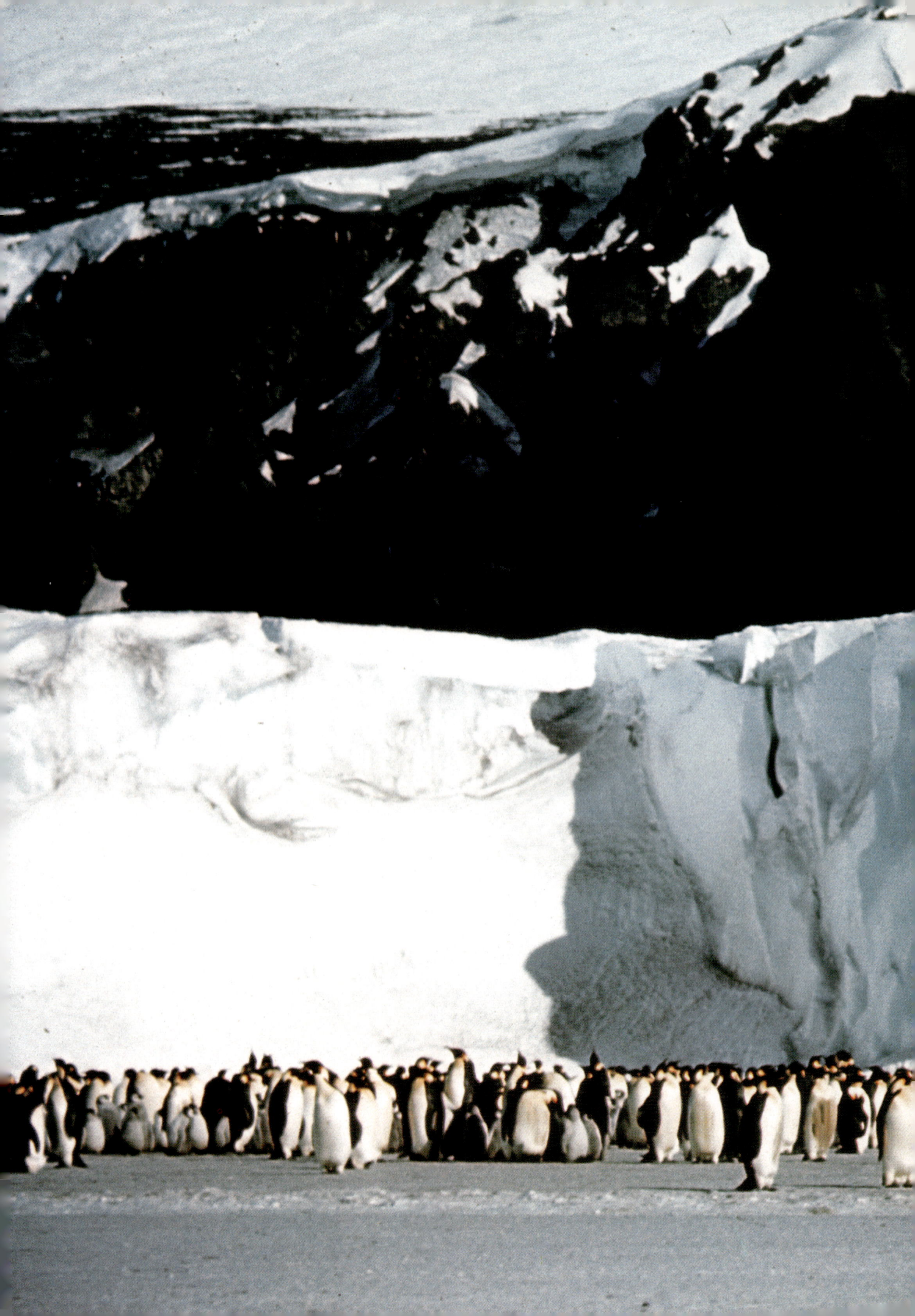

Sea sector seems excessive here compared with the continent as a whole because of the inclusion of the Balleny and Scott Islands off its northern edge. These islands contribute 1 penguin species, 3 petrels, and the southern skua to the regional total of 11 breeding species. Without them the Ross Sea sector has the more modest total of six breeding species.

Obviously the more travelling that is done over the Ross Sea or along the coast the more chance there is to see some of these birds. For those at Scott Base or McMurdo or travelling inland the only bird to be seen is the skua. For those lucky enough to get to Capes Royds, Bird, or Crozier, both Adélie penguins and skuas can be seen breeding. Although emperor penguins also breed at Crozier this occurs during winter and by the time the first visitors arrive in summer it is mostly over; by January the breeding site is deserted. Travelling by ship is the best way to see birds. Both Adélie and emperor penguins are commonly seen singly and in groups and they look much more attractive in this setting. Observations at sea also allow one to view the less common species, the local petrels and the vagrants from breeding grounds further north. These are not often seen over land.

Cape Crozier is a famous ornithological site (Plate 13.3). Its position facing the open sea is a good one for picking up vagrant species. Scientists working there have kept a detailed record of these over the years. Only eight vagrant species have been observed; of these chinstrap penguins, giant petrels, black-backed gulls, and southern skuas, all from breeding places much further north, are perhaps unexpected, and are all uncommon visitors.

There are numerous logs of the birds seen at sea from ships passing between New Zealand and McMurdo Sound. Their value depends on the skill of the observers in identifying flying birds at a distance, often in poor light. There is, however, one detailed study of the bird distribution and abundance in the Ross Sea. Peak densities of 60 birds per square kilometre were found along the oceanic continental slope and about the

Plate 13.4. *Emperor penguin and chicks huddle together to keep warm at Cape Crozier. Photo: Antarctic Division, DSIR.*

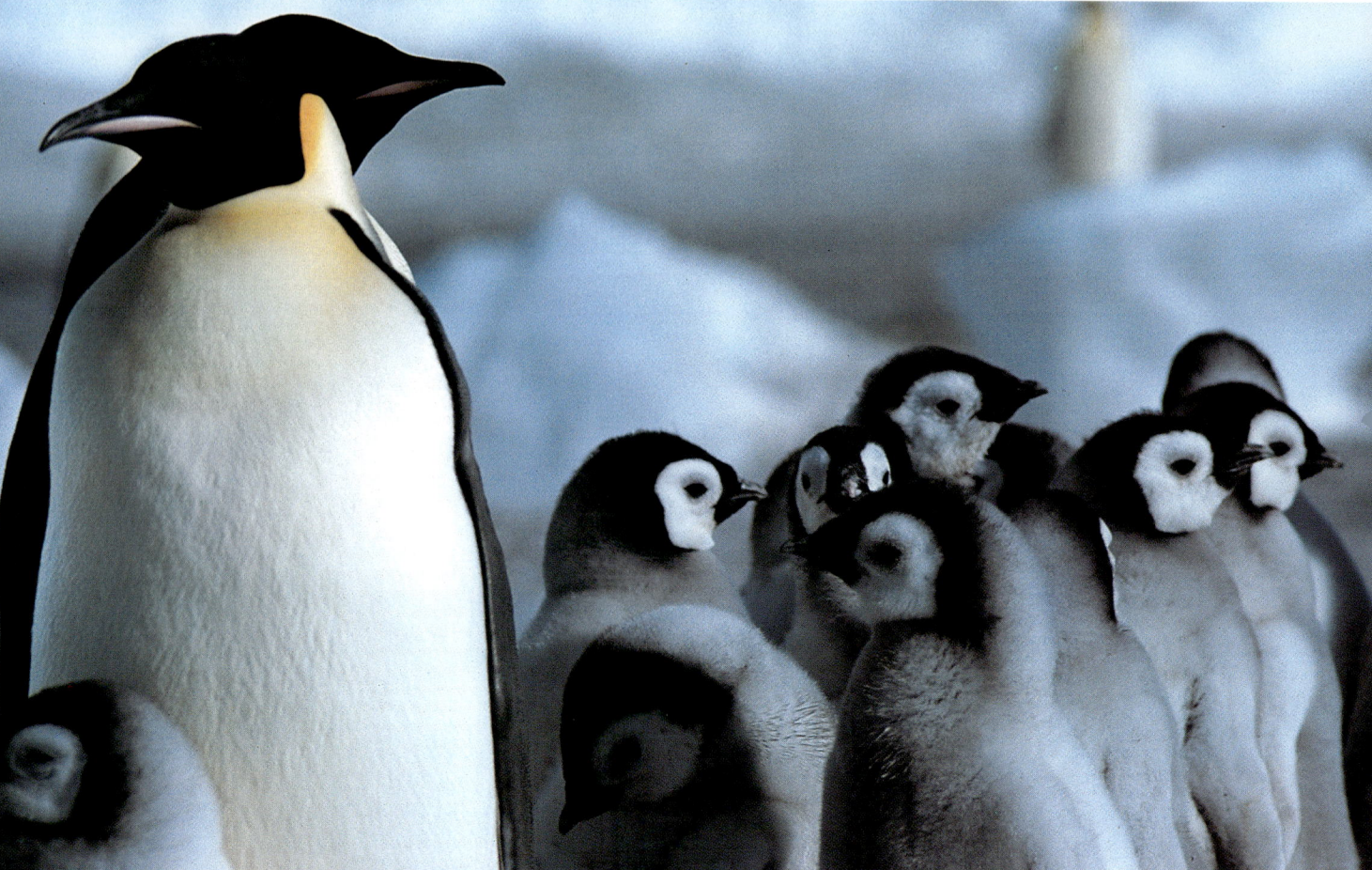

pack ice edge. Overall, however, densities were more modest with 4–25 birds per square kilometre and were generally lowest in ice-free areas. In the Ross Sea there is a pack ice community of the two penguins, and the snow and Antarctic petrels and an open-water community dominated by southern fulmar, cape pigeon, Wilson's storm petrel, mottled petrel, and Antarctic petrel—eight species in all. The two overlap along the ice edge and in broad, open leads. In part the division is determined by flight and foraging methods. Species depending on waves to obtain lift in flight, Wilson's petrel for example, are excluded from close pack where the waves are dampened. Six species occur typically in the loose pack of the southern Ross Sea in summer (emperor and Adélie penguins, snow and Antarctic petrels, Wilson's storm petrel, Antarctic skua). The most abundant species, and also contributing the greatest biomass, was the Antarctic petrel at 8.1 kilograms per square kilometre. Emperor penguins, in spite of their individual large size, occurred at lower densities and were second in importance at 7.5 kilograms and Adélie penguins and snow petrels each contributed 0.8 kilograms. Wilson's storm petrels contributed only 0.002 kilograms, reflecting both their small size and low numbers. No estimate was made for skuas.

Few terrestrial animals have adapted successfully to the harsh conditions of the Antarctic environment. On land a few primitive insect and mite species represent a true terrestrial fauna. At sea and in the pack ice there are four seals. On the margin of the land a small group of bird species breed successfully. This is a tiny fauna for such a huge area. The limitations are threefold. First, the isolation from other faunas precludes easy colonisation and adaptation. Secondly, the lack of food on land limits the range of species that can be supported. Thirdly, the harsh climate of intense winter cold with unbroken darkness and the very short summer excludes all but the most adapted forms. Even so, only the emperor penguin and the seals can survive through winter in the Ross Sea, relying on breaks in the pack ice for entry to the sea itself and its

Plate 13.5. *A creche of Adélie penguin chicks. Photo: E. Young.*

Plate 13.6. *A snow petrel at rest. Photo: Antarctic Division, DSIR.*

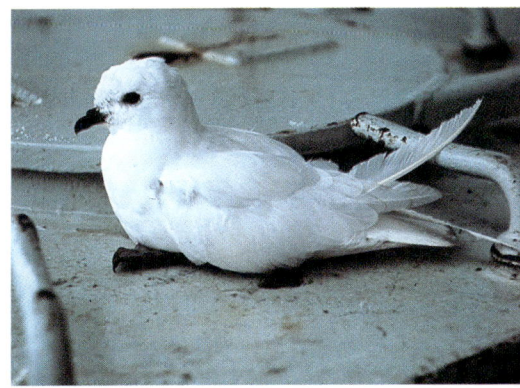

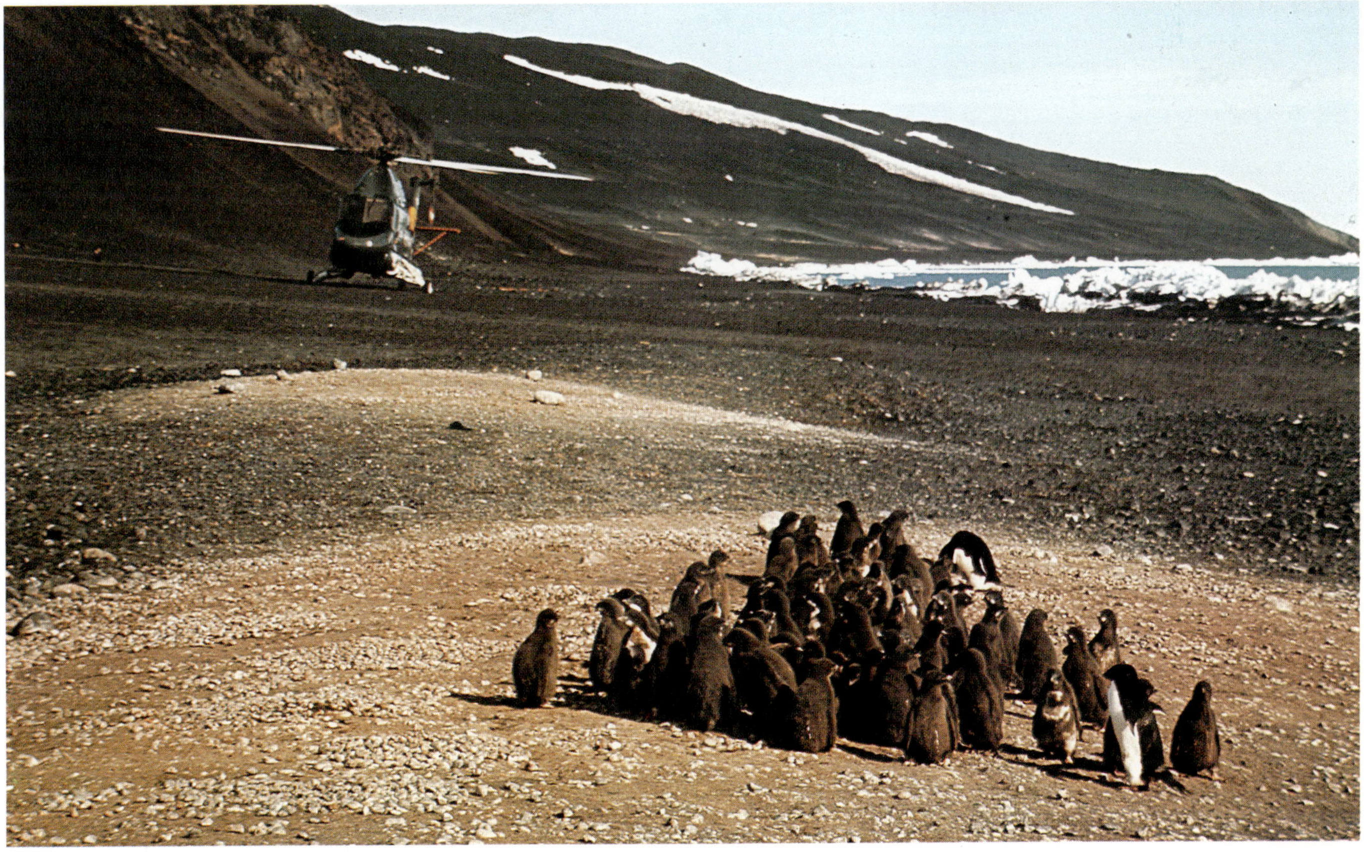

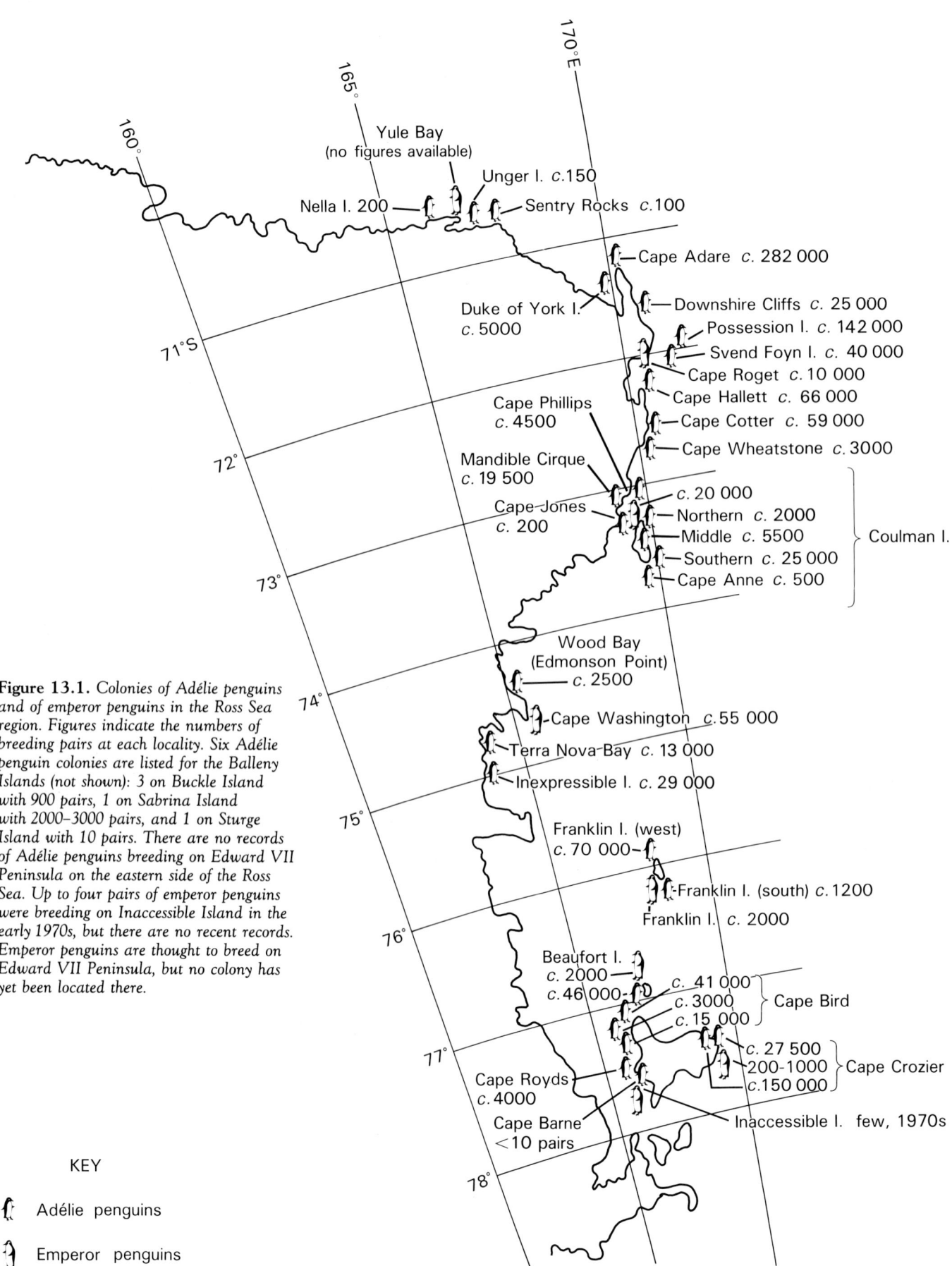

Figure 13.1. Colonies of Adélie penguins and of emperor penguins in the Ross Sea region. Figures indicate the numbers of breeding pairs at each locality. Six Adélie penguin colonies are listed for the Balleny Islands (not shown): 3 on Buckle Island with 900 pairs, 1 on Sabrina Island with 2000–3000 pairs, and 1 on Sturge Island with 10 pairs. There are no records of Adélie penguins breeding on Edward VII Peninsula on the eastern side of the Ross Sea. Up to four pairs of emperor penguins were breeding on Inaccessible Island in the early 1970s, but there are no recent records. Emperor penguins are thought to breed on Edward VII Peninsula, but no colony has yet been located there.

Yule Bay (no figures available)

Unger I. c.150

Nella I. 200 — Sentry Rocks c.100

Cape Adare c. 282 000

Duke of York I. c. 5000

Downshire Cliffs c. 25 000

Possession I. c. 142 000

Svend Foyn I. c. 40 000

Cape Roget c.10 000

Cape Hallett c. 66 000

Cape Phillips c. 4500

Cape Cotter c. 59 000

Cape Wheatstone c.3000

Mandible Cirque c. 19 500

c. 20 000

Cape Jones c. 200

Northern c. 2000

Middle c. 5500

Coulman I.

Southern c. 25 000

Cape Anne c. 500

Wood Bay (Edmonson Point) c. 2500

Cape Washington c.55 000

Terra Nova Bay c. 13 000

Inexpressible I. c. 29 000

Franklin I. (west) c. 70 000

Franklin I. (south) c. 1200

Franklin I. c. 2000

Beaufort I. c. 2000

c. 46 000

c. 41 000

c. 3000

Cape Bird

c. 15 000

c. 27 500

200-1000

Cape Crozier

c.150 000

Cape Royds c. 4000

Cape Barne <10 pairs

Inaccessible I. few, 1970s

KEY

Adélie penguins

Emperor penguins

food resources. The Weddell seal *(Leptonychotes weddelli)* living in the fast ice must work very hard to keep its access holes open, wearing down its teeth in the process. The remaining bird species migrate north in the autumn, to the pack ice fringe. Skuas, however, cross the Equator into the Northern Hemisphere.

We now have a good understanding of the adaptation to cold of these bird species. They are of course well insulated; the flying birds are thickly covered with down and feathers, the penguins with thick layers of subdermal fat (20 millimetres thick at the start of the breeding season) and with short, overlapping feathers that trap insulating air when the bird is in the water. These feathers are unique, they are the only ones known in which the insulation value *increases* in winds. Values of up to 180 percent of the still air value have been measured in penguins in winds of 15 metres per second (∼ 30 knots). The fat also smooths out the angularities of the penguin body to stream-line its shape for ease of movement through the water. The legs and feet are exposed to the cold, although as much as possible they are tucked up against the protecting warmth of the body. Body heat loss through these limbs is minimised through "heat exchangers" in which the veins and arteries are intertwined to transfer the warmth from the outward flow to the returning blood. The tissues of the legs and feet are extremely resistant to damage from cold and they take up the temperature of their surroundings. These superbly insulated birds are therefore well adapted to survive low temperatures when "loafing about" or incubating the eggs and protecting the chicks. In extreme conditions, emperor penguins huddle together to reduce heat loss (Plate 13.4). Adélie penguin chicks also huddle, forming creches (Plate 13.5). They do so when being harassed by skuas but more commonly it is a response to intense cold or wind.

At the other extreme of the temperature spectrum these birds must lose heat, when active and producing high levels of metabolic heat and when faced with hot, still conditions on the breeding grounds. Under these circumstances penguins radiate heat from the inside surface of their flippers. "Shunts" switch blood into a complex of surface blood vessels in heated birds, allowing the heat to escape across the nearly naked skin surface. It is easy to see when this is happening, the flipper glows pinkly. Skuas and petrels lack this feature and heat is lost through panting. On hot afternoons on the black rock and guano of the penguin rookeries both skuas and penguins are heat distressed, with gaping bills and a noisy rapid breathing.

The occurrence of diurnal cycles of activity in Antarctic birds during the continuous daylight of summer has long intrigued ornithologists. Numerous behavioural studies have attempted to demonstrate peaks and troughs of activity in relation to sun height and daily temperature cycles. Mostly records have been for specific behaviours at the colony, of sleep patterns, or the movements to and from the sea. The general conclusion from these studies is that there are no striking patterns of activity for Ross Island penguins, but they do not exclude more subtle, and as yet undetermined, rhythms. In part the conflicting evidence for cycles of activity is caused by the extreme variability in the daily weather and the vulnerability of both the penguins and skuas to heat stress. The colony may be quieter during the early hours of the morning, when it is both colder and darker, but it may also be quieter again during the mid-afternoon of warm, still days when the birds are struggling to keep cool. More recently the problem of diurnal periodicity in this summer environment has been studied by physiologists examining internal body cycles. At Cape Bird neither melatonin levels circulating in the blood nor body temperatures of the Adélie penguins

Plate 13.7. *Adélie penguins mate calling. Photo: Antarctic Division, DSIR.*

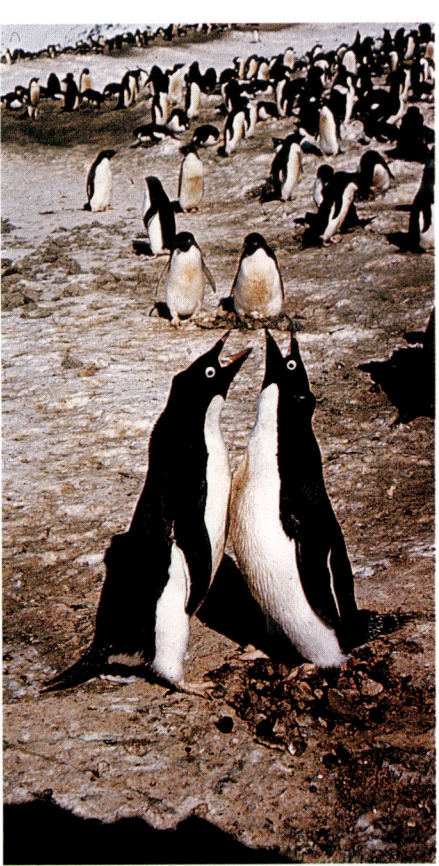

appear to have a 24-hour rhythmicity. Researchers are now attempting to determine whether the underlying "clock" controlling these rhythms is itself switched off during continuous daylight. So far it has been found that the cycles can be re-established by holding the penguins in alternating light and dark periods, simulating life under temperate conditions.

These problems, however, are minor compared with the difficulty of survival through winter in the Antarctic region. Not only is it much colder, with the regular occurrence of fierce and protracted storms, but it is at best a wan twilight and at worst completely dark. Little is known of the birds' lives at this time: their feeding behaviour, their response to severe weather, their movements. A life table of survival in relation to season, ice cover, and weather might point to the full significance of the overwintering period in the life of these birds.

Petrels

There is little information on the petrels of this region. Three sorts of information ought to be available. The first concerns the biology of the birds when at sea. Some of this is emerging through recent systematic surveys of bird and mammal densities in relation to pack ice and bottom topography in the Ross Sea. Antarctic petrels are the most abundant, snow petrels are the next most abundant, and Wilson's storm petrels occur at low densities.

Secondly, there should be information on breeding sites, where they are to be found, and how many breeding birds occur at each site. Petrels are difficult to survey because they nest in crevices on cliffs or on rocky outcrops and are too inconspicuous to be observed during routine travel. In general, it is not known where and in what numbers these birds are breeding in this region. There are records for the Balleny Islands of breeding by snow petrels, cape pigeons, Antarctic fulmars, and Wilson's storm petrels. Snow petrels (Plate 13.6) and Wilson's storm petrels certainly breed on the mountain cliffs on the western side of Edisto Inlet at Cape Hallett and at Crater Cirque (72°37'S, 169°20'E) on the southern margin of the Tucker Glacier, but there are no other published records of their breeding further south than this on the western side of the Ross Sea. Surprisingly, considering the isolation of Marie Byrd Land, there are good records of breeding Antarctic and snow petrels from several mountain ranges there. Breeding sites on Mount Paterson at 78°S are a little further south than Scott Base, but on the other side of the Ross Ice Shelf. This is, however, the farthest snow petrels and Antarctic petrels breed inland. The famous colony on the Theron Mountains at 79°S and 28°W, the southernmost breeding site for petrels and skuas on the continent, is 250 kilometres inland from the Weddell Sea coast. Snow petrels, either singly or in small groups, are commonly seen flying around Ross Island, and are the only petrels likely to be seen by visitors there. The possibility of their breeding on its cliffs should not be discounted.

The third kind of information needed is on breeding biology. Only one study has been carried out on the petrels of the region; this was on the snow petrels at Cape Hallett but it was so constrained by the difficult access to the breeding cliffs that much of the breeding cycle was missed. The birds occupied deep crevices from November, two eggs were laid, and fledgling chicks appeared in March.

Adélie penguins

More than a million pairs of Adélie penguins breed around the shores of the Ross Sea in 34 breeding sites. Most are in Northern Victoria Land but very large colonies also exist on the southern islands.

Plate 13.8. *An Adélie penguin feeding a chick. Photo: E. Young.*

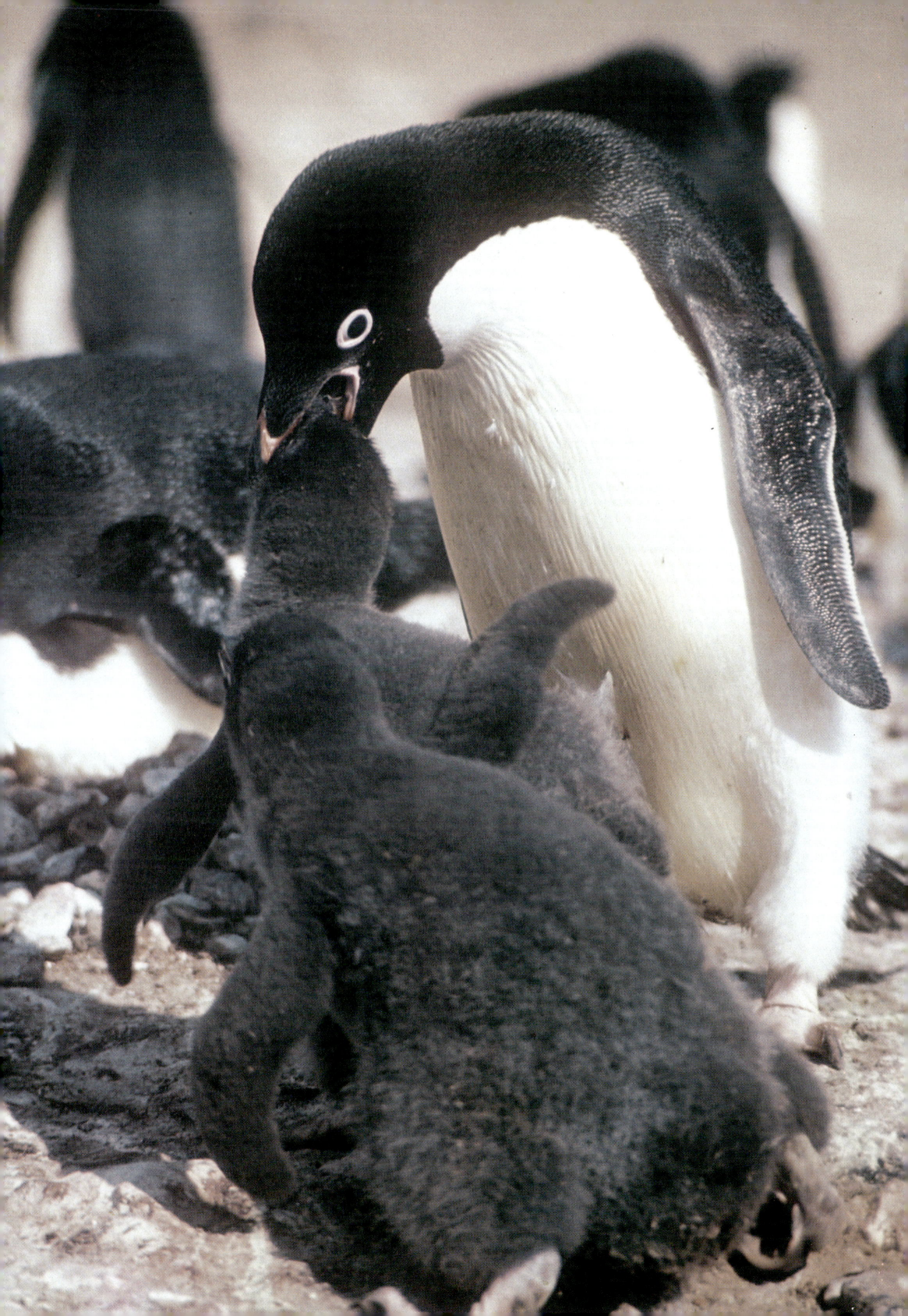

The closest one to Scott Base and McMurdo Station is at Cape Royds, but larger numbers also breed close at hand at Capes Bird and Crozier. The total numbers in the populations are, of course, much greater as this includes all the immature birds, and probably amounts to between 7 and 10 million birds.

Individual colonies vary in size from a few hundred pairs of birds to many thousands (Figure 13.1). The largest known is at Cape Adare (about 280 000 pairs); Cape Crozier has 175 000, but Cape Royds no more than 4000.

Several of the largest colonies, including Cape Adare and Cape Hallett, are on raised beaches, on the flat shelves formed as the coastal areas have lifted after the retreat of the ice cover. These areas are prime sites for penguins—and for the location of Antarctic bases. The siting of a base on the Cape Hallett colony is the most extreme example, so far, in this region of competition between penguins and people. The numbers of breeding penguins in this region have increased dramatically through the 1980s, with most sites recording a 50–100 percent increase in this decade. The numbers given in Figure 13.1 are the best estimates for December 1988.

In summer there is access from all breeding sites to the open sea. Although the birds can walk in to the rookery after covering long distances over the smooth ice during the spring migration, once the chicks have hatched the parents must be able to feed them easily from open sea nearby. Very severe egg and chick mortality results if the ice does not break out in summer or becomes very rough.

Breeding pairs have strong mate and site loyalties, often nesting year after year at the same nest site in the same colony (Plate 13.7). Few birds move to other colonies. The breeding cycle is perfect for summer research workers and visitors. There is no long drawnout preparation to breed once the birds come on to the land, as is common with subantarctic penguins, and the season cycles through briskly to fit into

Figure 13.2. *Breeding cycles of Adélie penguins, emperor penguins, and skuas on Ross Island.*

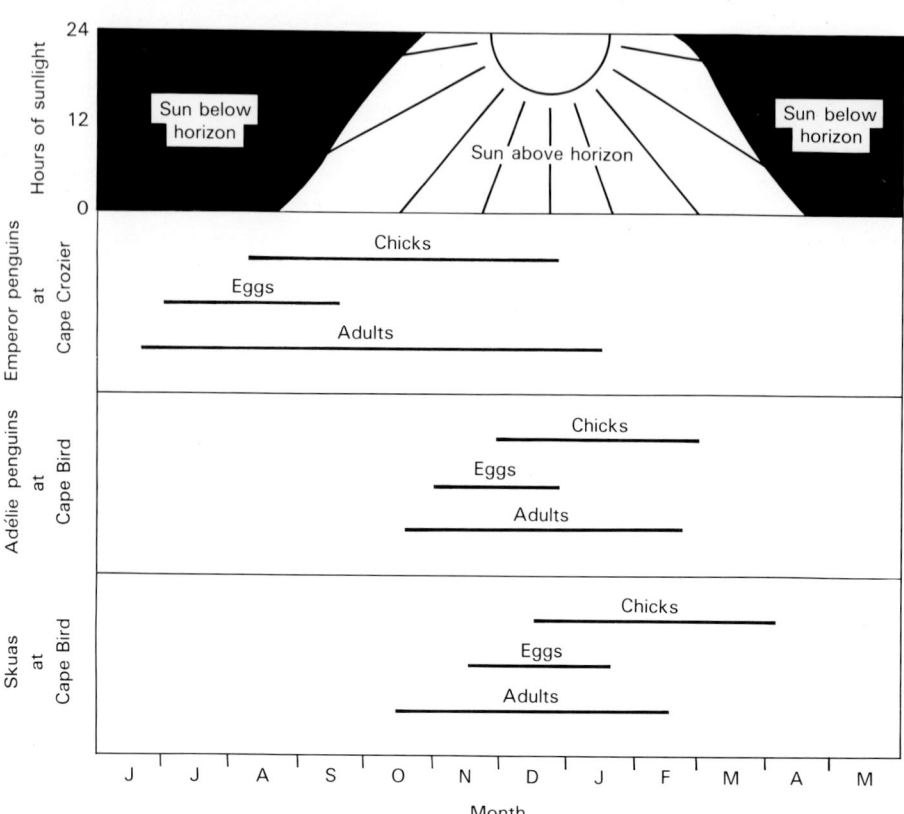

the very short summer. Mates that do not appear on time are soon forgotten in the imperative to breed and mate changing is common. Fierce fighting, causing great disturbance in the colony, follows if the original mates do arrive later.

The first penguins come ashore at the Ross Island colonies on about 20 October, the first eggs are laid in the first week of November, with a peak in about the middle of the month. Incubation lasts about 34 days (range 31–38), so that chicks appear in early December. They are ready to go to sea in late January. By mid-February the colonies are deserted apart from forlorn groups of moulting adults. This is a remarkably brisk cycle, much shorter than for penguins of comparable size in temperate habitats. Much of the shortening is, firstly, in the quickness with which eggs are laid after the birds arrive at the colony at the start of the season and, secondly, in the rate at which chicks mature to independence. The yellow-eyed penguin (Megadyptes antipodes) of New Zealand is much the same size as the Adélie penguin, yet its chick takes 14 weeks to reach independence compared with 8 weeks for the Adélie chick. Adélie chicks grow faster and become independent at a relatively lighter weight. The Antarctic summer is too short for a leisurely breeding regime, or for renesting.

The breeding cycle can begin so early in the season because the birds survive through the early weeks without feeding: for males not only during the long walk in from the sea but through the first 15–20 days or so of incubation; for females through the walk to the colony and egg laying. They have thus adapted to the vagaries of the ice cover and the difficulty of traversing broken ice to the open water by staying at the colony for weeks at a time. Adaptation to an aquatic life, with its need for compromises in body shape for swimming (essentially flying in water) and walking, has imposed constraints on the terrestrial life. Fifty kilometres is nothing to a flying bird but is an impossible distance across difficult terrain for a small walking one.

The sexes are identical and even the penguins have trouble sorting them out except through trial and error, or approach and response. Juvenile penguins differ from adults in having a white chin and a bluer plumage (Plate 13.8), but once the birds have moulted to adult plumage in their second year they can no longer be aged. This is a common problem for ornithologists. Population studies must begin, therefore, by banding chicks and waiting for them to grow up. In time there is a valuable stock of known-age adults to work with.

The general features of the breeding biology have been recorded many times and comprehensive accounts occur for this species. Only now, however, are the critical elements for breeding success being investigated. One of these is the timing of the return to the nest at the end of the long feeding and recovery periods at sea during the incubation period. There are three foraging trips during this period, of about 19, 13, and 3 days, respectively. The first is undertaken by the female immediately after egg laying. Getting this return timing correct is crucial to egg and chick survival. If the returning bird is late, the sitting bird may have to desert the nest to avoid starvation and the eggs would be lost. A late return after the chicks hatch would condemn them to starvation, as the sitting bird does not have any food left with which they can be fed.

The most comprehensive population study of the Adélie penguin was done at the Cape Crozier colony between 1961 and 1982. The first maturing adults observed at the colony were 2-year-olds (about 30 percent of the group turned up at this age) and all of the surviving banded chicks were back within 5 or 6 years. No birds bred as 2-year-olds but by 7 years 57 percent were breeding. The Adélie penguins at this colony were found to have a busy but short life: only one-half survived the first year, no more than 10 percent reached 10 years of age, and very few (4 percent) survived to the life span of the species at 20 years.

Young Adélie penguins are classed as wanderers when they first turn up at the colony, prospecting the breeding groups and marching through them and about the surrounding country. They are most likely to appear first during the re-occupation period in mid-December, when new birds and all those that failed to nest successfully earlier in the season flood into the colony and swell the numbers in the individual groups. This seems a strange time for this major event because it happens as the chicks are hatching and much fighting for mates and nest sites occurs over and around the sitting birds. However, for an individual bird there is no doubt advantage in forming pair bonds and claiming nest sites at this time in readiness for the following breeding season. Such preparation must shorten the time needed to lay the eggs at its start and ensure that the chicks are ready to leave before the ice begins to form across the autumn sea.

Early research on navigational ability of the Adélie penguin and skua involved taking birds out on to the Ross Ice Shelf out of sight of land and observing their direction of movement on release. Only limited information can be obtained from such experiments, although it was soon established that penguins at least could navigate by sun compass. An alternative approach has been to record the movements of free-ranging penguins fitted with radio transmitters whose signal can be received by direction-finding aerials on land. Cross plotting from two widely spaced aerials allows a bird's position to be recorded almost continuously. The most detailed studies have been carried out from Cape Bird, using one receiver on Mount Bird and a second on the far side of McMurdo Sound. This provides information both on the foraging behaviour of penguins at this colony and on navigation. Daily plots

Plate 13.9. *A pair of Antarctic skuas with chicks. Probably only the older chick will survive after it has driven the younger one away from the nest. Photo: C. Rudge, Antarctic Division, DSIR.*

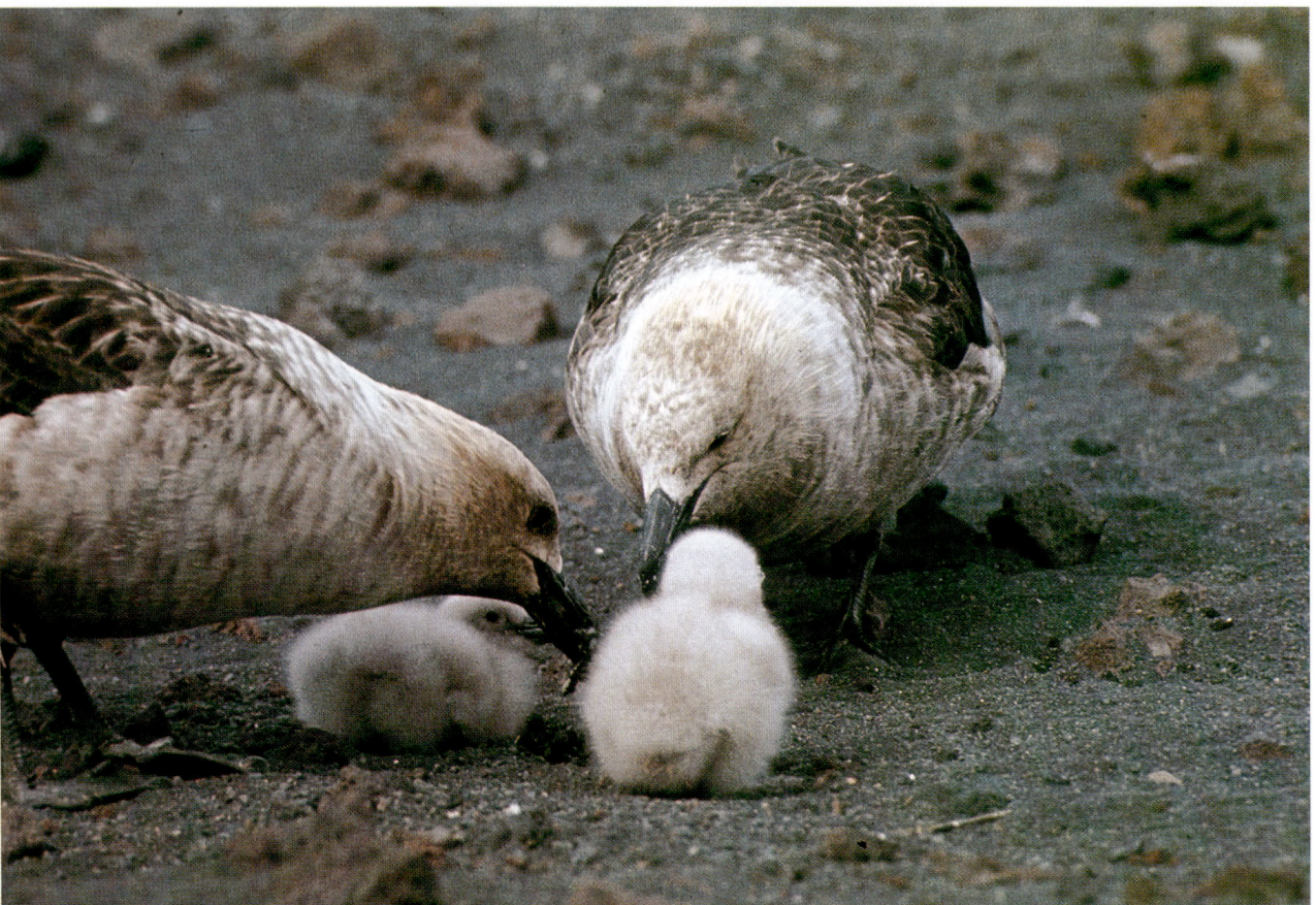

record both long-distance movements and times when little movement occurred.

The pattern of movement and foraging provides insight also into how the birds use the time available to them and this can be related to the timing of their return to the nest. Key questions are whether all the time away is spent foraging or whether it is spent "loafing" until it is time to return. Penguins from Cape Bird invariably travelled to the north-west for 30–40 kilometres into the centre of McMurdo Sound, where they appeared to begin feeding. But they then behaved variably, some continued towards the north past Beaufort Island, others moved to the east (against the current flow) around the north side of Ross Island. Even though the receivers were located on high points, birds often were lost over the radio horizon at about 150 kilometres. Return to the colony seemed very direct and the birds might travel up to 70 kilometres on the last day. These rates of travel clearly indicated that the birds knew where they were heading.

During the 1988–89 summer, transmitters relaying to a satellite were successfully tested and penguin movements throughout the foraging period were recorded. The use of satellites is a huge breakthrough in penguin biology and opens the way to tracking individual penguins over winter and as they leave and return to the colony. In addition, these records would provide insight into how the penguins navigate back to the colonies after the winter spent far to the north in the pack ice fringe.

Emperor penguins

These birds breed on sheltered fast ice close to land or ice cliffs in traditional sites. Seven sites are known in the Ross Sea (Figure 13.1). Others may exist, especially in the poorly explored coast of Marie Byrd Land. There are at least 90 000 breeding pairs in this region. The most famous site is at Cape Crozier—the place to which Wilson, Bowers, and Cherry-Garrard sledged in mid-winter on their memorable but scientifically modest expedition—but conditions there are so miserable during mid-winter breeding that this site is not to be recommended for studies. (A few birds have bred recently in some years at Inaccessible Island, near Cape Evans, but this is too fragile a breeding group to be disturbed.) For most of our knowledge of the breeding behaviour of emperor penguins we rely on French studies at Pointe Géologie, Adélie Land, where there is a rookery conveniently close to a winter base.

This penguin is so big and the physiological constraints on incubation and chick rearing, correlated with its large size, so severe that they are unable to fit breeding within the span of the summer months. The requirement for the newly independent chicks to leave the breeding area under the most favourable conditions for survival in early summer means that the entire breeding cycle has been advanced into the winter. At Cape Crozier, breeding adults congregate at the breeding site in March and April, the eggs are laid in May–June and the chicks hatch in July–August. The chicks become independent in December and all birds have left the rookery for the season by January. In 1983 an estimated 47 000 chicks were produced in the Ross Sea colonies.

The emperor penguin's closest relative is the king penguin which breeds on the subantarctic islands. This large penguin has a similar breeding cycle and shares the same habit of incubating the single egg while it is held on the feet against the belly. Neither species builds a nest. (Such large birds may be too clumsy to flop down on to eggs in a nest without breaking them.) Incubating birds shuffle about slowly, holding the eggs carefully. Transfer of eggs from parent to parent is obviously fraught with risk and any disturbance of birds by natural or

Plate 13.10. *Adélie penguins fleeing from a leopard seal on an ice floe. Photo: R. Thomson.*

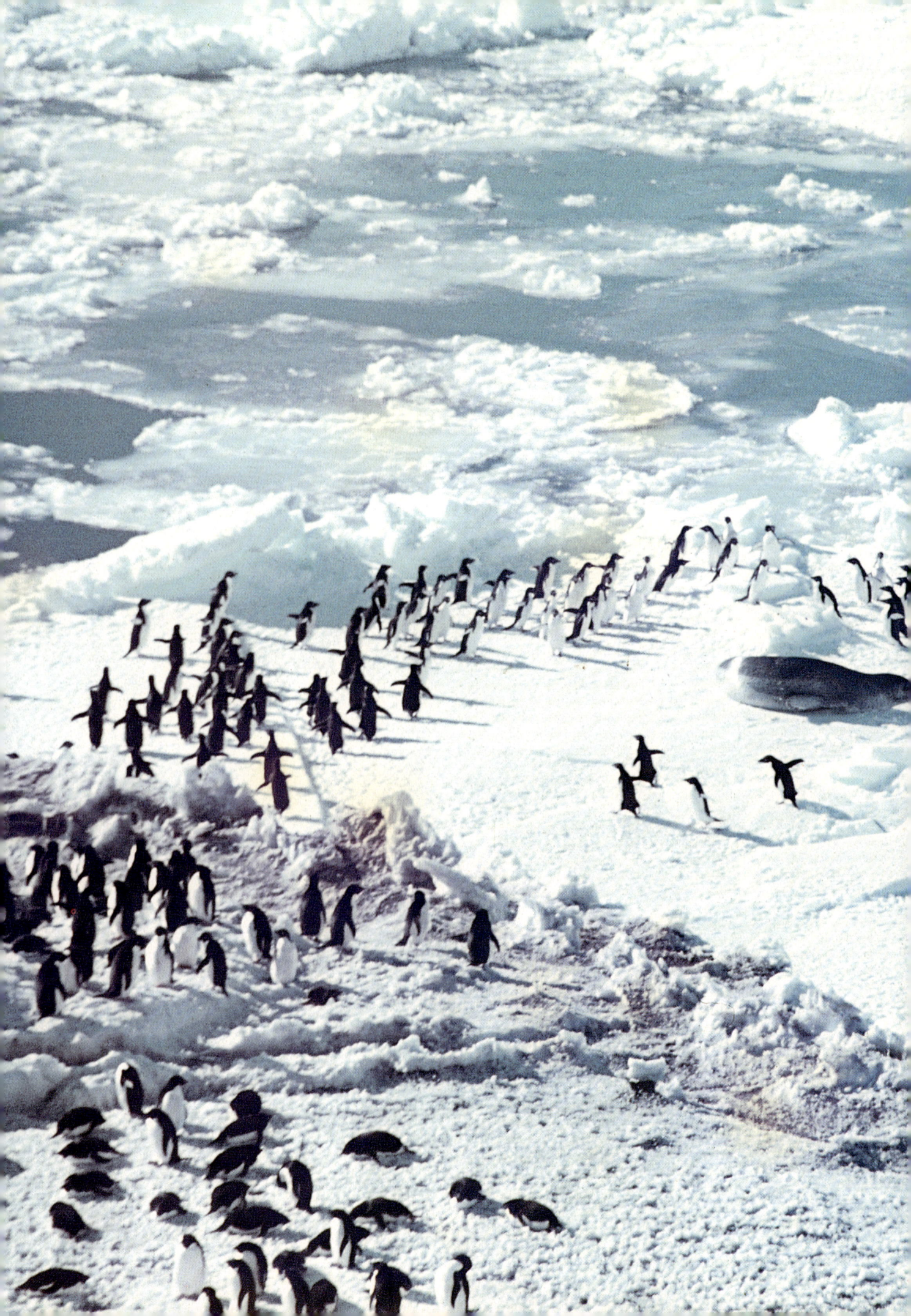

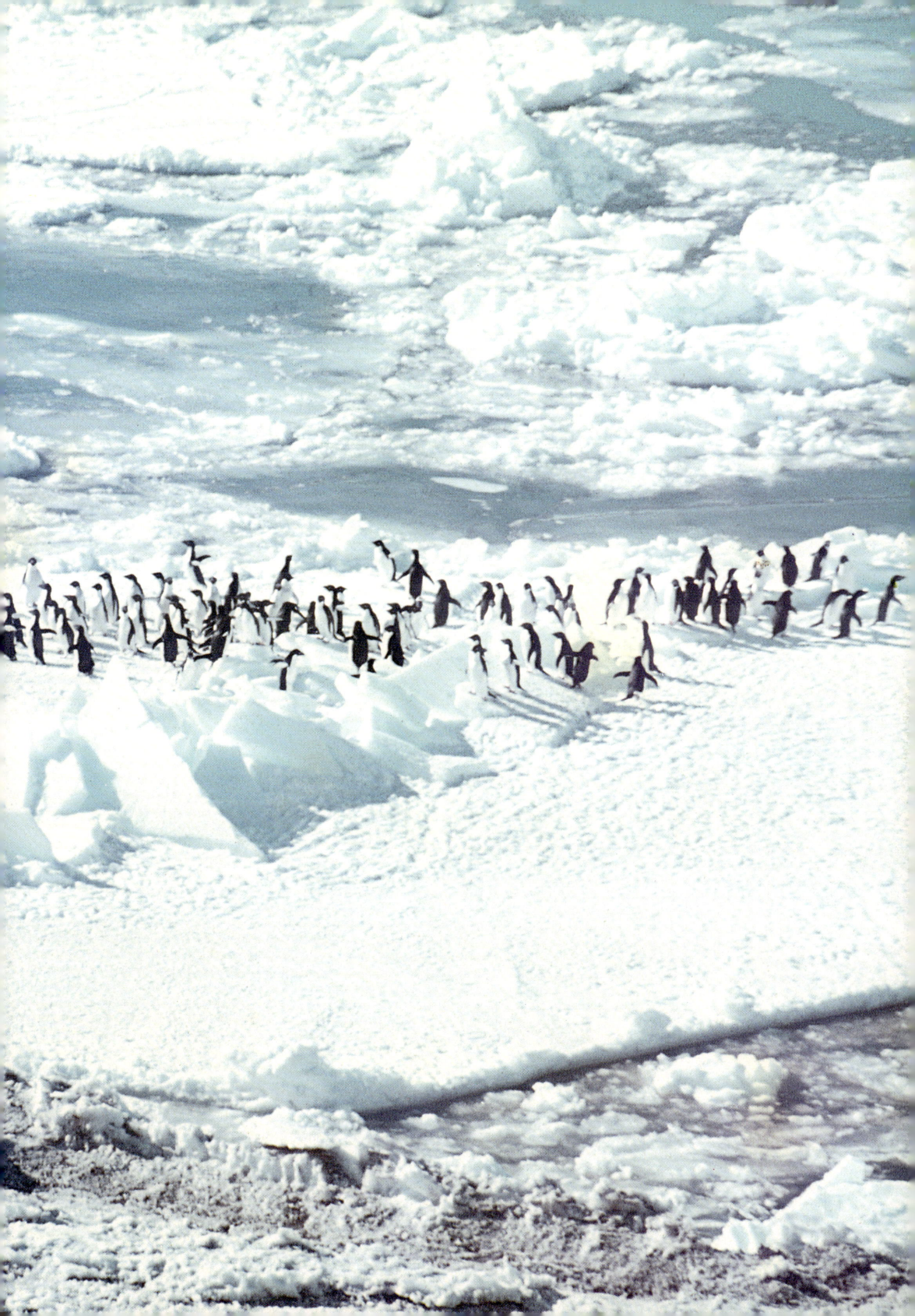

human events is potentially disastrous, as eggs may be dislodged and lost.

The tough breeding schedule running through the worst of winter is only possible because of fasting (Figure 13.2). Male birds fast for 3–4 months at the start of the cycle and may lose up to 45 percent of body weight. Energy reserves have to cover both the walk into the colony and the first stages of breeding and these are feasible only because individual birds can reduce the demands of maintaining body temperature in this chilling climate by huddling together. Later, the free-standing chicks also huddle tightly, forming solid groups of grey, snow-flecked, globular masses. The parents have to find their own chicks in these phalanxes and rely on voice recognition and response to ensure that they feed only their own chicks.

Under Antarctic winter conditions, breeding is at risk during all parts of the cycle. If the ice forms too solidly over the open water the birds cannot feed and breeding is abandoned. Conversely, the ice of the colony area may break out and the eggs and chicks lost. The parents may get lost, the egg may fall and break or not be recovered. Over all this is the universal spectre of cold and the possibility of hurricane-force winds. It is surprising that they manage to breed successfully; even more surprising that they so manifestly flourish.

Skuas

Visitors generally don't care much for skuas. The birds are drab compared with the striking black and white of the penguins. They congregate around the bases, scavenge at the rubbish dumps, attack and kill penguin chicks, scrabble messily over the birth remains of seals, and attack people noisily and fiercely if they enter breeding territories. Although they are big, tough, swaggering birds, paradoxically, they are more sensitive to disturbance than the penguins and need more protection. They are certainly not as tolerant of humans as the penguins appear to be and invariably stop all normal behaviour when people are in sight.

Skuas are a cold temperate, Southern Hemisphere group of birds with different species on Antarctica, the subantarctic islands, Chile, South America, the Falkland Islands, and Tristan da Cunha. However, one form now breeds in the north Atlantic. The identities and names of skuas are now so badly muddled or uncertain that there is little agreement among experts of how they should be grouped and identified. Their affinity to the smaller jaegers of the Northern Hemisphere is also unresolved. Fortunately most experts agree that the bird breeding on Antarctica is *maccormicki* but its genus and its common name is a matter of choice. It is known in New Zealand as the Antarctic skua, *Stercorarius maccormicki*, but it is also well known as the south polar skua or McCormick's skua and may be placed instead in the genus *Catharacta*.

Antarctic skuas leave the region entirely over winter (in contrast to the penguins). Evidence from the recovery of banded individuals indicates that Ross Sea birds cross the tropics to Japan, the North Pacific and the Alaskan coasts. They must be tough birds for they have a very high survival rate, probably live for 20–30 years, and do the long trip every year—returning to the same nesting site to mate each spring. This surely deserves our admiration. Such a trans-equatorial migration no doubt occurs also in the Atlantic and was probably how the recently established breeding populations in the North Atlantic first arose—by birds staying there instead of returning south. In a similar way Ross Sea birds may one day breed in the Aleutian Islands.

Skuas breed in large territories on terrain free from snow and ice within reasonable flying distance of open sea. Some of these places are also favoured by Adélie penguins, so that superficially the two birds

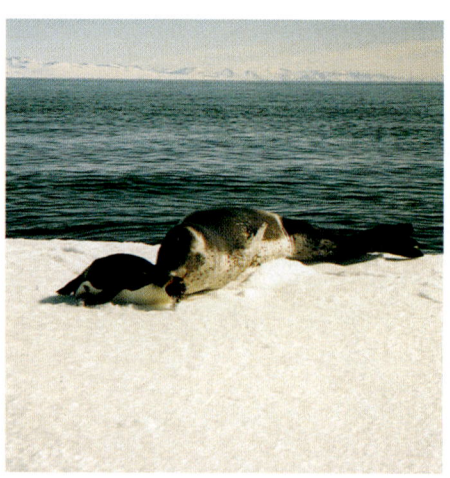

Plate 13.11. *A leopard seal with an emperor penguin it has just taken on the ice-edge. Such sightings are rare, and the seal has difficulty in dealing with such large prey. Photo: R. Thomson, Antarctic Division, DSIR.*

appear to be in a prey-predator association. There are, however, many more skuas breeding away from penguins in McMurdo Sound than with them (at least 1000 pairs breed south of Cape Royds) and most scientists now consider the association between the two birds fortuitous, arising from the same nesting site and breeding season requirements.

The first skuas arrive at the colonies on Ross Island in late October—the first eggs appear on about 15 November and the first chicks 1 month later on 15 December. The fledglings do not, however, leave the colonies until late March, even early April, but because of flight are less at risk from an early onset of sea ice than the penguins which have left much earlier. Few scientists have studied this last event in the breeding cycle for most have left their study areas and are already at home before the young skuas leave on what is assumed to be the first of their many flights to the Northern Hemisphere.

Antarctic skuas generally lay two eggs and usually both hatch. However, within a day or two of hatching the second chick is attacked by the older (and stronger) chick and chased from the nest area, to starve or be killed and eaten by other adult skuas (Plate 13.9). Parents seem unable to stop the siblings fighting; this behaviour is unique to Antarctic skuas. Subantarctic skuas seem to rear both chicks of the brood without conflict. Much research has already been done on this apparently maladaptive behaviour in the Antarctic species without producing a convincing explanation. Why do the skuas go to the effort of laying and caring for two eggs and not attempt to rear both chicks? Sibling hierarchial systems are common in birds and their intensity is generally found to be related to the current and predicted food availability for the brood.

Food and feeding of Antarctic birds

All the bird species in the Ross Sea region, indeed in Antarctica, directly or indirectly obtain their food from the sea: they are all sea

Plate 13.12. *A pair of skuas attack an Adélie penguin chick with no interference from the adults. Photo: E. Young.*

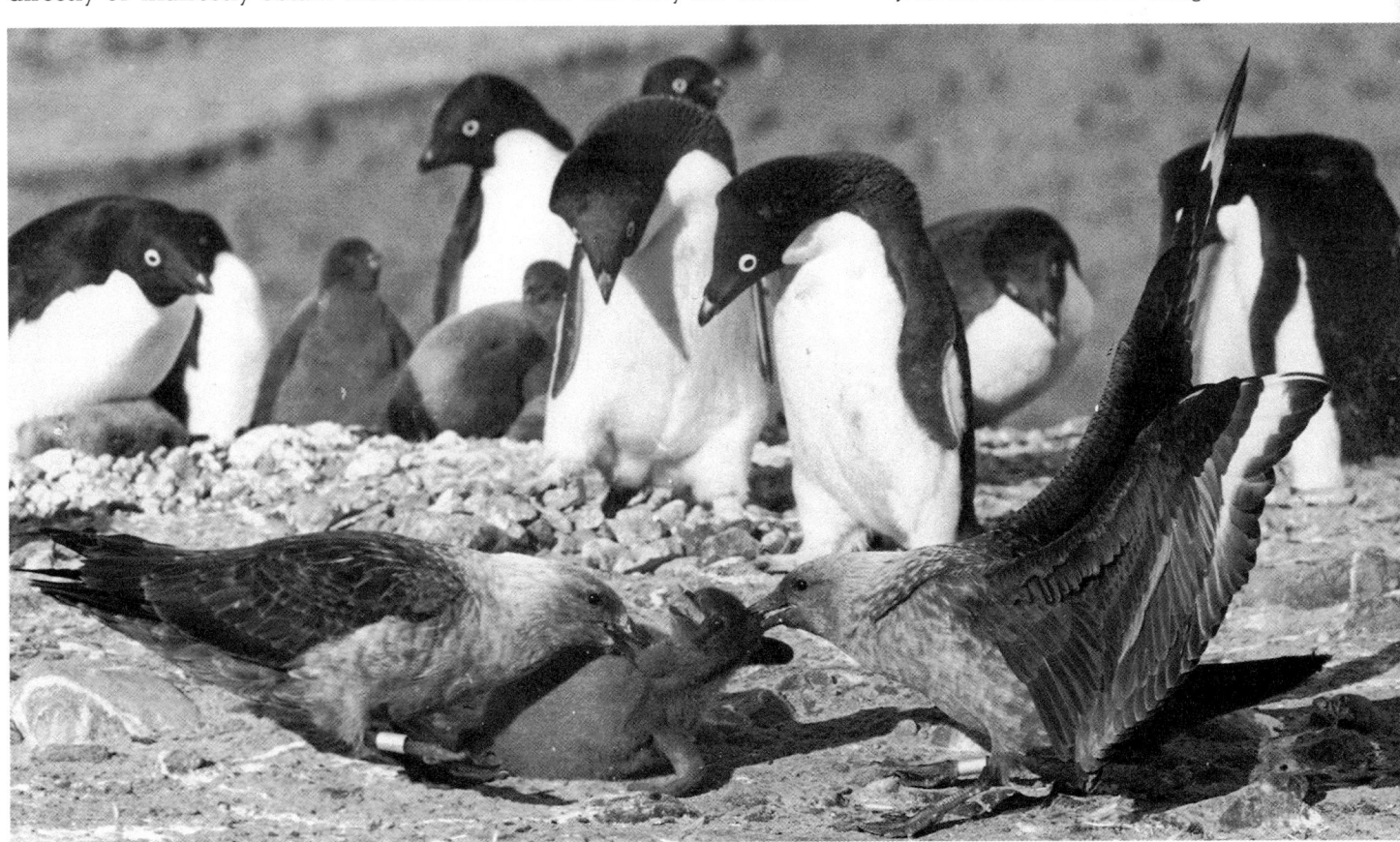

birds. They do, however, obtain this in different ways and appear to use different sectors of the marine food webs. There is no productive intertidal region here because of the sea ice and all birds feed on pelagic species—on plankton, squid, and fish—and on bottom fauna in shallow areas.

Penguins feed when diving, selecting prey individually. Adélie penguins can fish to 100 metres and emperor penguins may go to twice this depth. The diet of Adélie penguins in summer can easily be investigated (without harm) by using a small pump to suck food from the stomach of birds that have recently returned to the colony. Most of the food taken by birds at the Ross Island colonies is the coastal krill species, *Euphausia crystallorophias*. Stomach contents of 500–1000 grams might contain 10 000 krill, captured individually. These birds also take significant numbers of fish, mostly small Antarctic silverfish (*Pleurogramma antarcticum*). Further north the Antarctic krill (*E. superba*) is the dominant food and fish are less important.

An impressive volume of food is consumed by a colony of Adélie penguins over the breeding season. For example, at Cape Crozier the parents make about 40 feeding trips while raising their chicks. On average they bring back about 0.5 kilograms of food (krill and fish) from each foraging. For the total breeding population of 175 000 pairs an estimated 3500 tonnes of food is brought on to the colony during the season, producing about 875 tonnes of penguin chick.

Emperor penguins take more fish and squid but fewer krill. Their diet has not been studied at Ross Sea rookeries.

Little is known of the winter diet of these two penguin species. Both overwinter within the Antarctic ecosystem, when temperatures are most severe, light is minimal, and plankton abundance is lowest for the year. Foraging must be a desperate business. How does the emperor penguin survive so well, having to cater not only for its own breeding demands but also to feed the chick? Krill, indeed nearly all euphausids,

Plate 13.13. *Organised theft of an Adélie penguin's egg. In sequence: one skua attacks the penguin from the rear; the penguin responds, exposing the egg to the second skua; the second skua snatches the egg; and flies away with the egg to deal with it at its leisure. Photos: E. Young.*

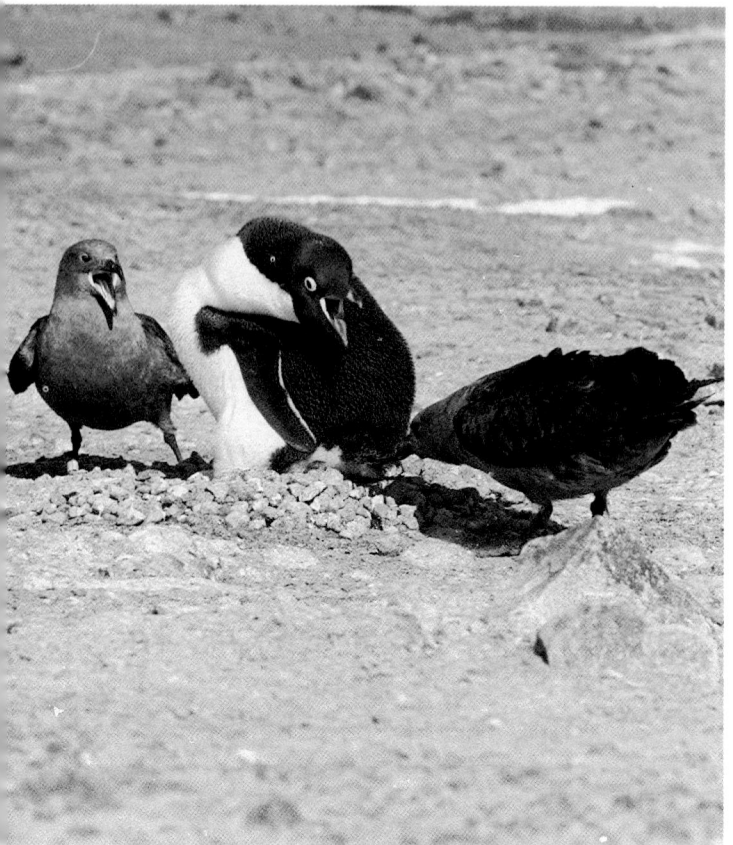
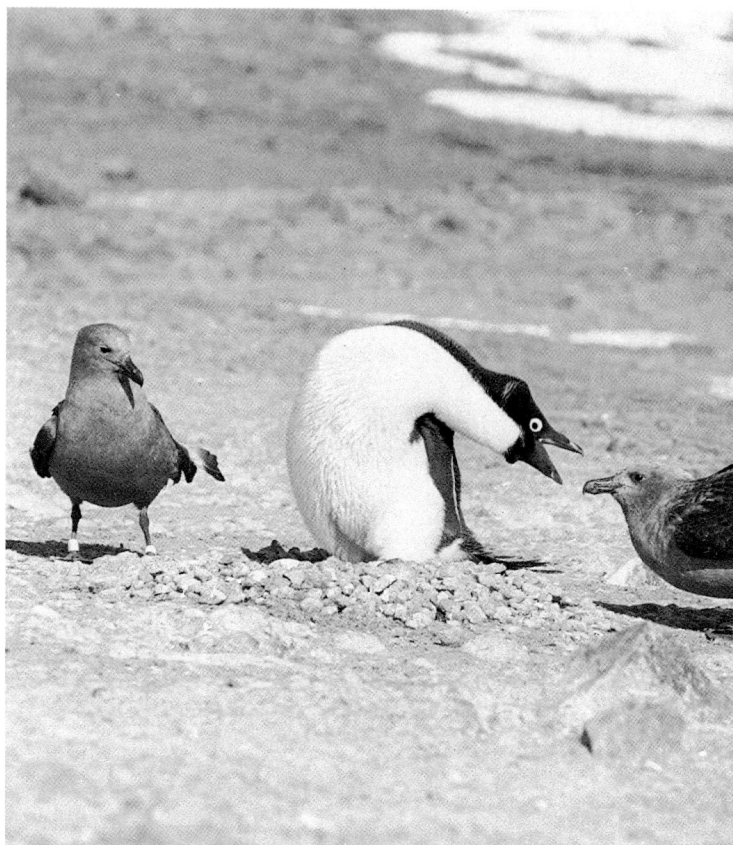

luminesce from organs called photophores at the base of the eye stalks and on some legs. Perhaps it is this luminescence which allows them to be found and caught by penguins at night and during winter.

Skuas feed by splashing down into the water to catch silverfish. They apparently do this very capably. Foraging flights from Cape Bird to the horizon and back need not take more than 30 minutes under good conditions. It is easy to see what the birds have been doing on really cold days: if fishing they return to the territory with so much frozen spray on their heads and breasts that they appear to be armoured.

The Antarctic silverfish is a pivotal species in the Ross Sea food web, yet little is known about it. Clearly it must be at the surface at times (because the skuas catch them) but it is rarely observed, is not taken commonly in the usual nets used in oceanography, and is not seen by SCUBA divers close to the island. Skuas head out to the north-west from Cape Royds and to the north and west from Cape Bird. Presumably this is where these fish are concentrated. Skuas are rarely seen fishing close inshore. Squid (Cephalopoda) are also important at this level in the food chain. Even less is known of this group of pelagic animals in the Ross Sea than is known of silverfish. They are abundant enough to form an important part of the diet of Weddell seals and emperor penguins but are difficult to catch from boats. Both silverfish and squid feed on krill and are themselves fed on by larger fish, birds, and seals.

Petrels everywhere feed on plankton and squid at the surface, the smaller birds patter across the sea, the larger ones pick up food from flight or when settled on to the water. There is no detailed account of the feeding behaviour of the petrels under different weather, ice, and water conditions.

Predators of Antarctic birds

The only significant predators of the penguins over summer at the

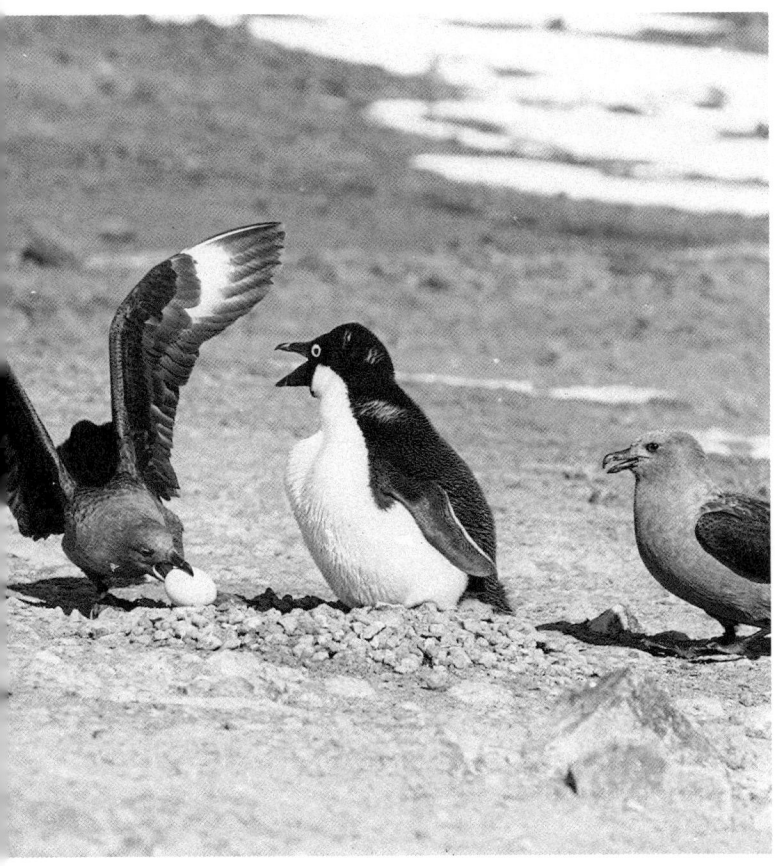
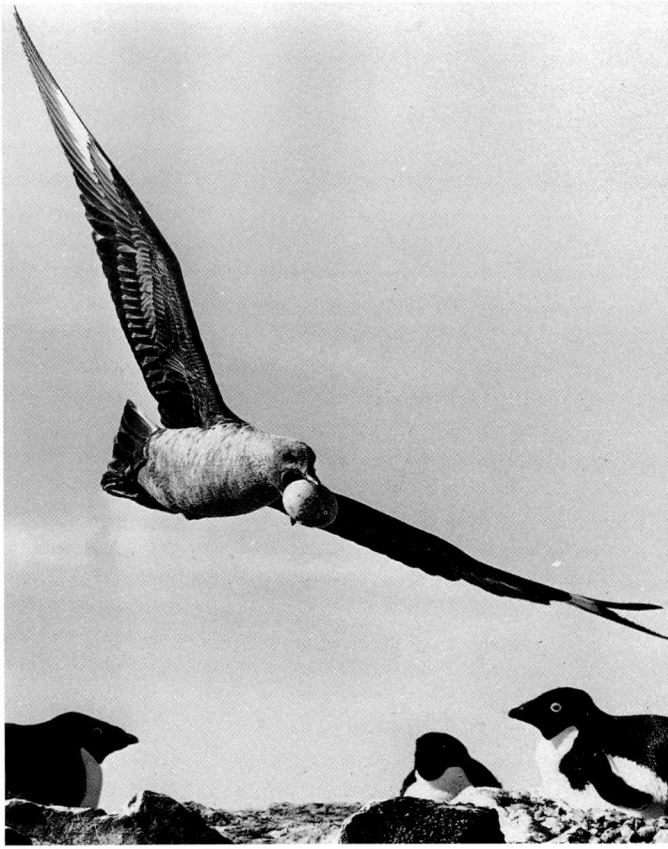

Ross Island colonies are the leopard seals (*Hydrurga leptonyx*) in the sea and on ice floes and the skua on land. Weddell seals only occasionally take penguins. At rookeries further north the giant petrel (*Macronectes giganteus*) takes emperor penguin eggs and chicks. Leopard seals take adult Adélie penguins during the breeding season and fledgling chicks at the end of the season, as they first swim away from the rookery. Although leopard seals chase penguins over ice they do not attempt to follow them on to shore (Plate 13.10). What chaos would occur if one ever entered a crowded colony! If they started doing so it would change the face of Antarctic ornithology. Leopard seals can also kill emperor penguin adults (Plate 13.11) and chicks, but these seals do not arrive at the southern colonies until early summer; by then most of the emperor penguin's breeding activity is over.

Where once penguins were considered to be the principal prey of leopard seals, and assumed to be crucial for their survival, more recent work has put the very obvious predation on Adélie penguins at colonies into perspective. The leopard seal's formidable array of teeth with their interlocking cusps is not for the capture and retention of large prey, as earlier thought, but for sieving krill. Most leopard seals live in the pack ice throughout the year. Those that patrol the penguin colony beaches and chase and catch penguins are the few opportunists that have discovered (and defend?) such a concentrated food source.

Skuas have achieved a similar, and equally undeserved, notoriety as penguin killers. This reputation is partly because they are such poor predators that killing large chicks is an agonisingly slow, messy process (Plate 13.12). It is this inefficiency as a predator that so horrifies observers and makes killing chicks such an obvious feature of the skuas' behaviour at the colony. They are sea birds well adapted for capturing and ingesting small fish. They have little more than their strength and agility to take on the role of predator in the penguin colonies. Large chicks are killed almost by exhaustion, the skuas seeming unable to deliver a rapid killing blow. Having finally killed a chick the skua has difficulty in feeding on it. The hooked beak is not used, as it would be by raptors (hawks and eagles), in conjunction with the feet to hold the prey. The skuas do not use their feet in this way, so that tearing prey apart is laborious unless two birds act together.

Their facility in manipulating and ingesting small chicks and their incompetence with large ones should have suggested from the earliest observations that the behaviour seen on the colony was uncharacteristic of the bird. A detailed study at Cape Bird found that only a few of the 130 skua pairs there with territories containing breeding penguins were skilled and aggressive predators. Many others scavenge, taking eggs (Plate 13.13) and chicks abandoned or outside the breeding groups. Some never bother with penguin food at any time even when it is readily available, preferring the more certain and safer sea foraging. At Cape Bird one expected impact of predator and prey is reversed: a higher proportion of skua eggs are damaged by penguins than are penguin eggs by skuas. Skua pairs nesting away from the penguins are the more successful breeders in most years.

Conclusion

At several places in this chapter reference has been made to human impact. The impact of gross disturbance by aircraft and visitors is now appreciated and sensible rules have been introduced throughout Antarctica to ameliorate this. The more subtle effects of disturbance are being evaluated—the insidious impact of continuing mild disturbance to breeding penguins, to the retention of territories by skuas, and to the selection of breeding places by the young birds of both species. The impact of visitors at Cape Royds and of the scientific station at Cape Hallett are early examples of gross impact.

It is important now to ensure that such situations do not reoccur. Of equal importance in the long term is to plan for the continued good health of the colony communities in the region, so that scientists (and visitors) can be assured that the community they are studying is not already degraded. As a first step, major science programmes should be located at different colonies from those used or likely to be used for behaviour or population research. Unless this is done, colony after colony will be taken over by groups of scientists intent on working with a "natural" system. If this is allowed to happen, there will soon be no natural colonies for anyone to study or visit (Plate 13.14).

Plate 13.14. *Brooding Adélie penguins on Inexpressible Island in a strong gale. Photo: K. Thompson.*

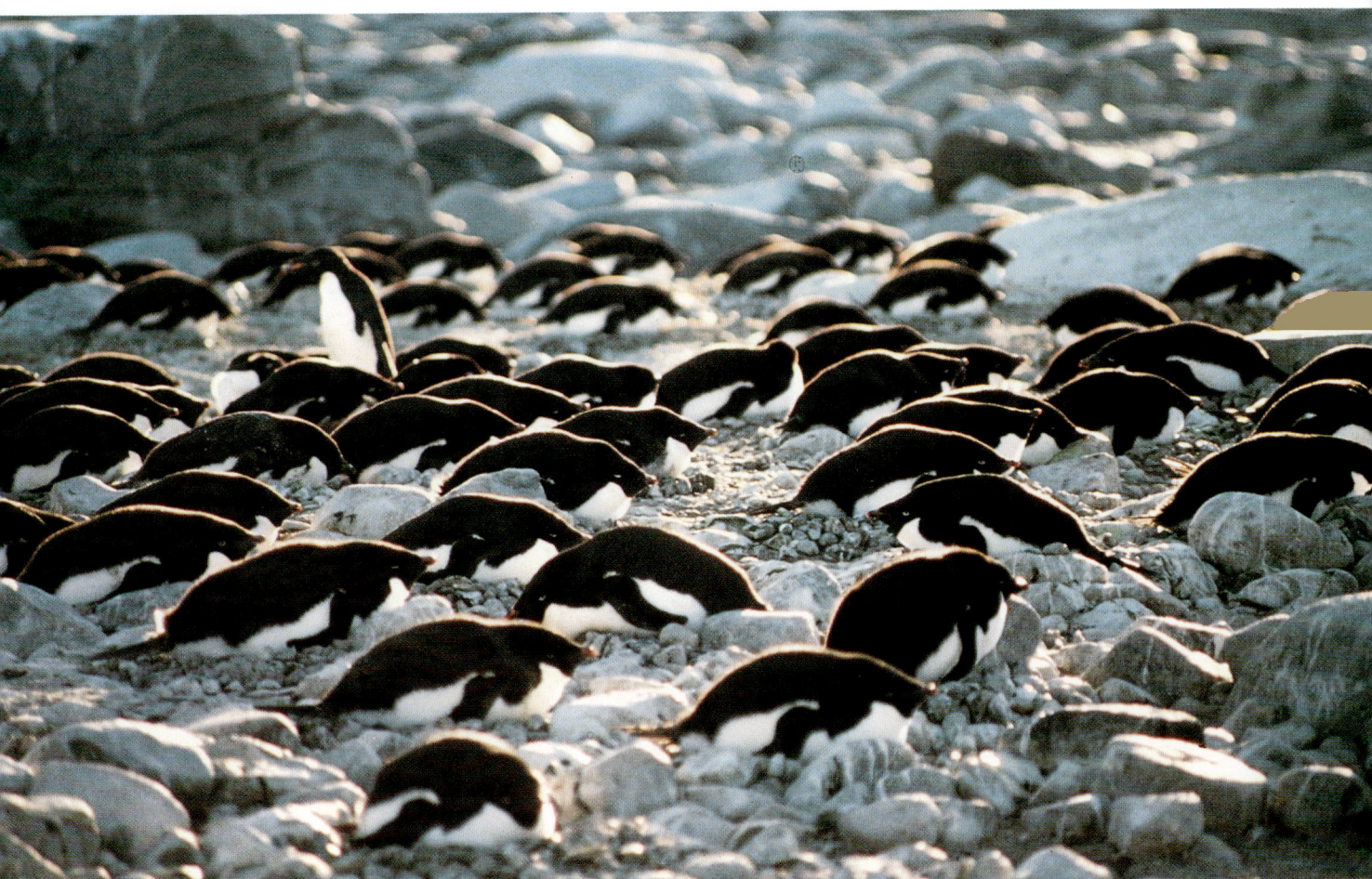

14. Animal adaptations to the Antarctic environment

Like the Antarctic Continent itself, its present animal inhabitants originated in more northerly, warmer climates. The ancestors of some animal groups have lived on Antarctica (or its submarine margins) since its separation from the Gondwanaland supercontinent, and, along with Antarctica, have drifted from temperate latitudes towards the South Pole. Other groups, however, have arrived more recently—blown by the winds, or migrating slowly across the sea floor. Animal migration to and from Antarctica is now limited by the great expanse of surrounding sea, the strong circumantarctic currents and winds, and the very sharp temperature change at the Antarctic Convergence.

The ancestors of today's Antarctic fauna must have been accustomed to much higher temperatures than the present cold environment. Thus, one of the most important questions to be asked about Antarctic animals is "How have they changed to adjust to polar conditions?".

Adaptations of Antarctic fishes

Adaptive radiation and buoyancy changes

Antarctic fishes are very unusual. Most belong to a single sub-order of bony fishes, the Notothenioidei, which appear to have evolved in Antarctica. The ancestors of this group were trapped on the continental shelf, and drifted with Antarctica as it separated from Gondwanaland and migrated to its present polar position. The notothenioids have diversified over about 50 million years, and now comprise five families (Figure 14.1):

- the Nototheniidae (Antarctic cod) which are the most generalised, and probably resemble the ancestral type (the true cod are members of the family Gadidae, based in the Northern Hemisphere);
- the Harpagiferidae (plunderfishes);
- the Bathydraconidae (dragonfishes);
- the Channichthyidae (icefishes, which are "white-blooded", and survive without a red blood pigment); and
- the Bovichthyidae (thornfishes) which are now found mainly to the north of Antarctica, on the southern coasts of Australia, New Zealand and South America.

None of these fishes possesses a gas-filled swimbladder, and so as adults most of them are heavy-bodied bottom dwellers, which is thought to represent their ancestral life style. However, some species have become secondarily pelagic, and now exploit the middle and surface waters. Lacking the flotation of a swimbladder, the free-swimming species have decreased their densities by incorporating oily lipid in their tissues, and reducing the amount of mineral in their bones so that their skeletons are mainly cartilaginous. (Lipids are compounds which include fats, oils, and waxes, and are generally less dense than water). Such changes in buoyancy have been particularly noted in members of the family Nototheniidae from McMurdo Sound, where one species, *Dissostichus mawsoni* (giant Antarctic cod) has neutral buoyancy; it weighs nothing in sea water. Another species, the herring-like *Pleuragramma antarcticum* (Antarctic silverfish) weighs in water only 0.6 percent of its normal weight in air. On the other hand, typical bottom-dwelling species such as *Trematomus bernacchii* and *T. centronotus* are about five times as dense as *P. antarcticum*.

These species probably achieved their pelagic, adult life styles by retaining juvenile characters, for the juvenile stages of many Antarctic fishes are free-swimming, incompletely ossified (boney), and contain much lipid.

The Antarctic fish fauna also contains several unrelated families of deep-water fishes, such as the eel-pouts (Zoarcidae) and sea-snails

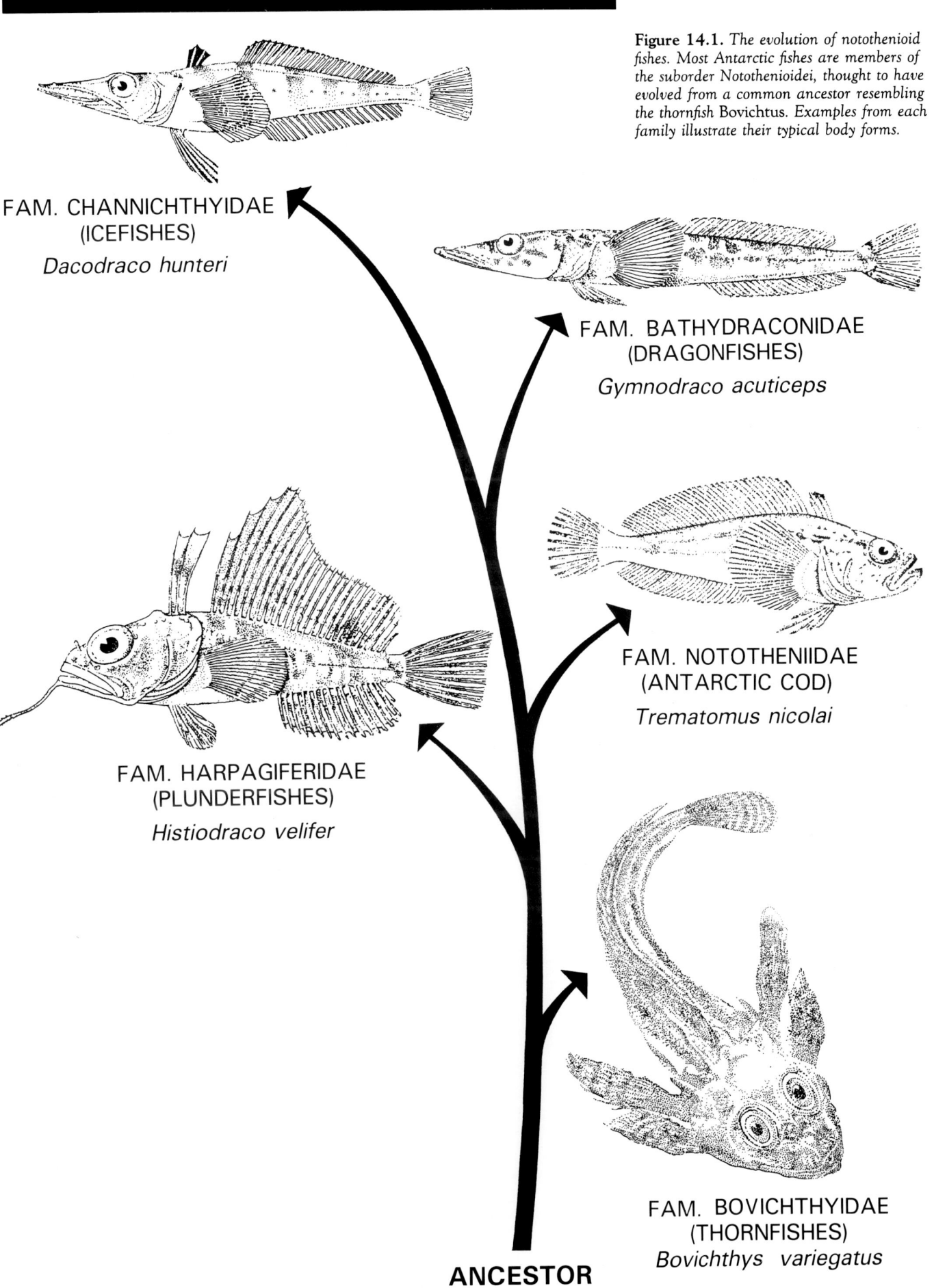

Figure 14.1. *The evolution of notothenioid fishes. Most Antarctic fishes are members of the suborder Notothenioidei, thought to have evolved from a common ancestor resembling the thornfish Bovichtus. Examples from each family illustrate their typical body forms.*

FAM. CHANNICHTHYIDAE
(ICEFISHES)
Dacodraco hunteri

FAM. BATHYDRACONIDAE
(DRAGONFISHES)
Gymnodraco acuticeps

FAM. NOTOTHENIIDAE
(ANTARCTIC COD)
Trematomus nicolai

FAM. HARPAGIFERIDAE
(PLUNDERFISHES)
Histiodraco velifer

FAM. BOVICHTHYIDAE
(THORNFISHES)
Bovichthys variegatus

ANCESTOR

(Liparidae), which presumably reached Antarctica by migrating across the sea floor. None of these more recent immigrants show the extreme diversification seen in the original inhabitants.

Temperature tolerance

During millions of years of evolution, Antarctic fishes have adjusted their body functions to cope with the low temperatures of Antarctic sea water.

Water temperatures in McMurdo Sound are extremely stable, and throughout the year do not vary more than a fraction of a degree from the freezing point of sea water, which is -1.86°C. This extraordinary constancy is due to the large amount of ice, which acts as a thermal buffer. As long as both ice and water are present, a loss or gain of heat will only change their relative proportions, but will not alter the temperature.

In such a stable environment, it is not surprising to find that Antarctic fishes are intolerant of temperature changes. Above 0°C they become agitated, seeking escape to colder water, and above 6°C they rapidly die.

Antifreeze compounds

Although low temperature and the constant presence of ice protect Antarctic fishes from temperature fluctuations, these conditions are also a potentially lethal combination. The concentration of salts in sea water is sufficient to depress the freezing point nearly 2°C below that of fresh water, but the blood of bony fishes contains only enough dissolved solids to lower the freezing point by about 1°C. Although some deep-dwelling species, such as those in Norwegian fjords, may live in a supercooled state below the freezing point of their body fluids, this is not possible for many Antarctic fishes, as even the smallest particle of ice will act as a seed crystal to trigger freezing of a supercooled solution. Thus, Antarctic

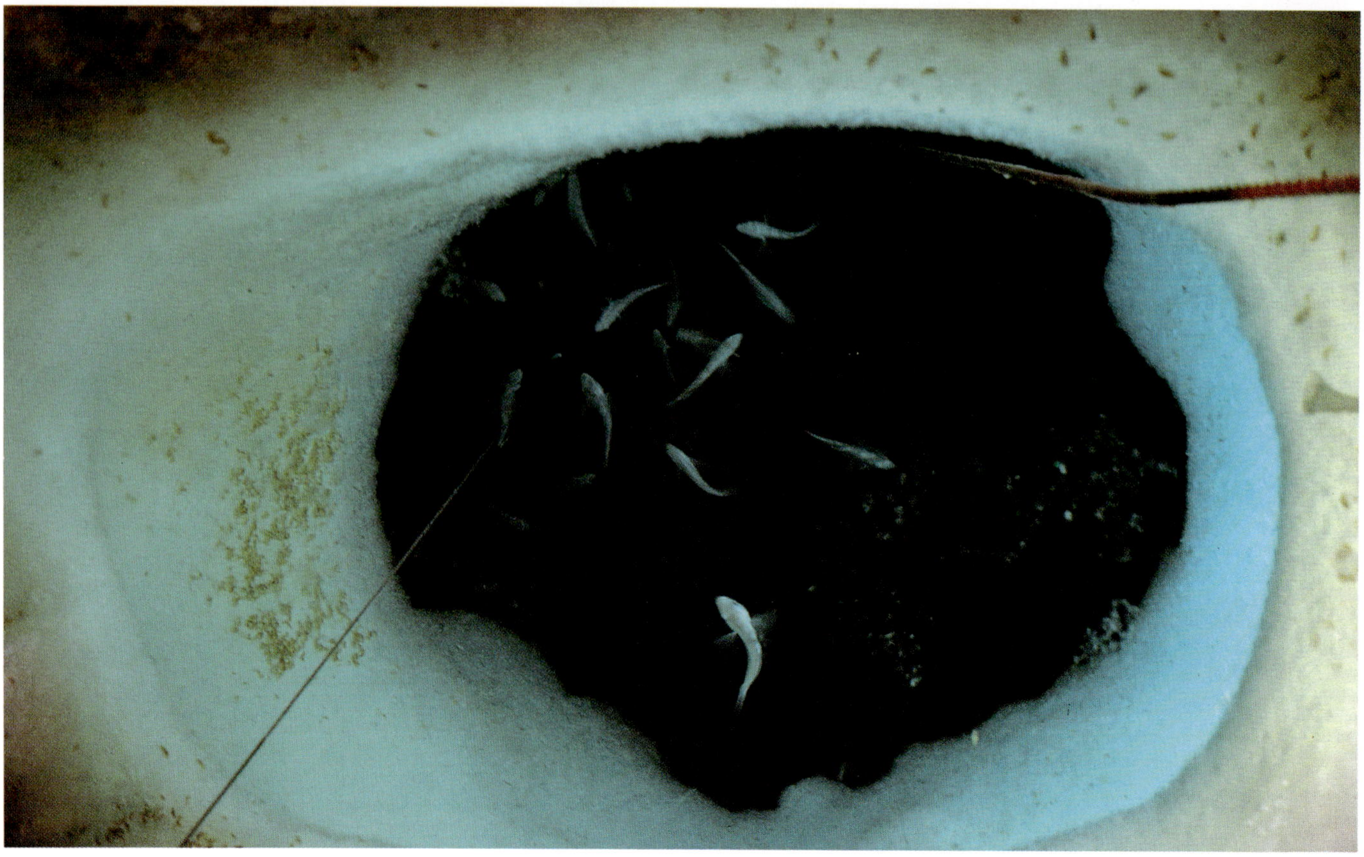

Plate 14.1. *An icy window into a watery world: silvery* Pagothenia borchgrevinki *swimming in a fishing hole. These small fish are easily caught on hand lines. A yellow-green film of diatoms is beginning to grow on the sides of the hole. Photo: J. Macdonald.*

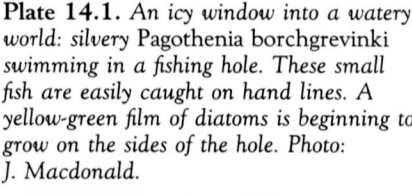

fishes must have some other defence against freezing. The nature of this protection is a remarkable molecular adaptation. Although familiar antifreezes rely on high concentrations of small molecules (salts, sugars, alcohols, glycerol, or ethylene glycol), the Antarctic notothenioids contain relatively low concentrations (about 3 percent by weight) of a large protein–sugar molecule (glycoprotein).

The regular spacing of sugar groups in the glycoprotein molecule matches the distance between water molecules in an ice crystal, so that the glycoprotein can act rather like an antibody, and readily attach (bind) to microscopic ice crystals. At its normal concentration, the bound glycoprotein inhibits further growth of ice crystals down to about −2.7°C. At lower temperatures in the presence of glycoprotein, ice crystals grow rapidly to form characteristic long spicules. The melting point of these spicules is significantly higher than their freezing point, and they persist until the temperature is raised to about −1°C, which is the normal freezing–thawing point predicted for conventional antifreezes (on the basis of total concentration of dissolved molecules). When freezing does occur in an Antarctic fish, the long ice spicules so damage cell membranes that the fish does not survive.

Antifreeze glycoprotein molecules occur in most body fluids, but some, such as urine, and fluids within the eye, lack antifreeze and apparently rely on the protection of surrounding tissues.

The kidneys of Antarctic fishes are specialised to conserve antifreeze, and prevent its loss from the blood. In most other fishes, urine is formed by pressure filtration through a ball of blood capillaries (the glomerulus), and only blood cells and very large proteins are retained in the blood. In such normal kidneys, smaller molecules such as sugars and the low molecular weight glycoproteins would pass from the blood into the urine. However, the kidneys of Antarctic notothenioids lack glomeruli, and urine is formed by active secretion through the walls of kidney tubules, so that the glycoproteins are retained.

The antifreeze glycoproteins are so effective in their protection against freezing that several Antarctic fishes have adopted masses of ice as their normal habitat, and as a refuge from predators. The naked dragon fish, *Gymnodraco acuticeps*, normally conceals itself on the bottom under anchor ice, lunging out to capture small passing fish. Another very successful species, *Pagothenia borchgrevinki* (Plate 14.1), inhabits the layer of loose ice platelets which form just beneath the solid sea ice, and often rests directly on masses of these crystals.

After their initial discovery in Antarctic notothenioids, ice-inhibiting molecules have been found in other groups of fishes, and seem to represent convergent evolution due to a problem common to most polar fish. For example, a protein antifreeze, lacking the sugar groups, is found in the McMurdo Sound eel-pout, *Rhigophila dearborni*, and analogous protein and glycoprotein antifreezes have been reported for Arctic fish species.

Blood and oxygen transport

All classes of vertebrates, from fishes to mammals, use the red respiratory pigment haemoglobin to carry oxygen to the tissues. Haemoglobin is contained in red blood cells, and combines temporarily with oxygen at high oxygen concentrations in the lungs or gills, later releasing the oxygen in the tissues where the oxygen level is low.

The amount of haemoglobin, and its affinity for oxygen, are lower in Antarctic fishes than in temperate fishes with similar life styles. The scarcity of haemoglobin is partly due to fewer red blood cells, but this varies between species, with more active species tending to have more blood cells. The number of blood cells in circulation can change rapidly, since many are stored in the spleen, and can be released suddenly to cope with stress. The concentration of haemoglobin within red blood cells is also lower in Antarctic fishes, but there does not seem to be any correlation with the activity of different species.

Both blood viscosity and oxygen solubility seem to play important roles in reducing haemoglobin in Antarctic fishes.

Viscosity, or "thickness" of blood, increases severalfold at low temperatures, and this in turn increases the work required to force blood through the blood vessels, particularly through small capillaries. Much of the viscosity of blood is due to its cellular components, and fewer red blood cells would be accompanied by lower viscosity and savings in energy requirements (Figure 14.2).

Low water temperatures also increase gas solubilities, and Antarctic waters generally contain much more dissolved oxygen than do temperate waters. This high ambient oxygen concentration partly compensates for the reduction in haemoglobin by assuring that the blood is saturated with oxygen. When Antarctic fishes are kept experimentally in water with artificially depleted oxygen, the number of red cells, the concentration of haemoglobin within the cells, and the affinity of the haemoglobin for oxygen all increase. The result is that the oxygen-carrying capacity of the blood is increased.

One family of Antarctic fishes, the Channichthyidae (icefish), is unique among the vertebrates as all its members lack haemoglobin (Plate 14.2). Channichthyids have a few blood cells which microscopically resemble red blood cells, except that they contain no haemoglobin. Oxygen is dissolved in the blood plasma, with its increased solubility (due to low temperature) partially offsetting the reduction in carrying capacity resulting from the lack of haemoglobin. The few, remaining non-pigmented "red cells" have apparently been retained to carry essential regulatory enzymes.

Several anatomical and physiological features help the channichthyids to function without a respiratory pigment. They have a large blood volume, which is circulated rapidly by an enlarged heart. Large-diameter capillaries and low blood viscosity (due to the absence of cells) both reduce the resistance of the circulatory system, so that less work is needed to maintain a high blood flow. Oxygen is extracted from the oxygen-rich water by large, well-developed gills, and also by exchange across the scaleless skin. Finally, their metabolic rates seem to be slightly lower than those of other Antarctic fishes, so that their requirements for oxygen are somewhat less. The channichthyids do not seem to be handicapped by the absence of a respiratory pigment; they are diverse and widespread around Antarctica, and some lead very active, predatory lives (Figure 14.3).

Food and energetics

All of an animal's vital functions are powered by the energy released from food. This release and its application to growth, protein synthesis, muscular contraction, etc., are known collectively as metabolism.

To understand an animal's ecology, and its place in an ecosystem, we must understand its metabolism. The fabled richness of southern waters is largely a myth, and at best the richness is a patchy phenomenon, so obtaining food is not always straightforward. In the continuous daylight of summer, food production may be high, resulting in large local populations of phytoplankton, sub-ice diatoms, and zooplankton, but even in summer there are some areas, under thick ice and snow cover, where little photosynthesis is possible, and food must drift in on currents from richer areas. Food abundance also fluctuates drastically with the seasons. Phytoplankton photosynthesis ceases during winter darkness, although enough algal material is trapped in sediments to maintain benthic (bottom-dwelling) invertebrates and fish through what would otherwise be a famine. Significant amounts of bottom sediments are also resuspended, and probably serve as a winter food reserve for animals higher in the water column.

Under such conditions, Antarctic fishes are generally opportunistic feeders. A minority specialise on particular prey such as plankton or nekton, but most seem to be omnivorous, devouring anything likely to supply metabolic energy. Analyses of stomach contents indicate that benthic species such as *Trematomus bernacchii* are indiscriminate feeders, taking bottom invertebrates, other fish, detritus, and even the occasional pebble! It would obviously be a great advantage for these fish to have an opportunistic metabolic system which could speed up to take advantage of abundance, but could also conserve energy when food is scarce.

In cold-blooded animals such as fish, metabolism is directly affected by environmental temperatures, so that we might expect it to be very low in Antarctic fishes. However, when animals adapt to cold environments they often increase their metabolic rates to compensate partially for the effects of low temperature, and this also seems to be true for Antarctic fishes. The metabolism of Antarctic fishes was first thought to be several times higher than it is, so that the Antarctic species at −1.9°C seemed to be almost totally temperature-compensated, with metabolic rates equivalent to those of fishes normally living at 15–20°C. More recent research indicates that the degree of compensation is much less, and that Antarctic fish do not show a unique degree of metabolic cold adaptation, but are only partially compensated, like other cold-adapted animals.

The most obvious use of metabolic energy is in locomotion. Most species of Antarctic fish propel themselves through the water with sculling movements of their pectoral (shoulder) fins, while holding the

Figure 14.2. *The giant Antarctic cod,* Dissostichus mawsonii, *showing biochemical, physiological, and anatomical features which have contributed to the success of red-blooded notothenioid fishes in Antarctica.*

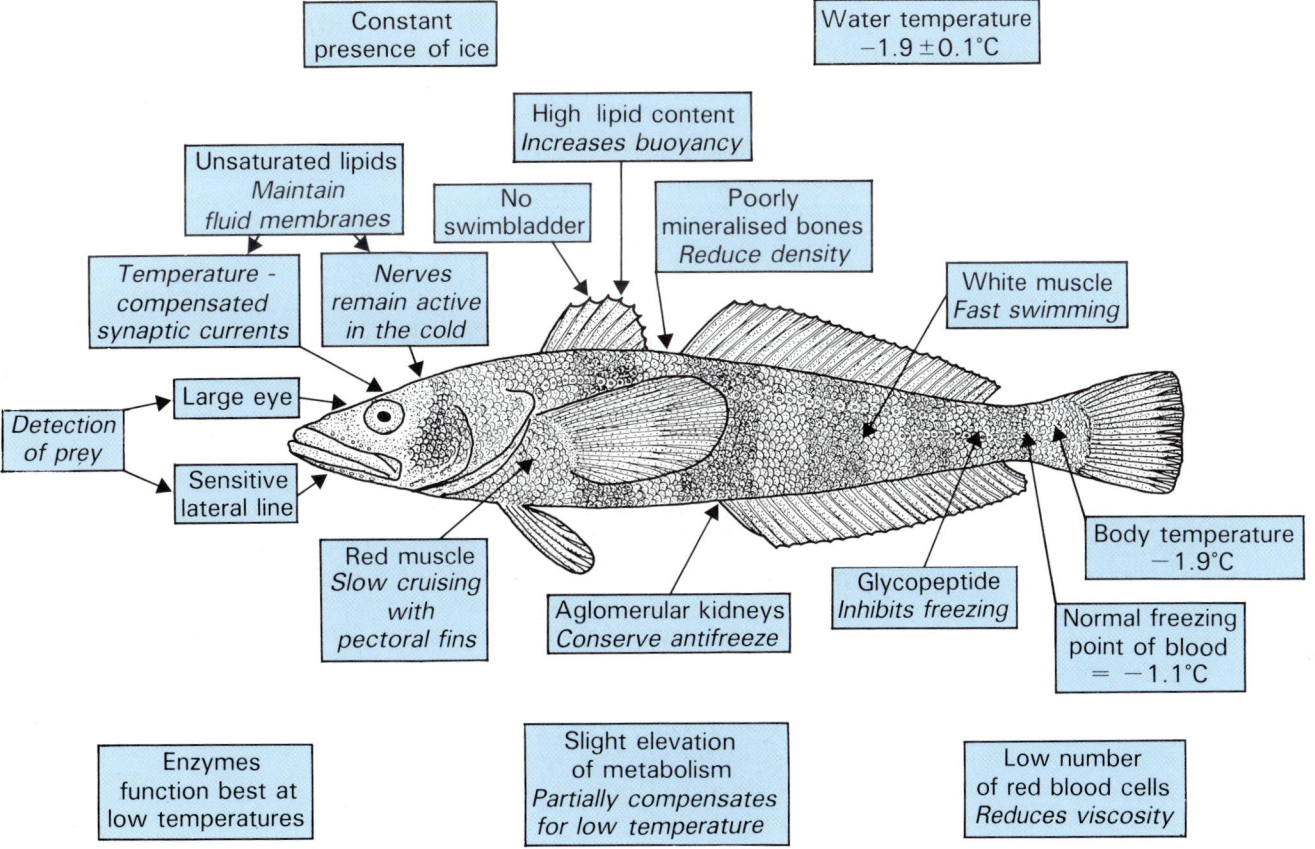

tail rigid as a rudder. This mode of swimming, known as labriform swimming, is particularly common among the notothenioids, and probably has been inherited from their bottom-dwelling ancestors. It is not a fast style of swimming, but it is well suited to a fish which spends much of its time lying on the bottom in wait for passing prey. In the neutrally buoyant midwater species, it can be maintained for long periods without rest. The pectoral fins are controlled by "red" muscles, which contain another oxygen-binding pigment, myoglobin, and rely on a continuous supply of oxygen to release metabolic energy (aerobic metabolism) for contraction.

Short rapid dashes driven by the caudal (tail) fin (subcarangiform swimming) can also be made to lunge at prey or to escape predators. The tail muscles, however, are "white", with no myoglobin, and can release energy very rapidly in the absence of oxygen (anaerobic metabolism). Anaerobic metabolism is much less efficient, and results in an "oxygen debt", so that rapid subcarangiform swimming can only be briefly sustained. Moreover, the oxygen debt must be repaid when strenuous activity ceases. In general, Antarctic fish seem to depend mainly on aerobic metabolism, and do not have a high capacity for anaerobic metabolism. Recovery from anaerobic oxygen debt may take them several days.

Another use of metabolic energy is in growth, and in the synthesis of large molecules such as proteins, where energy is required to bind smaller molecules together. Under experimental conditions, proteins may be manufactured by Antarctic fishes at rates equal to those of temperate fishes at warmer temperatures. The high rates of protein synthesis probably indicate an ability to use materials rapidly at the restricted times when they are available, whereas the slower overall growth rates reflect the availability of food averaged over the year, including the relative shortage during winter.

Molecular adaptations

Like all animals, the metabolism of Antarctic fishes is guided by enzymes, which act as molecular catalysts to control biochemical reactions. Enzymes of Antarctic fishes show features which seem to be related to their operation at low temperatures:

- Most enzymes function best at 20–40°C, but those of Antarctic fishes are most active near 0°C.
- Many Antarctic fish enzymes do not increase activity markedly at higher temperatures, which indicates that relatively little energy is needed to activate the enzymatic reaction.
- Enzymes and other proteins of Antarctic fishes tend to be less stable than those of warm-water fishes, and are easily denatured by both chemical agents and high temperatures.

One possible explanation for the enhanced reactivity and greater susceptibility of Antarctic enzymes lies in their structure. Although proteins are basically long chains of amino acids, they also coil and fold on themselves to assume complex three-dimensional shapes which are essential to their normal function. Antarctic enzymes seem to have an open configuration, with the reactive portions of the molecule being more readily exposed. Although these properties permit the fish to live at low temperatures, they also pose practical problems for cold storage. Frozen Antarctic fish should probably be stored at temperatures considerably lower than the usual deep-freeze temperature of –20°C.

In addition to changes in the reactivity of individual enzymes, there are changes in relative concentrations or activities of different enzymes, so that particular reactions or sequences of reactions are more favoured.

Figure 14.3. Pagetopsis macropterus *is an icefish or channichthyid, a family which lacks haemoglobin. Low temperature and anatomical specialisations permit these white-blooded fishes to flourish in Antarctica.*

For example, in many cold-adapted fishes the sugar glucose is metabolised mainly through a sequence known as the "pentose phosphate pathway" in preference to the "glycolytic pathway" favoured by many warm-adapted animals. The pentose phosphate pathway is closely linked to the biochemical pathways involved in lipid metabolism, which is also of greater importance in polar fishes.

Proteins are not the only molecules that have been remodelled to function better in the cold. Every living cell is bounded by a cell membrane, which is a thin film consisting of both protein and lipid. All materials entering or leaving the cell must pass through the membrane, and much of this transport seems to be controlled by the consistency of the lipid portion of the membrane. The lipids of cold-adapted plants and animals, including Antarctic fishes, tend to be more fluid at low temperatures than lipids from organisms adapted to warm temperatures.

The biochemical feature which seems to contribute most to increased lipid fluidity is chemical unsaturation of fatty acid hydrocarbon chains. (For example, compare the fluidity of an unsaturated cooking oil with that of a saturated animal fat such as butter or lard.) Unsaturation is achieved by reducing the number of hydrogen atoms attached to the carbon atoms of a chain, and substituting double, rather than single, bonds between the carbon atoms. Each double bond forms a rigid link in the carbon chain, interfering with free rotation of the carbon atoms and creating a permanent kink in the chain. A higher proportion of such unsaturated "kinked" chains prevents adjacent fatty acid chains from packing tightly together, and promotes freer movement between the chains.

As the cell membrane is extremely thin, about one-millionth of a centimetre, direct measurements of fluidity within the membrane are very difficult, but indirect measurements strongly suggest that the membranes of Antarctic fishes are more fluid than those of temperate fishes. In warm-adapted animals, spontaneous ion currents through the

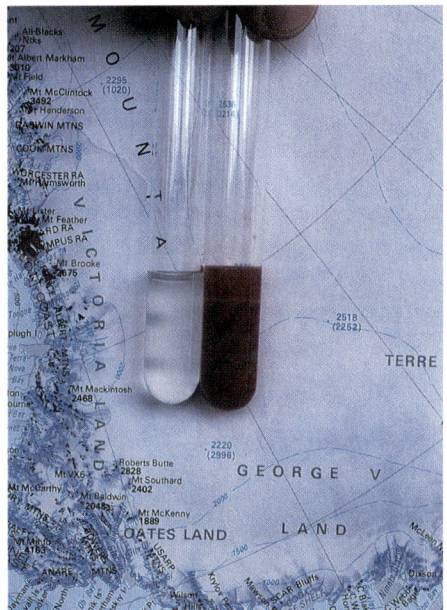

Plate 14.2. *Fresh red and "white" blood from Antarctic fishes. The left tube contains colourless whole blood from a "white-blooded" icefish, Chionodraco kathleenae (Family Channichthyidae), which lacks the red blood pigment haemoglobin. The right tube contains red blood from* Trematomus bernacchii *(Family Nototheniidae). Photo: J. Macdonald.*

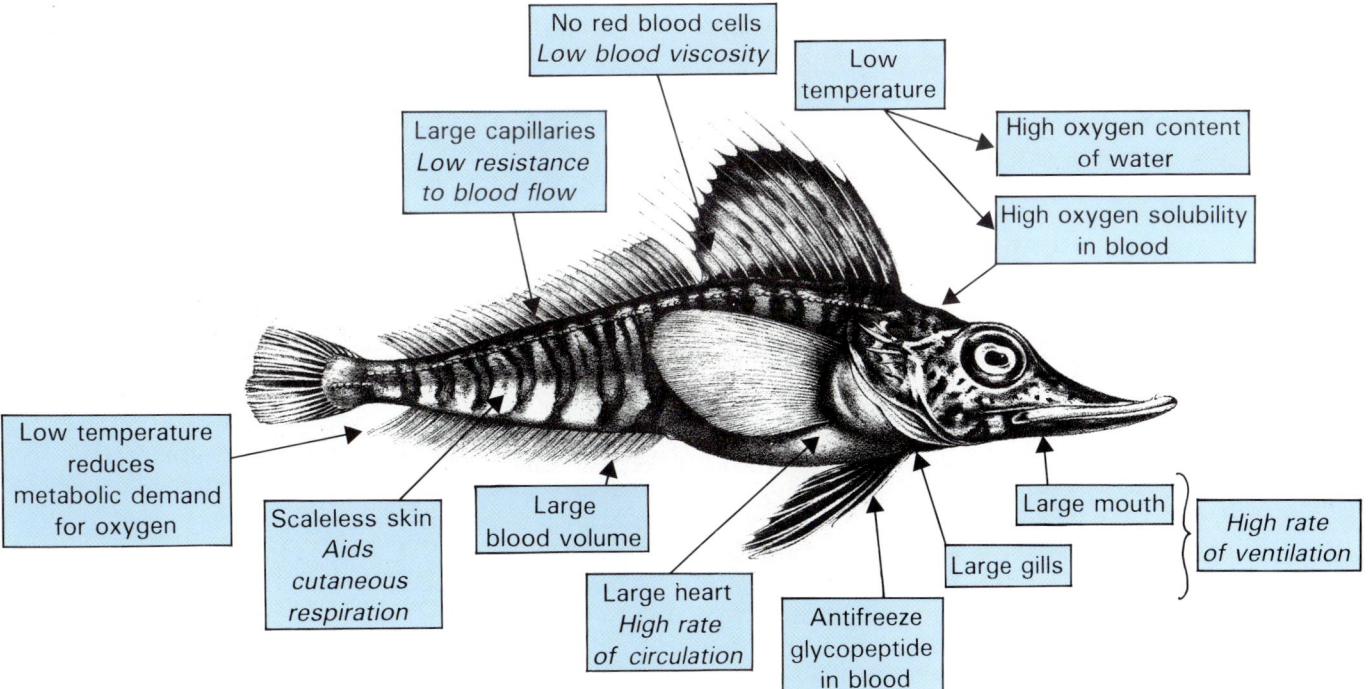

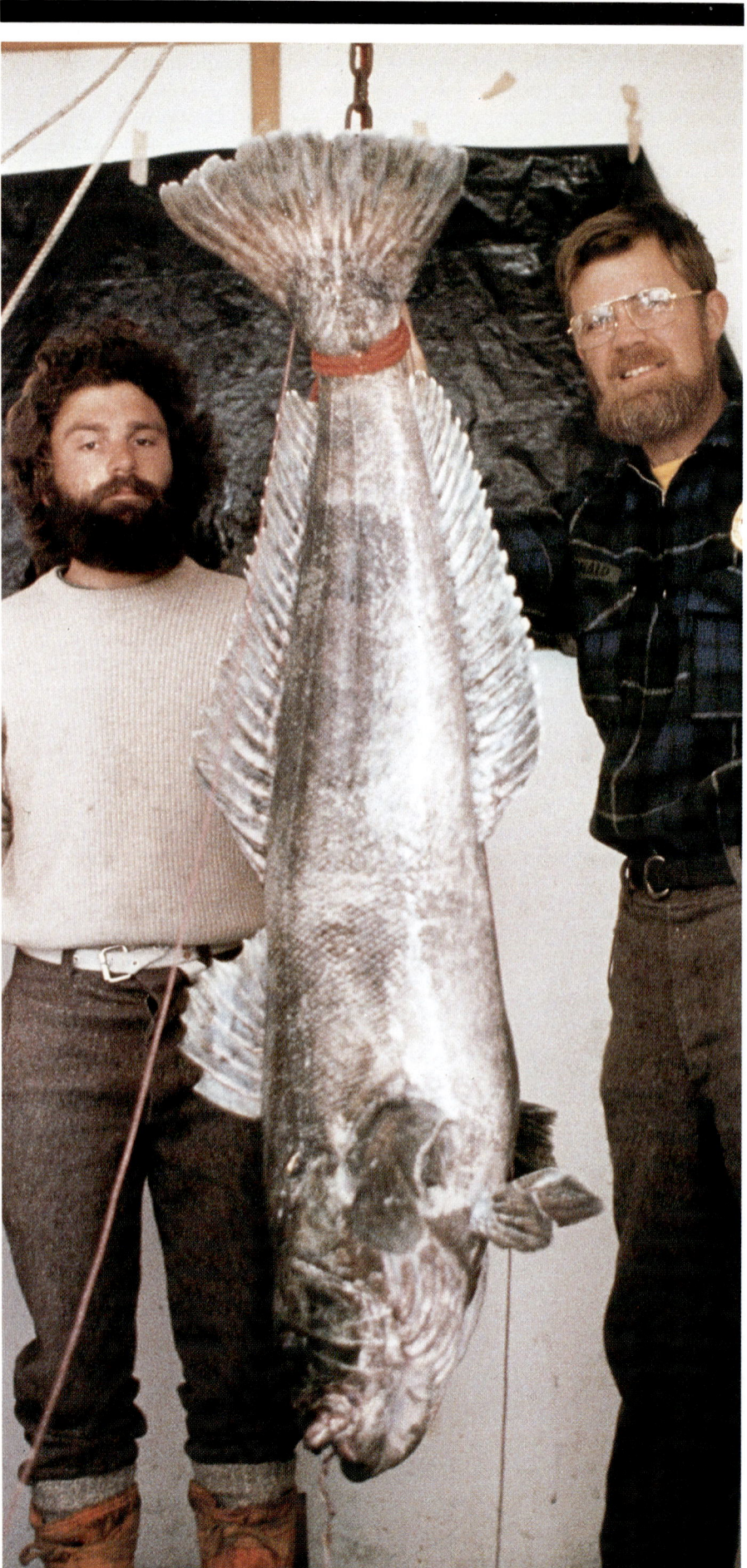

Plate 14.3. *The giant Antarctic cod,*
Dissostichus mawsonii, *is the largest of the*
fishes in Ross Sea waters. Photo:
D. Hayman.

membrane at junctions of nerves and muscles are greatly prolonged at low temperatures, whereas those of Antarctic fishes are of much shorter duration. The short duration is consistent with a high mobility of the molecular gates which control ion currents within the membrane.

Nerves and sense organs

One important consequence of the increased fluidity of cell membranes in Antarctic fishes is the ability of their nervous systems to function at subzero temperatures. When the nerves of warm-adapted animals are cooled, their ability to conduct nervous impulses usually fails above 0°C. But the nerves of Antarctic fishes continue to conduct impulses down to −5°C, and could probably function at even lower temperatures if the surrounding fluids were prevented from freezing. The upper temperature limits for most nervous processes are also reduced in Antarctic fishes to 15–20°C.

Sense organs of Antarctic fishes also seem to have adapted to low light levels. The eyes of at least one deep-dwelling species, the giant Antarctic cod (*Dissostichus mawsoni*) (Plate 14.3), are adapted to very low light levels. Adaptations include many light-detecting cells (photoreceptors), a high proportion of rod to cone receptors, and a high degree of convergence, so that the information gathered by many receptors is concentrated on fewer brain cells. In vertebrate eyes, rod-type photoreceptors are the most sensitive, and are responsible for black-and-white night-vision, whereas cones are used for high-resolution vision (and in some animals the perception of colour), but require much more light. Like nocturnal animals, these fishes can see in very dim light, but probably do not perceive much detail. As they feed on relatively large prey, the lack of resolution may not be a handicap.

Smaller shallow-water fishes such as *Trematomus bernacchii* and *Pagothenia borchgrevinki* feed on smaller prey, and in summer have enough light to locate their prey by sight. Their eyes do not show the adaptations to low light seen in the giant Antarctic cod (although they have only been examined in the summer). Without any retinal adaptations to the winter darkness, these fishes would have great difficulty finding food by sight alone, and it seems that they also employ another sensory system to assist in feeding.

All fishes possess lateral line sensory systems consisting of nerves running down the sides of the body, connected to end organs activated by water movements. The lateral line system of *Pagothenia borchgrevinki* is particularly well developed over the head and along the lower jaws. It is very sensitive to water-borne vibrations produced by several varieties of swimming planktonic crustaceans, which form much of their diet (Figure 14.4). Whether these fish actually use their lateral lines to locate and capture prey in the dark is a fascinating question.

Mammalian adaptations

Seals and whales are the only mammals native to Antarctica. Like the fishes, they must contend with the effects of cold, but they also need to overcome other problems, related to life as air-breathing animals. Generally, they show very different adaptations to those seen in Antarctic fishes. Although there has been much research on the biology of Antarctic whales, the emphasis of recent work in the Ross Sea region has been on seals.

Ross Sea seals—matching teeth with diet

Four species of seal, the Weddell, leopard, crabeater, and Ross seals, frequent the Ross Sea. All four are closely related members of the family of earless seals (Phocidae) which lack external ears (pinnae), in contrast to the eared seals (Otariidae) which possess small pinnae, and include sea lions and fur seals. The four Antarctic seals occupy distinct ecological niches, and differ greatly in both diet and behaviour. As in all mammals, their teeth reflect their diet, and are an excellent example of the relationship between form and function.

The Weddell seal (*Leptonychotes weddelli*) is found mainly on sea ice firmly attached to the shore ("fast ice"), and feeds primarily on fish. Like other fish-eating mammals, its molar and premolar teeth (cheek teeth) are simplified into a peglike form, mainly serving to grip the fish (Plate 14.4a). The canine and incisor teeth are modified for ice cutting—the second incisors of the upper jaw resemble canine teeth, and project forward along with the stout canines. The seal maintains breathing holes in the ice by shaking its head from side to side and sawing at the ice with its upper teeth. The canines and second incisors of mature Weddell seals are often extremely worn, and dental abcesses of these teeth are common. Tooth wear is probably important in limiting the life span of Weddell seals.

Although not the most abundant of Antarctic seals, the Weddell seal is probably the best known. Lacking any natural surface predators, Weddell seals are placid and easily approached; their behaviour differs significantly from that of Arctic seals which share a similar fast-ice habitat with predators such as the polar bear. In the summer Weddell seals spend many hours dozing on the sea ice.

The other three seal species are found mainly in the pack ice, but the leopard seal (*Hydrurga leptonyx*) is seen regularly off the Adélie penguin rookeries at Capes Bird and Crozier, and is the most dangerous of the Ross Sea seals. It is a generalised predator, with a powerful, heavy jaw, well-developed canine teeth, and distinctive trilobed cheek teeth. The cheek teeth apparently intermesh to form a sieve when the seal feeds on krill (*Euphausia*), which make up much of its diet (Plate 14.4b). Although some leopard seals eat well by devouring penguins off the

Figure 14.4. *Detection of a swimming crustacean by the anterior lateral line of Pagothenia borchgrevinki. The vibrations of an amphipod swimming near the fish, shown in the bottom record, stimulate sense organs in canals beneath the skin of the fish and produce the nerve spikes shown in the top record. The skin canals of the anterior lateral line, and the pores which communicate to the surface, are shown on the head of the fish.*

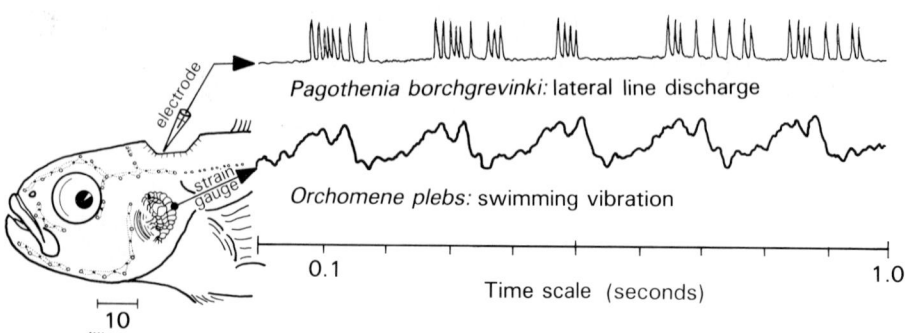

rookeries, many more are out amongst the pack ice, where fish, krill, and other invertebrates make up most of the diet. Even with its awesome mouth closed, the leopard seal may be clearly distinguished from the Weddell: firstly, it has a much longer head, with relatively small eyes (see Plate 11.11); and, secondly, it moves (very quickly) by writhing from side to side in a snake-like fashion (Plate 14.5), whereas the Weddell seal tends to hump itself along like a giant caterpillar. One should be very cautious about approaching a leopard seal!

In the McMurdo Sound region, a third species, the crabeater seal (*Lobodon carcinophagus*), is seen more often dead than alive (see Plate 11.10). Crabeater seals normally live in the pack ice, and only enter McMurdo Sound in late summer when the fast ice has broken out. They feed almost exclusively on krill, and have elaborately multilobed cheek teeth which are thought to serve as strainers like those of the leopard seal (Plate 14.4c). Also like leopard seals, crabeater seals can move rapidly across the ice by sideways undulations. Most of the dead seals in the Dry Valleys are juveniles, and probably died by becoming trapped and disoriented when winter ice begins to reform in the Sound. Very few carcasses of other seals are found.

The Ross seal (*Ommatophoca rossi*) is seldom seen, partly because of the inaccessibility of its usual habitat in dense pack ice. It feeds mainly on squid, plus a few fish; the cheek teeth are very small and relatively simple. The Ross seal has long acquired a reputation for a beautifully musical voice, but this may be more myth than reality.

A fifth species of true (earless) seal, the elephant seal (*Mirounga leonina*) occurs rarely in the Ross Sea as a stray. It is common in the subantarctic islands, where it feeds largely on squid. The cheek teeth are reduced to very simple blunt pegs.

Defences against the cold
Whereas Antarctic fish survive by adjusting each body cell to

Plate 14.4. *Skulls of the three most common seal species showing diagnostic features of dentition. (a); The Weddell seal's skull is somewhat flattened, with large, upward-directed eye sockets. Cheek teeth are relatively simple pegs, suited to gripping fish. Canine teeth and upper incisors are stout and directed forward to ream breathing holes through the sea ice. (b); Cheek teeth of the leopard seal are tri-lobed, serving as strainers when the seal feeds on krill, and also to grip fish. Canines are long and slender, suitable for feeding on large birds or fish. As with generalised predators the lower jaw is massive. (c); The crabeater seal's canine teeth, incisors, and the lower jaw are relatively weak, whereas the cheek teeth are elaborately lobed to serve as krill strainers. Photos: (a) G. Batt; (b), (c) J. Macdonald.*

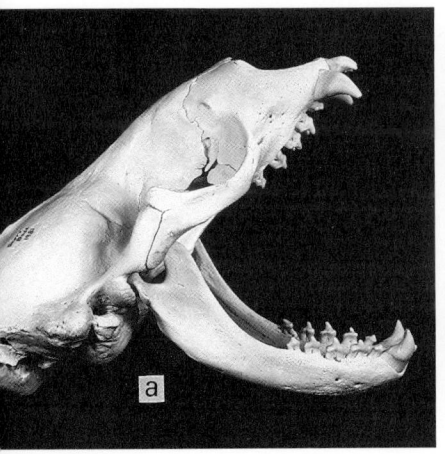

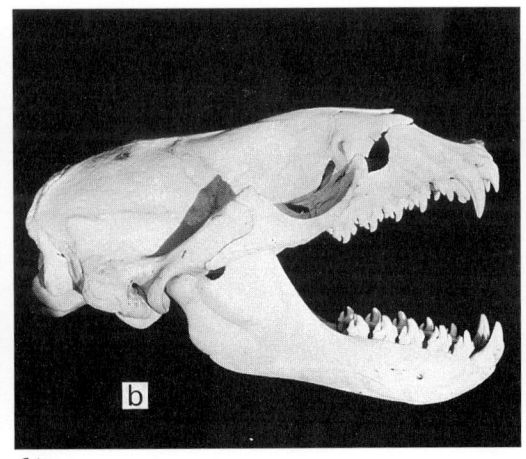

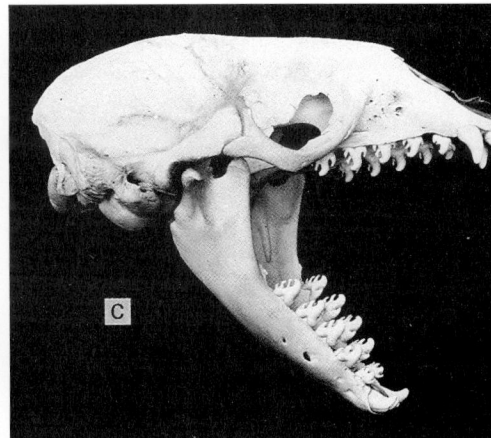

(a)   (b)   (c)

operate at subfreezing temperatures, seals and other mammals must maintain high internal temperatures. Most of the adaptive specialisations of Antarctic mammals result in ways for the brain and other vital organs to *avoid* being in direct contact with extreme environmental conditions. Their first defence, of course, is that they are aquatic animals and can avoid extremely low air temperatures by entering the comparatively warm water.

Like other marine mammals, Antarctic seals use an insulating blubber layer about 5 centimetres thick over the trunk to help retain body heat. The thermal conductance of fats is relatively low, and the blubber does not require a continuous high blood supply which would drain heat from the core of the animal.

The loss of heat from extremities, which are not protected by blubber, can be greatly reduced by an anatomical juxtaposition of blood vessels to form heat exchangers at the base of each limb. Each major artery running into a flipper is surrounded by a network of veins returning chilled blood from the surface. The cold venous blood absorbs heat from the arterial blood, so that most of the heat is returned to the body core, and very little is pumped out to the flipper and lost.

Blood supply to blubber, skin, and extremities can be controlled to regulate body temperature. When the seal is overheating, due to strenuous exercise or a balmy day on the surface, blood vessels leading to the skin open and blood bypasses the heat exchanger networks. Thus, more heat is pumped to the surface and lost to the environment (the same principle operates in humans when we become flushed in the face, hands, etc., when overheated). When the temperature falls, blood supply to blubber and skin are shut off, and the heat exchangers become fully functional.

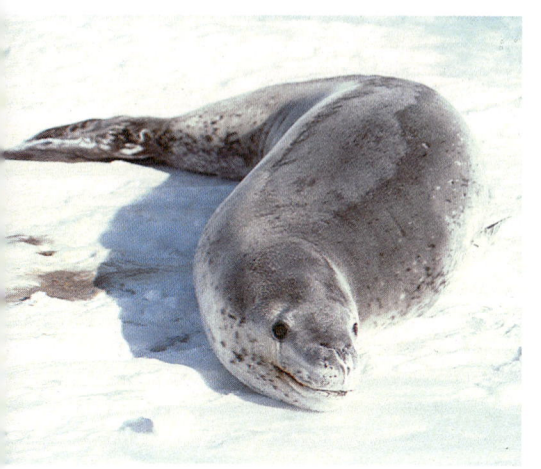

Plate 14.5. *A leopard seal,* Hydrurga leptonyx, *in its sinuous snake-like progress across across the sea ice. Photo: C. Monteath, Antarctic Division, DSIR.*

The size of an animal affects both the production and the loss of body heat. Large animals produce more heat, and tend to have proportionally less surface area from which to lose heat. Consequently, many marine mammals are large, with reduced extremities. Weddell seals are quite large seals, with adults weighing 300–450 kilograms. Large size may also confer other advantages, by increasing reserves of stored oxygen for prolonged dives, as well as increasing food reserves stored as blubber.

Because of their low body mass, newborn seal pups are extremely vulnerable. Weddell seals are born in November after a gestation period of about 11 months, with a body weight of about 30 kilograms. They grow rapidly on milk containing over 60 percent fat, doubling birth weight in 12 days, and reaching a weaning weight of over 100 kilograms at 5–6 weeks old. When the pups are born, they have only a few millimetres of blubber, but are covered with a lush fur which is suited for insulation in air, but not in water. This fur is shed as the subcutaneous blubber layer increases in thickness. At about 2 weeks old, pups begin to enter the water, when their blubber is about 20 millimetres thick.

"Weddell Seal: Consummate Diver"

This title of a book by G. L. Kooyman summarises the best-known feature of this extraordinary seal. Much of what we know about diving mammals has come from studies of the Weddell seal. It is a superb diver, submerging to depths of over 600 metres, or for up to 70 minutes. These records are exceeded only by sperm whales and elephant seals, which can dive to over 1100 metres, but temperament and ecology make the Weddell seal much easier to study. Not only is it a superb diver, with a placid nature, its greatest asset as a scientific subject is its habitat on and under the fast ice. If a hole is made in the sea ice several kilometres from other holes or cracks, captured Weddell seals can be forced to dive and return repeatedly to the same hole, so that their dives may be timed accurately, apparatus such as depth gauges can be recovered, and samples of blood or exhaled gas can be collected. With the aid of portable microcomputers glued to the seal's back, the physiological changes accompanying a voluntary dive made by a free-living wild animal can be measured.

Although the Weddell seal holds both depth and endurance records, it does not do both simultaneously. Deep feeding dives to 300 metres or more are relatively short (less than 20 minutes), whereas long exploratory dives, in which the seal may travel several kilometres from its breathing hole, and remain submerged for over 60 minutes, are comparatively shallow (about 50 metres) (Figure 14.5). The two kinds of dive are planned by the seal for different purposes.

Oxygen storage and consumption

Weddell seals can easily stop breathing for 15–20 minutes, which is a very long time by human standards. How do the seals accomplish this feat? Their initial advantage is undoubtedly their ability to store far more oxygen in their tissues. A Weddell seal contains twice as much blood for its body weight than a human, and the blood itself contains nearly twice as much haemoglobin as the same volume of human blood. If the weight of blubber, which is practically metabolically inert, were subtracted from the seal's weight, it would then have over five times more oxygen per unit weight stored in the blood than would a human. Another oxygen-binding compound, myoglobin, is found in some muscle cells, and gives red meat its characteristic colour (white meat has little or no myoglobin). Weddell seal muscle is very dark, almost black, and contains myoglobin concentrations 10 times that of human muscle, thus forming another large oxygen store for the exclusive use of muscle.

In common with other diving animals, Weddell seals can readjust their circulation during a dive. The spleen, which may contain up to 60

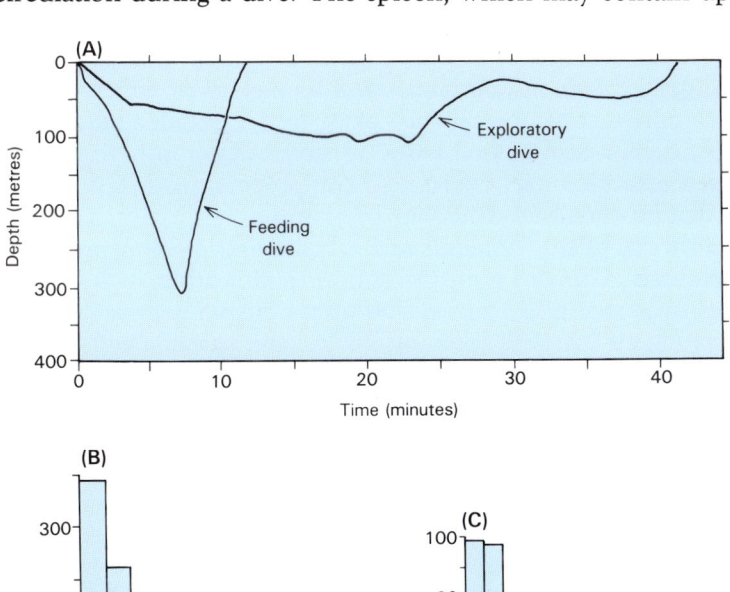

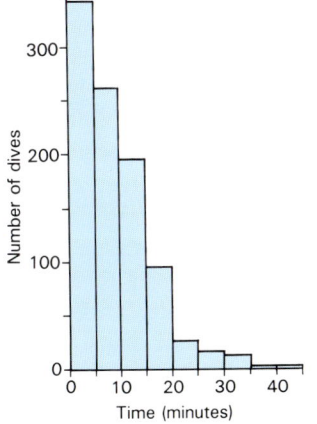

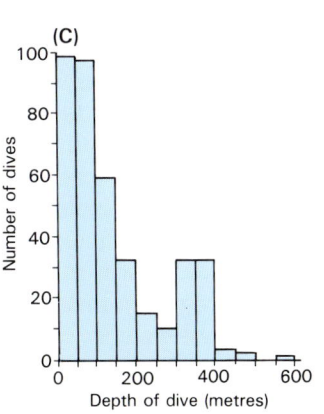

Figure 14.5. *Diving performance in Weddell seals. (a); Dive profiles, one relatively short and deep feeding dive, and one long, shallow exploratory dive are shown. (b); Dive times for 959 dives made by 31 seals. Most dives are shorter than 15 minutes, whereas dives longer than 20 minutes are rare. (c); Dive depths for 381 dives by 27 seals. Most dives were shallower than 200 metres, but some were much deeper.*

percent of the total red blood cells (25 litres!), contracts in a Weddell seal and adds enough oxygenated haemoglobin to the circulation to nearly double the haemoglobin concentration before the dive. Blood flow to other organs, such as the kidney, can be cut off forcing these organs to cease metabolism, or to operate without oxygen. Heart rate falls, and in extreme instances, blood flow is restricted to the heart and brain. Circulation to the muscles is normally cut off, and after exhausting its large store of myoglobin-bound oxygen, muscle functions without oxygen, producing a build-up of lactic acid in the muscle.

During short dives, up to about 20 minutes, seals do not produce lactic acid, and function aerobically. Their heart rate does not fall much, and they may recover and dive again within 5 minutes of surfacing.

For a long dive, however, the pattern is different from the beginning. Heart rate slows immediately, implying that the seal is planning ahead, and can voluntarily control some aspects of its circulation. At the end of a long dive, heart rate begins to increase just before the seal surfaces. A few minutes after surfacing, the lactic acid concentration in the blood increases, indicating that lactic acid is being flushed out of the muscle by restored circulation. Recovery, as indicated by a return to normal concentrations of lactic acid, may take well over an hour. Thus, extremely long dives are inefficient. Most dives are short, around 15 minutes (Figure 14.5).

Compensation for pressure

Extremely deep dives create other potential problems, related to high hydrostatic pressure. When an air-breathing animal dives, holding its breath, the chest is compressed, and internal gas pressure in the lungs and respiratory passages equals the external pressure. Some pockets of gas, such as those in the middle ear and frontal sinuses, are surrounded by rigid tissues and may fail to equalise pressure if the passages connecting them with the main airways are blocked. This can be extremely painful, and would usually end a human dive. Even with perfectly clear eustachian tubes and sinuses, human divers would be unable to compensate at depths beyond 60 metres, at which depth the chest is compressed to one-seventh of its fully expanded volume, and cannot be further compressed without the ribs breaking. Breathing compressed air, of course, offsets this compression, so that the air passages can be equalised to higher pressures.

The thorax of the Weddell seal is much more compressible than that of humans. The ribs are cartilaginous and flexible, as is the trachea. At depths over about 30 metres, the lungs collapse completely and force any remaining air into the bronchioles and the partially collapsed trachea. Pressure in the middle ear is equalised by soft-walled venous sinuses, which fill with blood and bulge into the ear to compress the air which is already there. Most diving mammals do not have open eustachian tubes, or frontal sinuses.

Even when breathing compressed air, human divers are at a disadvantage with Weddell seals. They may have extended their time underwater, and offset external water pressure with a high internal gas pressure, but they have acquired new problems. The solubility of gases in animal tissues increases with pressure. Air is nearly 80 percent nitrogen, and human tissues rapidly become saturated with dissolved nitrogen, which must come out of solution again when the pressure is reduced. At depths over 10 metres, the amount of nitrogen in the tissues can become so high that bubbles form in the tissues when the diver returns to the surface. The nitrogen bubbles can cause crippling pain, and, if they form in the nervous system, may produce paralysis or even death. This is a well-known occupational disease of human divers—caisson disease or

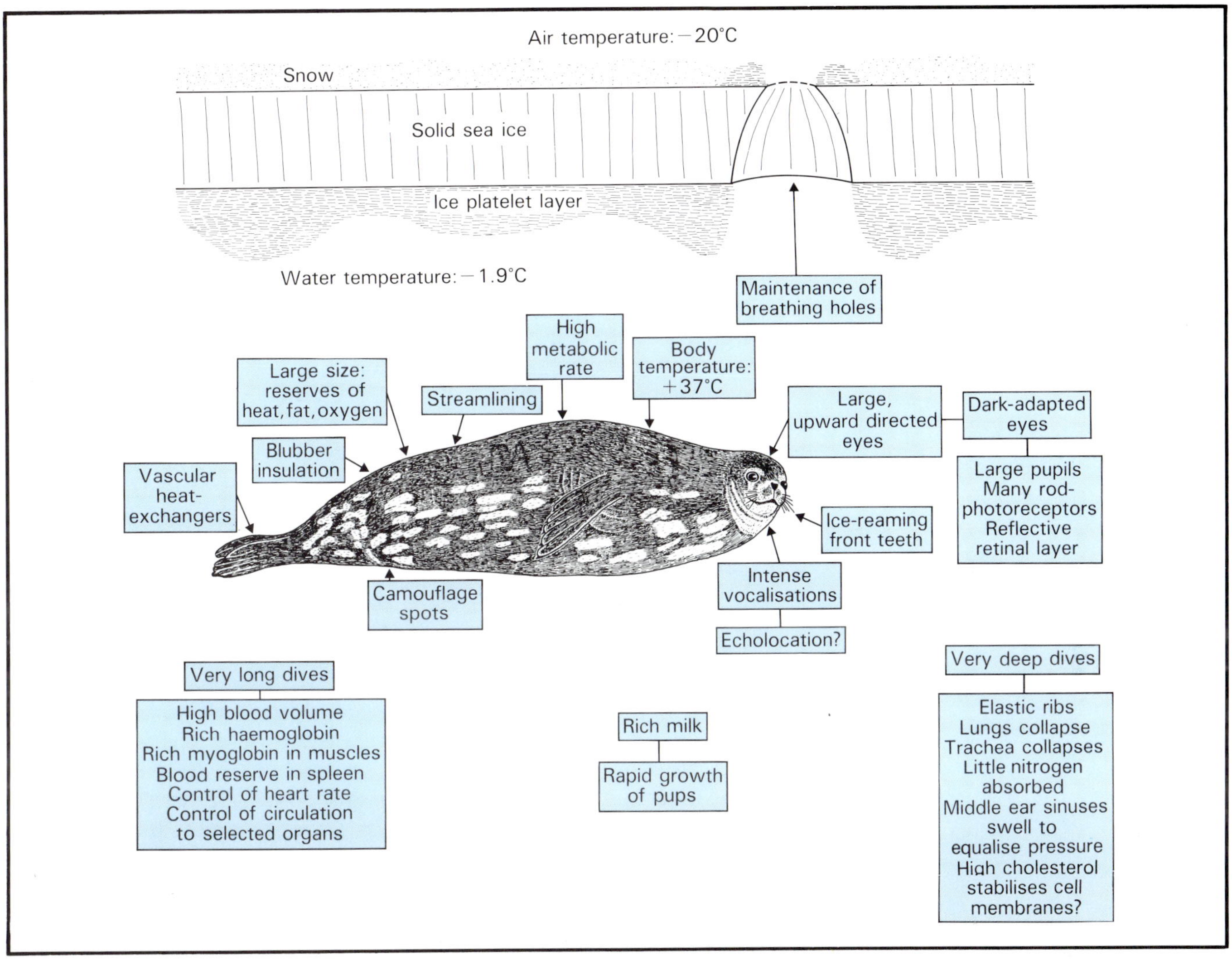

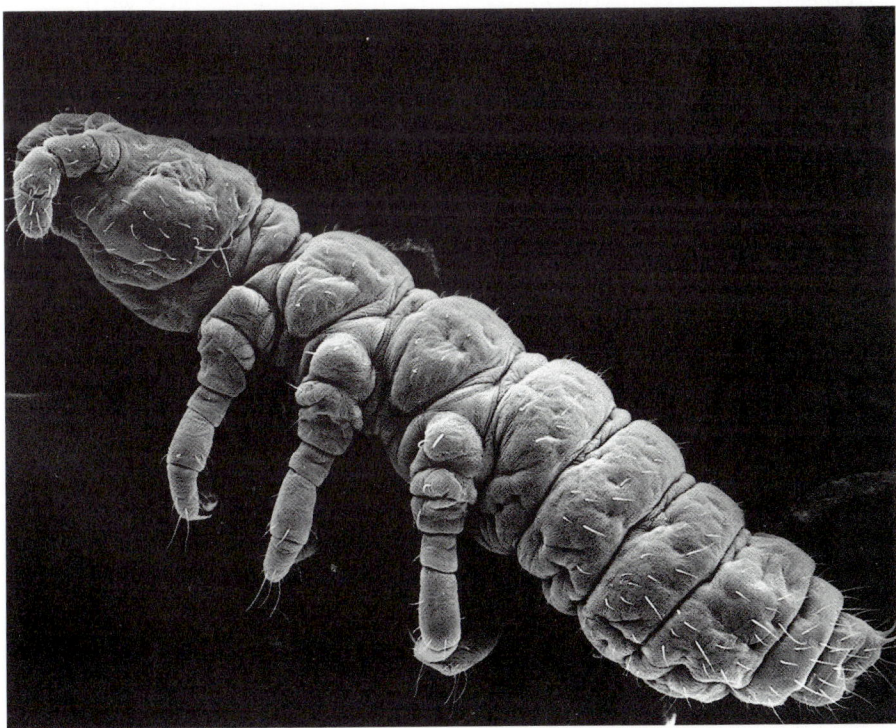

Air temperature: −20°C

Snow

Solid sea ice

Ice platelet layer

Water temperature: −1.9°C

Maintenance of breathing holes

High metabolic rate

Large size: reserves of heat, fat, oxygen

Streamlining

Body temperature: +37°C

Large, upward directed eyes

Dark-adapted eyes

Blubber insulation

Vascular heat-exchangers

Large pupils Many rod-photoreceptors Reflective retinal layer

Ice-reaming front teeth

Camouflage spots

Intense vocalisations

Echolocation?

Very long dives

High blood volume
Rich haemoglobin
Rich myoglobin in muscles
Blood reserve in spleen
Control of heart rate
Control of circulation
to selected organs

Rich milk

Rapid growth of pups

Very deep dives

Elastic ribs
Lungs collapse
Trachea collapses
Little nitrogen
absorbed
Middle ear sinuses
swell to
equalise pressure
High cholesterol
stabilises cell
membranes?

Figure 14.6. *Adaptations of the Weddell seal. Anatomical, physiological, and behavioural specialisations suit the Weddell seal to its habitat on and under the extensive sheets of sea ice bordering Antarctica.*

Plate 14.6. *The Antarctic springtail Gomphiocephalus hodgsoni, the largest terrestrial animal (about 1 millimetre long) in Southern Victoria Land, has high concentrations of polyalcohols in its body fluids. The polyalcohols inhibit freezing and enable some individuals to survive unfrozen through winter. Scanning electron micrograph: R. Eager.*

"the bends". Human divers must take special precautions, such as limiting the time spent at depths beyond 10 metres, to reduce the probability of suffering the bends. At depths of about 40 metres or more, the dissolved nitrogen has an additional direct effect on the nervous system, producing a sense of euphoria and loss of judgment known as nitrogen narcosis or "rapture of the deep".

Weddell seals suffer from neither nitrogen narcosis nor the bends. It seems that they avoid these problems in a straightforward way, by isolating the potentially lethal nitrogen from contact with the blood. Seals exhale before diving. Compression and collapse of the lungs forces any remaining gas, which is mainly nitrogen, into the bronchioles and trachea, which lack abundant blood vessels. So no further nitrogen is absorbed. Seals do not get the bends or nitrogen narcosis merely because they do not absorb large amounts of nitrogen.

Sight and sound beneath the ice

Very little light penetrates the sea ice to the waters below, which could create difficulties for Weddell seals in locating their prey. They seem to have overcome this potential problem by both anatomical and behavioural means.

The eye of the Weddell seal is very large, with a retina containing mainly rod-type photoreceptors, typical of eyes of nocturnal animals. Behind the retina, a reflective layer known as the tapetum lucidum ("bright curtain") acts much like a beaded cinema screen in reflecting light back through the retina, so the photoreceptors have a second chance at detecting unabsorbed light particles. Like the eyes of cats and possums, which also have reflective layers, those of a Weddell seal will shine brightly in torchlight or when photographed with a flash. In dim light the pupils can widen considerably, admitting more light, and also contributing to the seals' psychological appeal to humans.

Rather than being directed forward along the snout, the eyes of the Weddell seal are rotated upwards, so that the resting position of the seal's eyes is at about 15° to the axis of its head. The seal can easily look directly upwards without moving its head from the horizontal plane, or interfering with streamlining. Thus, a cruising seal is naturally designed to view the undersurface of the ice above, in contrast to a human diver, who looks downward. This upward vision may also explain why feeding dives are so deep. In the very clear water of early summer, seals are able to scan several hundred metres of water overhead to detect their prey silhouetted against the ice. From our own experience diving over deep water under ice, nothing is visible in the blackness below, although the ceiling of ice seems deceptively close and clear.

To supplement their vision, Weddell seals also have a remarkable underwater vocal repertoire, ranging from whistles and trills to chirps and grunts. Some of these sounds can be so loud that they are heard (or felt) through several metres of sea ice, and show as clear echoes on hydrophone recordings. Vocalisations are known to play a role in the seals' social interactions. Particularly during the mating season (mid summer), dominant male seals establish territories centred on breathing-exit holes near a group of females. They spend much of their time patrolling their underwater territories. Potential rivals are driven away with vocal warnings accompanied by bites to the rear flippers and genital region.

It has also been suggested that vocalisations may be used in echo-location, as in bats and porpoises, but so far we lack conclusive evidence. However, a lone seal could probably detect the sounds of a colony some distance away, and use them as a navigational beacon.

Figure 14.6 summarises the anatomical, physiological, and

behavioural adaptations of the Weddell seal to suit its habitat.

Terrestrial animals

Away from the stable environment of Antarctic sea water, conditions are much more severe. In the terrestrial environment, temperature and humidity, as well as light, vary seasonally, daily, and even hourly. The difficulties of life on dry land are reflected in very low productivity and low diversity of terrestrial ecosystems, in contrast to the richness of the marine environment. Microscopic animals such as ciliate protozoans, rotifers, tardigrades, and nematode worms are found in freshwater ponds and rivulets, and sometimes amongst unicellular plants in moist soil. However, the largest terrestrial animals on the Antarctic continent are arthropods, of which the most advanced are midges found only on the Antarctic Peninsula. In the McMurdo coastal region, arthropods are represented by three species: one primitive wingless insect, the collembolan (springtail) *Gomphiocephalus hodgsoni*, and two mites, *Stereotydeus mollis* and *Nanorchestes antarcticus*. The largest individuals of the springtail *Gomphiocephalus* are about 1.5 millimetres long, and the mites are only about one-third as big, so these are not spectacular animals and they are hard to find.

Other springtail and mite species are found further from the sea, in even harsher conditions in the South Victoria Land mountains; yet other species occur in North Victoria Land. The evolutionary origins of these discrete groups of species is a fascinating but unanswered question.

Despite their small sizes, these arthropods are complex animals, with nerves, muscles, sense organs, and other organ systems comparable to those found in fish, seals, and humans. Most of the work on Antarctic arthropods has dealt with their taxonomy, distribution, and ecology, but we are beginning to learn how the Antarctic springtails and mites cope with the rigours of their environment.

Avoidance of freezing

For much of the year, Antarctic air temperatures are well below freezing. Midwinter temperatures in the McMurdo region average -25 to -30°C, occasionally dipping to -40°C or lower. When faced with extremely low temperatures, terrestrial animals have two options. The first is to develop a tolerance to freezing, and to spend the winter literally frozen solid. To do this, some animals seem to be able to initiate freezing in selected sites, and to control ice crystal growth so that their cells are not damaged. A second option is to avoid freezing.

Freezing is lethal for both the McMurdo springtail and the mite *Stereotydeus mollis*. Many springtails die during winter, but some representatives of all life stages (eggs, nymphs, and adults) seem to survive winter, and thus must somehow avoid freezing. In favourable sites such as Cape Bird, there may be as many as 1000 springtails per square metre in early summer, rising to nearly 5000 per square metre in late summer. When tested in mid-summer, both the springtail and the mite show significant resistance to freezing, and many individuals can be cooled to -25 or -30°C. At these temperatures, both mites and springtails (Plate 14.7) are supercooled well below their normal freezing points. Supercooling is probably aided by high concentrations of polyalcohols such as glycerol, alcohol derivatives of sugars (e.g., mannitol), or sugars themselves (e.g., trehalose), which have been found to serve as supercooling agents in terrestrial arthropods from Signy Island (and also in some Arctic and alpine insects). Supercooling capability is reduced when animals are given water (which would dilute the supercooling agents), or when they have recently fed, which would create particles that would act as growth centres for ice in the gut. It seems likely that supercooling ability increases in winter.

Microhabitats

Although the McMurdo springtail can survive temperatures down to -30°C or lower, it is active only above 0°C, prefers temperatures between 10 and 12°C, and is adversely affected by temperatures above 17°C. Very little is known of temperature preferences for the few species living at even higher latitudes and/or elevations. The optimum temperature for the springtail is surprisingly high for one of the southernmost insects, and suggests that this species may be a recent immigrant, not yet completely adapted to conditions of high Antarctic latitudes.

The availability of water is very important to Antarctic terrestrial arthropods. Water, rather than temperature, is probably the dominant factor determining their distribution. Likely habitats for springtails and mites include north-facing slopes with perennial snowdrifts or ice faces furnishing a regular supply of meltwater, and areas with water-retaining clay substrates. The springtail is not found in saturated conditions, but in soils with moisture contents between 3 and 12 percent; mites prefer wetter conditions.

Both temperature and water problems can be alleviated by the selection of suitable microhabitats (Figure 14.7). Springtails are found most commonly near the air–soil boundary around large, partially buried stones (about 50 millimetres in diameter), in the top 25 millimetres of soil. The stone serves several functions. During the day it is a heat sink, preventing rapid temperature changes. Water vapour tends to condense on the stone's lower surface, locally raising the humidity. And, finally, the stone supplies physical stability by protecting the springtails from high winds. Springtails can be blown away, and this may be partly how they have become dispersed throughout the McMurdo region. In warm, still weather, both springtails and mites emerge from under their stones and appear on the surface, often in large numbers in patches of lichen or moss.

Parasitic arthropods

Other arthropods—fleas, lice, ticks, and mites—are known from Antarctica, as parasites of birds and seals. These arthropods have chosen warm and stable microenvironments on their host species. They are not free-living terrestrial animals, but their parasitic life style is yet another way of overcoming the extremes of the Antarctic climate.

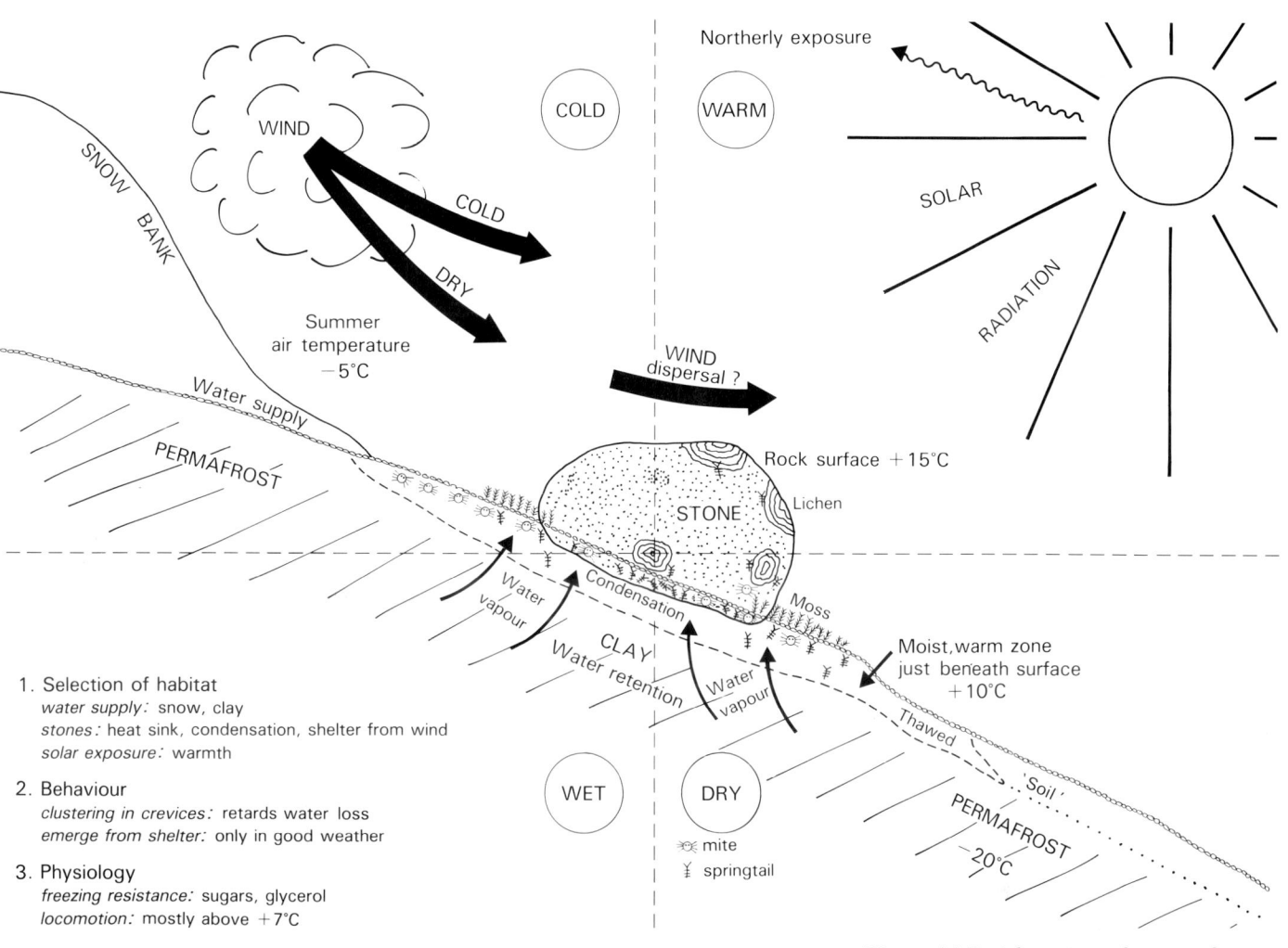

1. Selection of habitat
   water supply: snow, clay
   stones: heat sink, condensation, shelter from wind
   solar exposure: warmth

2. Behaviour
   clustering in crevices: retards water loss
   emerge from shelter: only in good weather

3. Physiology
   freezing resistance: sugars, glycerol
   locomotion: mostly above +7°C

Figure 14.7. *Adaptations of terrestrial Antarctic arthropods. Few terrestrial animals are known from high latitudes in Antarctica. Those that are known have taken advantage of favourable local conditions. Although they can survive low temperature and desiccation, they are only active for a brief period in summer when heat and moisture are available.*

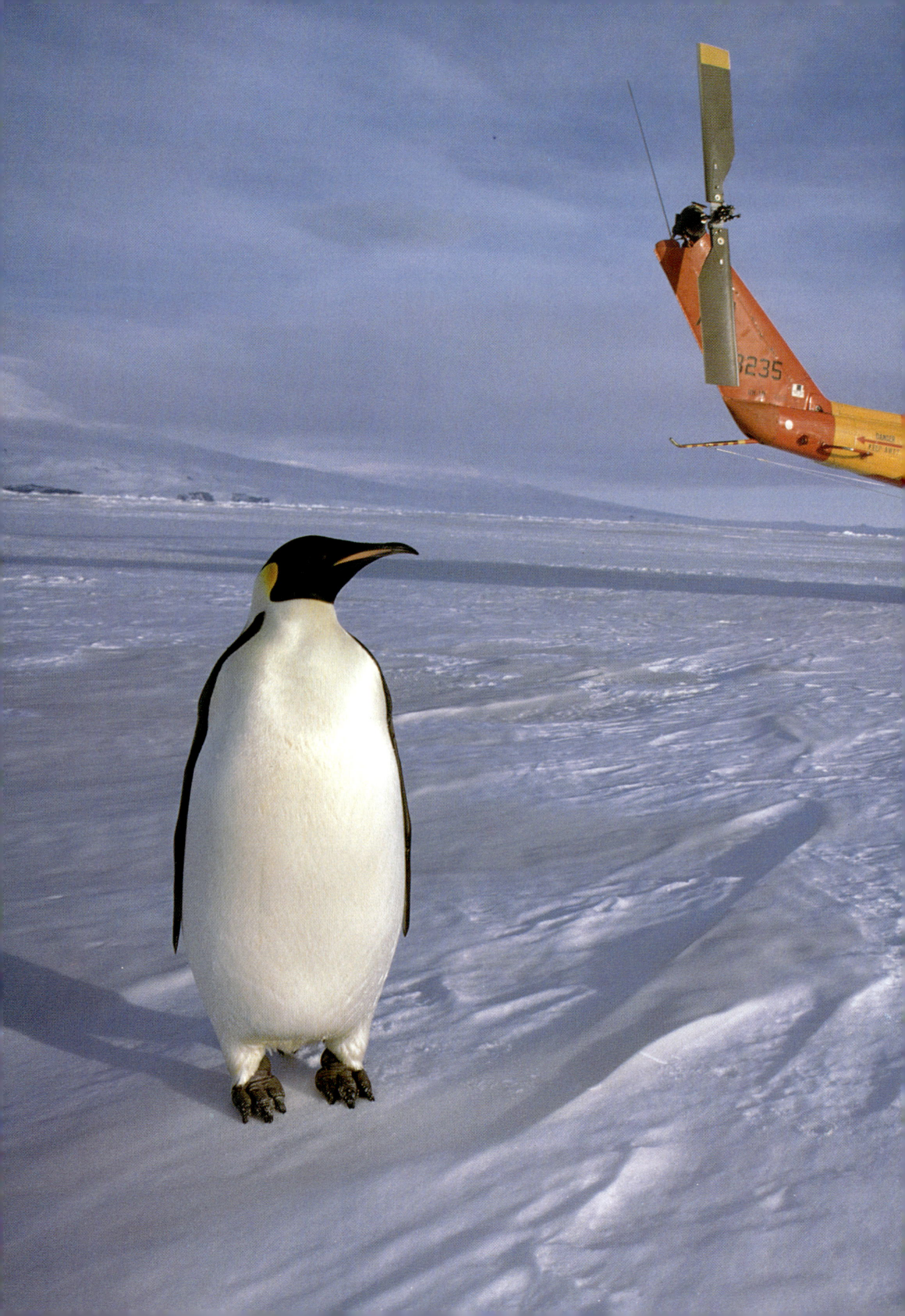

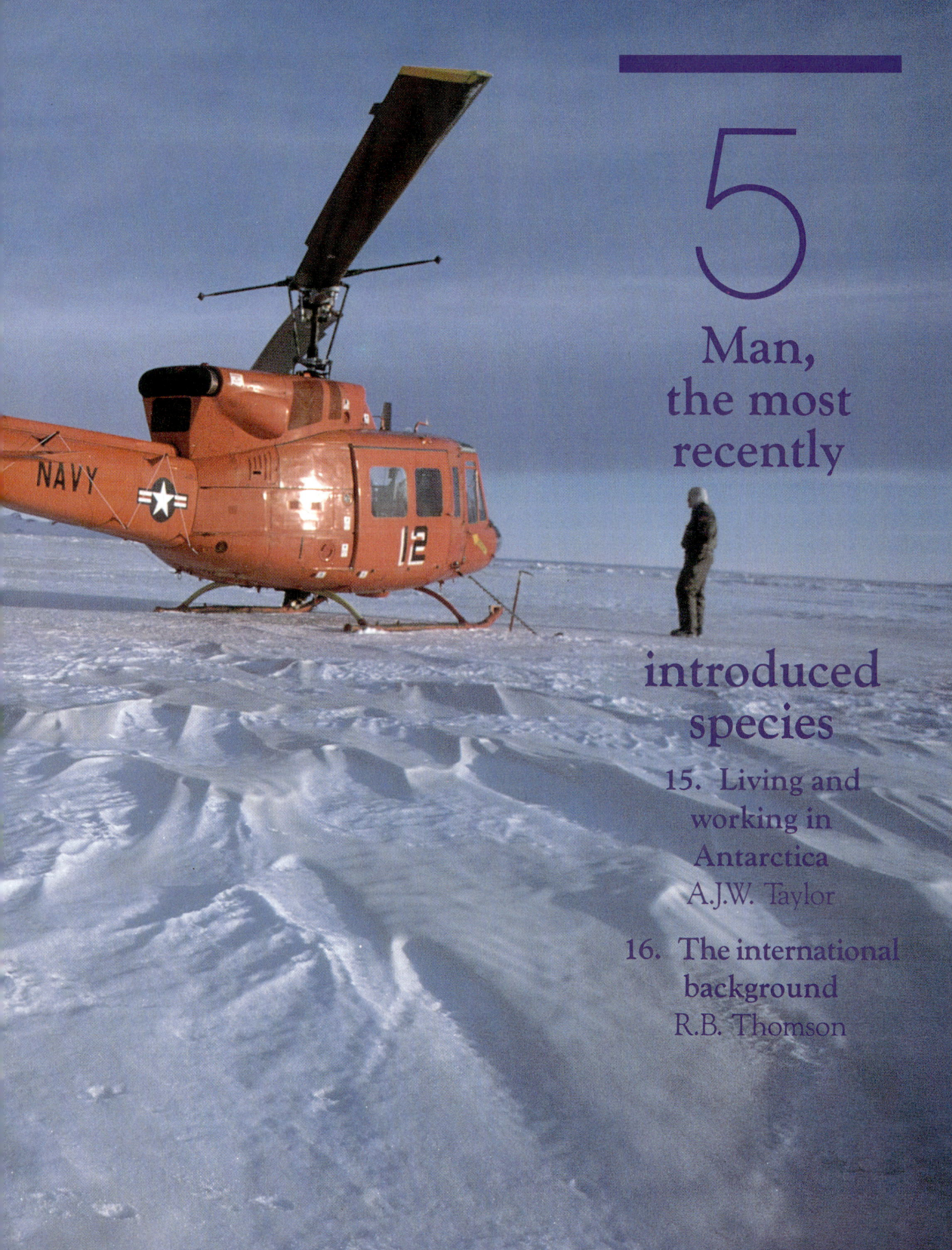

5

Man,
the most
recently

introduced
species

15. Living and working in Antarctica

Previous chapters have shown how the "White Continent" exerts a fascination over scientists and romantics alike (Plate 15.1). That was not always so. When on 30 January 1774, after a sojourn in the tropics, James Cook came closest to Antarctica, he wrote in his diary that "whoever has the resolution and perseverance to . . . [proceed] . . . further than I have done, I shall not envy him the honour of discovery, but I will be bold to say that the world will not be benefitted by it". He even doubted that anyone would be able to get further south, because he said "ambition leads me not only farther than any other man has been before, but as far as I think it possible for any man to go".

Obviously James Cook, the intrepid explorer, had gone as far as his logistics and technology would allow. Undoubtedly, he would have gone further if he had the same ships, communications, and supply bases of his successors. But the one factor that Cook had in common with those who immediately followed him was his character—a combination of ambition, determination, endurance, heroism, and selflessness. Those who came later were among the first to acknowledge that it was not so necessary for them to have an abundance of these qualities, because they had such superior access, accommodation, equipment, and transport.

Access to Antarctica in the early days entirely depended on the vagaries of sea ice. In some years these barriers were less extensive than in others, but in every year they yielded only in summer. Sailing ships went as close to land as they could get—unless like the *Belgica* they were set to drift firmly in the sea ice. Crews built temporary huts ashore for emergencies as well as to establish scientific laboratories and to relieve the monotony of shipboard life.

Typically, the crews would become familiar with their environment in the first summer, decrease their activities and do maintenance work in winter, and standby for their first real sortie in the following spring. Similarly, the scientists would have seasonal activities—first collecting specimens and setting up equipment, then working on their finds and planning experiments and field studies, before finally being free in spring to join exploration and mapping parties with which they would carry out their research.

Thus, 20 or so men would travel to the Antarctic and expect to be there for about 2 years before returning. Each man had a specific job (Plate 15.2) and each was dependent on helping the others to function efficiently and harmoniously (Plate 15.3). They committed themselves to danger, deprivations, discomfort (Plate 15.4), and hardship, in return for a sense of achievement, curiosity, duty, fame, fulfilment, honour, and notoriety.

Some of the early Antarctic explorers were not entirely ignorant of the life they faced, because they had either their own Arctic experience or the journals of others on which to draw — and their observations have a familiar ring today. For example, in 1835 while he was wintering in the far north, and before he went south into the sea that was named after him, James Clark Ross wrote a revealing account of his winter:

"We were much refreshed [at meeting Esquimaux] after continual self-occupation and the unvarying society . . . We had had our fill of anxiety and worry, of the sadness of frustrated hopes, and, above all, of the longing for our distant friends and home. What voyager can possibly be free from such moods? But there was something still harder to bear, and that never ceased. We were so terribly weary and dejected for lack of work, lack of variety, lack of mental occupation, lack of thought, and—why should I not say it?—lack of society. The everlasting monotony bore heavily on our spirits, and our minds grew weary for lack of incentive. The small events which occasionally happened were only a repetition of what we had so often experienced. Being imprisoned in one spot for so long, there was nothing more to interest us.

The very sight of the ice was to us an annoyance, a torture, as martyrdom, something vile, something to reduce us to desperation ... we hated the sight of it because we hated the consequences; and everything that concerned it, every thought connected with it, was an abomination.

We hoped and feared; we were nearly driven to despair, till we were able to foresee release and success."

To counter these reactions, Julius Payer, the Austrian Arctic explorer, declared in 1877 that in his opinion the entire

"... surrender of his individuality is an essential sacrifice that the Polar explorer must make to achieve his end. The job which has been apportioned to him admits of no party standpoint, still less of varying opinions as to taste and comfort. All irregularities of spirit, mind, or body must be subdued to a level of stoic apathy. He must become meek, keep his countenance at the daily reiteration of the dinner-table topics, the oft-described adventures, which never die, in spite of their staleness. His thoughts must be free from the dross of earthly desires; his self-consciousness must be buried under a mountain of respect towards the interests of peace ... Our own imagination was also crippled; even the most exciting story possessed nothing for us but the gravity and melancholy of a legend."

Yet another Arctic veteran, Lindhard, the Medical Officer to the 1910 Danish expedition in Northern Greenland, described the sleep problems and the inertia that he found during the winter:

"For many, sleep was difficult, broken and accompanied by dreams; one often lay for hours without being able to fall asleep. Heaviness followed in the morning, as if after a night on the watch; endeavours during the day to make up for the sleep lost resulted simply in a worse night afterwards ... Both the desire and the capacity for work were reduced, especially the latter and especially for brain-work ... The wish was present to begin work seriously; the pipe was lit and the pen filled, but one got no further ... One felt irritable and not disposed to be friendly, and the same qualities magnified were read into others. Conversation at the table had a tendency to become fragmentary or even to assume a pointed form ... There can be no doubt, however, that the polar night, the cold and the dark, especially the dark were indeed the true causes. The same discomforts were all present in the summer, though in less degree, and yet one's existence both outside and inside was quite different during the light period. We did not feel so overcome and crushed by the natural forces as in the dark period ... There was the desire to be active and we could work. We went unwillingly to bed, preferring to sleep just as the occasion offered; the sleep was short yet refreshing. One had the feeling of never resting and never tiring, which contrasted strongly with the physical helplessness of the dark period."

Subsequently, a whole generation of articulate and sensitive Antarctic explorers made similar first hand reports. Among the most illuminating was that written by Richard Byrd in 1934:

"When there is no growth or change outside, men are driven deeper and deeper inside themselves for materials of replenishment, and on these hidden levels of self-replenishment ... would depend the ability of any group of men to out-last such an ordeal and not

Plate 15.1. *Sunset from Razorback Island in McMurdo Sound. Photo: S. Sandblom.* ▶

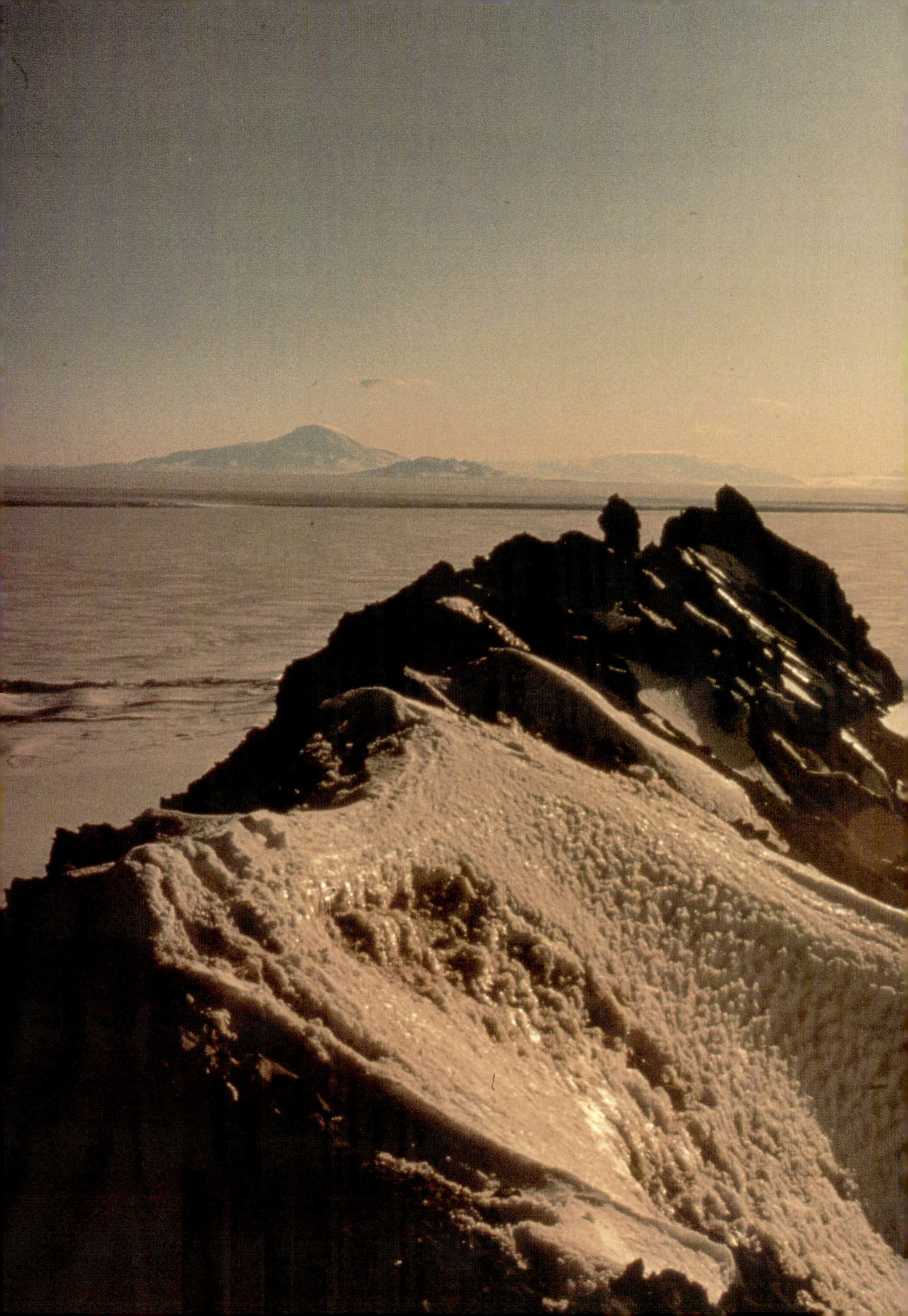

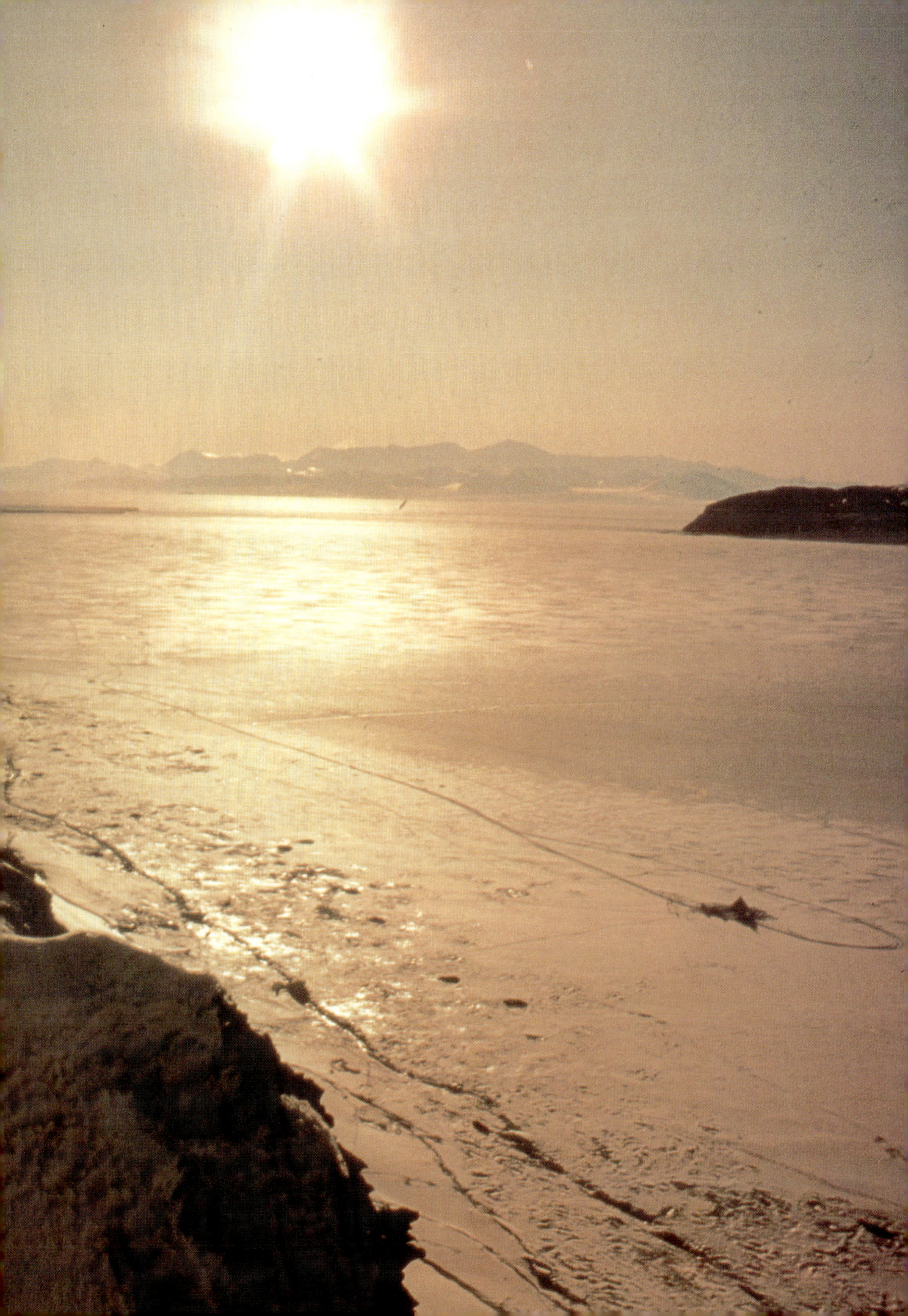

Plate 15.2. *A makeshift shelter, before the days of plastic domes, provides protection for Lieutenant Evans as he observes an occultation of Jupiter on 8 June 1911. Photo: Alexander Turnbull Library, Wellington.*

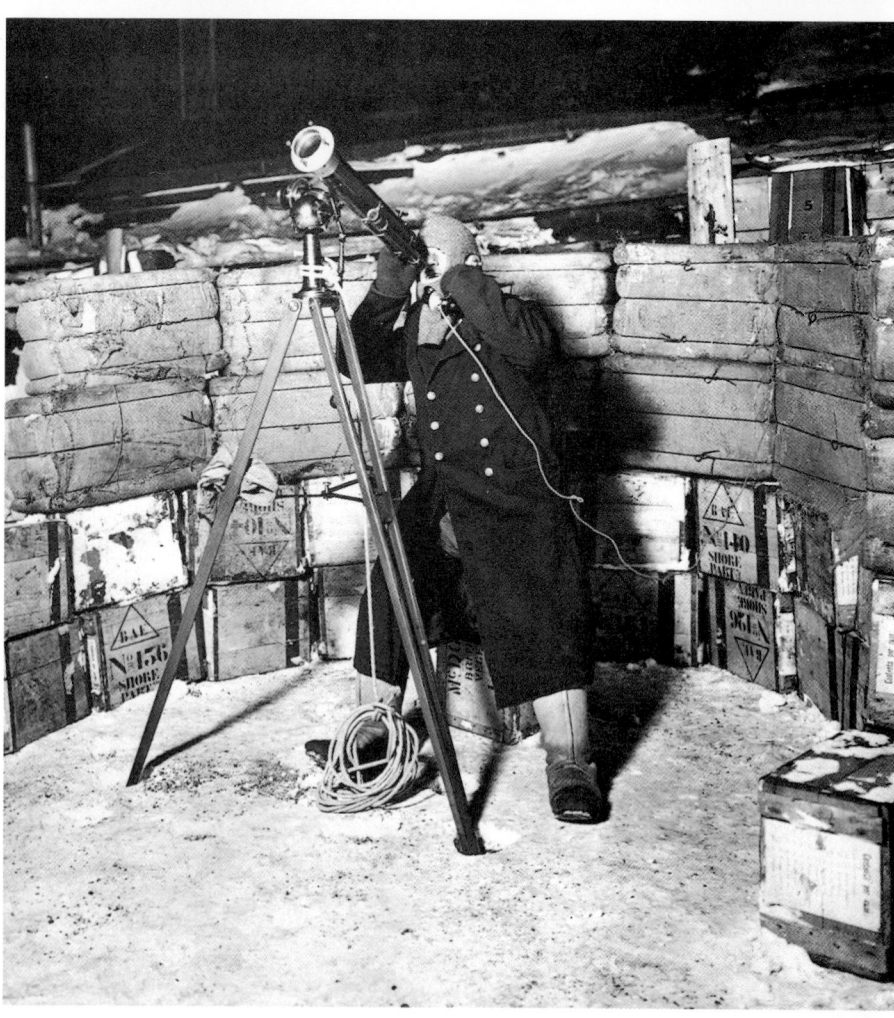

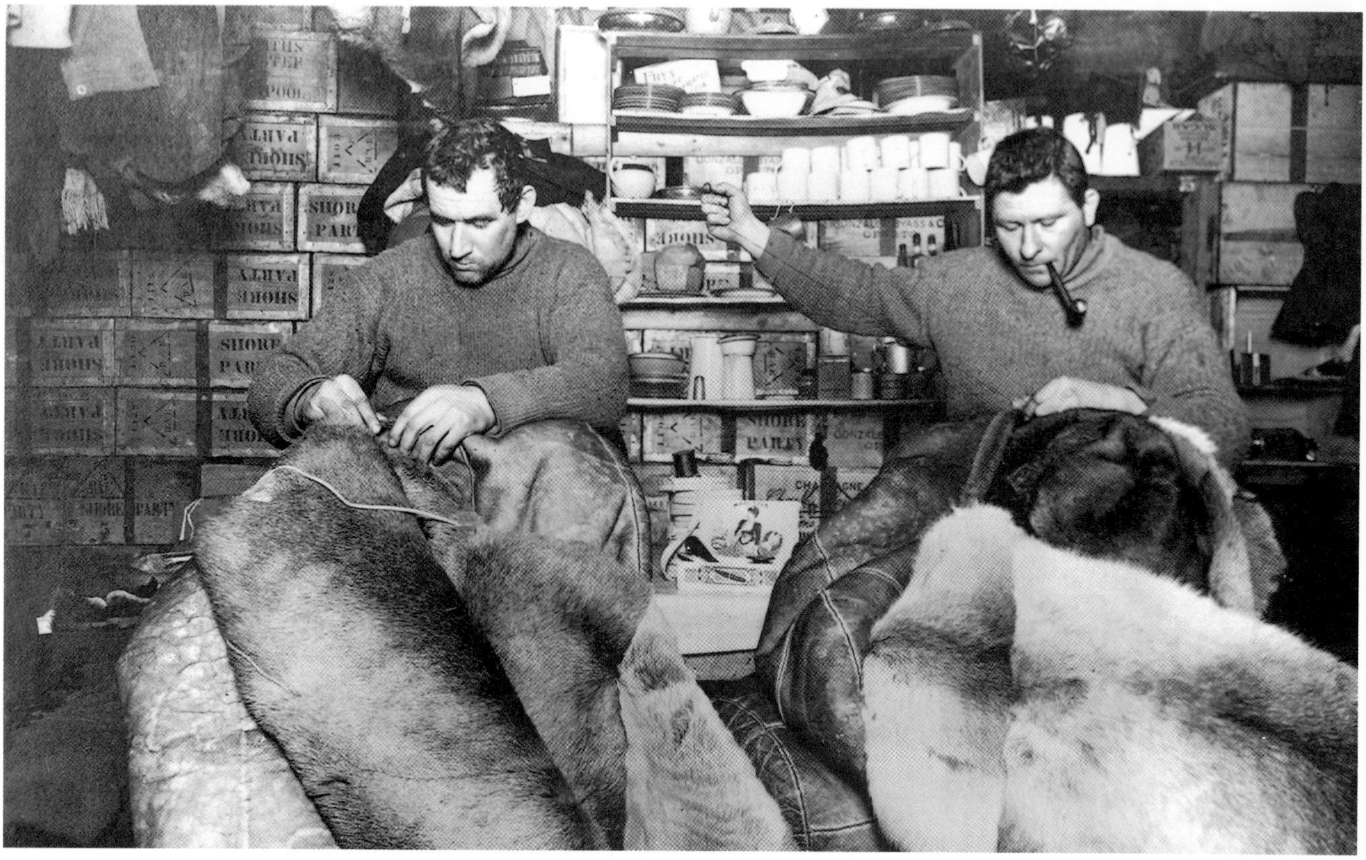

Plate 15.3. *Thomas Crean and Petty Officer Evans repairing sleeping bags, one of the many jobs which kept the early expeditioners busy during the winter months. Photo: Alexander Turnbull Library, Wellington.*

come to hate each other . . . For there is no escape anywhere. You are hemmed in on every side by your own inadequacies and the crowding pressure of your associates. The ones who survive with a measure of happiness are those who live profoundly off their intellectual resources as hibernating animals live off their fat."

Already by the time Byrd had written this there had been improvements in the technology used by the expeditioners. Ships were mechanised, radio communications were established, and aircraft had been introduced. Within another 25 years there were icebreakers, diesel tractors, and well designed, if still spartan, permanent bases on the continent.

These changes in access, movement, and living conditions were accompanied by changes in the selection of crews, in the duration of their assignments, and in the method of getting them to and from Antarctica.

For the first time a diversity of scientists from different specialities were able to go themselves for shorter periods to Antarctica, instead of getting a general scientist to work there on their behalf. They were now able to do specific short-term projects, make their observations, and depart to process the results. They were also able to have technicians gather data during the winter and send it to them in their home laboratories.

The growing number of summer field workers today, therefore, face few of the trials, uncertainties, and bleak long-term prospects of the early pioneers. They can still be confined to their tents for days by blizzards, and they still have to be wary of crevasses on their travels, as well as thin sea ice. But, mostly, they can look forward to short bursts of intensive field activity during 24 hours of daylight in which their only restraint will be the need to get adequate rest.

The increase of New Zealand based rather than Antarctic based scientists has affected the nature and purpose of Scott Base itself. No

Plate 15.4. *There was little privacy for the early expeditioners as this illustration of the "tenements" at Cape Evans in 1911 shows. Photo: Alexander Turnbull Library, Wellington.*

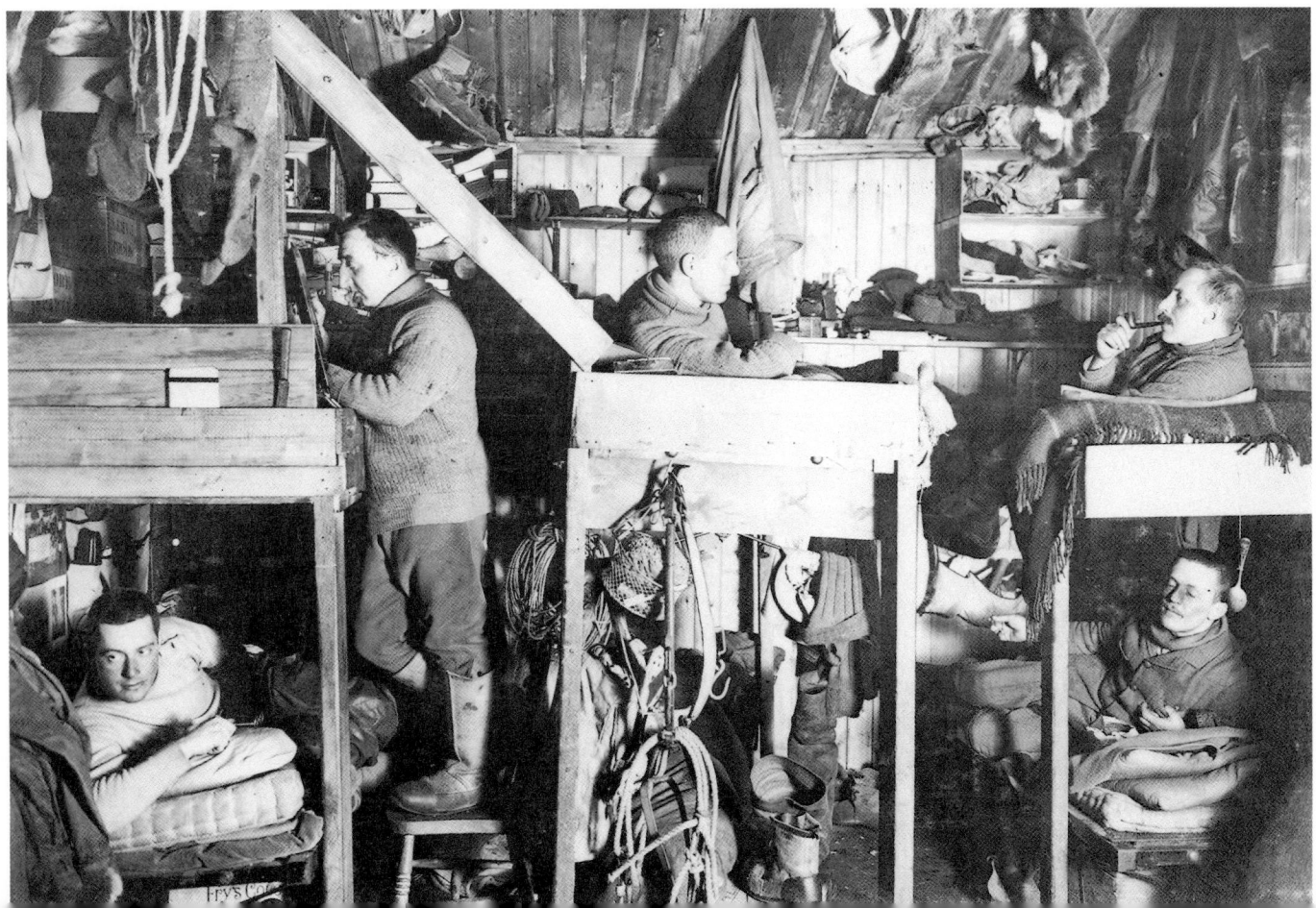

longer is it the home for a small isolated group for a long uncertain period. Instead, it is home only for the communications, laboratory, and maintenance staff for a full 12-month season, and a hostel for many more transient field staff. It is a staging post for the scientists and field workers between their laboratories back home and their areas of study. They use it as a place for learning to pack and stow their gear in sledges, to practice survival skills, and from which to spread out by skidoos, helicopters, and aircraft (Plate 15.5).

What sort of people went to Antarctica in the past, and what sort of people go there now?

Personnel selection

In the old days polar personnel were selected firstly with the appointment of competent and charismatic expedition leaders. The leaders were then free to use their own preferences for the type of person to accompany them. Nordenskjold, for example, looked for companions who would be "homogeneous and make living at close quarters tolerable". Amundsen sought highly experienced dog handlers and skiers who were nimble and mobile, "with a flamboyant fatalism to which risk appeals". Mawson, himself an academic geologist, selected graduates "with the dash and recuperative power of youth". Scott, in his first (1901–04) expedition, preferred a crew from the Royal Navy, because he felt sure that their sense of discipline would be an advantage and he had "grave doubts" about his ability to deal with civilians. Although he did not manage to have his way in the beginning, Scott found it "curiously satisfactory" and "a great matter for congratulation" when four merchant seaman volunteered to return home half way through the expedition on the relief ship with Shackleton, who was sick. When the time came later for Shackleton to select his own men, he chose those who were "qualified for the work . . . able to live together in harmony . . . have generally marked individuality." And when he advertised for

Plate 15.5. *Geological field party being "put in" to the Ohio Range area by ski-equipped C-130 Hercules aircraft. The field assistant is reporting their safe arrival back to base by radio. Photo: N. Roberts.*

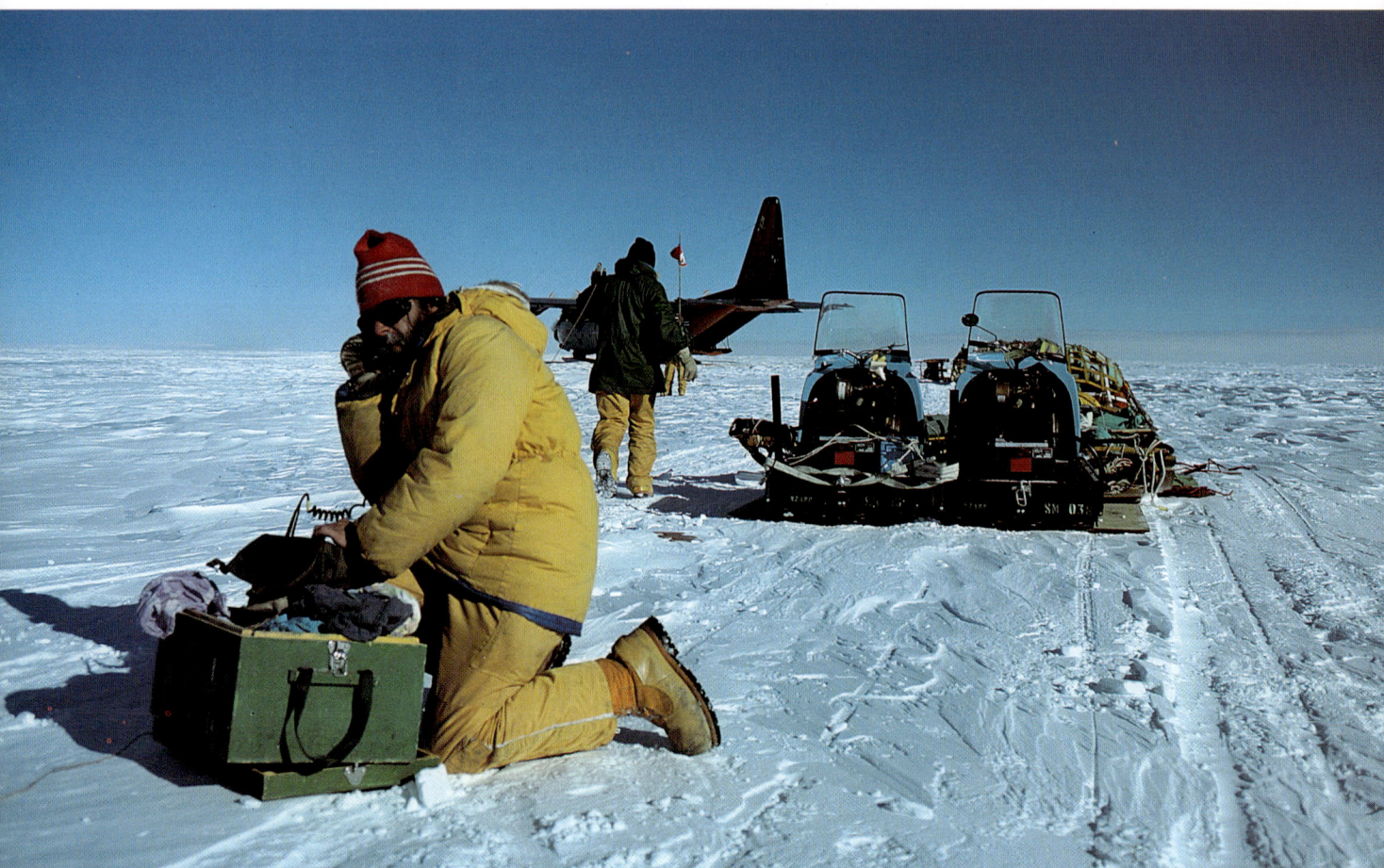

volunteers in 1913, Shackleton said simply " . . . men wanted for hazardous journey. Small wage, bitter cold, long months of complete darkness . . . safe return doubtful . . . and recognition in case of success." He had only 28 vacancies, and he drew 5000 applications for them!

Despite the improvements in the equipment and facilities that modern Antarcticans enjoy, and the changes in their type and period of work, they still need to have certain qualities of personality if they are to acquit themselves well. These qualities can be described as *ability*, *stability*, and *compatibility*, and selectors hope that these qualities would either be balanced within individual members or within their parties as a whole. *Ability* refers to the level of competence of a person at a given task and the versatility to overlap on other tasks when necessity demands. *Stability* concerns the degree of self awareness, self acceptance, and self control a person has which allows for the moderation and expression of impulses and needs without the loss of functional efficiency. *Compatibility* denotes the degree of respect, support, and tolerance a person has available to extend towards others in a group. Without compatibility there is insufficient group cohesion and group morale to counter adversity. With too much compatibility there is too great an emphasis on the group at the expense of an individual. As Schopenhauer said, we are like porcupines . . . we need to be sufficiently far apart to preserve our individuality but sufficiently close for warmth and security.

Good selection is not the only safeguard for satisfactory performance. Personnel still need to be given information about Antarctica and to be encouraged to counter the continent's adverse mental and social effects. They need to be encouraged to plan personal programmes of hobbies and activities as "food for the mind" in times of impoverishment (Plate 15.6). They need to work out informal procedures for ventilating grievances and generating mutual support.

Today, before they depart for Antarctica, New Zealand personnel

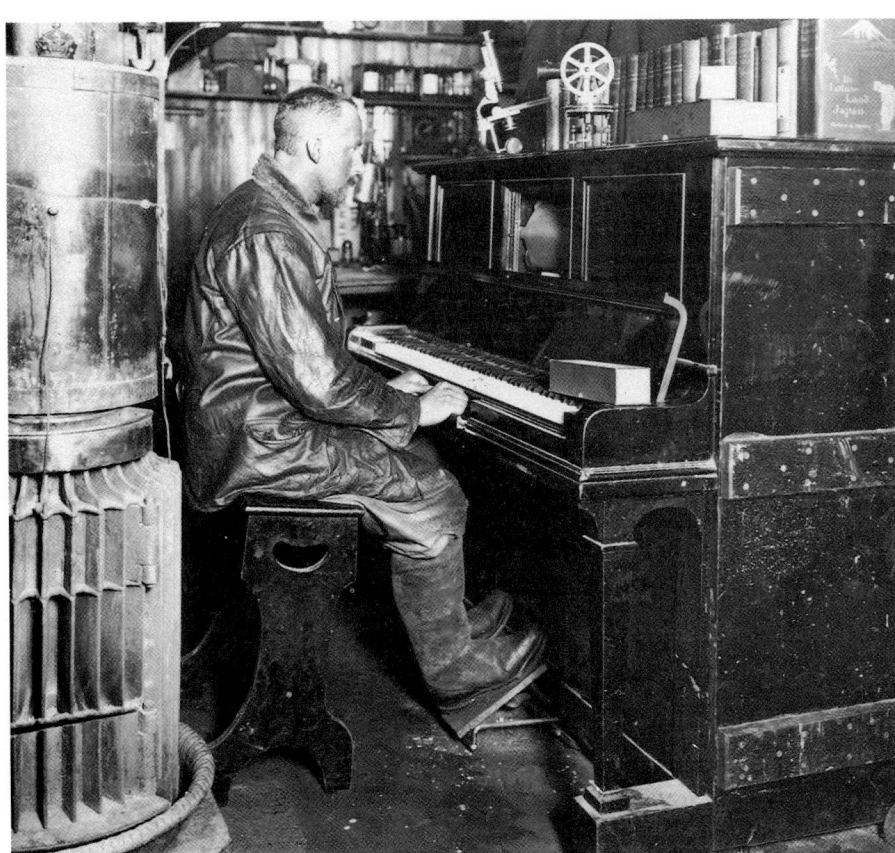

Plate 15.6. *In the early expeditions men had to make their own music. Here Meares plays the pianola at Cape Evans in 1912; the instrument was returned to its makers in London at the end of the expedition. Photo: Alexander Turnbull Library, Wellington.*

(a)

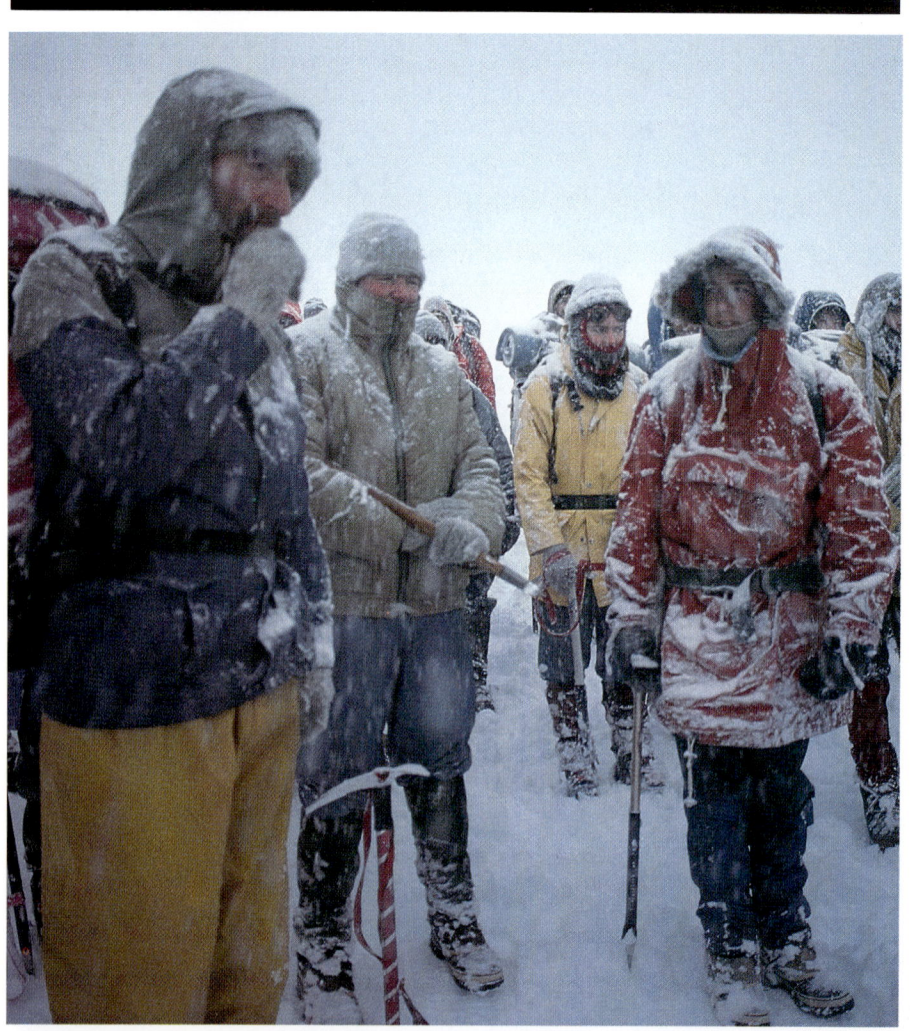

Plate 15.7(a), (b). *Antarctic personnel undergo field training at a camp on Mount Stevenson in New Zealand. Photos: C. Rudge, Antarctic Division, DSIR.*

(b)

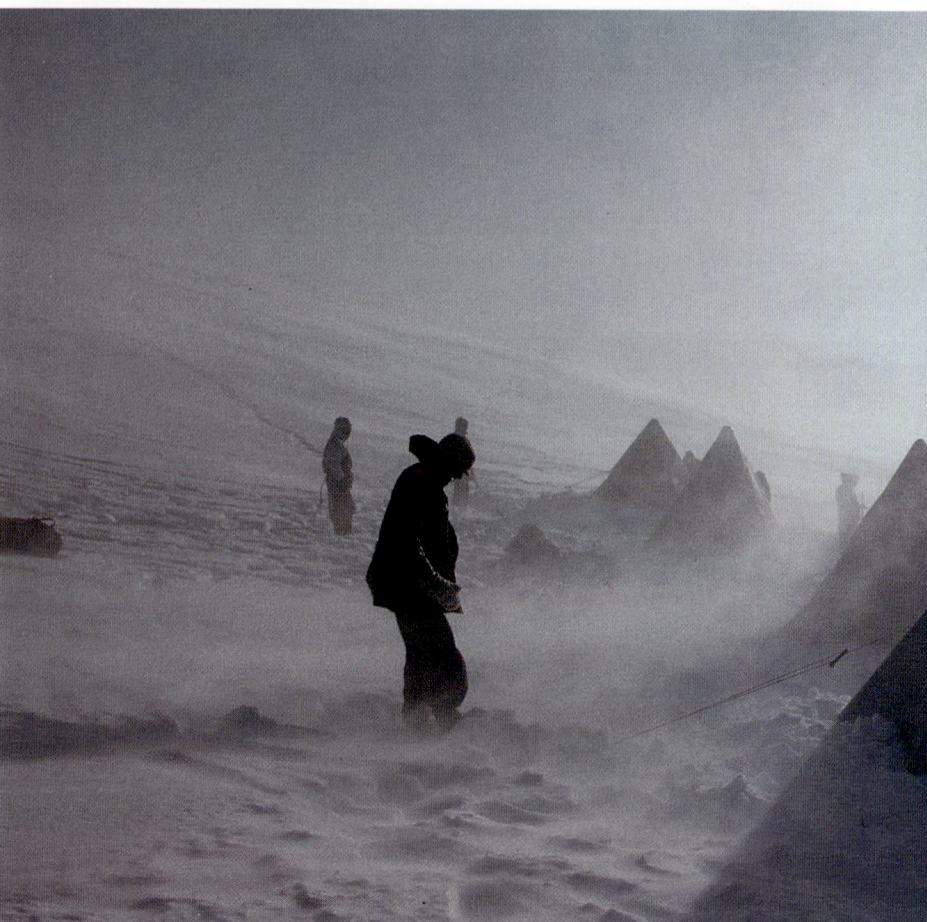

are given extensive field training in New Zealand so that they are accustomed to living and working in the hazardous environment (Plate 15.7a, b). In addition, all members of the N.Z. and U.S. Antarctic research programmes in the Ross Sea region likely to be involved in any field activity have to undertake a survival training course; this is carried out amongst ice cliffs and crevasses near Scott Base.

Leaders these days can no longer rely on the authority and charisma of Scott, Shackleton, Amundsen, and Byrd. They must have managerial skills and a knowledge of group dynamics, so that they can maintain objectives and bring unity out of any discord which might arise. Not that the earlier leaders were lacking in these qualities but their tasks may have been made easier by the nature of their participatory amusement activities. In contrast with the present availability of movies and videos and other such passive forms of entertainment in non-working hours, earlier expeditions produced substantial, imaginative, and beautifully illustrated newspapers and books such as the 120-page "Aurora Australia" of Shackleton's 1907–09 expedition, and the *South Polar Times* (Plate 15.8) of Scott's two expeditions. The *Discovery* personnel also held concerts and dramatic performances in the building at Hut Point. Hockey on the ice was a great diversion for them until it ceased in April for lack of light. "Sometimes the officers play the men, sometimes we divide by age limit, and sometimes in other ways. Today it has been "Married and Engaged v. Single" and as the former side lacked numbers we had to include in it those who were accused of being engaged, in spite of protest. The match was played in a temperature of –40°[C] and it was odd to see the players rushing about with clouds of steam about their heads and their helmets sparkling with frost." These days outdoor sport in the vicinity of Scott Base is invariably rugby (Plate 15.9), although various more individualistic sports are available.

Now as before, the winter is broken by special occasions such as birthdays (Plates 15.10, 15.11). Midwinter's Day (June 21 or 22 according

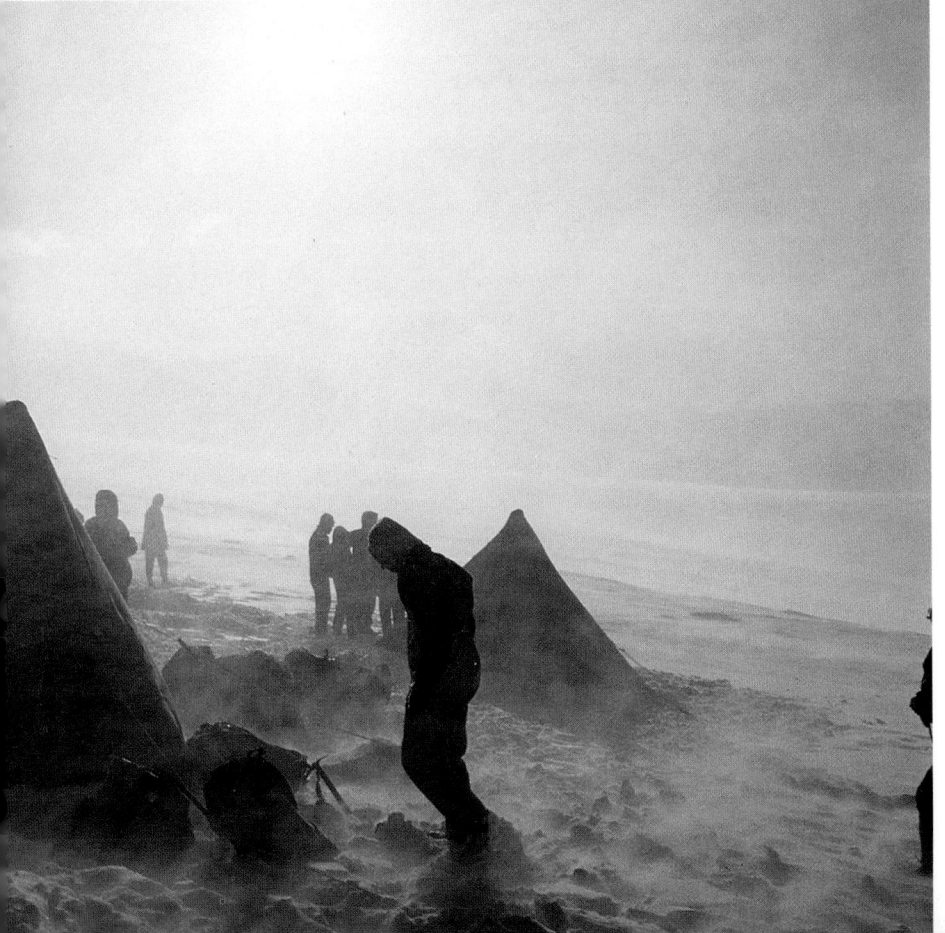

252

Plate 15.8. *A page from the* South Polar Times *produced in winter throughout both Scott's expeditions.*

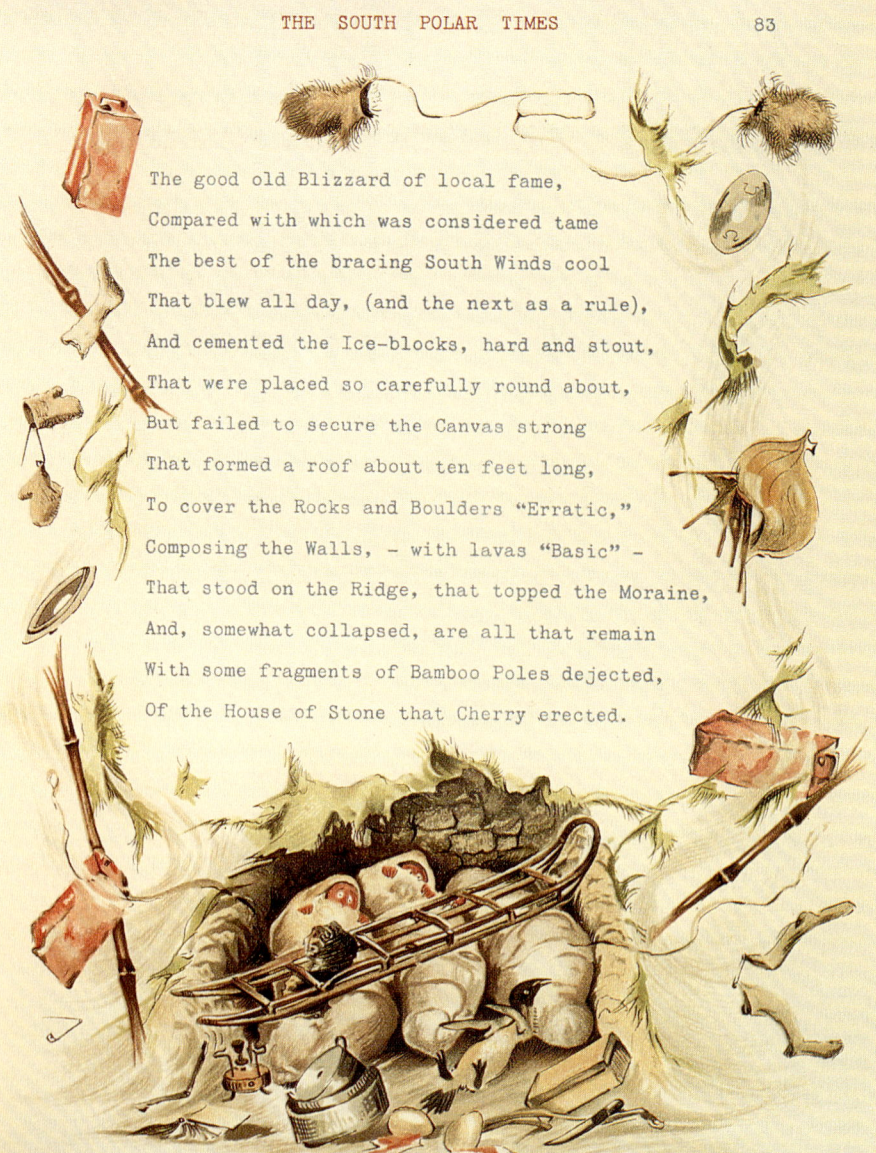

The good old Blizzard of local fame,
Compared with which was considered tame
The best of the bracing South Winds cool
That blew all day, (and the next as a rule),
And cemented the Ice-blocks, hard and stout,
That were placed so carefully round about,
But failed to secure the Canvas strong
That formed a roof about ten feet long,
To cover the Rocks and Boulders "Erratic,"
Composing the Walls, – with lavas "Basic" –
That stood on the Ridge, that topped the Moraine,
And, somewhat collapsed, are all that remain
With some fragments of Bamboo Poles dejected,
Of the House of Stone that Cherry erected.

Plate 15.9. *A game of rugby, "Scott Base" versus "Kiwi Caterers". Photo: L. Harrison.*

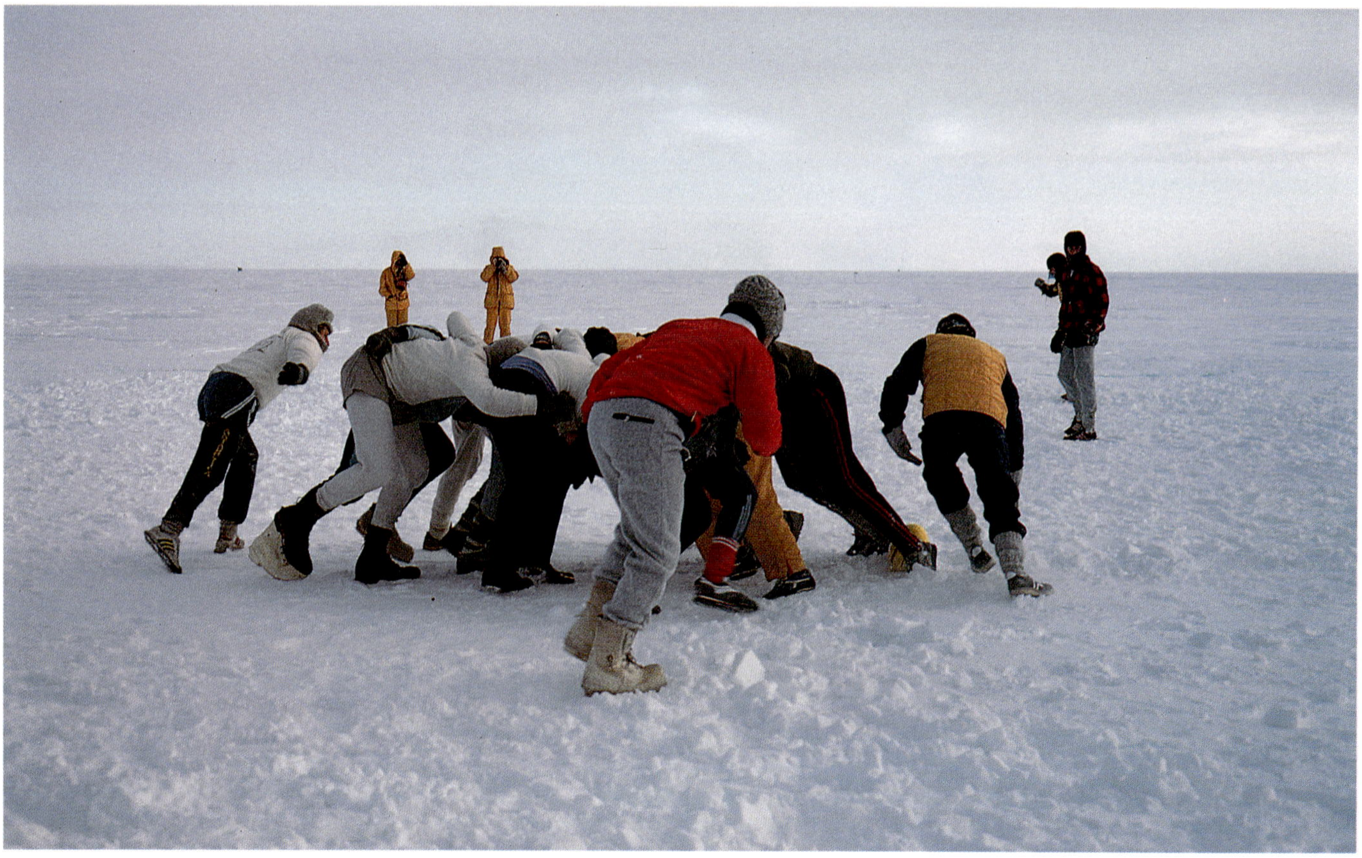

to inclination) is another event which is celebrated, and it leads to annual reunions of wintering groups for the rest of their lives. The first appearance of the sun in late August, after 4 months' absence, is another cause of rejoicing (Plate 15.12). "I am not sure that a polar night is not worth living through for the mere joy of seeing the day come back", wrote Scott after his first winter in McMurdo Sound.

There have been a few extreme instances of isolation by accident or choice. For example, Scott's northern party of six men, after wintering at Cape Adare in 1911, were landed by ship at Terra Nova Bay during the succeeding summer, with a further rendezvous planned with the ship to be withdrawn 6 weeks later. March passed with no sign of the ship and by mid-April it was obvious that they would have to survive somehow through an Antarctic winter with no food supplies and no hut. They were in an appalling situation, but they were not appalled (Plate 15.13). They dug a cave out of solid ice and killed and stored all the seals they could find, but even so they were half starved. For 9 months they could not wash or change their clothes. But on 30 September 1912 they set out to trek down the coast to McMurdo Sound, reaching Hut Point on 6 November after a journey of 400 kilometres. As L.B. Quartermain wrote: "Few parties under anything approaching these conditions can ever have preserved such high morale and such unruffled good humour as did this curiously mixed group consisting of two naval officers, a scientist, two petty officers and a seaman". Byrd, on the other hand, *chose* to winter alone during 1934 at the Bolling Advanced Weather Base some 200 kilometres polewards from Little America. He went there on 22 March, and for over 2 months all went well. But on the last day of May he collapsed, probably due to fumes from his generator, and he never really recovered his former physical and mental state. His deteriorating condition was diagnosed at Little America from the nature of his radio transmissions, and, after several attempts to relieve him under very difficult winter conditions, he was finally reached on 11 August.

Few Antarcticans are likely to be so isolated and exposed as Scott's Northern Party and as Byrd, but accidents sometimes happen, and rescue and recovery might be delayed. More generally, modern Antarcticans are likely to show some mild affects of isolation that can be insidious. The effects may be seen in a group of tired people whose reserves of interest, information, and energy are drained. Their monotonous and unstimulating environment can induce lethargy and withdrawal; it lowers motivation, morale, and efficiency. But Antarctic isolation can also have its positive effects, and there are some who thrive on such experience. It often brings a sense of personal discovery, a sense of achievement, the fulfilment of ambition, and the membership of a rather special, informal club. To minimise the adverse effects and to maximise the beneficial effects is the aim of the Antarctic selectors, administrators, and party leaders.

"The proper study of mankind is man" (Alexander Pope)

While scientists have been studying the Antarctic, they have themselves been studied. Between 1967 and 1982 all New Zealanders who wintered-over participated in isolation studies of a systematic nature that used clinical, observational, psychometric, and experimental methods. There were 204 subjects in the study. Their ages ranged from 20 to 49 with a median age of 26, most were single, and all but one were men. About one-half had tertiary education and about 20 percent had previously worked in isolated places.

The experience of wintering-over was satisfactory for most, an endurance for many, depressing for some, and produced symptoms of disorder in a few. Fifty percent thought that they had slowed in their responses and 60 percent declared that they had experienced unusual sleep difficulties. Their symptoms were consistent with those described by the early explorers, but experimental testing showed that none had lost the capacity to respond quickly and accurately as and when required to do so. It seemed as if they had slowed their normal pace of activity to match the unstimulating and monotonous world around them during their long Antarctic winter. Similarly, their sleep difficulties were related to the absence of the normal 24-hour rhythm of daylight and dark. When they got home they soon regained their normal level of mental and physical functioning.

In these days of ongoing, annual occupation of the bases, the changeover is a critical time for both the outgoing and incoming groups. Those going out have established their territorial rights over the base as a whole, over their separate working and cabin areas (Plate 15.14), and over their specific places in the dining and recreational areas. Then, at changeover they are obliged to share their public and private places with relative strangers and to relinquish their hold on the base that has been their home for so long. Such preservation of territorial rights was held firmly by the earlier expeditioners over the use of the "abandoned" huts and even localities. Scott, for example, wanted assurance from Shackleton that he would not use Ross Island for the location of his 1907–09 expedition base when the area was unoccupied and there was no immediate prospect of his using it.

The small New Zealand contingent at Scott Base is fortunate in having a much larger group of Americans as neighbours at McMurdo Station, only 3 kilometres away. It gives them an opportunity both to work as a cohesive group for indoor sports competitions, and an opportunity to mix as individuals to ameliorate their own inter-personal and inter-group tensions with well-disposed outsiders. People in both places with shared vocational interests can forge an instant bond whether they are chefs or radio technicians. And, until recently when the last

Plate 15.10. *Captain Scott's birthday dinner at Cape Evans, 6 June 1911. Photo: Alexander Turnbull Library, Wellington.*

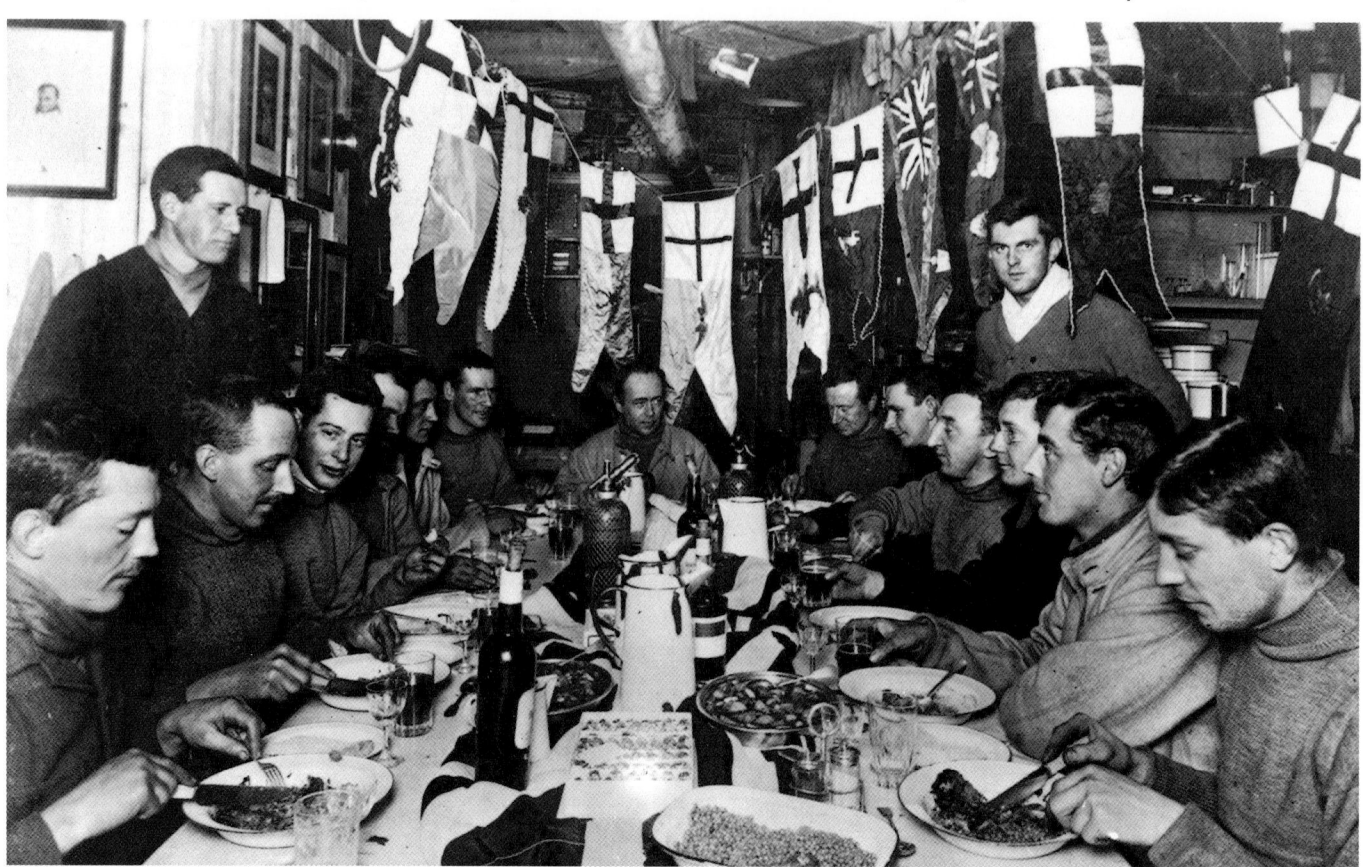

dogs were withdrawn, the huskies at Scott Base provided opportunities for recreation in exercising the dogs, a chance for the dog handlers to identify more closely with the earlier expeditioners as they exercised themselves, and a "third party" with which base staff could also communicate (Plate 15.15).

The question arises in Antarctic staffing as to whether some of the adverse effects of isolation might be minimised if the gender compositions of the bases were closely to resemble that of the communities from which the Antarcticans come. In recent years Argentina and Chile have provided for partners and families at two of their bases, but there are no data yet on their effects. Although a sexually balanced group has the potential for providing a broader range of inter-personal experience than an unbalanced one, it has to be accepted that tensions of a different kind could arise that might have the same disruptive or even violent outcome as arises in normal life. It is a moot point whether the benefits of a mixed group would outweigh the disadvantages, but in this matter it may well be that the benefits will far outweigh the risks for all concerned. Certainly, there has been a marked improvement in the tone and quality of discourse at the bases since the advent of women, and also the female scientists have made substantial contributions to Antarctic research (Plate 15.16).

The call of the Antarctic

Clinical questionnaires have shown that wintering-over expeditioners are drawn to Antarctica more for personal challenge than monetary or occupational reasons. But in the secure and relatively luxurious environment of a modern Antarctic base they might be denied this personal challenge. A winter period might not require them to be motivated, stimulated, energetic , and efficient. In reality, survival in isolation might require them to exercise the "disciplined lethargy" to which Priestley referred after he lived in an ice cave on Inexpressible

Plate 15.11. *A Kiwi-style barbeque in progress at Scott Base. Photo: C. Rudge, Antarctic Division, DSIR.*

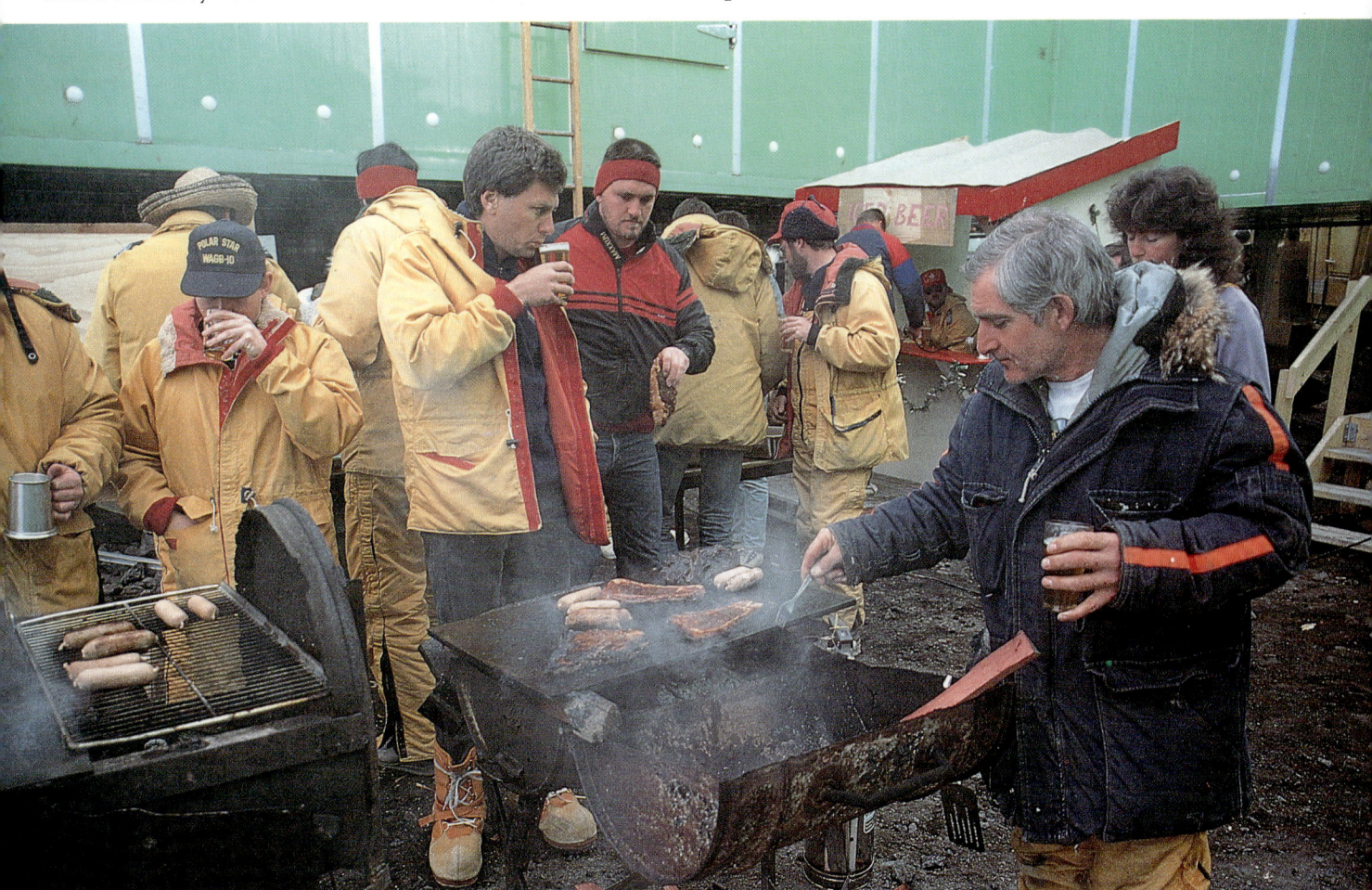

Island in 1912:

> "One of the greatest surprises of our behaviour during this winter was the unexpected way in which the whole party settled down to the inert, vegetating existence without fretting or protest. Four of us, at any rate, are usually active people, and cannot bear to be unoccupied in normal life, while I myself could never have believed that I could have been happy without something to read. Yet during the greater part of this inactive life we were certainly happy, as was witnessed by the seraphic state of our tempers, and, far from pining for books, I can remember many times when I could have been reading . . . and I preferred to lie and let my thoughts wander at their own sweet will. It had shown to all of us, I think, without doubt with how little it is possible to be content, and it has been a most decisive proof that in many cases the luxuries of civilisation only fulfil the wants they create. I do not say for a moment that any of us would care to repeat that winter; indeed I believe that another similar experience would kill most of us, or drive us mad, but it is certain that our pleasures during the hardest winter that any of us are likely ever to experience again were as acute as our pains . . . It is probable, indeed, that here we have another of the clearest notes that together make up that elusive something we term, for want of a better name 'the Call of the Antarctic'."

There are those who consider that a response to this call can no longer be made under the banner of science, and that close communication with the great forces of nature is no longer possible from within the restrictive protection of the modern Antarctic base. They are seen as a growing number of small, private adventure expeditions such as the "Footsteps of Scott Expedition" and the "90° South Expedition". Before long these expeditions, together with a seemingly inevitable tourist expansion, could change the face of the present, almost exclusively "scientific" profile of Antarctica.

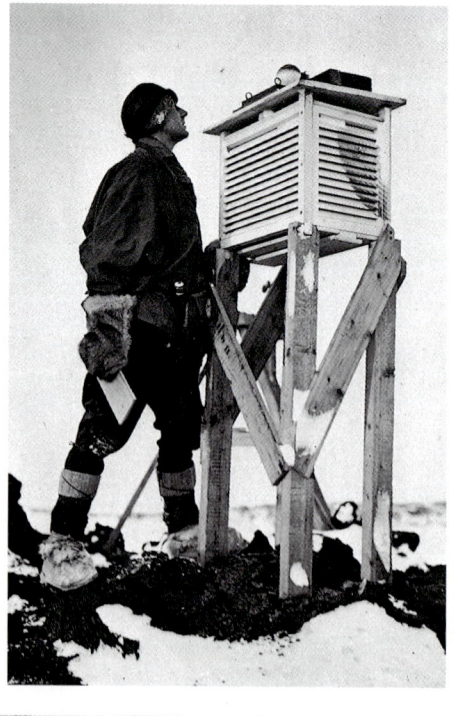

Plate 15.12. *Edward Wilson preparing to record the return of the sun to Cape Evans, 26 August 1911. After Midwinter's Day, this is the most keenly anticipated event in the Antarctic calendar. Photo: Alexander Turnbull Library, Wellington.*

Plate 15.13. *A probably never to be repeated winter isolation was that of Scott's Northern Party, three of the members of which are seen here in camp. Photo: Canterbury Museum, Christchurch, N.Z.*

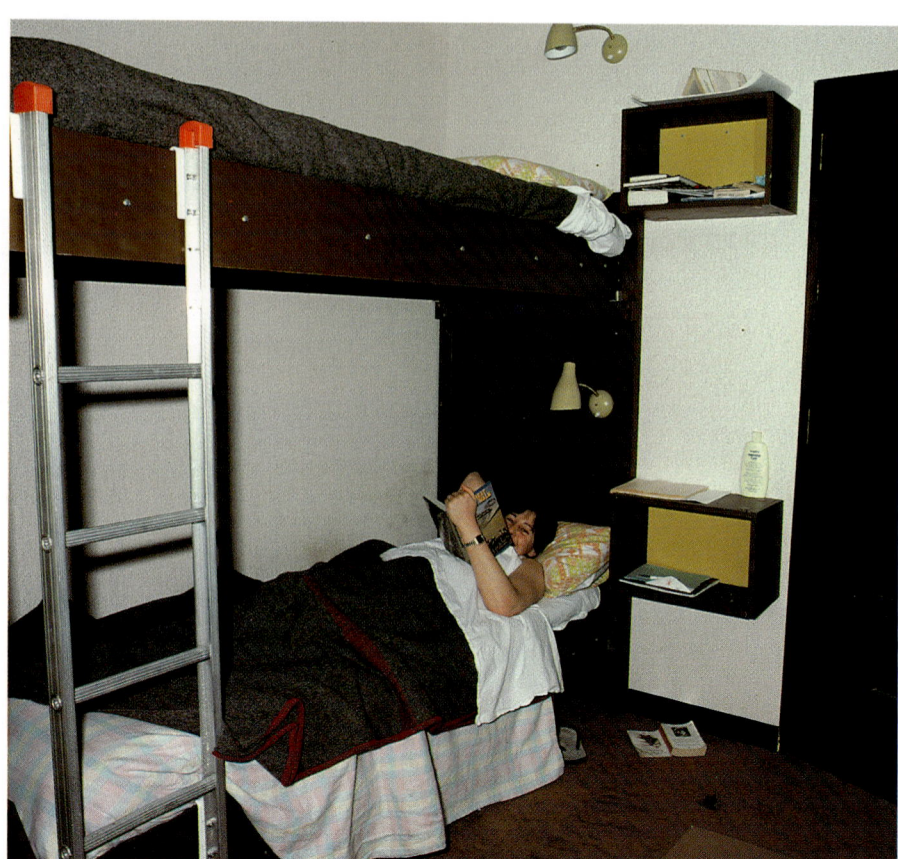

Plate 15.14. *Modern over-winterers usually have complete privacy in their sleeping quarters. The second bunk is only used in summer. Photo: K. Westerskov, Antarctic Division, DSIR.*

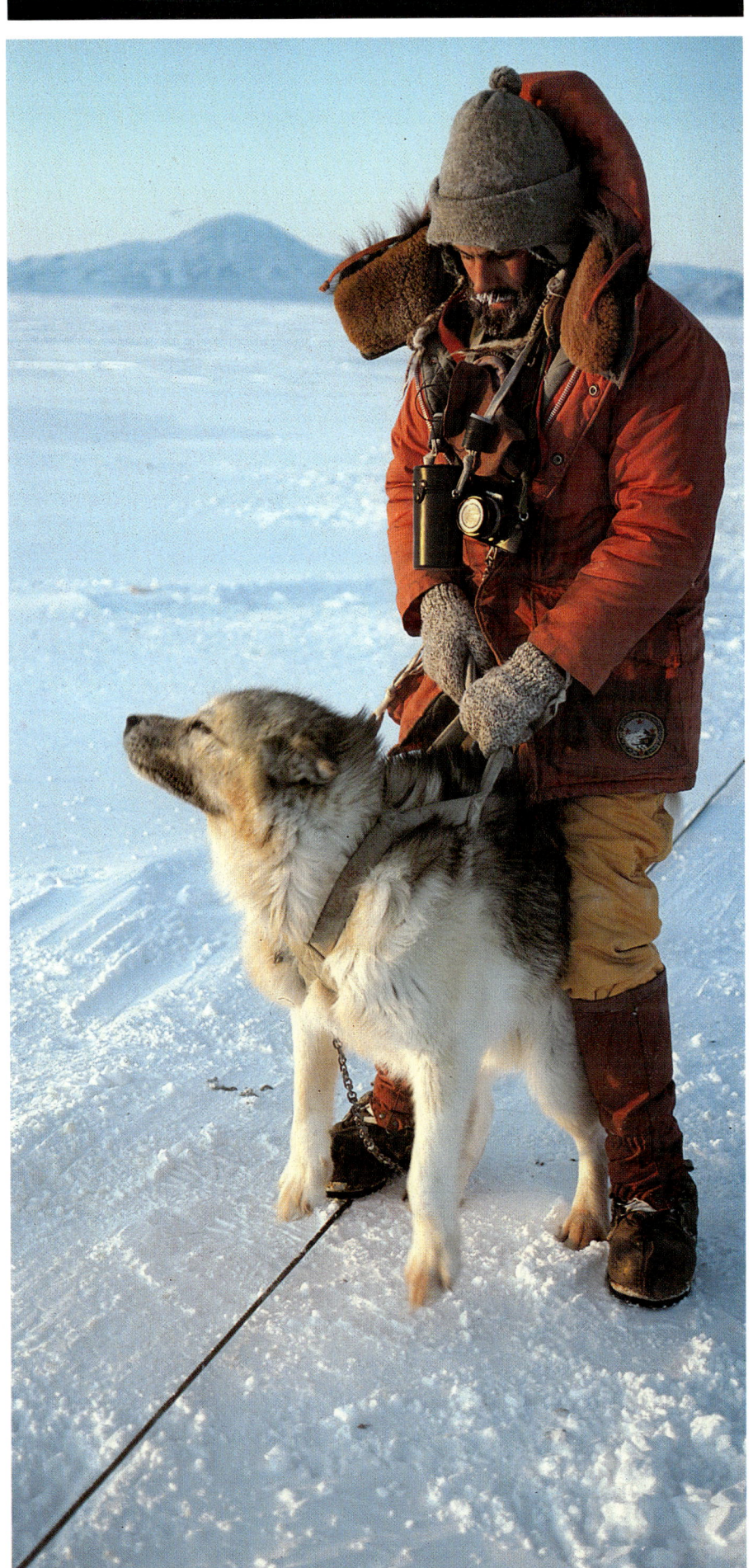

Plate 15.15. *Strong bonds used to develop between huskies and doghandlers. Photo: Kim Westerskov, Antarctic Division, DSIR.*

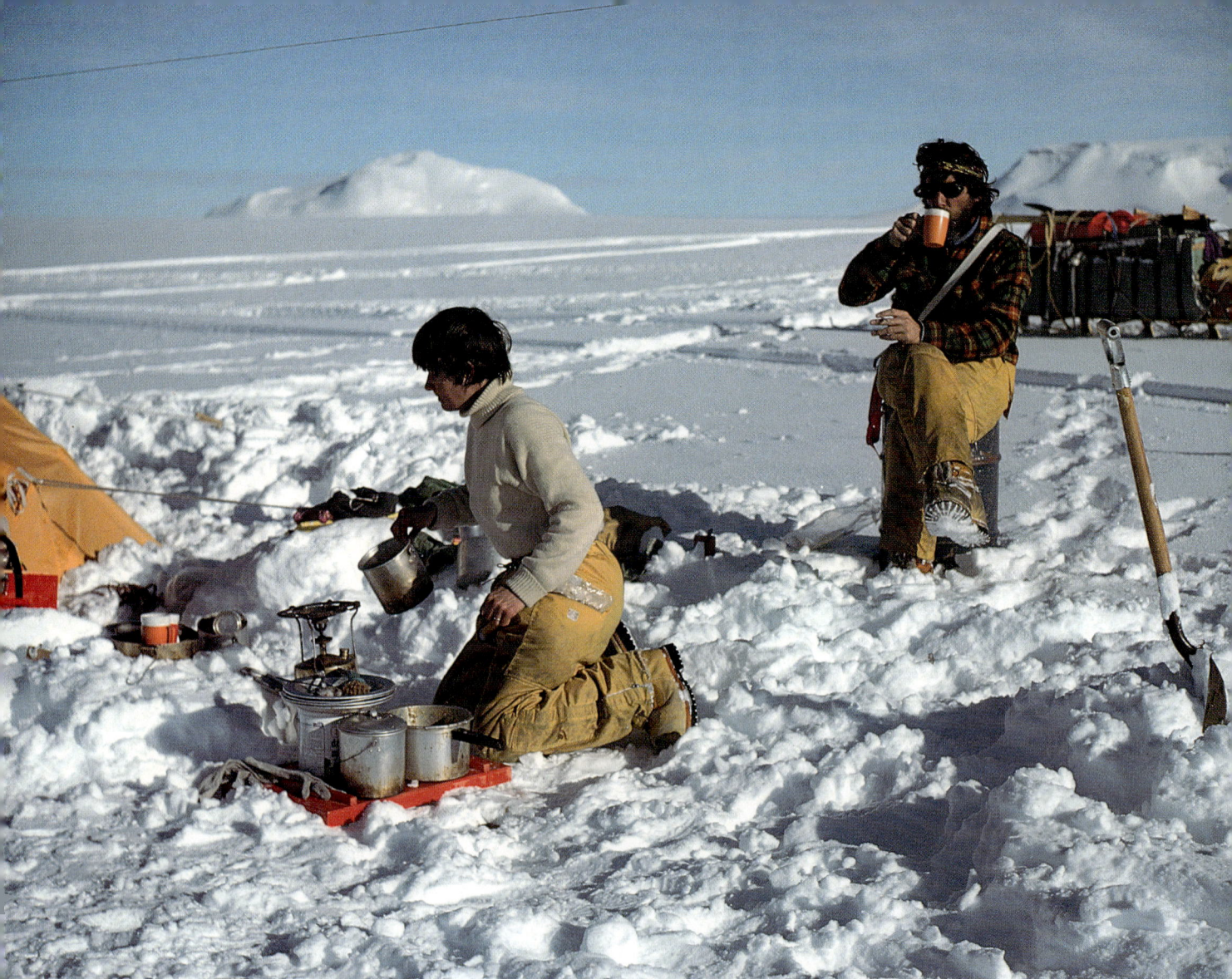

Plate 15.16. *Woman geologist preparing dinner on a fine summer "evening" on the plateau about 500 kilometres from the South Pole. Photo: M. Bradshaw.*

T he legal status of Antarctica was of little practical concern to the early explorers, who had more challenging interests. For some the challenge was adventure itself whereas for others national prestige was paramount and also, typical of so many of the nineteenth century explorers, the discovery and annexation of new territory. In 1908 the United Kingdom proclaimed sovereignty in the Weddell Sea and Antarctic Peninsula areas to the south of South America. Similar proclamations followed in 1923 when the United Kingdom established the Ross Dependency (a claim entrusted to the Governor-General of New Zealand), and the Australian Antarctic Territory in 1933. France and Norway also asserted jurisdiction over sections of East and West Antarctica. Argentina and Chile responded to the British assertion of sovereignty over the Antarctic Peninsula with conflicting claims of their own. In all, seven states claimed sovereignty in Antarctica (Figure 16.1). Their claims were not recognised by the United States or the Soviet Union, although the United States reserved the right to make a claim of its own.

During the late 1940s and the 1950s the political situation in Antarctica worsened as competition increased and attempts to consolidate sovereignty in some areas were matched by equally vigorous attempts to rebut those claims by some non-claimants. Moreover, the conflicting claims also produced international tension. British and Argentinian warships engaged in several incidents; bases were ransacked and burnt, a variety of disputes occurred and even armed conflict was reported.

This was the scene then in Antarctica in the mid 1950s: no agreed legal system, and an international political problem of serious proportions. Superimposed on this scene were the tensions of the outside world which was experiencing the height of the Cold War and there was real danger that these tensions, if extended to Antarctica, would have serious consequences for any stability that existed there.

The International Geophysical Year, 1957–58

Fortunately at this time, far seeing and imaginative scientists from many countries were planning the greatest scientific event in history— the International Geophysical Year (IGY) of 1957–58. This worldwide programme gave priority to upper atmosphere and polar research. Thus, it was imperative that several scientific stations were established in Antarctica. This would require the combined efforts of several countries. Forty-seven research stations operated in Antarctica during the IGY, with a further eight stations on subantarctic islands. Nations co-operated freely in these endeavours and good relationships became established between the logistics groups and scientists in the field.

To consolidate the advances made in international science during the IGY the International Council of Scientific Unions (ICSU) established the Special (changed to "Scientific" in 1961) Committee on Antarctic Research (SCAR) in 1957. This Committee was charged with the initiation, promotion and co-ordination of scientific activity in the Antarctic, to frame and review scientific programmes of circumpolar scope and significance. The membership of SCAR includes representatives from "other international bodies", as well as a delegate from each of those countries active in Antarctic research programmes.

The IGY created a new spirit of co-operation in Antarctica. It brought a breathing space in the form of an understanding that scientific activity carried out during the IGY would not prejudice the positions of the various countries regarding claims. Also it paved the way for a period of stewardship in Antarctic affairs.

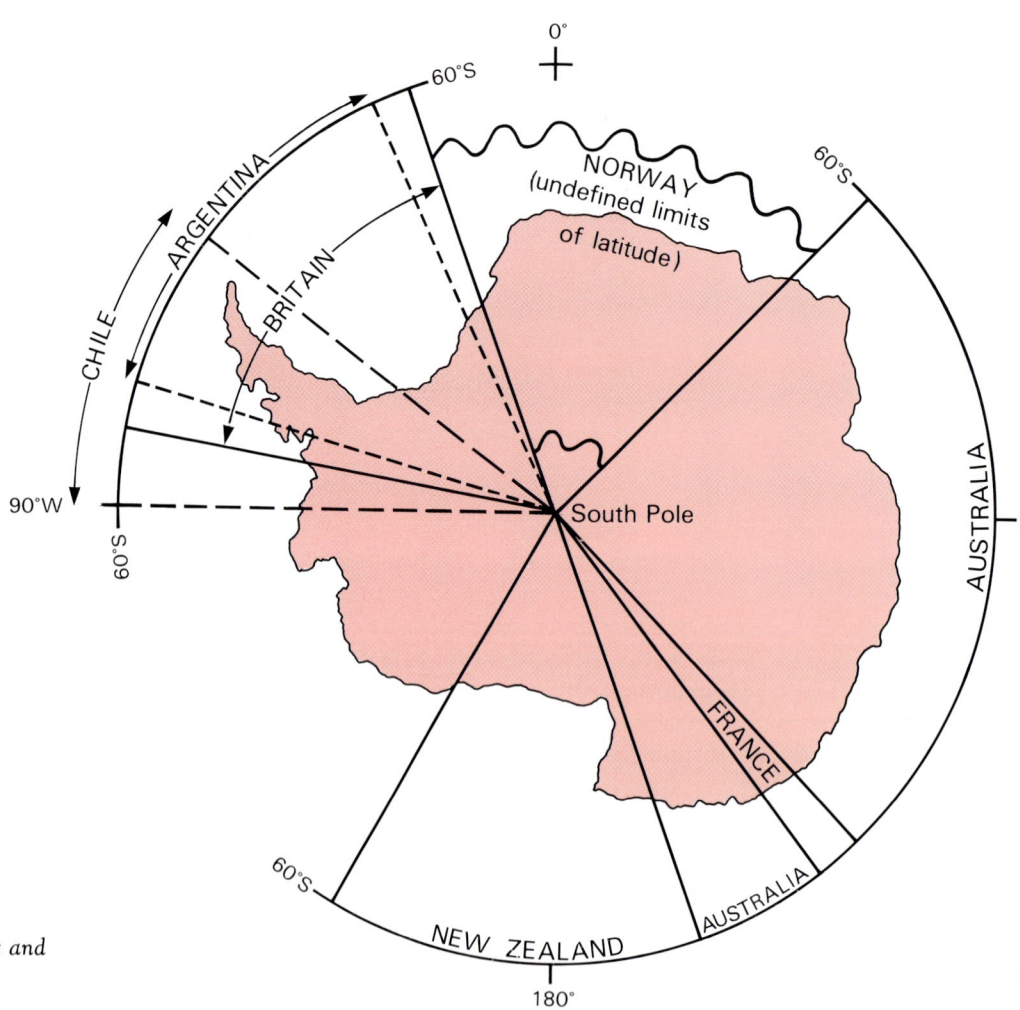

Figure 16.1. *Map of Claimant States and table of Antarctic Treaty signatories.*

TABLE OF ANTARCTIC TREATY SIGNATORIES
(at November 1989)
with dates of deposit or accession

Consultative Parties
Argentina* 1961; Australia* 1961; Belgium* 1960; Brazil 1983;
Chile* 1961; China 1985; Finland 1989; France* 1960;
Germany, Federal Republic 1981; Germany, Democratic Republic 1987;
India 1983; Italy 1987; Japan* 1960; New Zealand* 1960;
Norway* 1960; Peru 1989; Poland 1977; South Africa* 1960;
South Korea 1989; Spain 1988; Sweden 1988; United Kingdom* 1960;
United States* 1960; Uruguay 1985; U.S.S.R.* 1960.

Acceding Parties
Austria 1987; Bulgaria 1978; Canada 1988; Colombia 1989; Cuba 1984;
Czechoslovakia 1962; Denmark 1965; Ecuador 1987; Greece 1987;
Hungary 1984; Korea, Democratic People's Republic 1987;
Netherlands 1967; Papua New Guinea 1981; Romania 1971;
Switzerland 1987.

*Original signatories, 1 December 1959

The Antarctic Treaty

As the IGY drew to a close it became clear that the IGY type of activity would continue for several years. Indeed, the scientific community headed by SCAR had already held a series of discussions where it had been agreed that "co-operation in science in Antarctica should be expanded and continued for an indefinite period". The scientists had shown how international co-operation in science in Antarctica could be easily achieved to the benefit of all. Now it was over to the politicians to ensure that Antarctica should never again become a place for international discord that could threaten world peace.

In late 1957 and early 1958 various agencies of interested governments frequently consulted and it was generally agreed that a more permanent special regime would be desirable. The aims were to continue to foster co-operation in scientific research and to resolve, so far as possible, the many differences regarding territorial sovereignty.

Accordingly, on 2 May 1958 the United States took the initiative in proposing that a conference be convened to conclude an Antarctic Treaty. The broad objectives of the United States note, which was sent to all 12 countries active in Antarctica during the IGY, were stated in the third paragraph.

"The International Geophysical Year comes to a close at the end of 1958. The need for coordinated scientific research in Antarctica, however, will continue for many years into the future. Accordingly, it would appear desirable for those countries participating in the Antarctic programme of the International Geophysical Year to reach agreement among themselves on a program to assure the continuation of the fruitful scientific cooperation referred to above. Such an arrangement could have the additional advantage of preventing unnecessary and undesirable political rivalries in that continent, the uneconomic expenditure of funds to defend individual national interests, and the recurrent possibility of international misunderstanding. It would appear that if harmonious agreement can be reached among the countries directly concerned in regard to friendly cooperation in Antarctica, there would be advantages not only to those countries but to all other countries as well."

Following further negotiations during a series of preparatory talks held from June 1958 and continued for over one year, the Conference on Antarctica convened on 15 October 1959 and closed on 1 December 1959 with the signing of the Antarctic Treaty on that day.

The Treaty, which came into force on 23 June 1961, is a remarkable and unique international instrument. It formalised and guaranteed the kind of free access and research rights that had spelt success for the IGY and established a framework for all countries, both claimant and non-claimant, to work together for the common cause. That it managed to balance these complex and competing interests in a simple yet practical way is perhaps the most remarkable aspect of the Treaty. It also required that Antarctica should be used for peaceful purposes only, thus prohibiting military activities in what could be described as as disarmament regime which is still the most comprehensive in the world and the only one which specifically provides for on-site inspection. The Treaty also prohibits nuclear explosions and the dumping of nuclear waste. It requires that each Contracting Party shall annually provide advance notice, to the other Contracting Parties, of expeditions planned, stations occupied, and military personnel or equipment involved in a support role.

The 60th South parallel, which runs through open water, was

262

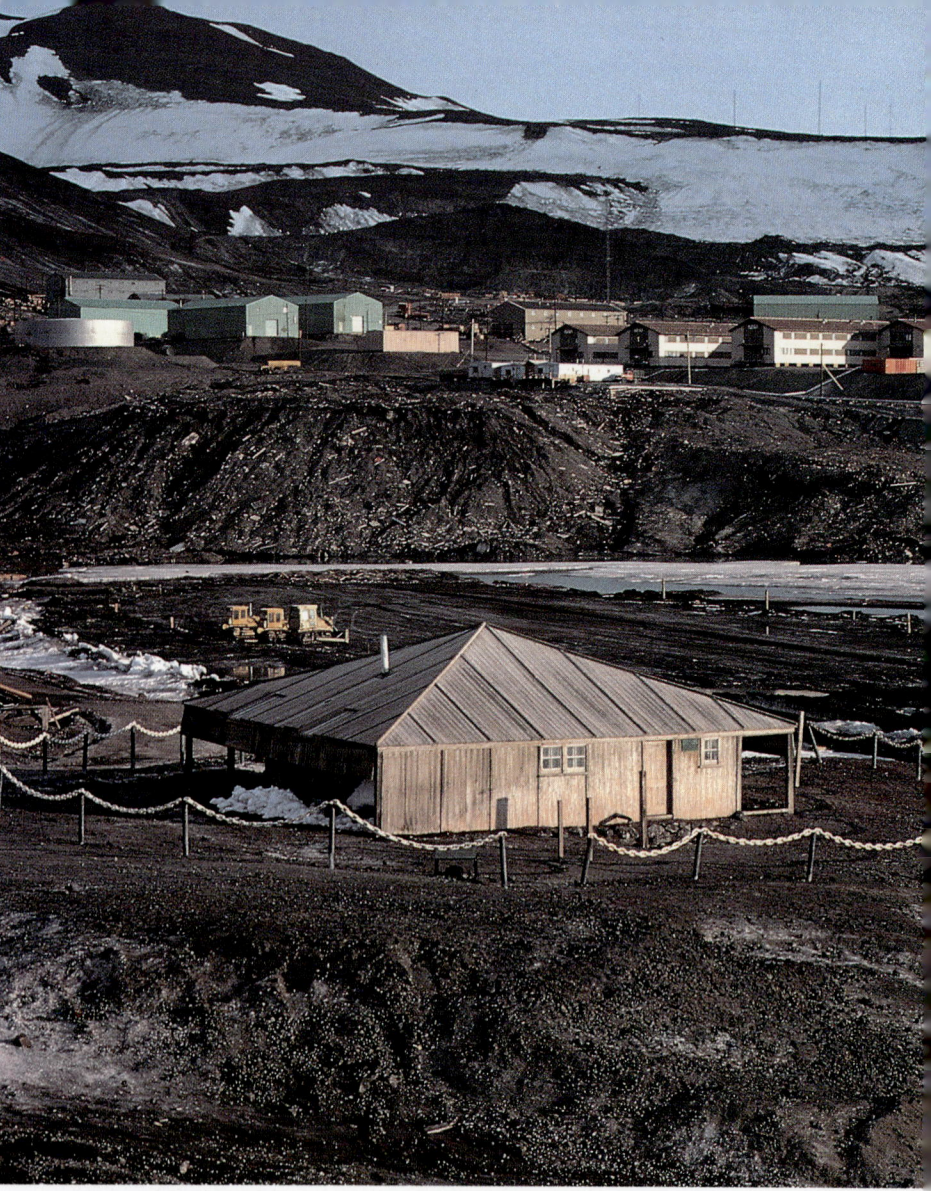

Plate 16.1. *Scott's "Discovery Hut", with McMurdo Station in the background. Photo: K. Westerskov, Antarctic Division, DSIR.*

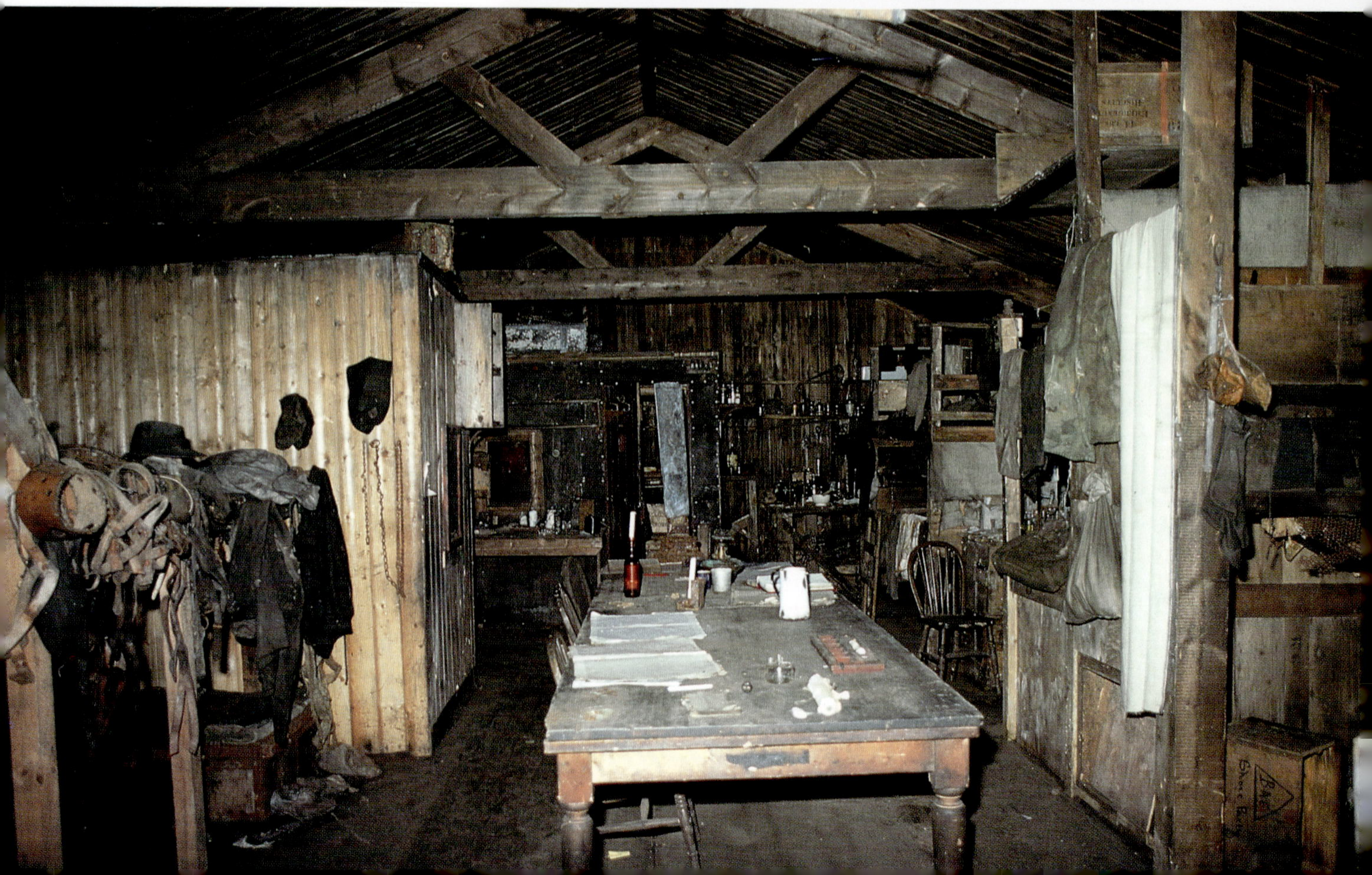

Plate 16.2. *The interior of Scott's Cape Evans Hut after excavation and restoration. Note the officer's dining table (see Plate 15.10). Photo: C. Rudge, Antarctic Division, DSIR.*

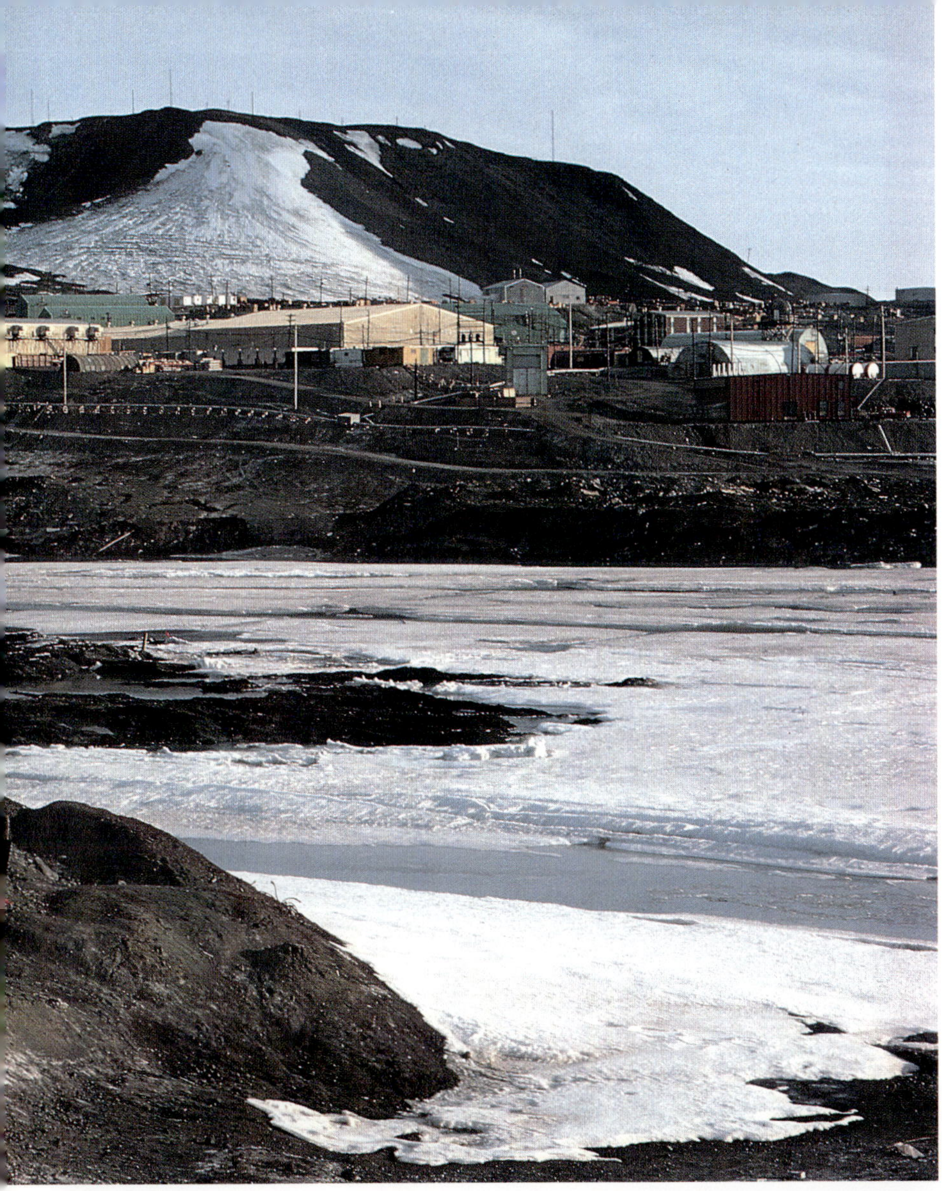

Plate 16.3. *The restored interior of Shackleton's Hut at Cape Royds, Ross Island. Photo: C. Rudge, Antarctic Division, DSIR.*

Plate 16.4. *Cross erected on Observation Hill by survivors of Scott's last expedition. Beyond the McMurdo Ice Shelf is White Island. Photo: C. Rudge, Antarctic Division, DSIR.*

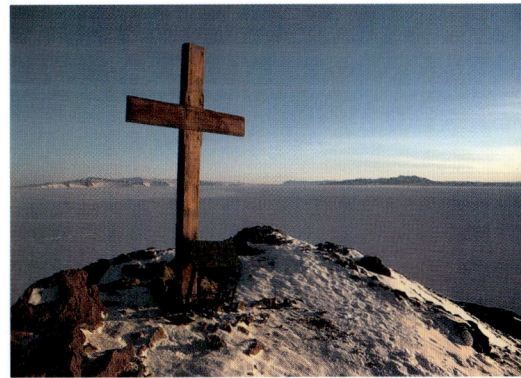

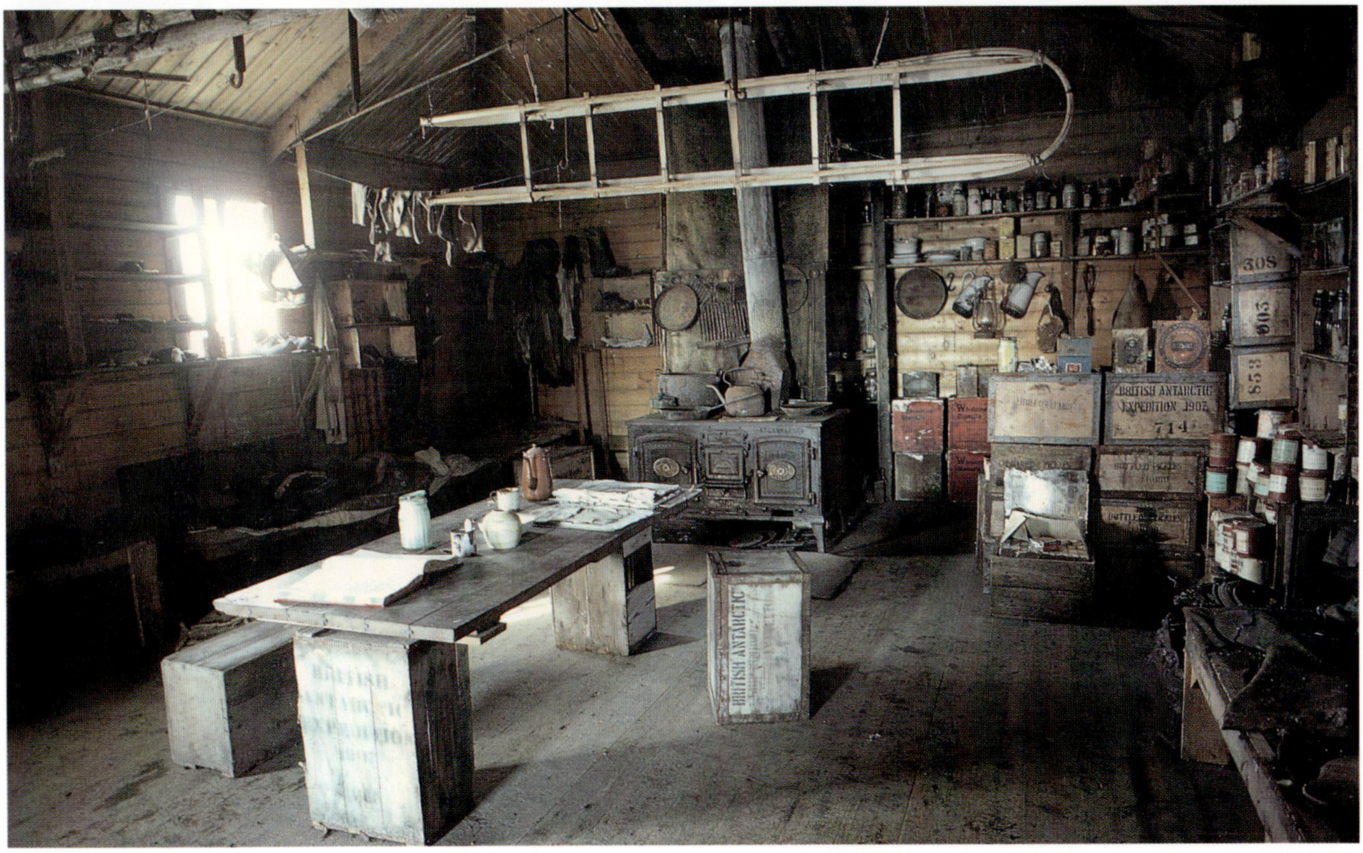

selected as the Antarctic Treaty's border line. It delimits a unique international accommodation, in a combination of geographical and political circumstances found nowhere else. The importance of the Treaty is attested by its growth from 12 original signatories to 38 acceding states representing about 90 percent of the world's population (see Figure 16.1).

A major feature of the Treaty is the consultative system which it established. The record is impressive. In the 28 years since 1961 when the Treaty came into force, 197 Recommendations have been adopted from the 14 Consultative Meetings held, most dealing with aspects of conservation and environmental protection.

Environmental protection

Attention to environmental matters arose initially from an increasing concern for the conservation and preservation of living resources in Antarctica. Although the Treaty was founded on the principle of conservation, specific measures to regulate human activities needed to await further experience and observations to better identify those activities more likely to adversely affect the environment.

The first "restricted" areas in Antarctica were established jointly by New Zealand and the United States and covered the Hut Point Peninsula of Ross Island and identified historic sites (Plates 16.1, 16.2, 16.3, 16.4), sites of special scientific interest, central station complexes, outlying science areas, and areas for future scientific or logistical development. A set of rules was agreed on which included prohibiting low flying aircraft over penguin rookeries and restricting human activities near and within the Adélie penguin colony at Cape Royds.

The McMurdo Land Management and Conservation Board was then established to oversee and co-ordinate the management plans and it has continued to meet over the years whenever there has been a need to discuss any important question relating to land management of the Hut

Figure 16.2. *Specially Protected Areas (SPA's) and Sites of Special Scientific Interest (SSSI's) in the Ross Sea region.*

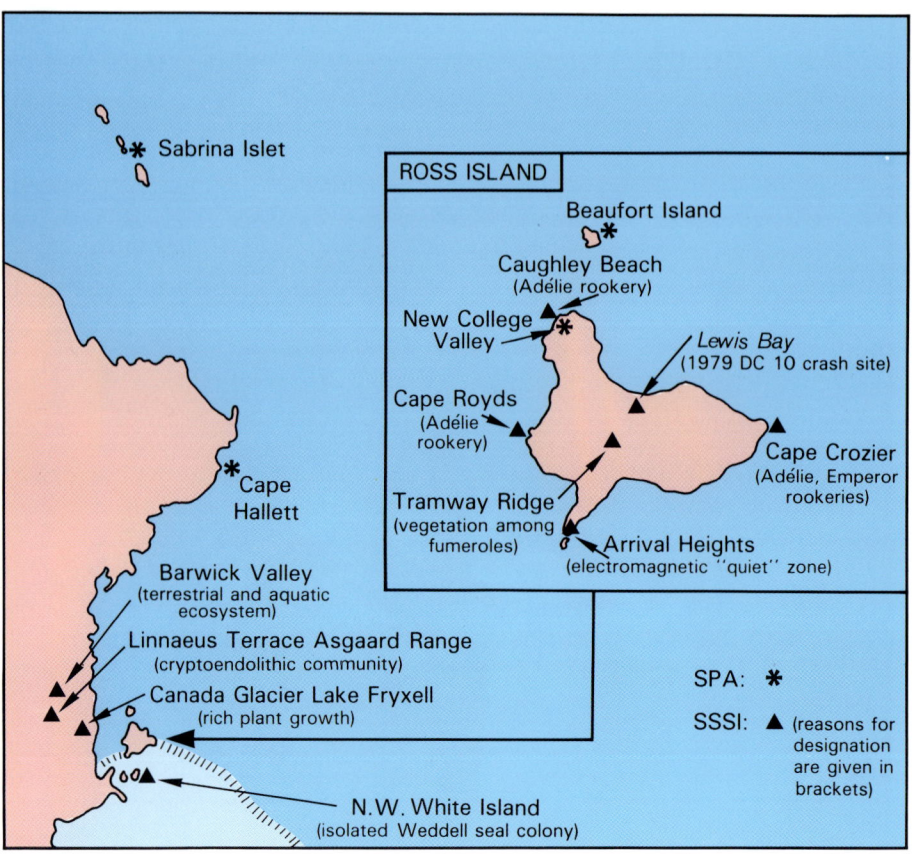

Point Peninsula. Of great importance too was the experience gained by both the United States and New Zealand programmes in working within the restrictions imposed by Antarctica's first land management plan. This experience was put to good use in the following years in formulating proposals for Specially Protected Areas and prompting the concept of Sites of Special Scientific Interest (Figure 16.2), restricting the use of radio-isotopes in Antarctica, and developing a Code of Conduct for Antarctic expeditions and station activities (including waste disposal and many other matters related to concern for the environment, Plate 16.5).

The most comprehensive and important set of rules designed to minimise harmful interference and preserve and protect living resources was "The Agreed Measures for the Conservation of Fauna and Flora". This was adopted by the Treaty Nations in 1964 as a result of the initiative taken by SCAR scientists in 1960 who developed some "general rules of conduct for the preservation and conservation of living resources in Antarctica".

Within the 1964 "Agreed Measures" provision was made for:
- prohibition of the killing, wounding, capturing, or molesting of any native mammal or native bird except in accordance with a permit;
- such permits to be issued only for certain restricted purposes;
- the designation of Specially Protected Species;
- the designation of Specially Protected Areas;
- regulating the importation into Antarctica of non-indigenous species, parasites and diseases;
- minimising harmful interference with the normal living conditions of Antarctic mammals and birds;
- exchange of information between consultative parties as to actions they have permitted.

The New Zealand Antarctica Act 1960 was amended in 1970 to include regulations for the Conservation of Antarctic Fauna and Flora in accordance with the "Agreed Measures".

Antarctic resources—minerals

The text of the Antarctic Treaty signed in 1959 makes no reference to commercial activities such as the exploration and exploitation of marine and mineral resources. Indeed it appears to have deliberately avoided specific mention of sensitive matters that could have prejudiced the delicate negotiations taking place which led to final agreement and the achievements of the Treaty in other respects.

The question of resources has posed the greatest threat to the Antarctic Treaty so far. Therefore, it is important to understand how these concerns originated and what has been accomplished to ensure that the fullest protection to the Antarctic environment is maintained in any agreement that may provide for the wise use of a resource.

By the late 1960s the geology of Antarctica became better known and there was reasonable agreement amongst geologists as to its age, its place in the ancient super continent of Gondwanaland, and the general acceptance of this hypothesis to account for many similarities in the continents of the Southern Hemisphere. Some geologists were bold enough to predict Antarctica's likely mineral wealth based on a few occurrences only and analogies they made with the surrounding continents from their reconstructions of Gondwanaland. The late M. G. Ravich, a leading Soviet polar geologist, wrote in 1960: "East Antarctica, which is part of the ancient Gondwana continent may, like Australia, Africa, Brazil and India, contain valuable minerals including diamonds, gold, mica and others, and it possesses significant coal resources. West Antarctica like the South American Andes may contain deposits of lead, tin, copper and gold".

Thus, the question of what would happen if somebody found minerals in Antarctica which seemed economically viable and so started exploiting them was raised. The immediate reaction of the Consultative Parties was a wish to not discuss it and hope that such a question would go away. But this immediate negative response was soon followed by a more positive realisation by some Treaty Members that this was a question of considerable importance. The first opportunity for discussion was at the Sixth Antarctic Treaty Consultative Meeting in 1970 when New Zealand raised the question informally.

At about this time the first commercial interest in Antarctic mineral resources became apparent when three large exploration companies sought individual mining rights in Antarctica from the Australian, United Kingdom, and New Zealand Governments. This spurred these Governments into giving more attention to this question which they now saw as being much more urgent. They had to provide an appropriate response to the inquiries, and also meet the demands of increasing international interest in Antarctic mineral resources and the many questions on who owned any resources and what restrictions would be placed on any commercial activities (Plate 16.6).

Formal discussions on mineral exploration and exploitation continued to be restricted when the Consultative Parties met for their seventh meeting in 1972. Those present could only agree (the Treaty requires unanimity) to discussing minerals under an agenda item "Effects of Mineral Exploration". Thus few, if any, of the difficult legal and political questions could be discussed nor could all the many activities likely to be associated with exploitation be considered. At least it was a start. Formal discussions concentrated around environmental concerns, for these were seen as the most likely effects of mineral exploration, while many of the leading questions (especially those with legal–political overtones) were discussed informally.

Plate 16.5. *Retrieval of waste metal at Scott Base. Photo: C. Rudge, Antarctic Division, DSIR.*

Similar restrictions were imposed at the 1975 Consultative Meeting. But the Parties did adopt the Recommendation proposed by New Zealand that: "the subject Antarctic Resources—The Question of Mineral Exploration and Exploitation be fully stated in all its aspects in relation to the Treaty and be the subject of consultation among them with a view to convening a special preparatory meeting during 1976, the terms of reference of which will be determined precisely through diplomatic channels; the special preparatory meeting to report to the Ninth Consultative Meeting". Other important progress at this Consultative Meeting included Recommendations "to study the environmental implications of mineral resource activities in the Antarctic . . . " and "to invite SCAR to study the likely environmental impact of resource-related activities" and to "continue to co-ordinate national geological and geophysical programmes in the Antarctic Treaty Area with the aim of obtaining fundamental scientific data on the geological structure of Antarctica" and to "consider what further scientific programmes are necessary in pursuit of these objectives".

The Special Preparatory Meeting was held in Paris in 1976. Two Working Groups were formed, one to consider the legal–political questions and the other, the scientific-technical questions. Both Groups produced reports identifying the main areas requiring further attention and importantly, noted that "offshore oil and gas deposits are a far more likely subject for mineral exploration in the Antarctic than land-based minerals". This conclusion was based on the high increases in the price of energy at that time and, consequently, that exploration companies were looking globally for new areas of hydrocarbon deposits.

In 1977 the Antarctic Treaty Parties agreed to "urge their nations and other States to refrain from all exploration and exploitation of mineral resources while making progress towards the timely adoption of an agreed regime concerning Antarctic mineral resource activities". Mineral exploration and exploitation has been an agenda item at all subsequent Antarctic Treaty Consultative meetings. Also, several Special Consultative Meetings have been held since June 1982 to develop a regime dealing with all aspects of this question.

The last meeting of this series was held in Wellington, New Zealand, in May–June 1988. All the 20 Consultative Parties to the Treaty, together with 3 Contracting Parties, continued the earlier work of reviewing, drafting issues in the text of the Convention, and ensuring the concordance of the text in the official languages of the Antarctic Treaty.

At the conclusion of the session the Representatives of the Consultative Parties adopted by consensus the Convention on the Regulation of Antarctic Mineral Resource Activities (CRAMRA). The Meeting agreed that the Convention would be open for signature at Wellington on 25 November 1988.

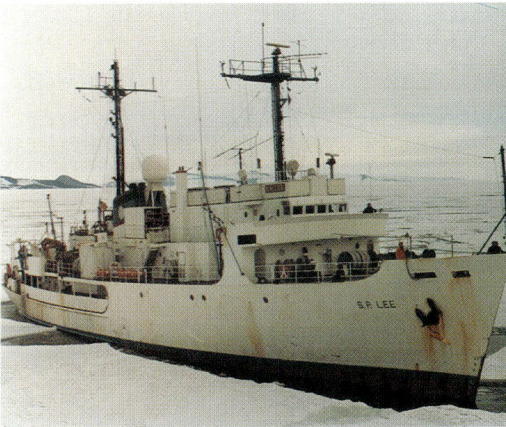

Plate 16.6. *A geophysical exploration ship in the Ross Sea. Photo: F. Davey.*

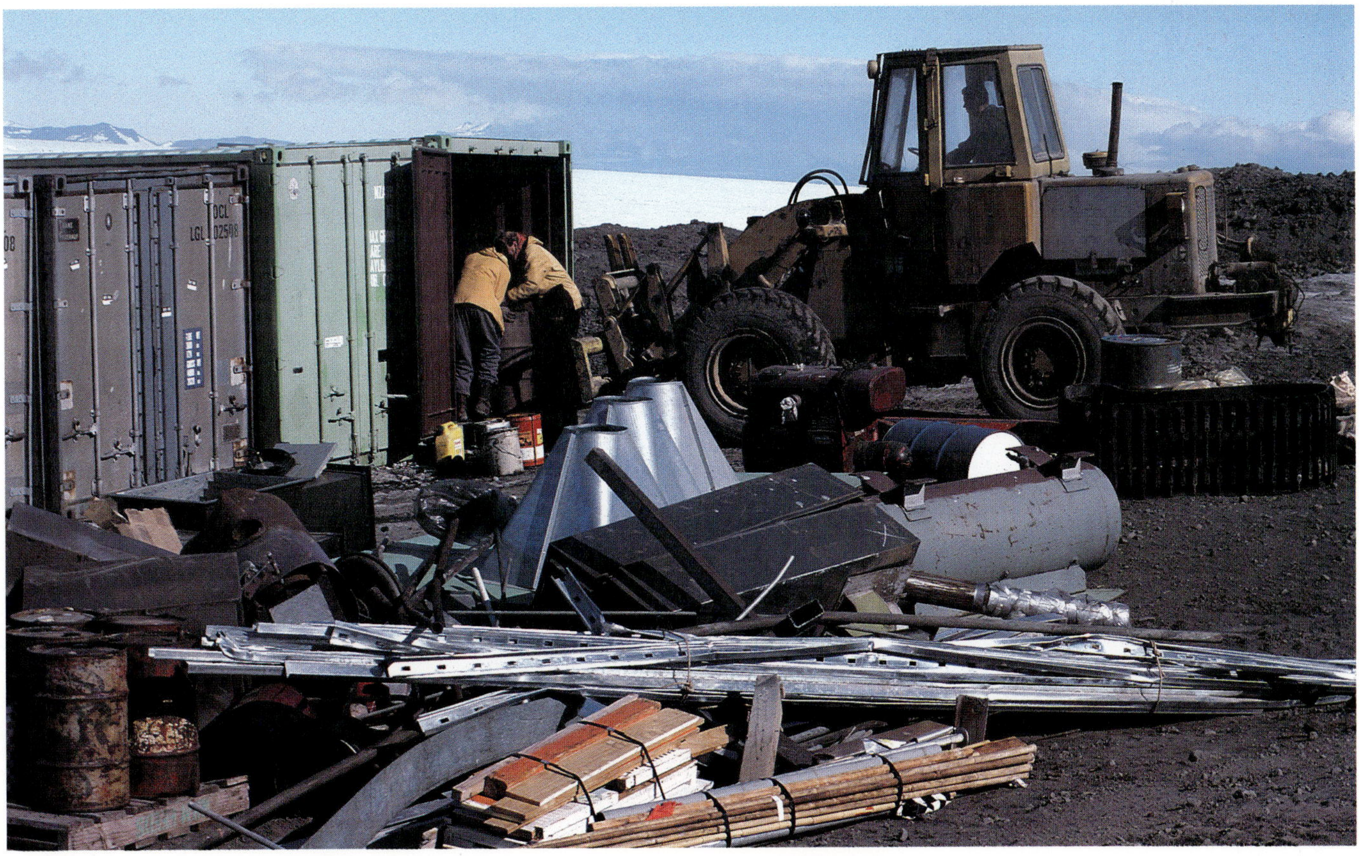

The Convention takes into account the special legal and political status of Antarctica and places highest priority on ensuring that the fullest protection of the Antarctic environment and of its dependent or associated ecosystems must be a basic consideration in decisions taken on all possible mineral resources activities.

To be fully effective in all its work and responsibilities, CRAMRA will establish five separate institutions: the Commission; the Special Meeting of States Parties; the Scientific, Technical, and Environmental Advisory Committee; different Regulatory Committees for selected geographical areas; and a Secretariat.

However, there is a general consensus amongst Antarctic geologists and solid-earth geophysicists that no economic development of mineral resources in Antarctica is possible for several decades, if ever. No minerals in any quantity have so far been found, exploration would be risky and extremely expensive, and exploitation of any resource would be more expensive again when the additional high risks and transportation costs are included.

Antarctic resources—biological

Meanwhile, another matter arose of great concern to the Treaty Parties, that of marine living resources. This problem differed from that of minerals because some exploitation was already taking place; thus priority was given to negotiating a Convention on the Conservation of Marine Living Resources.

Possibly resulting from the decrease in whales, krill (their main food) became noticeable in larger quantities in the mid 1960s; many countries experimented with processing krill to make it palatable for human consumption. Methods of catching, identifying the best areas for fishing, and gaining a better overall knowledge of this resource were also studied.

The conclusion of the Convention for the Conservation of Antarctic Seals in 1972, which places strict limits on the taking of some Antarctic seals and prohibits the taking of all others, opened the way to a consideration of the question that would be posed by the potential large-scale exploitation of krill. At the Eighth Antarctic Treaty Consultative Meeting the Treaty partners agreed on the need to "promote and achieve within the framework of the Antarctic Treaty the objectives of protection, scientific study and rational use of (Antarctic) marine living resources". This Recommendation went on to focus attention on scientific study as the necessary basis for protection and rational use and remitted certain questions to SCAR for consideration. SCAR responded by holding a meeting of its Biology Working Group in 1976, where a research programme, the Biological Investigations of Marine Antarctic Systems and Stocks (BIOMASS), was developed. The

Plate 16.7. *Tourists unloading from the* Lindblad Explorer *for shore visits. Photo: Antarctic Division, DSIR.*

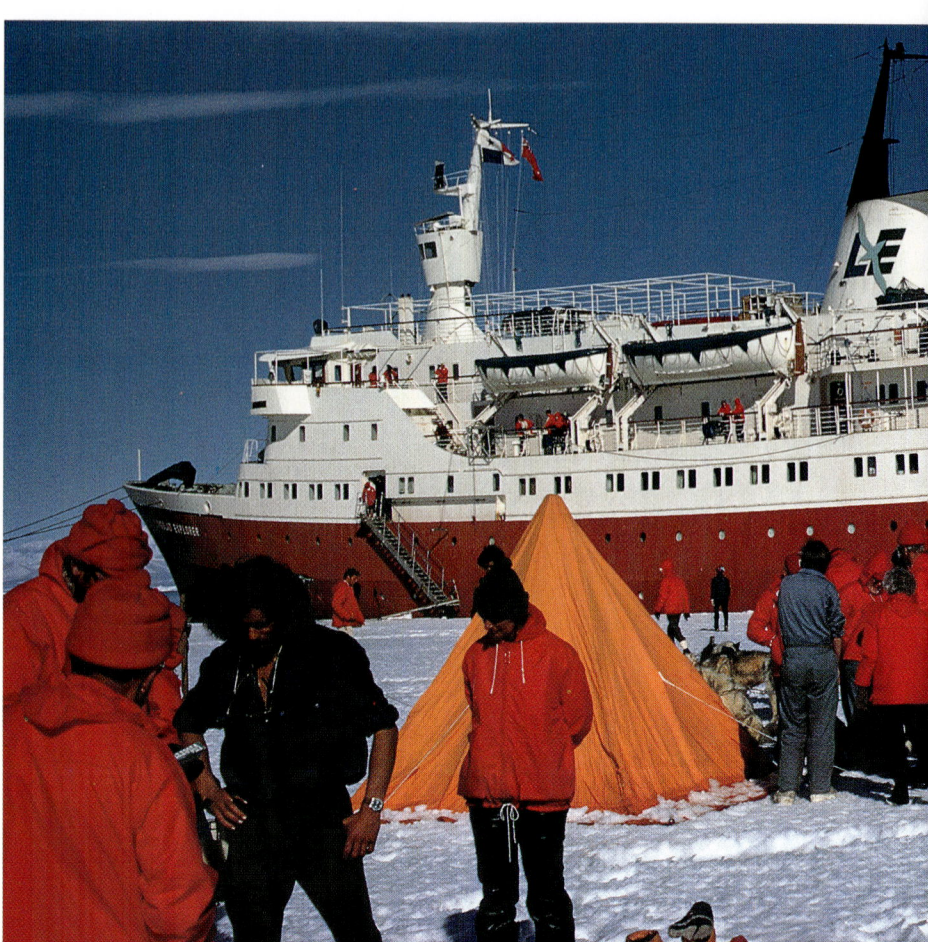

BIOMASS programme, which is continuing, highlighted the importance of krill in the Antarctic marine ecosystem. It noted that irrational, large-scale exploitation of krill could have severe repercussions on the birds, seals, and whales of the Antarctic which depend on krill for their food.

In May 1980 a Convention for the Conservation of Antarctic Marine Living Resources (CCAMLR) was concluded in Canberra. The Convention is unique for, although it was developed under the auspices of the Antarctic Treaty, the area of application of the Convention is larger than the Antarctic Treaty area. The boundary of the Antarctic marine ecosystem is the Antarctic Convergence, a major circum-Antarctic biogeographical boundary where the cold northerly-moving waters dip beneath warmer southerly-moving sub-tropical waters. South of the Antarctic Convergence and within the Antarctic marine ecosystem, krill is dominant in the food web. It was therefore decided that the northern boundary of the Convention's area of application should coincide within the Antarctic Convergence, not latitude 60°S as is applicable to the Antarctic Treaty. The Convention also broke new ground in that, firstly, it was concluded before the taking of the crucial species (krill) had got much beyond the experimental phase and, secondly, it was based on the concept of "conservation" rather than "exploitation".

CCAMLR provides the essential legal framework for management of the living resources of the Southern Ocean and it has established both a Commission (based in Hobart) and a Scientific Committee to provide information and management advice about the complex Antarctic ecosystem. To assist the Commission in this work, the Scientific Committee holds annual meetings to which representatives of appropriate international organisations such as SCAR are invited. Its success so far has provided a lead for those developing the Convention on the Regulation of Antarctic Mineral Resources Activities (CRAMRA).

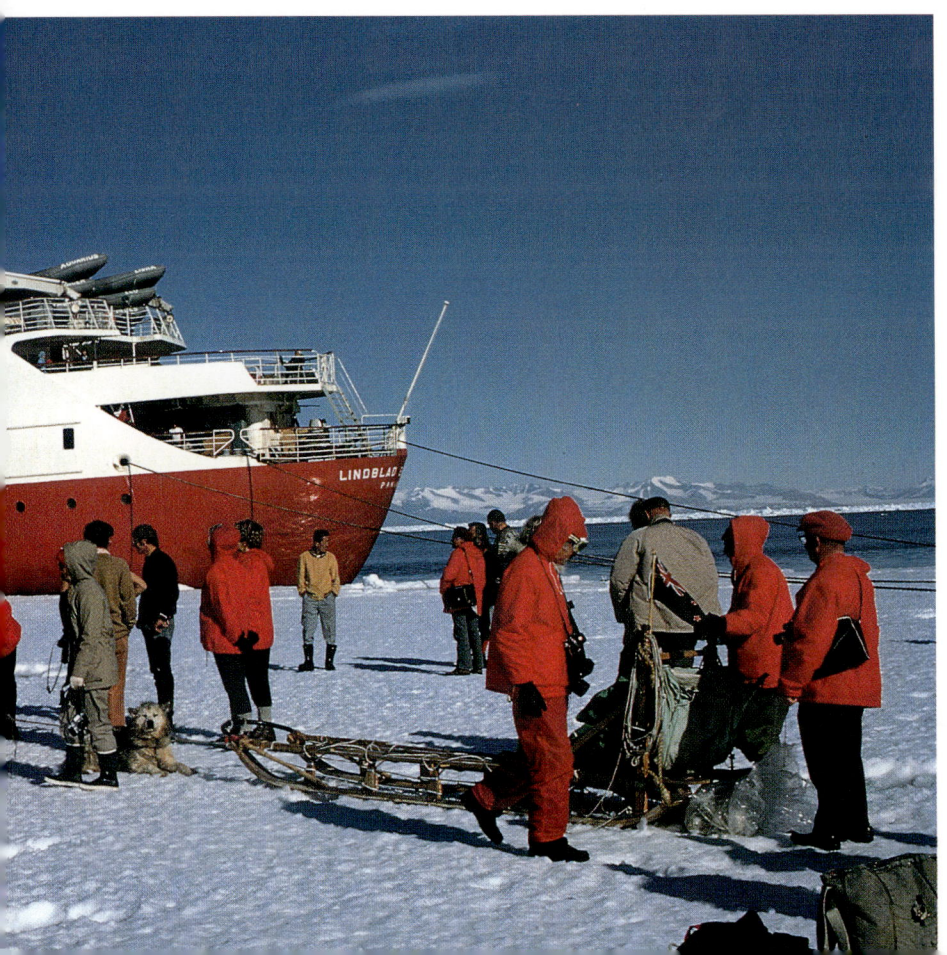

Antarctic resources—tourism

Another commercial activity, tourism, is likely to be one of Antarctica's major industries. Antarctic tourism has already started but Antarctica is still the most underdeveloped tourist area in the world. This is not due to Antarctica being unattractive, for the natural assets are there—such as breathtaking beauty, uniqueness, a new frontier, and adventure. There are thousands who would like to go there at a price, but the isolation of Antarctica makes travel times very lengthy and therefore very expensive; a not very attractive combination to most tour organisers or tourists.

The first tourist visits to Antarctica began in 1958 aboard an Argentine vessel but regular ship tours did not commence until 1966 (Plate 16.7). The Ross Sea region was identified as one of the most interesting areas for tourists, particularly because of the historic huts of Scott and Shackleton on Ross Island. It thus became necessary to establish what conditions, if any, should be applied to visits to bases or historic sites in the region.

The Antarctic Treaty Parties had discussed the "Effects of Antarctic Tourism" when they met in 1966. But their Recommendation did not suggest any conditions that might be applicable when governments were faced with granting permission for tourist groups to visit their Antarctic stations.

Consequently, New Zealand and the United States took the initiative in preparing policy statements on conditions of access for tourist expeditions to their stations. These points became incorporated at the 1970 Consultative Meeting in an agreed Recommendation which applied to private expeditions as well as tourists. About one tourist ship visits the Ross Sea region each year. A ranger-guide is provided by New Zealand to acquaint passengers with the Antarctic Treaty and New Zealand policies, to lecture on Antarctica, and to supervise shore visits.

Plate 16.8. *Jonathon Chester's private "Australian Bicentennial Antarctic Expedition" to climb Mount Minto camps for the second night during the hazardous journey. They achieved the first ascent of the 4163-metre peak in the Ross Dependency. However, gales badly damaged the expedition's schooner. Photo: J. Chester, Extreme Images.*

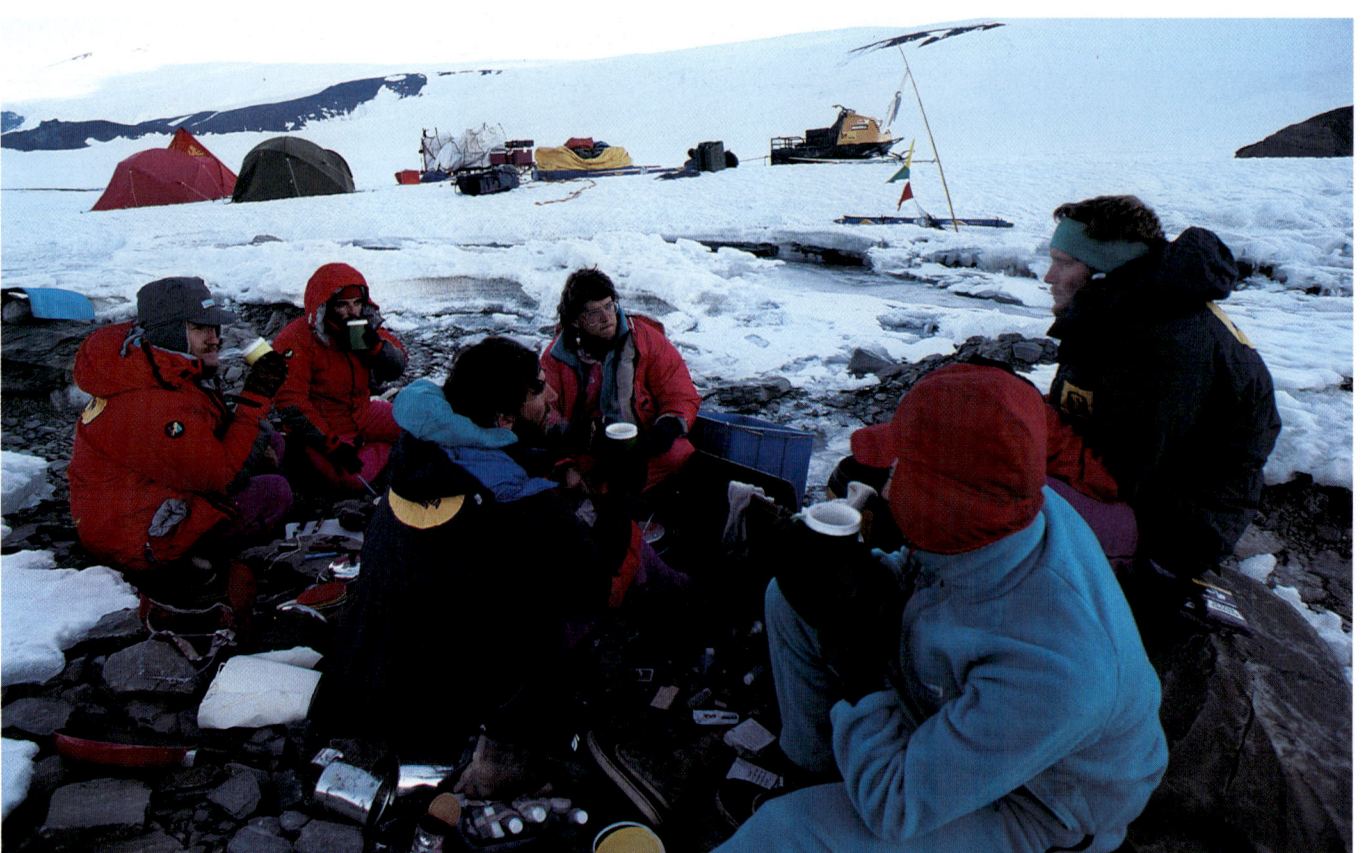

In this way, adverse impact by tourists has been minimised and no major problems have been encountered.

Regular airborne tourism began in 1977 with aircraft originating from Australia and New Zealand. These overflights did not land in Antarctica but provided sightseeing of a section of Antarctica (usually the Ross Sea region) from comfortable armchairs. These terminated soon after the tragic crash of the Air New Zealand DC10 on Mount Erebus, Ross Island, on 28 November 1979 with the loss of 257 lives (see Plate 6.6).

Up to that time airborne tourists had provided few problems for Antarctic administrators for they did not land and so did not require guidance and surveillance. Overflights may be reintroduced in the future and could be the best way for most to at least see something of the marvellous scenery of Antarctica—but without touching!

Most Antarctic Treaty countries regard tourism as one of the legitimate uses of Antarctica. But they are concerned to ensure that tours are operationally self-sufficient, are conducted safely with minimal impact on the environment and scientific research, and that tour organisers are well acquainted with all relevant provisions of the Antarctic Treaty.

Private expeditions

Non-government (private) expeditions also began in Antarctica in 1966. They are distinguished from government expeditions by being inspired by an individual and having little more than notional support from the government of the country in which they are organised (Plate 16.8). They are usually adventure or publicity seeking expeditions which often find their way to Antarctica in small ships or yachts not well suited to polar seas.

Private expeditions differ from tourist expeditions also in that they usually operate with restricted finances; thus their resources are limited and few could claim to be self-sufficient to the degree desirable for carrying out any activity in Antarctica. Indeed, many private expeditions have been undertaken with the thought that should they get into any great difficulty government expeditions would come to their aid. This aspect has become an increasing concern for government agencies because of the diversion of resources, disruption to scientific programmes, and risk to personnel and equipment in emergencies or rescues.

The Treaty has not been idle in giving attention to the problems private expeditions may cause. Several Recommendations have been agreed to by the Treaty Partners in recent years. But some governments have not always been able to respond to their responsibilities as required under all these Recommendations, especially in "providing detailed information in advance of any private expedition being organised in or proceeding from its territory".

Antarctica in the satellite age

International science in Antarctica, which was initiated during the IGY, has for the past 30 years or so developed and expanded its research in many scientific disciplines. Much of this work has been devoted to discovering more about the history of the continent and the development of its life forms, and also in using the Antarctic Continent as a giant observing platform to study the rest of our globe, the atmosphere above, and beyond to outer space.

To provide the necessary facilities for this research, numerous scientific stations were constructed by participating countries. Many have since been upgraded or rebuilt in recent years in keeping with demands for better accommodation and laboratory space (Plate 16.9 a, b).

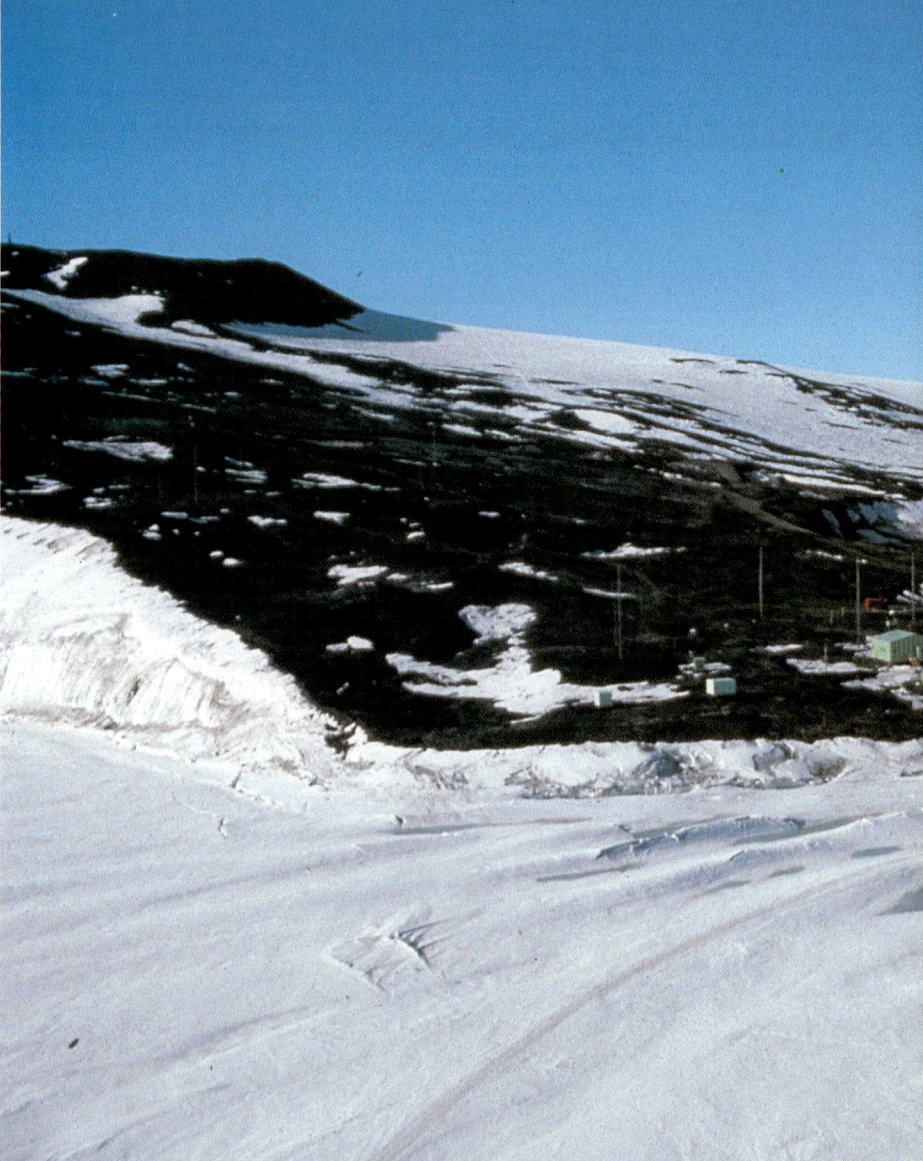

(b)

Plate 16.9. *The new look in Antarctic bases.*
(a) A huge geodesic dome covers the buildings of the United States base at the South Pole. Photo: I. Paterson.
(b) Scott Base today. Photo: G. Varcoe, Antarctic Division, DSIR. (c) The Federal Republic of Germany's Station "Gondwana Hutte" at Terra Nova Bay. Photo: A. Mortimer.

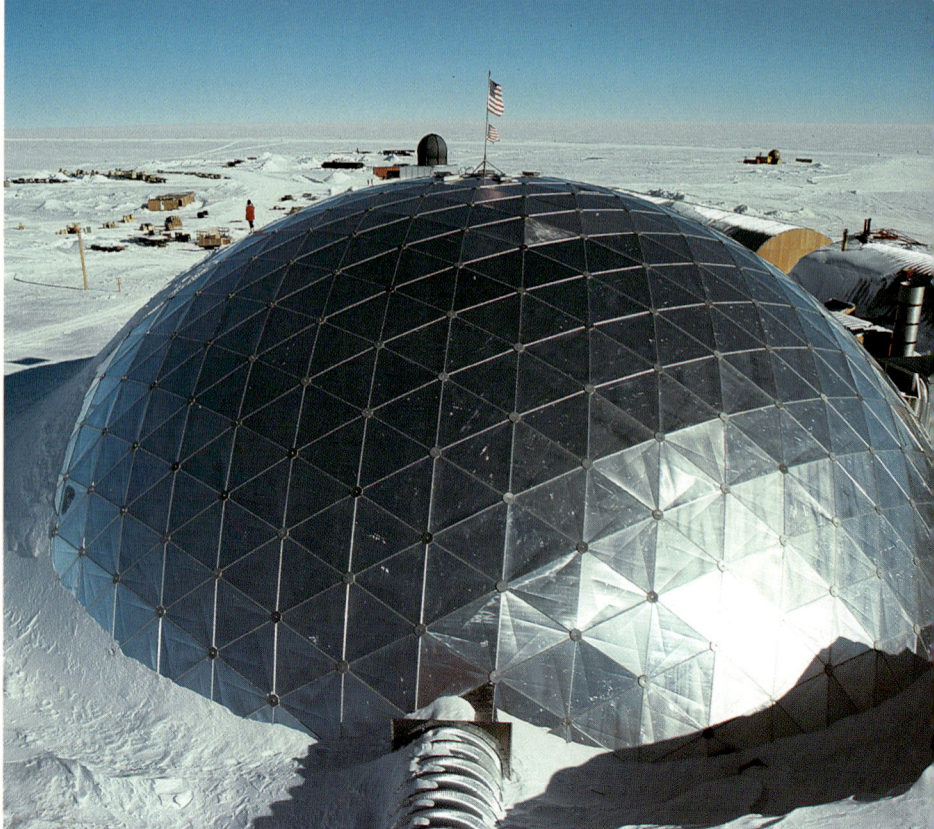

(a)

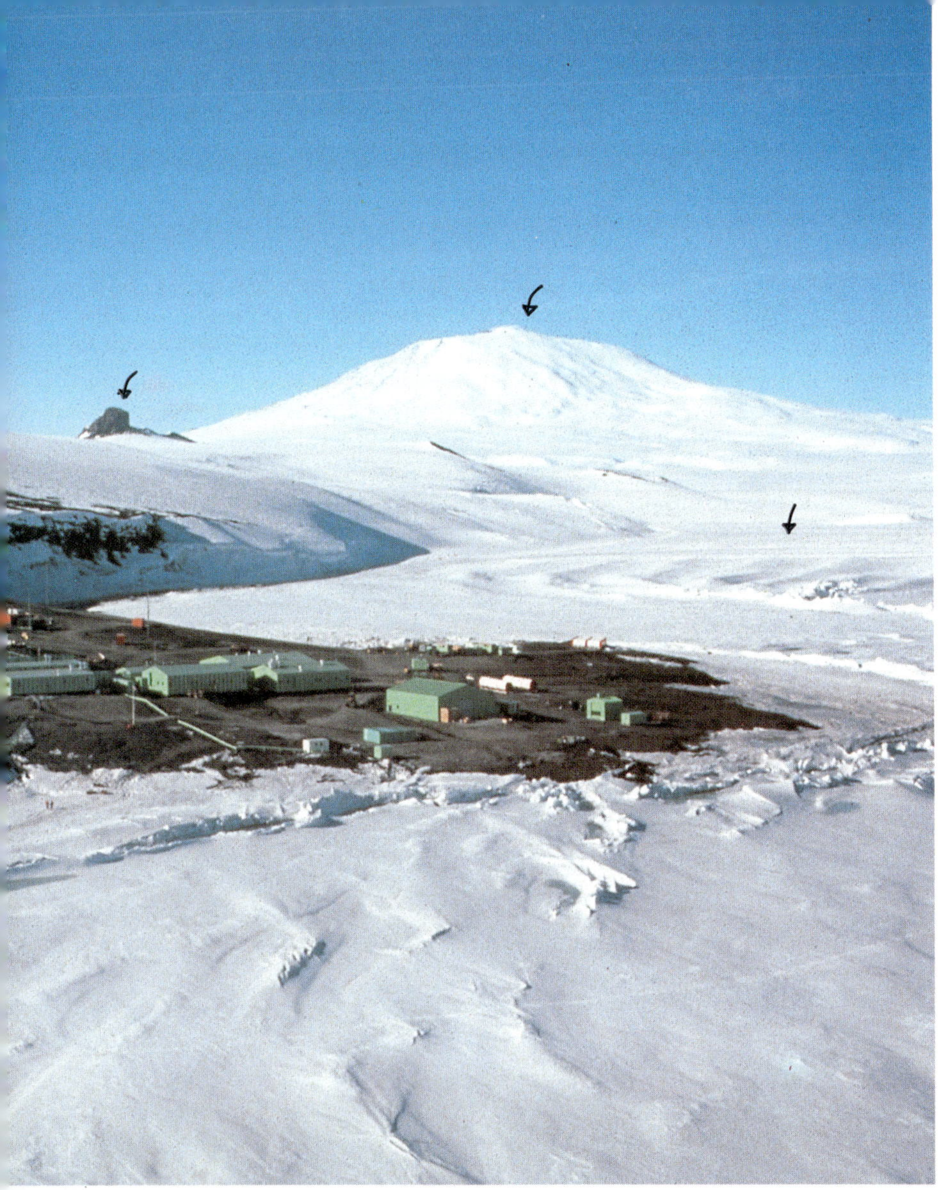

(c)

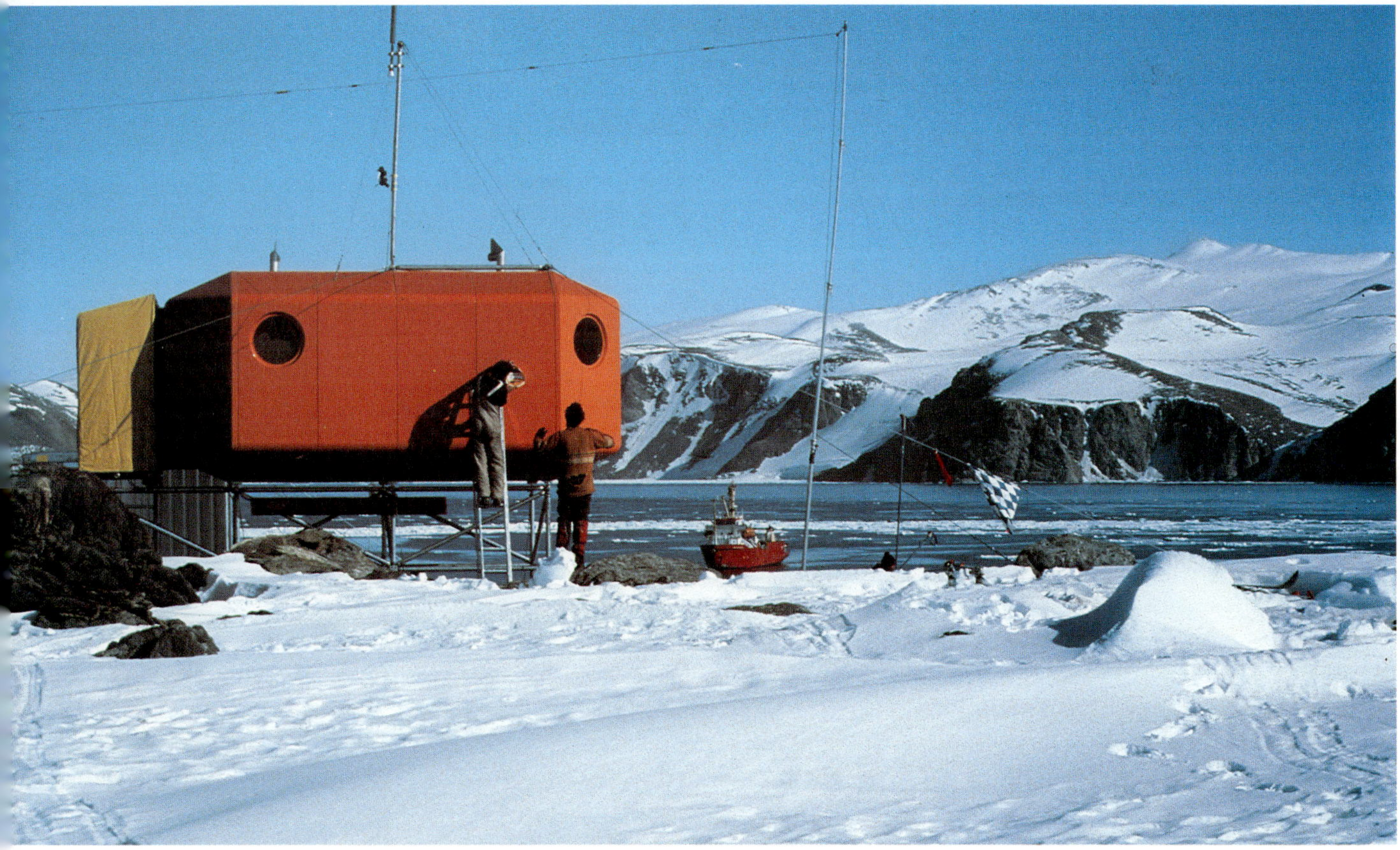

In brief, Antarctica has experienced three distinct phases to date: the exploration phase around the turn of the century, the introduction of international science and the construction of bases for the IGY in the late 1950s, followed by the subsequent expansion of scientific activities together with the consolidation of permanent bases over the last three decades. We are now rapidly approaching a fourth phase, which could be described as "Antarctic science in the satellite age" (Plate 16.10).

This new era will bring major changes to Antarctic research and to the Antarctic scene generally. The first impact will result from the Polar Platform (PF) component of the National Aeronautical and Space Administration (NASA) Space Station project scheduled for launch in 1994. A co-operative programme has been agreed to between NASA and the European Space Agency (ESA), known as the "Columbus Programme", with additional co-operation from Japan and Canada to further develop the equipment and research programmes. The PF will provide the scientific community with a permanent astronaut-tended vantage point in low earth orbit from which the Earth's atmosphere, oceans, land masses, and space environment will be monitored. As such, the PF can become the home for remote-sensing instruments that are carried on current polar space craft or those scheduled to fly within the next few years. The operational payload of the PF will monitor the Earth's magnetic field, atmospheric temperatures and water vapour, ozone, aerosols, outgoing radiation, precipitation, sea surface temperature, sea ice, ocean chlorophyll, surface winds, wave height, ocean circulation, snow cover, land use, vegetation, crops, volcanoes, and the hydrologic cycle.

Communications equipment proposed for the PF will provide for routine interrogation of multi-disciplinary environmental gauges, the potential to improve significantly the high-frequency telecommunications systems currently used within Antarctica and to the outside world, and the ability to monitor emergency transmissions from ships, vehicles, and

Plate 16.10. *Satellite test antenna at Arrival Heights, Hut Point Peninsula.* Photo: S. O'Doud.

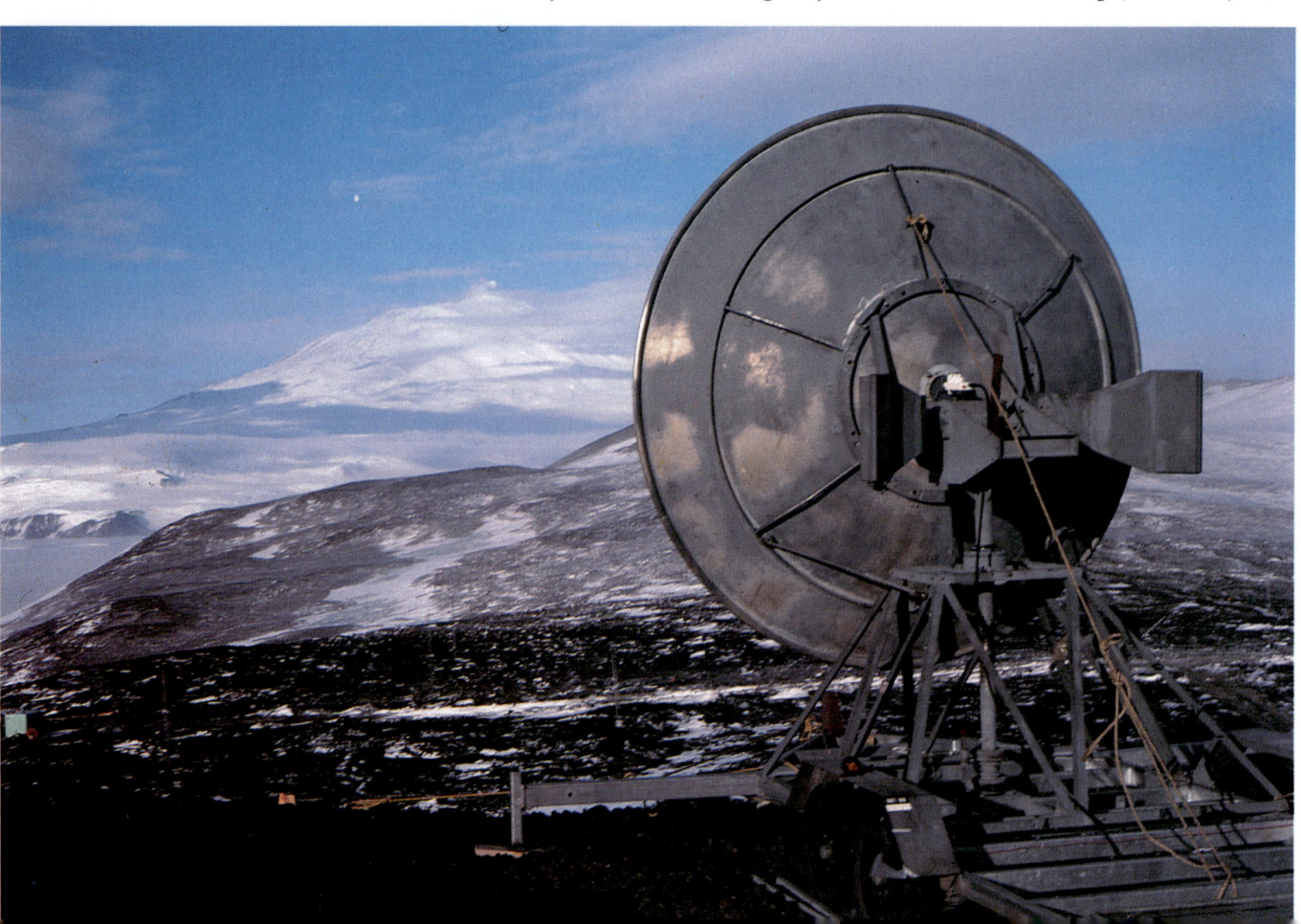

aircraft. Additional instrumentation being developed should eventually permit routine monitoring of atmospheric chemistry, global winds, and space plasma, and even be able to identify various rock types and minerals from space. It is envisaged there will be eight PFs in orbit by the year 2000, each with an operation life expectancy of 20 years.

PFs and associated satellites will, by the mid 1990s, do much of the current work being carried out in Antarctica. Especially with environmental monitoring, such as providing weather predictions and ice forecasts, they will do it even better. So what work will remain to be done at Antarctic bases and by field parties?

Even allowing for long-term development of technology in future space programmes there will remain a need for detailed surface and sub-surface studies to be done by humans. There will be a move away from many of the traditional monitoring programmes in earth and upper atmosphere studies to life sciences. Greater emphasis will be placed on marine biology, involving diving programmes, deep drilling programmes both into the ice and bedrock in search of more details covering past climatic and geological events, and the exploration of natural resources. As most if not all field activity is likely to continue to be done during summer only, the question of the need for maintaining personnel in Antarctica to "winter-over" will have to be addressed. Those programmes that need to be operated on the ground year-round (such as seismology) may not require continuous human overseeing for data collection and maintenance.

Internationalisation of Antarctica has brought stability to the region and Antarctica has been promoted as a "continent for science". It is very difficult to predict the ultimate future of Antarctica for this will likely be determined by events elsewhere in the world. Nevertheless, scientific knowledge has been the main export of Antarctica for much of this century and, considering all other factors, it seems likely that science will remain its biggest industry well into the next century.

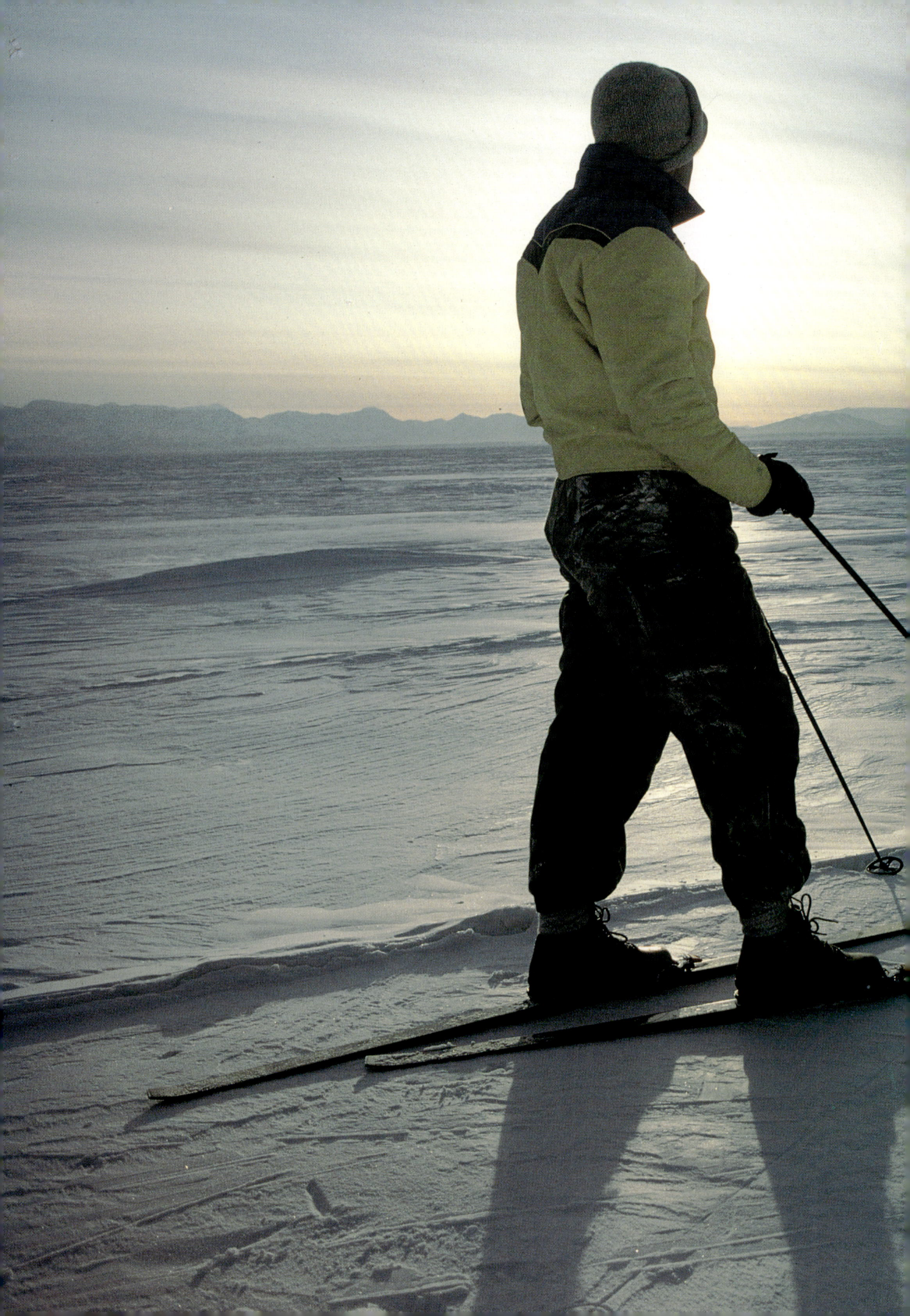

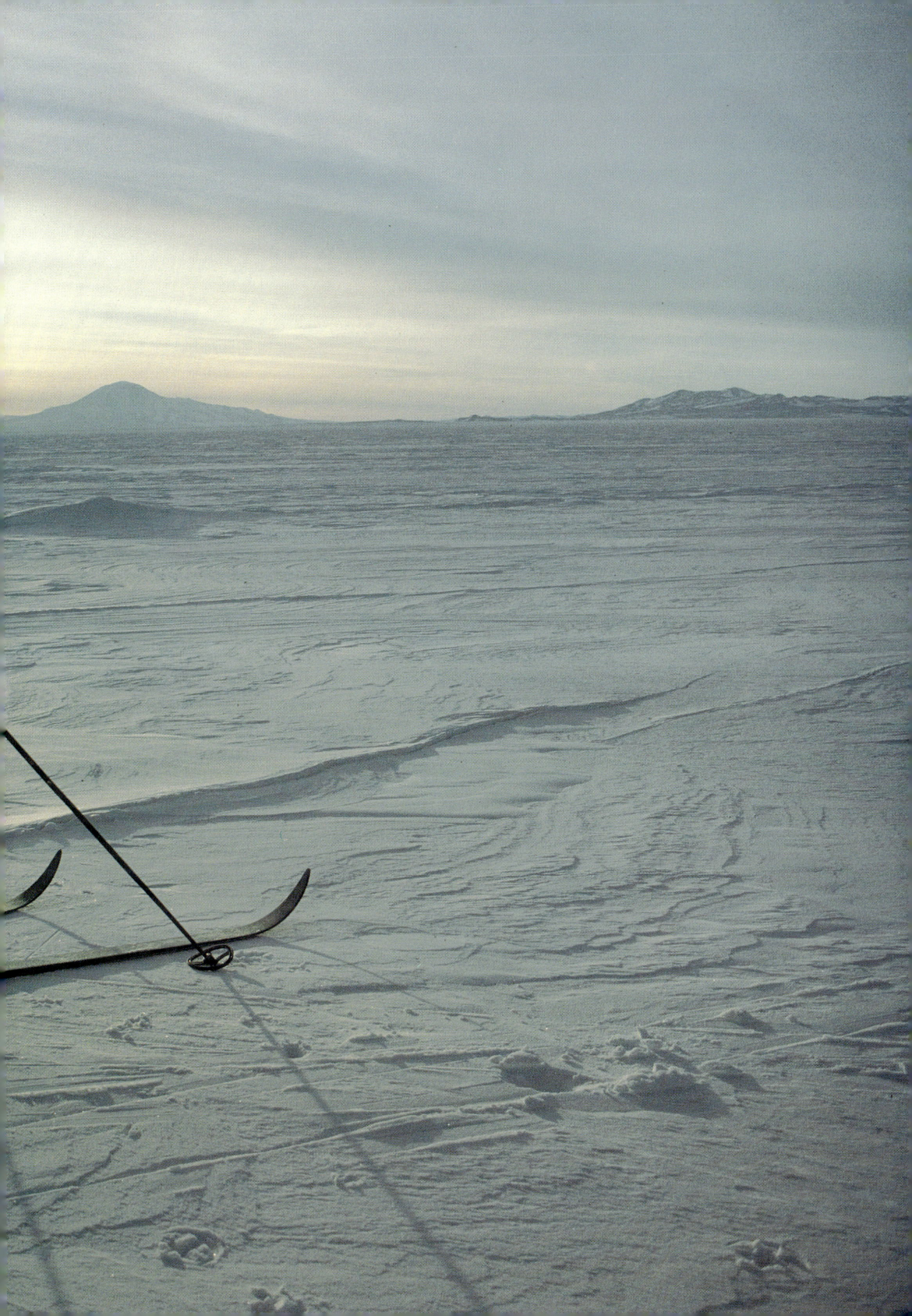

Locality map

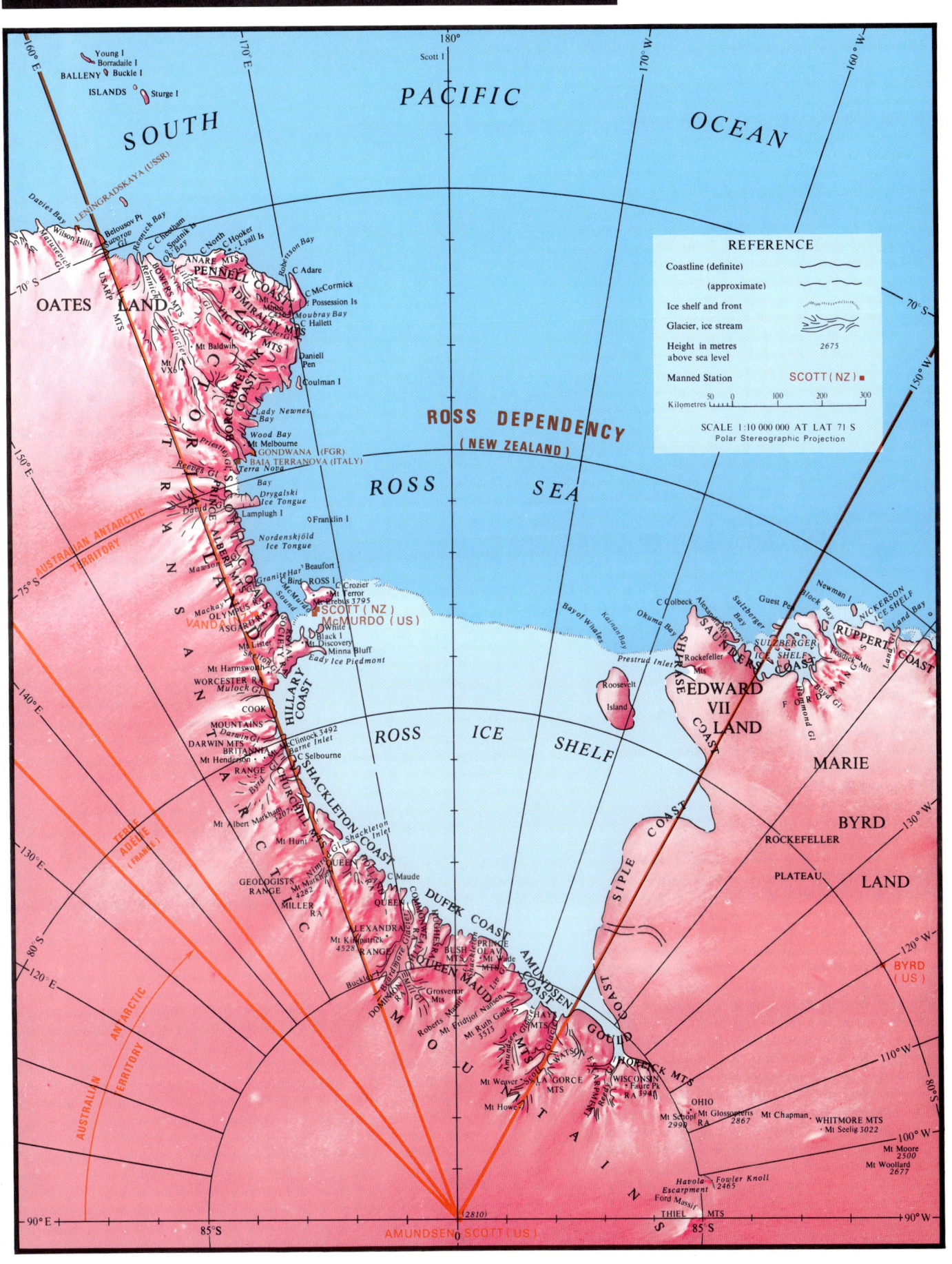

Editor. **Trevor Hatherton**, formerly Director of DSIR's Geophysics Division, first visited Antarctica in 1955 and wintered-over with New Zealand's first expedition in 1957. He has been involved in New Zealand's Antarctic activities ever since, serving on the Ross Dependency Research Committee for 30 years, the last 5 as Chairman.

Chapter 1. **David Harrowfield** has written on the early expeditions, particularly their huts, stores, and artefacts, and he is a leading proponent of their preservation and conservation. (see Suggested Reading)

Chapter 2. **Hugh Logan** is Manager of Antarctic Division, DSIR, the New Zealand Antarctic planning and logistics group. He was awarded the Q.S.M. for his work at the Mount Erebus disaster site, 1979.

Chapter 3. **Margaret Bradshaw** is Principal Curator of Canterbury Museum, Christchurch. As a geologist and paleontologist she has accompanied five field parties to Antarctica since 1975.

Chapter 4. **Michael Selby**, Professor of Earth Sciences and Deputy Vice-Chancellor of the University of Waikato, is a geomorphologist, with particular interests in how rock slopes develop.

Chapter 5. **Graeme Claridge** and **Ian Campbell**, both formerly with N.Z. Soil Bureau, DSIR, have worked together on Antarctic soils for 25 years and have written a major monograph on this topic. (see Suggested Reading)

Chapter 6. **Ray Dibble** is Reader in Geophysics at Victoria University of Wellington. Since 1962 he has spent 12 summers in the Ross Sea region, principally working on the volcano-physics of Mount Erebus.

Chapter 7. **Murray Gregory** is a Senior Lecturer at the Department of Geology, University of Auckland. He combines an interest in modern coastal processes with environmental and pollution studies. **Bob Kirk**, a Senior Lecturer at the Department of Geography, University of Canterbury, is particularly interested in coastal landforms and sea level history.

Chapter 8. **Brett Mullan** and **Mark Sinclair** are both research meteorologists with the N.Z. Meteorological Service. Brett's interest is the influence of the Antarctic on the Southern Hemisphere's atmospheric circulation and Mark is concerned with the synoptic meteorology of high latitudes.

Chapter 9. **Harry Keys** is with N.Z.'s Department of Conservation. He has worked on a variety of earth science projects since his first visit to Antarctica in 1972 but recently has concentrated on ice studies and environmental problems. (see Suggested Reading)

Chapter 10. **Trevor Chinn**, hydrologist and glaciologist with the N.Z. Geological Survey, DSIR, has studied the glaciers, lakes, and rivers of the Dry Valleys for the past 20 years.

Chapter 11. **Brian Foster**, Associate Professor of Zoology at the University of Auckland, is particularly interested in the analysis of Antarctic marine ecosystems, especially zooplankton and planktivory.

Chapter 12.   **Paul Broady** is a Lecturer in the Department of Plant and Microbial Sciences, University of Canterbury. He has worked with British, Australian, and New Zealand expeditions for over 18 years, studying Antarctic algae in terrestrial and freshwater habitats. **Warwick Vincent**, a scientist with the Division of Water Sciences, DSIR, has worked on, and written widely about, the microbial ecosystems of Antarctica. (see Suggested Reading)

Chapter 13.   **Euan Young**, Professor of Zoology, University of Auckland, has studied Antarctic and Subantarctic birds for 30 years. His wife Pam was the first New Zealand woman scientist to work in the Ross Sea region.

Chapter 14.   **John Macdonald**, whose interests are in the biology, physiology, biochemistry, and evolution of Antarctic fishes, has visited the Ross Sea region on 13 occasions and wintered-over at McMurdo Station in 1964. He is a Senior Lecturer in the Department of Zoology, University of Auckland. **John Montgomery**, an Associate Professor in the same department, shares his co-authors' interests in Antarctic fishes, particularly their neurophysiology.

Chapter 15.   **Tony Taylor**, Professor of Clinical Psychology at Victoria University of Wellington, has been psychological consultant to the New Zealand Antarctic Research Programme for over 20 years. He has consolidated the results of his study of the effects of Antarctic isolation into a book. (see Suggested Reading)

Chapter 16.   **Bob Thomson** was the Director of Antarctic Division, DSIR, for 25 years and was responsible for implementing the research programme. He has wintered-over at Hallett and Wilkes Stations and from the latter he led a major overland traverse to Vostok Station and back, receiving the O.B.E. for his contributions to Australian–Antarctic activities.

General

Anon. 1985: Antarctica: great stories from the frozen continent. Sydney, Readers Digest. 319 p.

May, J. (Ed.) 1988: Greenpeace book of Antarctica. Auckland, Macdonald. 192 p.

Sugden, D. 1982: Arctic and Antarctic. Cambridge, Blackwell. 478 p.

Walton, D. W. H. (Ed.) 1987: Antarctic science. Cambridge, Cambridge University Press. 280 p.

Historical

Harrowfield, D. 1981: Sledging into history. Auckland, Macmillan. 119 p.

Huntford, R. 1979: Scott and Amundsen. London, Hodder & Stoughton. 665 p.

Quartermain, L. 1967: South to the Pole. London, Oxford University Press. 481 p.

Quartermain, L. 1981: The forgotten men. Wellington, Millwood Press. 192 p.

Wildlife

Bonner, W. N.; Walton, D. W. H. (Eds) 1985: Antarctica. Oxford, Pergamon. 381 p. (In: Key Environments Series)

Eastman, J. T.; DeVries, A. L. 1986: Antarctic fishes. *Scientific American* 255: 96–103.

Kooyman, G. L. 1981: Weddell seal—consummate diver. Cambridge, Cambridge University Press. 135 p.

Laws, R. M. (Ed.) 1984: Antarctic ecology. London, Academic Press. 2 v.

Petersen, R. 1989: Penguins. Boston, Houghton Mifflin. 238 p.

Stonehouse, B. 1972. Animals of the Antarctic. New York; Holt, Rinehart, & Winston. 171 p.

Vincent, W. F. 1988: Microbial ecosystems of Antarctica. Cambridge, Cambridge University Press. 304 p.

Environmental aspects

Barnes, J. N.; Porter, E. 1982: Let's save Antarctica. Richmond, Va., Greenhouse Publications. 96 p.

Benninghoff, W. S.; Bonner, W. N. 1985: Man's impact on the Antarctic environment. Cambridge, Scientific Committee on Antarctic Research. 56 p.

Campbell, I. B.; Claridge, G.G.C. 1987: Antarctica: soils, weathering processes and environment. Amsterdam, Elsevier. 368 p.

Keys, J. R. 1984: Antarctic marine environments and offshore oil. Wellington, Commission for the Environment. 168 p.

Effects of isolation

Taylor, A. J. W. 1988: Antarctic psychology. Wellington, Science Information Publishing Centre, Department of Scientific and Industrial Research. 145 p.

International and legal aspects

Anon. 1986: Antarctic treaty system: an assessment. Washington, D.C., National Academy Press. 435 p.

Auburn, F. M. 1982: Antarctic law and politics. London, C. Hurst. 361 p.

A page number set in *italic* indicates that a figure or plate occurs on that page.
A title set in *italic* indicates a ship.